Handbook of
Advanced Multilevel Analysis

The European Association of Methodology (EAM) serves to promote research and development of empirical research methods in the fields of the Behavioural, Social, Educational, Health and Economic Sciences as well as in the field of Evaluation Research.

Homepage: http://www.eam-online.org

The purpose of the EAM book series is to advance the development and application of methodological and statistical research techniques in social and behavioral research. Each volume in the series presents cutting-edge methodological developments in a way that is accessible to a broad audience. Such books can be authored, monographs, or edited volumes.

Sponsored by the European Association of Methodology, the EAM book series is open to contributions from the Behavioral, Social, Educational, Health and Economic Sciences. Proposals for volumes in the EAM series should include the following: (1) Title; (2) authors/editors; (3) a brief description of the volume's focus and intended audience; (4) a table of contents; (5) a timeline including planned completion date. Proposals are invited from all interested authors. Feel free to submit a proposal to one of the members of the EAM book series editorial board, by visiting the EAM website http://eam-online.org. Members of the EAM editorial board are Manuel Ato (University of Murcia), Pamela Campanelli (Survey Consultant, UK), Edith de Leeuw (Utrecht University) and Vasja Vehovar (University of Ljubljana).

Volumes in the series include

Hox/Roberts: Handbook of Advanced Multilevel Analysis, 2011

De Leeuw/Hox/Dillman: International Handbook of Survey Methodology, 2008

Van Montfort/Oud/Satorra: Longitudinal Models in the Behavioral and Related Sciences, 2007

Handbook of Advanced Multilevel Analysis

Edited by

Joop J. Hox
Utrecht University

J. Kyle Roberts
Southern Methodist University

Routledge
Taylor & Francis Group
New York London

Routledge
Taylor & Francis Group
270 Madison Avenue
New York, NY 10016

Routledge
Taylor & Francis Group
27 Church Road
Hove, East Sussex BN3 2FA

Printed in the United States of America on acid-free paper
10 9 8 7 6 5 4 3 2 1

International Standard Book Number: 978-1-84169-722-2 (Hardback)

Library of Congress Cataloging-in-Publication Data

Handbook of advanced multilevel analysis / editors, Joop J. Hox, J. Kyle Roberts.
p. cm.
Includes bibliographical references and index.
ISBN 978-1-84169-722-2 (hardcover : alk. paper)
1. Social sciences--Statistical methods. 2. Multilevel models (Statistics) 3. Regression analysis. I. Hox, J. J. II. Roberts, J. Kyle.

HA29.H2484 2011
519.5'36--dc22 2010021673

Visit the Taylor & Francis Web site at
http://www.taylorandfrancis.com

and the Psychology Press Web site at
http://www.psypress.com

Contents

Preface

As statistical models become more and more complex, there is a growing need for methodological instruction. In the field of multilevel and hierarchical linear modeling, there is a distinct need to not only continue the development of complex statistical models, but also to illustrate their specific applications in a variety of fields. Although multilevel modeling is a relatively new field introduced first by Goldstein (1987) and then by Bryk and Raudenbush (1992), this field has enjoyed a large collection of published articles and books in just the last few years. In addition, statistical software has become more powerful, providing substantive researchers with a new set of analytic choices. Based on the explosion of research in this methodological field, the editors felt a need for a comprehensive handbook of advanced applications in multilevel modeling.

The benefit of this book to the broader research community is twofold. First, in many current texts, space is largely devoted to explaining the structure and function of multilevel models. This book is aimed at researchers with advanced training in multivariate and multilevel analysis. Therefore, the book immediately turns to the more difficult complexities of the broader class of models. Although the chief concern for the handbook is to highlight advanced applications, the initial chapter written by the editors, discusses the broad idea of multilevel modeling in order to provide a framework for the later chapters. Second, some of the leading researchers in the field have contributed chapters to this handbook. Thus, the later chapters are introduced and discussed by authors who are actively carrying out research on these advanced topics.

The handbook is divided into five major sections: introduction; multilevel latent variable modeling; multilevel models for longitudinal data; special estimation problems; and specific statistical issues. Section I, the Introduction, describes the basic multilevel regression model and multilevel structural equation modeling. Section II encompasses topics such as multilevel structural equation modeling, multilevel item response theory, and latent class analysis. Section III primarily covers panel modeling and growth curve analysis. Section IV devotes attention to the difficulties involved in estimating complicated models, including the analysis of ordered categorical data, generalized linear models, bootstrapping, Bayesian estimation, and multiple imputations. The latter half of Section IV is devoted to explaining variance, power, effect sizes, model fit and selection, and optimal design in multilevel models. Section V and final section covers centering issues, analyzing cross-classified models, and models for dyadic data.

The primary audience for this handbook is statisticians, researchers, methodologists, and advanced students. The handbook is multidisciplinary; it is not limited to one specific field of study. Educational researchers may use these models to study school effects; while researchers in medicine will use the techniques to study genetic strains; and economists will use the methods to register market, national, or even global trends. We will assume that the primary audience has a good working knowledge of multilevel modeling; therefore, the book aims at more advanced readers, but not necessarily readers with a lot of mathematical statistics. The book should also be useful to researchers looking for a comprehensive treatment of the best practices when applying these models to research data. This handbook would be ideal for a second course in multilevel modeling.

Supplementary materials for the handbook, such as data sets and program setups for the examples used in the chapters, are hosted on http://www.HLM-Online.com/.

We thank the chapter authors for their commitment in contributing to our book. We also thank the multilevel community, members of the multilevel and semnet discussion lists, and participants in the International multilevel conferences for the many lively discussions that have inspired the editors to compile this handbook. We appreciated the feedback received from the reviewers: Ronald H. Heck, the University of Hawaii, Manoa; Noel A. Card, University of Arizona; and Scott L. Thomas, Claremont Graduate University. Their input was instrumental in helping us finalize the overall plan for the book. We also thank Debra Riegert at Routledge/Taylor & Francis for her continued support on this project. Without her help and encouragement, we might never have seen this project to completion!

Joop J. Hox
J. Kyle Roberts

REFERENCES

Bryk, A. S., & Raudenbush, S. W. (1992). *Hierarchical linear models*. Newbury Park, CA: Sage Publications, Inc.

Goldstein, H. (1987). *Multilevel models in educational and social research*. London, UK: Charles Griffin & Company Ltd.

Section I

Introduction

1

Multilevel Analysis: Where We Were and Where We Are

Joop J. Hox
Department of Methodology and Statistics,
Utrecht University, Utrecht, The Netherlands

J. Kyle Roberts
Annette Caldwell Simmons School of Education and Human
Development, Southern Methodist University, Texas

1.1 INTRODUCTION

Hierarchical or multilevel data are common in the social and behavioral sciences. The interest in analyzing and interpreting multilevel data has historical roots in educational and sociological research, where a surge in theoretical and statistical discussions occurred in the 1970s. Although sociology, by definition, studies collective phenomena, the issue of studying relationships between individuals and the contexts in which they exist traces back to Lazarsfeld and Menzel (1961) and Galtung (1969). Lazarsfeld and Menzel developed a typology to describe the relations between different types of variables, defined at different levels. Galtung (1969) developed this scheme further, including levels within individuals. A simplified scheme is presented by Hox (2002):

Level	**1**		**2**		**3**		**Etc.**
Variable type	Global	⇒	Analytical				
	Relational	⇒	Structural				
	Contextual	⇐	Global	⇒	Analytical		
			Relational	⇒	Structural		
			Contextual	⇐	Global	⇒	
					Relational	⇒	

In this scheme, the lowest level (level 1) is usually formed by the individuals. However, this is not always the case. Galtung (1969), for instance, defines roles within individuals as the lowest level, and in longitudinal designs one can define repeated measures within individuals as the lowest. At each level, several types of variables are distinguished. Global variables refer only to the level at which they are defined, without reference to any other units or levels. Relational variables also refer to one single level, but they describe the relationships of a unit with other units at the same level. Sociometric indices, for example the reciprocity of relationships, are of this kind. Analytical and structural variables are created by aggregating global or relational variables to a higher level. They refer to the distribution of a global or relational variable at a lower level, for instance to the mean of a global variable from a lower level. Contextual variables, on the other hand, are created by disaggregation. All units at a lower level receive the value of a variable for the super unit to which they belong at a higher level.

The advantage of this typology is mainly conceptual; the scheme makes it clear to which level the measurements properly belong, and how related variables can be created by aggregation or disaggregation. Historically, the problem of analyzing data from individuals nested within groups was "solved" by moving all variables by aggregation or disaggregation to one single level, followed by some standard (single-level) analysis method. A more sophisticated approach was the "slopes as outcomes" approach, where a separate analysis was carried out in each group and the estimates for all groups were collected in a group level data matrix for further analysis. A nice introduction to these historical analysis methods is given by Boyd and Iverson (1979). All these methods are flawed, because the analysis either ignores the different levels or treats them inadequately. Statistical criticism of these methods was expressed early after their adoption, for example by Tate and Wongbundhit (1983) and de Leeuw and Kreft (1986). Better statistical methods were already available, for instance Hartley and Rao (1967) discuss estimation methods for the mixed model, which is essentially a multilevel model, and Mason, Wong, and Entwisle (1984) describe such a model for multilevel data, including software for its estimation. A nice summary of the state of the art around 1980 is given by van den Eeden and Hüttner (1982). The difference between the 1980 state of the art and the present (2010) situation is clear from its contents: there is a lot of discussion of (dis)aggregation and the "proper" level for the analysis, and of multiple regression tricks such as slopes as outcomes and other two-step procedures. There is no mention of statistical models as such, statistical dependency, random coefficients, or estimation methods. In short, what is missing is a principled statistical modeling approach.

Current statistical modeling approaches for multilevel data are listed under multilevel models, mixed models, random coefficient models, and hierarchical linear models. There are subtle differences, but the similarities are greater. Given these many labels for similar procedures, we simply use the term multilevel modeling or multilevel analysis to indicate the application of statistical models for data that have two or more distinct hierarchical levels, with variables at each of these levels, and research interest in relationships that span different levels. The prevailing multilevel model is the multilevel linear regression model, with explanatory variables at several levels and an outcome variable at the lowest level. This model has been extended to cover nonnormal outcomes, multivariate outcomes, and cross-classified and multiple membership structures. There is increasing interest in multilevel models that include latent variables at the distinct levels, such as multilevel structural equation models and multilevel latent class models. Although multiple regression is just a specific structural equation model, and multilevel modeling can be incorporated in the general structural equation framework (Mehta & Neale, 2005), the differences in the typical application and the software capacities are sufficiently large that it is convenient to distinguish between these two varieties of multilevel modeling. Hence, in the next two sections we will introduce multilevel regression and multilevel structural equation modeling briefly, and in the final section we will provide an overview of the various chapters in this book.

What seems to be lost in the current surge of statistical models, estimation methods, and software development is the interest in multilevel theories that is evident in the historical literature referred to earlier. Two different theoretical approaches are merged in multilevel research: the more European approach of society as a large structure that should be studied as a whole, and the more American approach of viewing society as primarily a collection of individuals. Multilevel theories combine these approaches by focusing on the questions of how individuals are influenced by their social context, and of how higher level structures emerge from lower level events. Good examples of a contextual theory are the variety of reference theories formulated in educational research to explain the effect of class and school variables on individual pupils. One such is Davis's (1966) frog-pond theory. The frog-pond theory poses that pupils use their relative standing in a group as a basis for their self-evaluation, aspirations, and study behavior. It is not the absolute size of the frog that matters, but the relative size given the pond it is in. Erbring and Young (1979) elaborate on this model by taking the interaction structure in a school class into account to predict school success. In brief, they state that the outcomes of pupils that are, in the sociometric sense, close to a specific pupil, affect the aspiration level and hence the success of that pupil. In their endogeneous feedback model the success of individual pupils becomes a group level determinant of that same success, mediated by the sociometric structure of the group. Such explicit multilevel theories appear more rare today. Certainly, theory construction is lagging behind the rapid statistical developments.

1.2 MULTILEVEL REGRESSION MODELS

Although Robinson (1950) was arguably one of the first individuals to recognize the need for multilevel analysis through his studies in ecological processes, no great progress was made in this area until the 1980s due to lack of statistical power available in computers. Lindley and Smith (1972) were the first to use the term *hierarchical linear models* for the method by which to analyze such data through Bayesian estimation. However, prior to the development of the EM algorithm (Dempster, Laird, & Rubin, 1977), the analysis of hierarchically structured data could prove magnanimous. Recently, however, the development of multiple statistical packages makes this type of analysis more accessible to those researchers who wish to examine the hierarchical structure of data.

Texts by both Goldstein (1995) and Raudenbush and Bryk (2002) were the initial texts that led to the rise of multilevel analysis. Although similar in their approach to handling hierarchically structured data, each had their own notation to describe these models. For example, Raudenbush and Bryk would notate the variance of the intercepts as τ_{00} whereas Goldstein would notate σ^2_{u0}. To begin describing the multilevel regression model, we will first consider a model in which no covariates are added to the model. This is sometimes referred to as the null model or the multilevel ANOVA. We model this as:

$$y_{ij} = \gamma_{00} + u_j + e_{ij}, \tag{1.1}$$

where y_{ij} represents the score for individual i in cluster j, γ_{00} is the grand estimate for the mean of y_{ij} for the population of j clusters, u_{0j} is the unique effect of cluster j on y_{ij} (also called the cluster-level error term), and e_{ij} is the deviation of individual i around their own cluster mean (also called the individual-level error term). In this case, we assume that $e_{ij}\tilde{}N(0,\sigma^2_e)$ and that $u_{0j}\tilde{}N(0,\sigma^2_u)$. This may also be written in matrix notation as:

$$\mathbf{y}_j = \mathbf{X}_j\gamma + \mathbf{Z}_j\mathbf{U}_j + \mathbf{e}_j, \tag{1.2}$$

where $\mathbf{y}_j$ is a $n_j \times 1$ response vector for cluster j, $\mathbf{X}_j$ is a $n_j \times p$ design matrix for the fixed effects, γ is a $p \times 1$ vector of unknown fixed parameters, $\mathbf{Z}_j$ is a $n_j \times r$ design matrix for the random effects, $\mathbf{U}_j$ is the $r \times 1$ vector of unknown random effects $\tilde{}N(0, \sigma_u)$, and $\mathbf{e}_j$ is the $n_j \times 1$ residual vector $\tilde{}N(0, \sigma_e)$.

For the null model, the matrix notation for a single cluster could be represented as:

$$\begin{bmatrix} Y_{1j} \\ Y_{2j} \\ \cdots \\ Y_{n_j j} \end{bmatrix} = \begin{bmatrix} 1 \\ 1 \\ \cdots \\ 1 \end{bmatrix} \begin{bmatrix} \gamma_{00} \end{bmatrix} + \begin{bmatrix} u_{oj} \end{bmatrix} + \begin{bmatrix} e_{1j} \\ e_{2j} \\ \cdots \\ e_{n_j j} \end{bmatrix}. \tag{1.3}$$

Likewise, a model with a single individual-level covariate (say "math") would take on the following matrix model form for cluster j:

$$\begin{bmatrix} Y_{1j} \\ Y_{2j} \\ \cdots \\ Y_{n_j j} \end{bmatrix} = \begin{bmatrix} 1 & \text{math}_{1j} \\ 1 & \text{math}_{2j} \\ \cdots & \cdots \\ 1 & \text{math}_{n_j j} \end{bmatrix} \begin{bmatrix} \gamma_{00} \\ \gamma_{10} \end{bmatrix} + \begin{bmatrix} u_{oj} \end{bmatrix} + \begin{bmatrix} e_{1j} \\ e_{2j} \\ \cdots \\ e_{n_j j} \end{bmatrix} \tag{1.4}$$

Were a random effect for math now modeled, the matrix model form for cluster j would now be:

$$\begin{bmatrix} Y_{1j} \\ Y_{2j} \\ \cdots \\ Y_{n_j j} \end{bmatrix} = \begin{bmatrix} 1 & \text{math}_{1j} \\ 1 & \text{math}_{2j} \\ \cdots & \cdots \\ 1 & \text{math}_{n_j j} \end{bmatrix} \begin{bmatrix} \gamma_{00} \\ \gamma_{10} \end{bmatrix} + \begin{bmatrix} u_{oj} & u_{1j} \end{bmatrix} + \begin{bmatrix} e_{1j} \\ e_{2j} \\ \cdots \\ e_{n_j j} \end{bmatrix} \tag{1.5}$$

where

$$\begin{bmatrix} u_{oj} \\ u_{1j} \end{bmatrix} \sim N\left(\begin{bmatrix} 0 \\ 0 \end{bmatrix}, \begin{bmatrix} \sigma^2_{u_0} & \sigma_{u_0 u_1} \\ \sigma_{u_0 u_1} & \sigma^2_{u_1} \end{bmatrix} \right). \tag{1.6}$$

Adding a cluster-level variable (say "schsize") to the model would make the matrix model form for cluster j take on the form:

$$\begin{bmatrix} Y_{1j} \\ Y_{2j} \\ \cdots \\ Y_{n_j j} \end{bmatrix} = \begin{bmatrix} 1 & \text{math}_{1j} & \text{schsize}_j \\ 1 & \text{math}_{2j} & \text{schsize}_j \\ \cdots & \cdots & \cdots \\ 1 & \text{math}_{n_j j} & \text{schsize}_j \end{bmatrix} \begin{bmatrix} \gamma_{00} \\ \gamma_{10} \\ \gamma_{01} \end{bmatrix}$$

$$+ \begin{bmatrix} u_{oj} & u_{1j} \end{bmatrix} + \begin{bmatrix} e_{1j} \\ e_{2j} \\ \cdots \\ e_{n_j j} \end{bmatrix}. \tag{1.7}$$

Finally, we could add a cross-level interaction effect between math and schsize making the matrix model form for cluster j:

$$\begin{bmatrix} Y_{1j} \\ Y_{2j} \\ \cdots \\ Y_{n_i j} \end{bmatrix} = \begin{bmatrix} 1 & \text{math}_{1j} & \text{schsize}_j & \text{schsize}_j\text{*math}_{1j} \\ 1 & \text{math}_{2j} & \text{schsize}_j & \text{schsize}_j\text{*math}_{2j} \\ \cdots & \cdots & \cdots & \\ 1 & \text{math}_{n_i j} & \text{schsize}_j & \text{schsize}_j\text{*math}_{n_i j} \end{bmatrix} \begin{bmatrix} \gamma_{00} \\ \gamma_{10} \\ \gamma_{01} \\ \gamma_{11} \end{bmatrix} + \begin{bmatrix} u_{oj} & u_{1j} \end{bmatrix} + \begin{bmatrix} e_{1j} \\ e_{2j} \\ \cdots \\ e_{n_i j} \end{bmatrix}. \quad (1.8)$$

It should be noted that the above model may take on different notations. For example, Raudenbush and Bryk (2002) would notate this as a hierarchical linear model with the following form for the full model:

$$y_{ij} = \gamma_{00} + \gamma_{10}\,\text{math}_{ij} + \gamma_{01}\,\text{schsize}_j + \gamma_{11}\,\text{math}_{ij}\,\text{schsize}_j + u_{1j}\,\text{math}_{ij} + u_{0j} + r. \quad (1.9)$$

This model could also be represented in the Raudenbush and Bryk form as a level 1 model:

$$y_{ij} = \beta_{0j} + \beta_{1j}\,\text{math}_{ij} + r \quad (1.10)$$

and level 2 model:

$$\beta_0 = \gamma_{00} + \gamma_{01}\,\text{schsize}_j + u_0 \quad (1.11)$$

$$\beta_1 = \gamma_{10} + \gamma_{11}\,\text{schsize}_j + u_1, \quad (1.12)$$

with random effects:

$$\begin{bmatrix} u_{oj} \\ u_{1j} \end{bmatrix} \sim N\left(\begin{bmatrix} 0 \\ 0 \end{bmatrix}, \begin{bmatrix} \tau_{00} & \tau_{01} \\ \tau_{10} & \tau_{11} \end{bmatrix}\right). \quad (1.13)$$

Goldstein (1995) would notate the same model as:

$$y_{ij} = \gamma_{00} + \gamma_{10}\,\text{math}_{ij} + \gamma_{01}\,\text{schsize}_j + \gamma_{11}\,\text{math}_{ij}\,\text{schsize}_j + u_{1j}\,\text{math}_{ij} + u_{0j} + e_{ij}, \quad (1.14)$$

with random effects:

$$\begin{bmatrix} u_{oj} \\ u_{1j} \end{bmatrix} \sim N\left(\begin{bmatrix} 0 \\ 0 \end{bmatrix}, \begin{bmatrix} \sigma^2_{u_0} & \sigma_{u_0 u_1} \\ \sigma_{u_0 u_1} & \sigma^2_{u_1} \end{bmatrix}\right). \quad (1.15)$$

1.3 MULTILEVEL STRUCTURAL EQUATION MODELS

Multilevel structural equation modeling assumes sampling at two levels, with both within group (individual level) and between group (group level) variation and covariation. In multilevel regression modeling, there is one dependent variable and several independent variables, with independent variables at both the individual and group level. At the group level, the multilevel regression model includes random regression coefficients and error terms. In the multilevel structural equation model, the random intercepts are second level latent variables, capturing the variation in the means. Conceptually, the issue is whether the group level covariation can be explained by a theoretical model. Statistically, the model used is often a structural equation model, which explains the covariation among the cluster level variables by a model containing latent variables, path coefficients, and (co)variances. Some of the group level variables may be random intercepts of slopes, drawn from the first level model, other group level variables may be variables defined at the group level, which are nonexistent at the individual level. In the typology presented above, the random intercepts and slopes are analytical variables (being a function of the lower level variables) and the group level variables proper are global variables. The first useful estimation method for multilevel structural equation models was a limited information method named MUML by Muthén (1989, 1994). The MUML approach follows the conventional notion that structural equation models are constructed for the covariance matrix with added mean vector, which are the sufficient statistics when data have a multivariate normal distribution. Thus, for a confirmatory factor model, the covariance matrix Σ is modeled by:

$$\Sigma = \Lambda\Psi\Lambda + \Theta, \qquad (1.16)$$

where Λ is the matrix of factor loadings, Ψ is the covariance matrix of the latent variables and Θ is the vector with residual variances. The MUML method distinguishes between the within groups covariance matrix Σ_w and between groups covariance matrix Σ_B, and specifies a structural equation model for each. As Muthén (1989) shows, the pooled within groups sample matrix S_{PW} is the maximum likelihood estimator of Σ_w, but the between groups sample matrix $\mathbf{S}_B^*$ is the maximum likelihood estimator of the composite $\Sigma_w + c\Sigma_B$, with scale factor c equal to the common group size n:

$$\mathbf{S}_{PW} = \overline{\Sigma_W}, \qquad (1.17)$$

and

$$\mathbf{S}_B^* = \overline{\Sigma_W + c\Sigma_B}. \qquad (1.18)$$

Originally, the multigroup option of conventional SEM software was used to carry out a simultaneous analysis at both levels, which leads to complicated software setups (cf. Hox, 2002). More recently, software implementations of the MUML method hide all the technical details, and allow direct specification of the within and the between model. However, the main limitation of the MUML is still there: MUML assumes a common group size, and the fact that groups are generally not equal is simply ignored by using an average group size. Simulations (e.g., Hox & Maas, 2001) have shown that this works reasonably well, and analytical work (Yuan & Hayashi, 2005) shows that accuracy of the standard errors increases when the number of groups becomes large and the amount of variation in the group sizes decreases. A second, probably more important limitation is that the MUML approach models only group level variation in the intercepts, group level slope variation cannot be included.

A more advanced approach is to use Full Information Maximum Likelihood (FIML) estimation for multilevel SEM. The FIML approach defines the model and the likelihood in terms of the individual data. The FIML minimizes the function (Arbuckle, 1996)

$$F = \sum_{i=1}^{N} \log|\Sigma_i| + \sum_{i=1}^{N} \log(\mathbf{x}_i - \mu_i)'\Sigma_i^{-1}(\mathbf{x}_i - \mu_i), \qquad (1.19)$$

where the subscript i refers to the observed cases, $\mathbf{x}_i$ to the variables observed for case i, and μ_i and Σ_i contain the population means and covariances of those variables that are observed for case i. Since the FIML estimation method defines the likelihood on the basis of the set of data observed for each specific individual, it is a very useful estimation method when there are missing data. If the data are incomplete, the covariance matrix is no longer a sufficient statistic, but minimizing the FIML likelihood for the raw data provides the maximum likelihood estimates for the incomplete data.

Mehta and Neale (2005) link multilevel SEM to the FIML fit function given above. By viewing groups as observations, and individuals within groups as variables, they show that models for multilevel data can be specified in the full information SEM framework. Unbalanced data, for example, unequal numbers of individuals within groups, are handled the same way as incomplete data in standard SEM. So, in theory, multilevel structural equation models can be specified in any SEM package that supports FIML estimation for incomplete data. In practice, specialized software routines are used that take advantage of specific structures of multilevel data to achieve efficient computations and good convergence of the estimates. Extensions of this approach include extensions for categorical and ordinal data, incomplete data, and adding more levels. These are described in detail by Skrondal and Rabe-Hesketh (2004).

Asparouhov and Muthén (2007) describe a limited information Weighted Least Squares (WLS) approach to multilevel SEM. In this approach, univariate Maximum Likelihood (ML) methods are used to estimate the vector of means $\boldsymbol{\mu}$ at the between group level, and the diagonal elements of $\boldsymbol{\Sigma}_W$ and $\boldsymbol{\Sigma}_B$. Next, the off-diagonal elements of $\mathbf{S}_W$ and $\mathbf{S}_B$ are estimated using bivariate ML methods. The asymptotic covariance matrix for these estimates is obtained, and the multilevel SEM is estimated for both levels using WLS. This estimation method is developed for efficient estimation of multilevel models with nonnormal variables, since for such data it requires only low-dimensional numerical integration, while ML requires generally high-dimensional numerical integration, which is computationally very demanding. However, multilevel WLS can also be used for multilevel estimation with continuous variables. With continuous variables, WLS does not have a real advantage, since ML estimation is very well possible and should be more efficient. A limitation to the WLS approach is that it, like MUML, does not allow for random slopes.

Muthén and Muthén (2007) and Skrondal and Rabe Hesketh (2004) have suggested extensions of the conventional graphic path diagrams to represent multiple levels and random slopes. The two-level path diagram in Figure 1.1 uses the Muthén and Muthén notation to depict a two-level regression model with an explanatory variable X at the individual level and an explanatory variable Z at the group level. The within part of the model in the lower area specifies that Y is regressed on X. The between part of the model in the upper area specifies the existence of a group level variable Z. There are two latent variables represented by circles. The group level latent variable Y represents the group level variance of the intercept for Y. The group level latent variable XYslope represents the group level variance of the slope for X and Y, which is on the group level regressed on Z. The black circle in the within part is a new symbol, used to specify that this path coefficient is assumed to have random variation at the group level. This variation is modeled at the group level using group level variables. The path diagram in Figure 1.2 uses the Skrondal and Rabe-Hesketh notation to depict the same

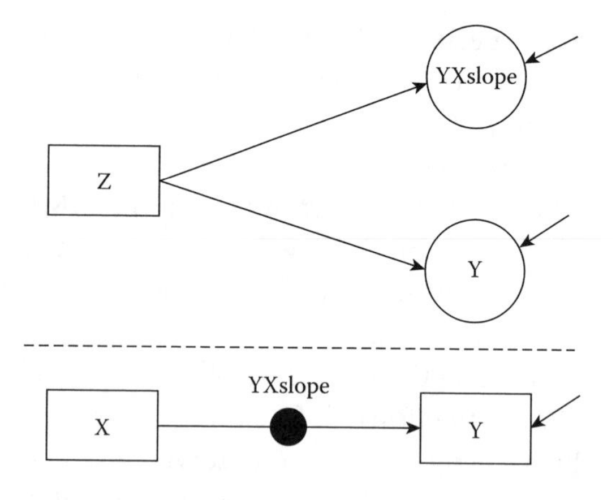

FIGURE 1.1
Path diagram for a two-level regression model, Muthén and Muthén style.

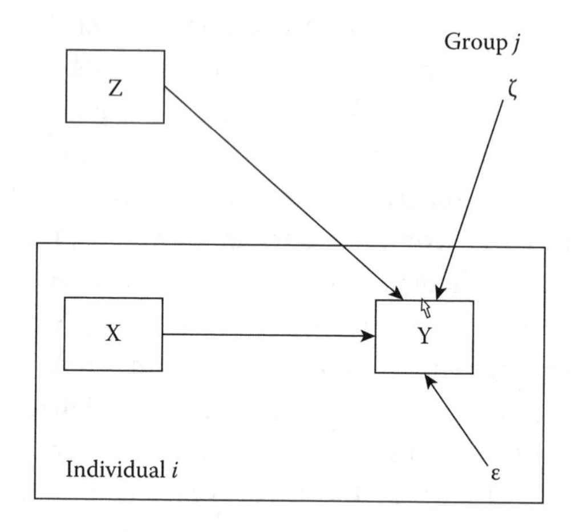

FIGURE 1.2
Path diagram for a two-level regression model, Skrondal and Rabe-Hesketh style.

two-level regression model. This notation is a little more complicated, but is more easily extended to complex models, for example, with a partial nesting structure.

1.4 CONTENTS OF THIS BOOK

This book is divided into five sections. The first section is an introduction by the authors about multilevel analysis then and now. The second section is about multilevel latent variable models and contains chapters by Muthén and Asparouhov on general multilevel latent variable modeling, by Kamata and Vaughn on multilevel item response theory, and by Vermunt on multilevel mixture modeling. The third section focuses on longitudinal modeling, with chapters by Hox on panel modeling and by Stoel and Galindo Garre on growth curve analysis. The fourth section focuses on special estimation problems, including a chapter by Hedeker and Mermelstein on ordinal data, Hamaker and Klugkist on Bayesian estimation, Goldstein on bootstrapping, Van Buuren on incomplete data, and Kim and Swoboda on omitted variable bias. Also included in the fourth section is a chapter by Roberts, Monaco, Stovall and Foster on model fit, power, and explained variance, Hamaker, van Hattum, Kuiper, and Hoijtink on model selection, and by Moerbeek and Teerenstra on optimal design. The fifth and final section discusses a selection of special problems, including a chapter by Algina and Swaminathan on centering, Beretvas on cross-classified models, and Kenny and Kashy on dyadic data analysis.

REFERENCES

Arbuckle, J. L. (1996). Advanced structural equation modeling: Issues and techniques. In G. A. Marcoulides & R. E. Schumacker (Eds.), *Structural equation modeling* (pp. 243–277). Mahwah, NJ: Erlbaum.

Asparouhov, T., & Muthén, B. (2007, August). Computationally efficient estimation of multilevel high-dimension latent variable models. In *Proceedings of the joint statistical meeting,* Salt Lake City, UT.

Boyd, L. H., & Iverson, G. R. (1979). *Contextual analysis: Concepts and statistical techniques.* Belmont, CA: Wadsworth.

Davis, J. A. (1966). The campus as a frog pond: An application of the theory of relative deprivation to career choices of college men. *American Journal of Sociology, 72,* 17–31.

De Leeuw, J., & Kreft, I. G. G. (1986). Random coefficient models. *Journal of Educational Statistics, 11*(1), 55–85.

Dempster, A. P., Laird, N. M., & Rubin, D. B. (1977). Maximum likelihood from incomplete data via the em algorithm. *Journal of the Royal Statistical Society, Series B, 39,* 1–8.

Erbring, L., & Young, A. A. (1979). Contextual effects as endogenous feedback. *Sociological Methods and Research, 7,* 396–430.

Galtung, J. (1969). *Theory and methods of social research.* New York, NY: Columbia University Press.

Goldstein, H. (1995). *Multilevel statistical models.* London, UK: Edward Arnold.

Hartley, H. O., & Rao, J. N. K. (1967). Maximum likelihood analysis for the mixed analysis of variance model. *Biometrika, 54,* 93–108.

Hox, J. J., & Maas, C. J. M. (2001). The accuracy of multilevel structural equation modeling with pseudobalanced groups with small samples. *Structural Equation Modeling, 8,* 157–174.

Hox, J. J. (2002). *Multilevel analysis.* Mahwah, NJ: Erlbaum.

Lazarsfeld, P. F., & Menzel, H. (1961). On the relation between individual and collective properties. In A. Etzioni (Ed.), *Complex organizations: A sociological reader.* New York, NY: Holt, Rhinehart and Winston.

Lindley, D. V., & Smith, A. F. M. (1972). Bayes estimates for the linear model. *Journal of the Royal Statistical Society, Series B, 34,* 1–41.

Mason, W. M., Wong, G. Y., & Entwisle, B. (1984). Contextual analysis through the multilevel linear model. In S. Leinhardt (Ed.), *Sociological methodology.* San Francisco, CA: Jossey-Bass.

Mehta, P. D., & Neale, M. C. (2005). People are variables too: Multilevel structural equations modeling. *Psychological Methods, 10,* 259–284.

Muthén, B. (1989). Latent variable modeling in heterogeneous populations. *Psychometrika, 54,* 557–585.

Muthén, B. (1994). Multilevel covariance structure analysis. *Sociological Methods and Research, 22,* 376–398.

Muthén, L. K., & Muthén, B. (2007). *Mplus. the comprehensive modeling program for applied researchers* (5th ed.) [Manuel de logiciel]. Los Angeles, CA: Authors.

Raudenbush, S. W., & Bryk, A. S. (2002). *Hierarchical linear models: Applications and data analysis methods* (2nd ed.). Thousand Oaks, CA: Sage.

Robinson, W. S. (1950). Ecological correlations and the behavior of individuals. *Sociological Review, 15,* 351–357.

Skrondal, A., & Rabe-Hesketh, S. (2004). *Generalized latent variable modeling: Multilevel, longitudinal and structural equation models.* Boca Raton, FL: Chapman and Hall/CRC.

Tate, R. L., & Wongbundhit, Y. (1983). Random versus nonrandom coefficient models for multilevel analysis. *Journal of Educational Statistics, 8,* 103–120.

Van den Eeden, P., & Huttner, H. J. M. (1982). Multilevel research. *Current Sociology, 30*(3), 1–181.

Yuan, K. H., & Hayashi, K. (2005). On Muthén's maximum likelihood for two-level covariance structure models. *Psychometrika, 70,* 147–167.

Section II

Multilevel Latent Variable Modeling (LVM)

2

Beyond Multilevel Regression Modeling: Multilevel Analysis in a General Latent Variable Framework

Bengt Muthén
Graduate School of Education & Information Studies,
University of California, Los Angeles, California

Tihomir Asparouhov
Muthen & Muthen, Los Angeles, California

2.1 INTRODUCTION

Multilevel modeling is often treated as if it concerns only regression analysis and growth modeling (Raudenbush & Bryk, 2002; Snijders & Bosker, 1999). Furthermore, growth modeling is merely seen as a variation on the regression theme, regressing the outcome on a time-related covariate. Multilevel modeling, however, is relevant for nested data not only with regression analysis but with all types of statistical analyses including

- Regression analysis
- Path analysis
- Factor analysis
- Structural equation modeling
- Growth modeling
- Survival analysis
- Latent class analysis
- Latent transition analysis
- Growth mixture modeling

This chapter has two aims. First, it shows that already in the traditional multilevel analysis areas of regression and growth there are several new modeling opportunities that should be considered. Second, it gives an overview with examples of multilevel modeling for path analysis, factor analysis, structural equation modeling, and growth mixture modeling. Due to lack of space, survival, latent class, and latent transition analysis are not covered. All of these topics, however, are covered within the latent variable framework of the M*plus* software, which is the basis for this chapter. A technical description of this framework including not only multilevel features but also finite mixtures is given in Muthén and Asparouhov (2008). Survival mixture analysis is discussed in Asparouhov, Masyn, and Muthén (2006). See also examples in the M*plus* User's Guide (Muthén & Muthén, 2008). The user's guide is available online at http://www.statmodel.com.

The outline of the chapter is as follows. Section 2.2 discusses two extensions of two-level regression analysis, Section 2.3 discusses two-level path analysis and structural equation modeling, Section 2.4 presents an example of two-level exploratory factor analysis (EFA), Section 2.5 discusses two-level growth modeling using a two-part model, Section 2.6 discusses an unconventional approach to three-level growth modeling, and Section 2.7 presents an example of multilevel growth mixture modeling.

2.2 TWO-LEVEL REGRESSION

One may ask if there really is anything new that can be said about multilevel regression. The answer, surprisingly, is yes. Two extensions of conventional two-level regression analysis will be discussed here, taking into account measurement error in covariates and unobserved heterogeneity among level 1 subjects.

2.2.1 Measurement Error in Covariates

It is well-known that measurement error in covariates creates biased regression slopes. In multilevel regression a particularly critical covariate is the level 2 covariate $\bar{x}_{.j}$, drawing on information from individuals within clusters to reflect cluster characteristics, as, for example, with students rating the school environment. Based on relatively few students such covariates may contain a considerable amount of measurement error, but this fact seems to not have gained widespread recognition in multilevel regression modeling. The following discussion draws on Asparouhov and Muthén (2006) and Ludtke et al. (2008). The topic seems to be rediscovered every two decades given earlier contributions by Schmidt (1969) and Muthén (1989).

Raudenbush and Bryk (2002, p. 140, Table 5.11) considered the two-level, random intercept, group-centered regression model

$$y_{ij} = \beta_{0j} + \beta_{1j}(x_{ij} - \bar{x}_{.j}) + r_{ij}, \qquad (2.1)$$

$$\beta_{0j} = \gamma_{00} + \gamma_{01}\,\bar{x}_{.j} + u_j, \qquad (2.2)$$

$$\beta_{1j} = \gamma_{10}, \qquad (2.3)$$

defining the "contextual effect" as

$$\beta_c = \gamma_{01} - \gamma_{10}. \qquad (2.4)$$

Often, $\bar{x}_{.j}$ can be seen as an estimate of a level 2 construct that has not been directly measured. In fact, the covariates $(x_{ij} - \bar{x}_{.j})$ and $\bar{x}_{.j}$ may be seen as proxies for latent covariates (cf. Asparouhov & Muthén, 2006),

$$x_{ij} - \bar{x}_{.j} \approx x_{ijw}, \qquad (2.5)$$

$$\bar{x}_{.j} \approx x_{jb}, \qquad (2.6)$$

where the latent covariates are obtained in line with the nested, random effects

ANOVA decomposition into uncorrelated components of variation,

$$x_{ij} = x_{jb} + x_{ijw.} \tag{2.7}$$

Using the latent covariate approach, a two-level regression model may be written as

$$y_{ij} = y_{jb} + y_{ijw} \tag{2.8}$$

$$= \alpha + \beta_b x_{jb} + \varepsilon_j \tag{2.9}$$

$$+ \beta_w x_{ijw} + \varepsilon_{ij}, \tag{2.10}$$

defining the contextual effect as

$$\beta_c = \beta_b - \beta_w. \tag{2.11}$$

The latent covariate approach of Equations 2.9 and 2.10 can be compared to the observed covariate approach Equations 2.1 through 2.3. Assuming the model of the latent covariate approach of Equations 2.9 and 2.10, Asparouhov and Muthén (2006) and Ludtke et al. (2008) show that the observed covariate approach introduces a bias in the estimation of the level 2 slope γ_{01} in Equation 2.3,

$$E(\hat{\gamma}_{01}) - \beta_b = \frac{(\beta_w - \beta_b)\psi_w/c}{\psi_b + \psi_w/c}$$

$$= (\beta_w - \beta_b)\frac{1}{c}\frac{1-icc}{icc+(1-icc)/c}, \tag{2.12}$$

where c is the common cluster size and icc is the covariate intraclass correlation ($\psi_b/(\psi_b + \psi_w)$). In contrast, there is no bias in the level 1 slope estimate $\hat{\gamma}_{10}$. It is clear from Equation 2.12 that the between slope bias increases for decreasing cluster size c and for decreasing icc. For example, with $c = 15$, $icc = 0.20$, and $\beta_w - \beta_b = 1.0$, the bias is 0.21.

Similarly, it can be shown that the contextual effect for the observed covariate approach $\hat{\gamma}_{01} - \hat{\gamma}_{10}$ is a biased estimate of $\beta_b - \beta_w$ from the latent covariate approach. For a detailed discussion see Ludtke et al. (2008), where the magnitudes of the biases are studied under different conditions.

As a simple example, consider data from the German Third International Mathematics and Science Study (TIMSS, 2003). Here there are $n = 1980$ students in 98 schools with average cluster (school) size = 20. The dependent variable is a math test score in Grade 8 and the covariate is student-reported disruptiveness level in the school. The intraclass correlation for disruptiveness is 0.21. Using maximum-likelihood (ML) estimation for the latent covariate approach to two-level regression with a random intercept in line with Equations 2.9 and 2.10 results in $\hat{\beta}_b = -1.35$ ($SE = 0.36$), $\hat{\beta}_w = -0.098$ ($SE = 0.03$), and contextual effect $\hat{\beta}_c = -1.25$ ($SE = 0.36$). The observed covariate approach results in the corresponding estimates $\hat{\gamma}_{01} = -1.18$ ($SE = 0.29$), $\hat{\gamma}_{10} = -0.097$ ($SE = 0.03$), and contextual effect $\hat{\beta}_c = -1.08$ ($SE = 0.30$).

Using the latent covariate approach in M*plus*, the observed covariate *disrupt* is automatically decomposed as $disrupt_{ij} = x_{jb} + x_{ijw}$. The use of M*plus* to analyze models under the latent covariate approach is described in Chapter 9 of the user's guide (Muthén & Muthén, 2008).

2.2.2 Unobserved Heterogeneity Among Level 1 Subjects

This section reanalyzes the classic High School & Beyond (HSB) data used as a key illustration in Raudenbush and Bryk

(2002; RB from now on). The HSB is a nationally representative survey of U.S. public and Catholic high schools. The data used in RB are a subsample with 7185 students from 160 schools, 90 public, and 70 Catholic. The RB model presented on pages 80–83 is considered here for individual i in cluster (school) j:

$$y_{ij} = \beta_{0j} + \beta_{1j}\,(ses_{ij} - mean\_ses_j) + r_{ij}, \quad (2.13)$$

$$\beta_{0j} = \gamma_{00} + \gamma_{01}\, sector_j + \gamma_{02}\, mean\_ses_j + u_{0j}, \quad (2.14)$$

$$\beta_{1j} = \gamma_{10} + \gamma_{11}\, sector_j + \gamma_{12}\, mean\_ses_j + u_{1j}, \quad (2.15)$$

where *mean_ses* is the school-averaged student *ses* and *sector* is a 0/1 dummy variable with 0 for public and 1 for Catholic schools. The estimates are shown in Table 2.1. The results show for example that, holding *mean_ses* constant, Catholic schools have significantly higher mean math achievement than public schools (see the γ_{02} estimate) and that Catholic schools have significantly lower *ses* slope than public schools (see the γ_{12} estimate).

What is overlooked in the above modeling is that a potentially large source of unobserved heterogeneity resides in variation of the regression coefficients between groups of individuals sharing similar but unobserved background characteristics.

TABLE 2.1

High School & Beyond Two-Level Regression Estimates

Log-likelihood				−23,248
Number of parameters				10
BIC				46,585

Parameter	Estimate	*SE*	Est./*SE*	Two-Tailed P-Value
Within level				
Residual variance				
math	36.720	0.721	50.944	0.000
Between level				
math (β_{0j}) ON				
sector (γ_{01})	1.227	0.308	3.982	0.000
mean_ses (γ_{02})	5.332	0.336	15.871	0.000
s_ses (β_{1j}) ON				
sector (γ_{11})	−1.640	0.238	−6.905	0.000
mean_ses (γ_{12})	1.033	0.333	3.100	0.002
math WITH				
s_ses	0.200	0.192	1.041	0.298
Intercepts				
math (γ_{00})	12.096	0.174	69.669	0.000
s_ses (γ_{10})	2.938	0.147	19.986	0.000
Residual variances				
math	2.316	0.414	5.591	0.000
s_ses	0.071	0.201	0.352	0.725

Source: Raudenbush, S.W., & Bryk, A.S., *Hierarchical linear models: Applications and data analysis methods* (2nd ed.). Newbury Park, CA: Sage Publications, 2002.

It seems possible that this phenomenon is quite common due to heterogeneous subpopulations in general population surveys. Such heterogeneity is captured by level 1 latent classes. Drawing on Muthén and Asparouhov (2009), these ideas can be formalized as follows.

Consider a two-level regression mixture model where the random intercept and slope of a linear regression of a continuous variable y on a covariate x for individual i in cluster j vary across the latent classes of an individual-level latent class variable C with K categories labeled $c = 1, 2, \ldots, K$,

$$y_{ij|C_{ij}=c} = \beta_{0cj} + \beta_{1cj}\, x_{ij} + r_{ij}, \qquad (2.16)$$

where the residual $r_{ij} \sim N(0, \theta_c)$ and a single covariate is used for simplicity. The probability of latent class membership varies as a two-level multinomial logistic regression function of a covariate z,

$$P(C_{ij} = c \mid z_{ij}) = \frac{e^{a_{cj} + b_c\, z_{ij}}}{\sum_{s=1}^{K} e^{a_{sj} + b_s\, z_{ij}}}. \qquad (2.17)$$

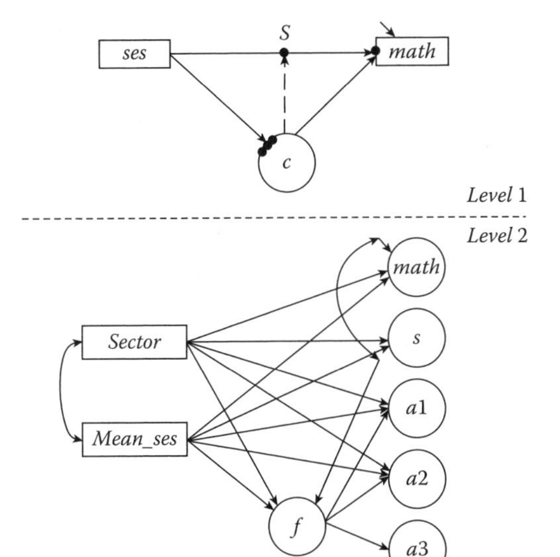

FIGURE 2.1
Model diagram for two-level regression mixture analysis.

The corresponding level-2 equations are

$$\beta_{0cj} = \gamma_{00c} + \gamma_{01c}\, w_{0j} + u_{0j}, \qquad (2.18)$$

$$\beta_{1cj} = \gamma_{10c} + \gamma_{11c}\, w_{1j} + u_{1j}, \qquad (2.19)$$

$$a_{cj} = \gamma_{20c} + \gamma_{21c}\, w_{2j} + u_{2cj}. \qquad (2.20)$$

With K categories for the latent class variable there are $K-1$ equations (Equation 2.20). Here, w_{0j}, w_{1j}, and w_{2j} are level-2 covariates and the residuals u_{0j}, u_{1j}, and u_{2j} are $(2 + K - 1)$-variate normally distributed with means zero and covariance matrix Θ_2 and are independent of r_{ij}. In many cases $z = x$ in Equation 2.17. Also, the level 2 covariates in Equations 2.18 through 2.20 may be the same as is the case in the HSB example considered below, where there is a common $w_j = w_{0j} = w_{2j}$. To reduce the dimensionality, a continuous factor f will represent the random intercept variation of Equation 2.20 in line with Muthén and Asparouhov (2009).

Figure 2.1 shows a diagram of a two-level regression mixture model applied to the HSB data. A four-class model is chosen and obtains a log-likelihood value of 22,812 with 30 parameters, and BIC = 45,891. This BIC value is considerably better than the conventional two-level regression BIC value of 46,585 reported in Table 2.1 and the mixture model is therefore preferable. The mixture model and its ML estimates can be interpreted as follows. Since this type of model is new to readers, Figure 2.1 will be used to understand the estimates rather than reporting a table of the parameter estimates for Equations 2.16 through 2.20.

The latent class variable *c* in the level 1 part of Figure 2.1 has four classes. As indicated by the arrows from *c*, the four classes are characterized by having different intercepts for *math* and different slopes for *math* regressed on *ses*. In particular, the *math* mean changes significantly across the classes. An increasing value of the *ses* covariate gives an increasing odds of being in the highest math class that contains 31% of the students. For three classes with the lowest *math* intercept, *ses* does not have a further, direct influence on *math*: the mean of the random slope *s* is only significant in the class with the highest *math* intercept, where it is positive.

The random intercepts of *c*, marked with filled circles on the circle for *c* on level 1, are continuous latent variables on level 2, denoted *a*1 – *a*3 (four classes gives three intercepts because the last one is standardized to zero). The (co-)variation of the random intercepts is, for simplicity, represented via a factor *f*. These random effects carry information about the influence of the school context on the probability of a student's latent class membership. For example, the influence of the level 2 covariate *sector* (public = 0, Catholic = 1) is such that Catholic schools are less likely to contribute to students being in the lower *math* intercept classes relative to the highest *math* intercept class. Similarly, a high value of the level 2 covariate *mean_ses* causes students to be less likely to be in the lower *math* intercept classes relative to the highest *math* intercept class.

The influence of the level 2 covariates on the random slope *s* is such that Catholic schools have lower values and higher *mean_ses* schools have higher values. The influence of the level 2 covariates on the random intercept *math* is insignificant for *sector* while positive significant for *mean_ses*. The insignificant effect of *sector* does not mean, however, that *sector* is unimportant to math performance given that *sector* had a significant influence on the random effects of the latent class variable *c*.

It is interesting to compare the mixture results to those of the conventional two level regression in Table 2.1. The key results for the conventional analysis is that (a) Catholic schools show less influence of *ses* on *math*, and (b) Catholic schools have higher mean math achievement. Neither of these results are contradicted by the mixture analysis. But using a model that has considerably better BIC, the mixture model explains these results by a mediating latent class variable on level 1. In other words, students' latent class membership is what influences math performance and latent class membership is predicted by both student-level ses and school characteristics. The Catholic school effect on math performance is not direct as an effect on the level 2 math intercept (this path is insignificant), but indirect via the student's latent class membership. For more details on two-level regression mixture modeling and a math achievement example focusing on gender differences, see Muthén and Asparouhov (2009).

2.3 TWO-LEVEL PATH ANALYSIS AND STRUCTURAL EQUATION MODELING

Regression analysis is often only a small part of a researcher's modeling agenda. Frequently a system of regression equations is specified as in path analysis and structural equation modeling (SEM). There have been recent developments for path analysis and SEM in multilevel data and a brief overview of new kinds of models will be presented in this section. No data analysis is done, but focus is instead on modeling ideas.

Consider the left part of Figure 2.2 where the binary dependent variable *hsdrop*, representing dropping out by Grade 12, is related to a set of covariates using logistic regression. A complication in this analysis is that many of those who drop out by Grade 12 have missing data on *math*10, the mathematics score in Grade 10, where the missingness is not completely at random. Missingness among covariates can be handled by adding a distributional assumption for the covariates, either by multiple imputation or by not treating them as exogenous. Either way, this complicates the analysis without learning more about the relationships among the variables in the model. The right part of Figure 2.2 shows an alternative approach using a path model that acknowledges the temporal position of *math*10 as an intervening variable that is predicted by the remaining covariates measured earlier. In this path model, "missing at random" (MAR; Little & Rubin, 2002) is reasonable in that the covariates may well predict the missingness in *math*10. The resulting path model has a combination of a linear regression for a continuous dependent variable and a logistic regression for a binary dependent variable.

Figure 2.3 shows a two-level counterpart to the path model. The top part of Figure 2.3 shows the within-level part of the model for the student relationships. Here, the filled circles at the end of the arrows indicate random intercepts. On the between level these random intercepts are continuous latent variables varying across schools. The two random intercepts are not treated symmetrically, but it is hypothesized that increasing *math*10 intercept decreases the *hsdrop* intercept in that schools with good mean math performance in Grade 10 tend to have an environment less conducive to dropping out. Two school-level covariates are used as predictors of the random intercepts, *lunch*, which is a dummy variable used as a

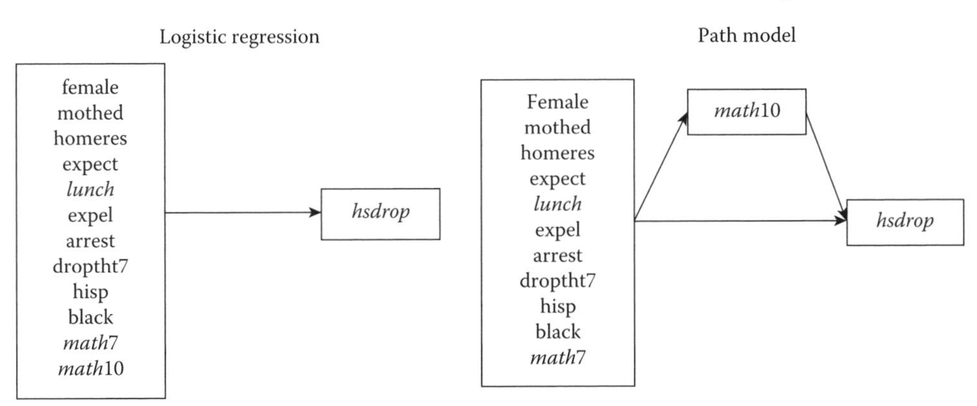

FIGURE 2.2
Model diagram for logistic regression path analysis.

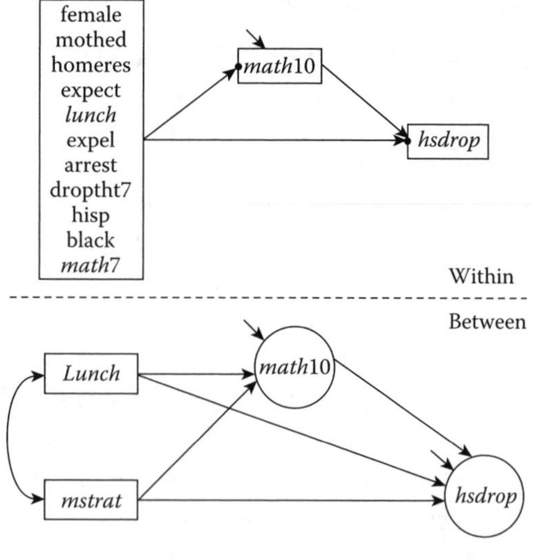

FIGURE 2.3
Model diagram for two-level logistic regression path analysis.

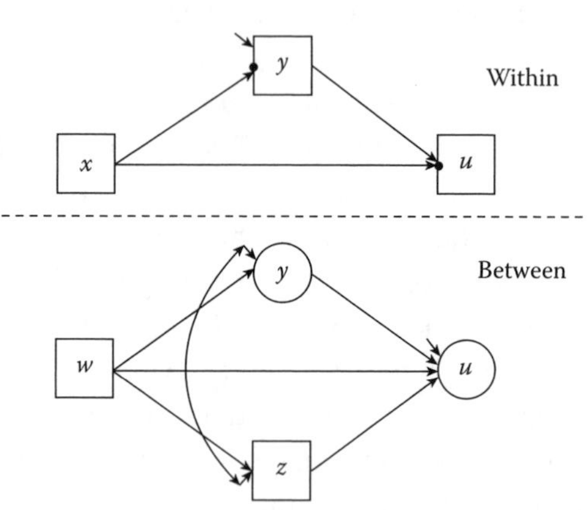

FIGURE 2.4
Model diagram for path analysis with between-level dependent variable.

poverty proxy and *mstrat*, measuring math teacher workload as the ratio of students to full-time math teachers.

Another path analysis example is shown in Figure 2.4. Here, u is again a categorical dependent variable and both u and the continuous variable y have random intercepts. Figure 2.4 further illustrates the flexibility of current two-level path analysis by adding an observed between-level dependent variable z that intervenes between the between-level covariate w and the random intercept of u. Between-level variables that play a role as dependent variables are not used in conventional multilevel modeling.

Figure 2.5 shows a path analysis example with random slopes a_j, b_j, and c_j'. This illustrates a two-level mediational model. As described in Bauer, Preacher, and Gil (2006) for example, the indirect effect is here $\alpha \times \beta + Cov(a_j, b_j)$, where α and β are the means of the corresponding random slopes a_j and b_j.

Figure 2.6 specifies a MIMIC model with two factors *fw*1 and *fw*2 for students on the within level. The filled circles at the binary indicators $u1 - u6$ indicate random intercepts that are continuous latent variables on the between level. The between level has a single factor *fb* describing the variation and covariation among the random intercepts. The between level has the unique feature of also adding between-level indicators $y1 - y4$ for a between-level factor *f*, another example

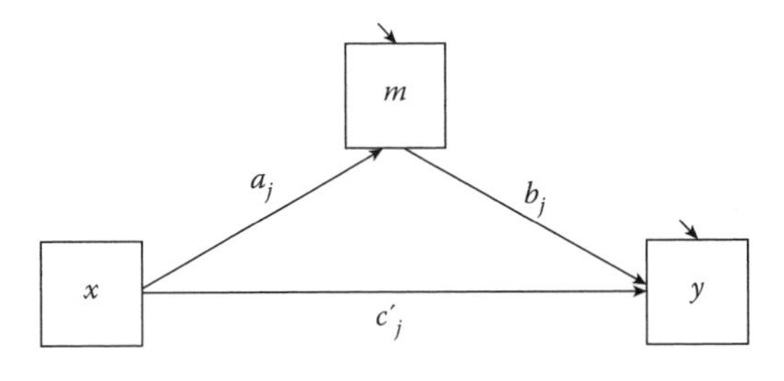

FIGURE 2.5
Model diagram for path analysis with mediation and random slopes.

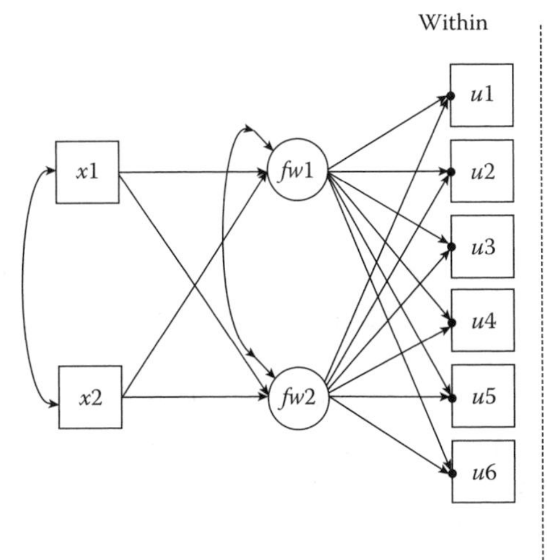

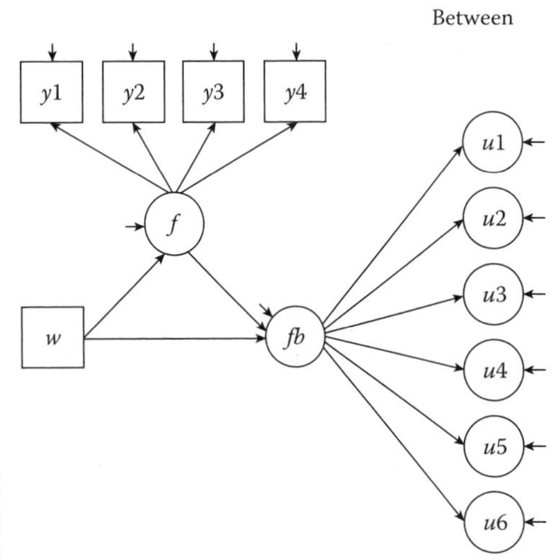

FIGURE 2.6
Model diagram for two-level SEM.

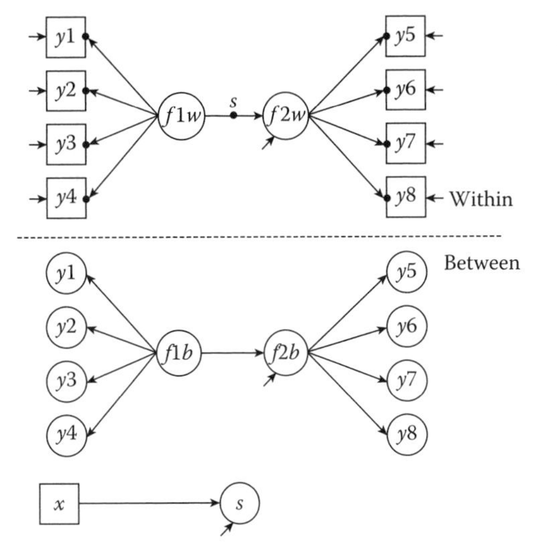

FIGURE 2.7
Model diagram for two-level SEM with a random structural slope.

of between-level dependent variables. Two-level factor analysis will be discussed in more detail in Section 2.4.

Figure 2.7 shows a structural equation model with an exogeneous and an endogenous factor that has both within-level and between-level variation. The special feature here is that the structural slope s is random. The slope s is regressed on a between-level covariate x.

2.4 TWO-LEVEL EXPLORATORY FACTOR ANALYSIS

A recent multilevel development concerns a practical alternative to ML estimation in situations that would lead to heavy ML computations (cf. Asparouhov & Muthén, 2007). Heavy ML computations occur when numerical integration is needed, as for instance with categorical outcomes. Many models, including factor analysis models, involve many random effects, each one of which adds a dimension of integration. The new estimator uses limited information from first- and second-order moments to formulate a weighted least squares approach that reduces multidimensional integration into a series of one and two-dimensional integrations for the uni- and bivariate moments. This weighted least squares approach is particularly useful in EFA where there are typically many random effects due to having many variables and many factors.

Consider the following EFA example. Table 2.2 shows the item distribution for a set of 13 items measuring aggressive-disruptive behavior in the classroom among 363 boys in 27 classrooms in Baltimore public schools. It is clear that the variables have very skewed distributions with strong floor effects so that 40–80% are at the lowest value. If treated as continuous outcomes, even nonnormality robust standard errors and χ^2 tests of model fit would not give correct results in that a linear model is not suitable for data with such strong floor effects. The variables will instead be treated as ordered polytomous (ordinal). The 13-item instrument is hypothesized to capture three aspects of aggressive–disruptive behavior: property, verbal, and person. Figure 2.8 shows a model diagram with notation analogous to two-level regression. On the within (student) level the three hypothesized factors are denoted $fw1 - fw3$. The filled circles at the observed items indicate random measurement intercepts. On the between level

TABLE 2.2

Distributions for Aggressive-Disruptive Items

Aggression Items	Almost Never (Scored as 1)	Rarely (Scored as 2)	Sometimes (Scored as 3)	Often (Scored as 4)	Very Often (Scored as 5)	Almost Always (Scored as 6)
Stubborn	42.5	21.3	18.5	7.2	6.4	4.1
Breaks rules	37.6	16.0	22.7	7.5	8.3	8.0
Harms others and property	69.3	12.4	9.40	3.9	2.5	2.5
Breaks things	79.8	6.60	5.20	3.9	3.6	0.8
Yells at others	61.9	14.1	11.9	5.8	4.1	2.2
Takes others' property	72.9	9.70	10.8	2.5	2.2	1.9
Fights	60.5	13.8	13.5	5.5	3.0	3.6
Harms property	74.9	9.90	9.10	2.8	2.8	0.6
Lies	72.4	12.4	8.00	2.8	3.3	1.1
Talks back to adults	79.6	9.70	7.80	1.4	0.8	1.4
Teases classmates	55.0	14.4	17.7	7.2	4.4	1.4
Fights with classmates	67.4	12.4	10.2	5.0	3.3	1.7
Loses temper	61.6	15.5	13.8	4.7	3.0	1.4

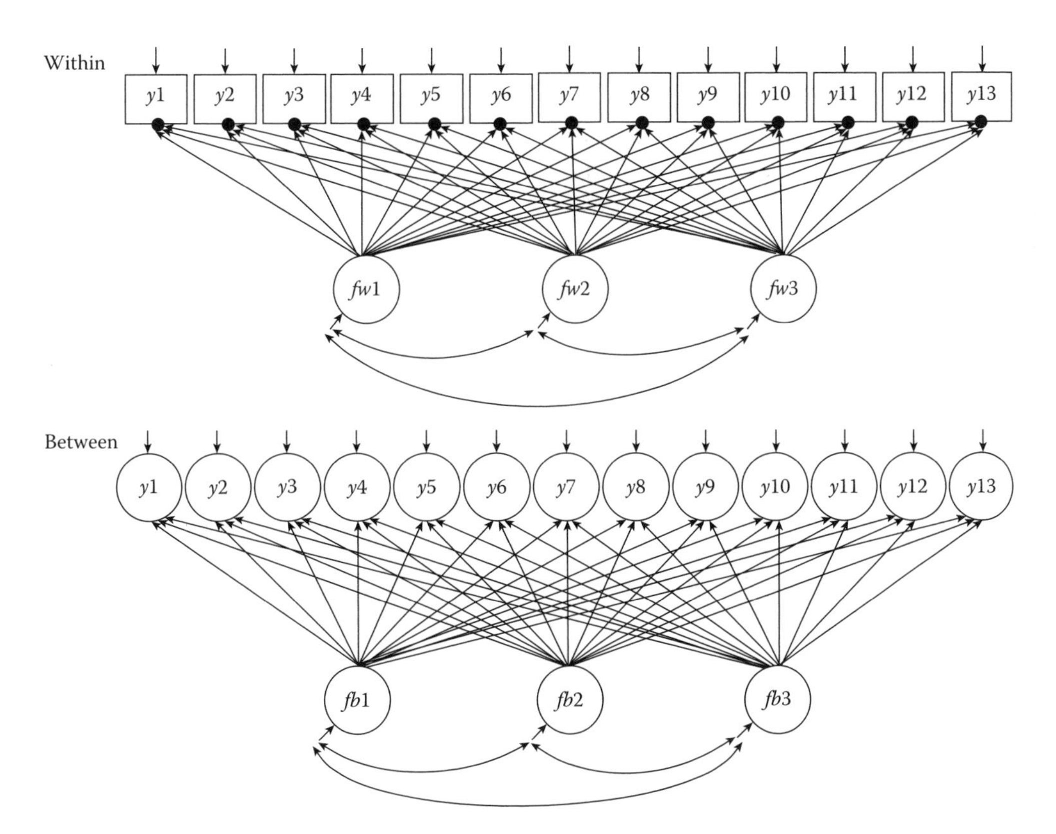

FIGURE 2.8
Two-level factor analysis model.

these random intercepts are continuous latent variables varying over classrooms, where the variation and covariation is represented by the classroom-level factors *fb*1 – *fb*3. The meaning of the student-level factors *fw*1 – *fw*3 is in line with regular factor analysis. In contrast, the classroom-level factors *fb*1 – *fb*3 represent classroom-level phenomena for which a researcher typically has less understanding. These factors require new kinds of considerations as follows. If the same set of three within-level factors (property, verbal, and person) are to explain the (co-)variation on the between level, classroom teachers must vary in their skills to manage their classrooms with respect to all three of these aspects. That is, some teachers are good at controlling property-oriented, aggressive-disruptive behavior and some are not, some teachers are good at controlling verbally oriented, aggressive-disruptive behavior and some are not, and so on. This is not very likely and it is more likely that teachers simply vary in their ability to manage their classrooms in all three respects fairly equally. This would lead to a single factor *fb* on the between level instead of three factors.

As shown in Figure 2.8, ML estimation would require 19 dimensions of numerical integration, which is currently an impossible task. A reduction is possible if the between-level, variable-specific residuals are zero, which is often a good approximation. This makes for a reduction to six dimensions of integration, which is still a very difficult task. The Asparouhov and Muthén (2007) weighted least squares approach is suitable for such a situation and will be used here. The approach assumes that the factors are normally distributed and uses an ordered probit link function for the item probabilities as functions of the factors. This amounts to assuming multivariate normality for continuous latent response variables underlying the items in line with using polychoric correlations in single-level analysis. Rotation of loadings on both levels is provided along with standard errors for rotated loadings and resulting factor correlations.

Table 2.3 shows a series of analyses varying the number of factors on the within and between levels. To better understand how many factors are needed on a certain level, an unrestricted correlation model can be used on the other level. Using an unrestricted within-level model it is clear that a single between-level factor is sufficient. Adding within-level factors shows an improvement in fit going up to four factors. The 4-factor solution, however, has no significant loadings for the additional, fourth factor. Also, the 3-factor solution captures the three hypothesized factors. The factor solution is shown in Table 2.4 using Geomin rotation (Asparouhov & Muthén, 2009) for the within level. Factor loadings with

TABLE 2.3

Two-Level EFA Model Test Result for Aggressive–Disruptive Items

Within-Level Factors	Between-Level Factors	Df	Chi-Square	CFI	RMSEA
Unrestricted	1	65	66 (p = 0.43)	1.000	0.007
1	1	130	670	0.991	0.107
2	1	118	430	0.995	0.084
3	1	107	258	0.997	0.062
4*	1	97	193	0.998	0.052

*4th factor has no significant loadings.

TABLE 2.4

Two-Level EFA of Aggressive–Disruptive Items Using WLSM and Geomin Rotation

	Within-Level Loadings			
Aggression Items	Property	Verbal	Person	Between-Level Loadings General
Stubborn	0.00	**0.78***	0.01	**0.65***
Breaks rules	0.31*	0.25*	0.32*	**0.61***
Harms others and property	**0.64***	0.12	0.25*	**0.68***
Breaks things	**0.98***	0.08	–0.12*	**0.98***
Yells at others	0.11	**0.67***	0.10	**0.93***
Takes others' property	**0.73***	**–0.15***	0.31*	**0.80***
Fights	0.10	0.03	**0.86***	**0.79***
Harms property	**0.81***	0.12	0.05	**0.86***
Lies	**0.60***	0.25*	0.10	**0.86***
Talks back to adults	0.09	**0.78***	0.05	**0.81***
Teases classmates	0.12	0.16*	**0.59***	**0.83***
Fights with classmates	–0.02	0.13	**0.88***	**0.84***
Loses temper	–0.02	**0.85***	0.05	**0.87***

asterisks represent loadings significant on the 5% level, while bolded loadings are the more substantial ones. The loadings for the single between-level factor are fairly homogeneous supporting the idea that there is a single classroom management dimension.

2.5 GROWTH MODELING (TWO-LEVEL ANALYSIS)

Growth modeling concerns repeated measurement data nested within individuals and possibly also within higher-order units (clusters such as schools). This will be referred to as two- and three-level growth analysis, respectively. Often, two-level growth analysis can be performed in a multivariate, wide data format fashion, letting the level 1 repeated measurement on y over T time points be represented by a multivariate outcome vector $y = (y_1, y_2, \ldots, y_T)'$, reducing the two levels to one. This reduction by one level is typically used in the latent variable framework of M*plus*. More common, however, is to view growth modeling as a two-level model with features analogous to those of two-level regression (see, e.g., Raudenbush & Bryk, 2002). In this case, data are arranged in a univariate, long format.

Following is a simple example with linear growth, for simplicity using the notation of Raudenbush and Bryk (2002). For time point t and individual i, consider

y_{ti}: individual-level, outcome variable

a_{ti}: individual-level, time-related variable (age, grade)

x_i: individual-level, time-invariant covariate

and the two-level growth model

$$\textit{Level } 1: y_{ti} = \pi_{0i} + \pi_{1i}\, a_{ti} + e_{ti}, \quad (2.21)$$

$$\textit{Level } 2: \begin{cases} \pi_{0i} = \gamma_{00} + \gamma_{01}\, x_i + r_{0i}, \\ \pi_{1i} = \gamma_{10} + \gamma_{11}\, x_i + r_{1i}, \end{cases} \quad (2.22)$$

where π_0 is a random intercept and π_1 is a random slope. One may ask if there really is anything new that can be said about (two-level) growth analysis. The answer, surprisingly, is again yes. Following is a discussion of a relatively recent and still underutilized extension to situations with very skewed outcomes similar to those studied in the above EFA. Here, the example concerns frequency of heavy drinking in the last 30 days from the National Longitudinal Survey of Youth (NLSY), a U.S. national survey. The distribution of the outcome at age 24 is shown in Figure 2.9, where a majority of individuals did not engage in heavy drinking in the last 30 days. Olsen and Schafer (2001) proposed a two-part or semicontinuous growth model for data of this type, treating the outcome as continuous but adding a special modeling feature to take into account the strong floor effect.

The two-part growth modeling idea is shown in Figure 2.10, where the outcome is split into two parts, a binary part and a continuous part. Here, *iy* and *iu* represent random intercepts π_0, whereas *sy* and *su* represent random linear slopes π_1. In addition, the model has random quadratic slopes *qy* and *qu*. The binary part is a growth model describing for each time point the probability of an individual experiencing the event, whereas for those who experienced it the continuous part describes the amount, in this case the number of heavy drinking occasions in the last 30 days. For an individual who does not experience the event, the continuous part is recorded as missing. A joint growth model for the binary and the

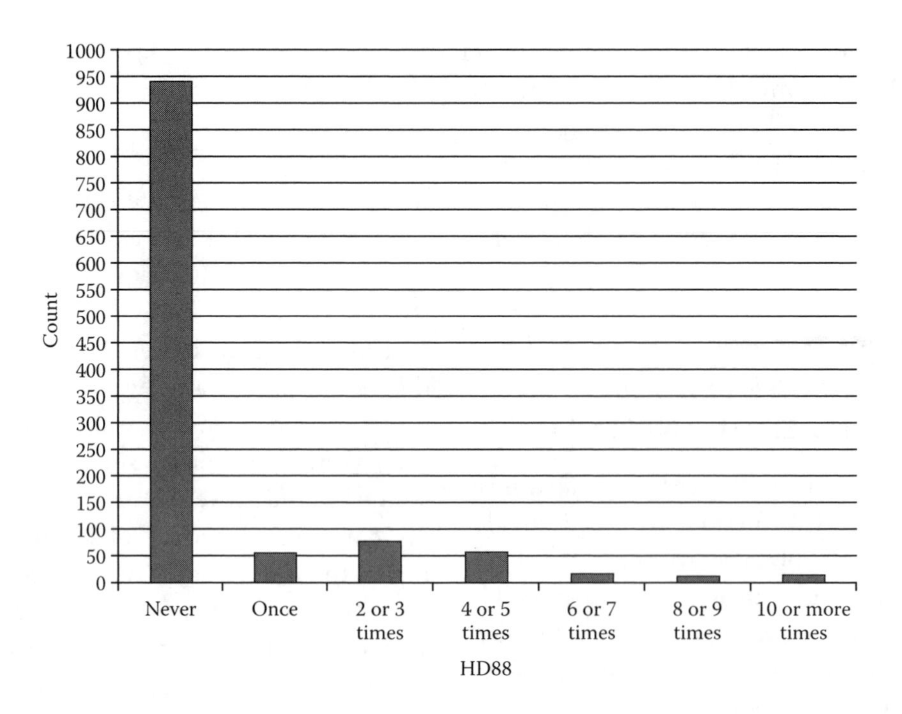

FIGURE 2.9
Histogram for heavy drinking at age 24.

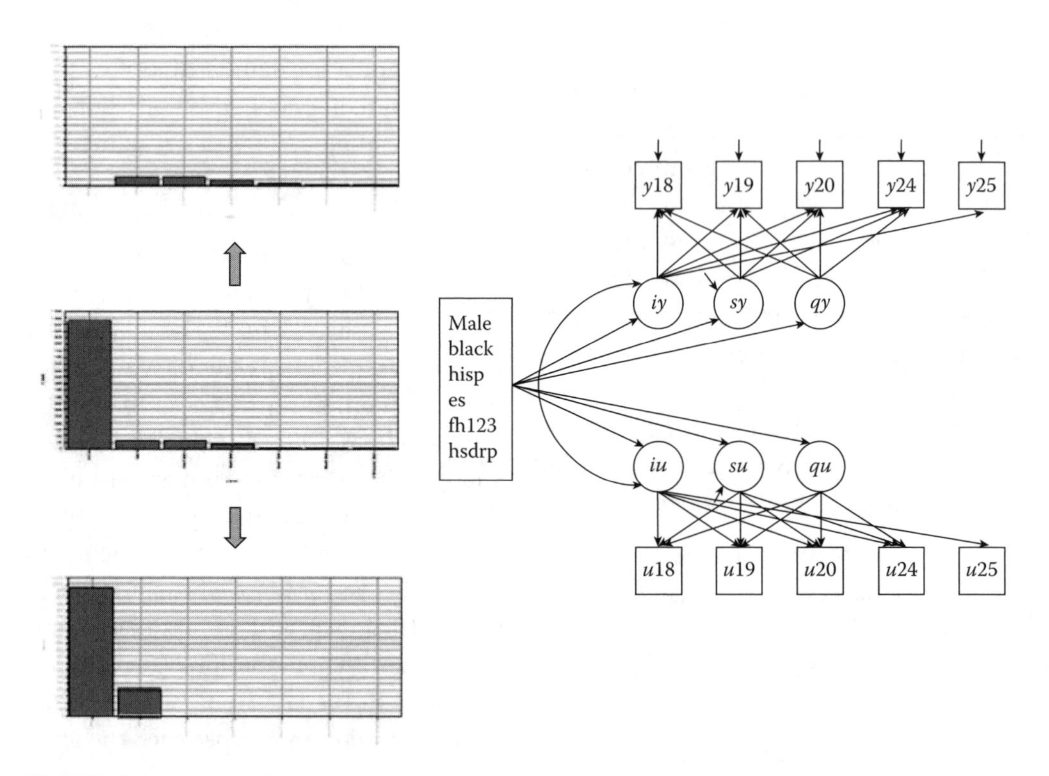

FIGURE 2.10
Two-part growth model for heavy drinking.

continuous process scored in this way represents the likelihood given by Olsen and Schafer (2001).

Nonnormally distributed outcomes can often be handled by ML using a nonnormality robust standard error approach, but this is not sufficient for outcomes such as shown in Figure 2.9 given that a linear model is unlikely to hold. To show the difference in results as compared two-part growth modeling, Table 2.5 shows the M*plus* output for the estimated growth model for frequency of heavy drinking ages 18 to 25. The results focus on the regression of the random intercept *i* on the time-invariant covariates in the model. The time scores are centered at age 25 so that the random intercept refers to the systematic part of the growth curve at age 25. It is seen that the regular growth modeling finds all but the last two covariates significant. In contrast, the two-part modeling finds several of the covariates insignificant in one part or the other (the two parts are labeled *iy ON* for the continuous part and *iu ON* for the binary part. Consider as an example, the covariate *black*. As is typically found being black has a significant negative influence in the regular growth

TABLE 2.5

Two-Part Growth Modeling of Frequency of Heavy Drinking Ages 18–25

Parameter	Estimate	*SE*	Est./*SE*	Std	StdYX
Regular growth modeling, treating outcome as continuous. Nonnormality robust ML (MLR)					
***i* ON**					
male	0.769	0.076	10.066	0.653	0.326
black	−0.336	0.083	−4.034	−0.286	−0.127
hisp	−0.227	0.103	−2.208	−0.193	−0.071
es	0.291	0.128	2.283	0.247	0.088
fh123	0.286	0.137	2.089	0.243	0.075
hsdrp	−0.024	0.104	−0.232	−0.0240	−0.008
coll	−0.131	0.086	−1.527	−0.111	−0.052
Two-part growth modeling					
***iy* ON**					
male	0.262	0.052	5.065	0.610	0.305
black	−0.096	0.059	−1.619	−0.223	−0.099
hisp	−0.130	0.066	−1.963	−0.111	−0.111
es	0.082	0.062	1.333	0.191	0.068
fh123	0.213	0.076	2.815	0.495	0.152
hsdrp	0.084	0.065	1.289	0.195	0.078
coll	−0.015	0.053	−0.280	−0.035	−0.016
***iu* ON**					
male	2.041	0.176	11.594	0.949	0.474
black	−1.072	0.203	−5.286	−0.499	−0.222
hisp	−0.0545	0.234	−2.331	−0.254	−0.093
es	0.364	0.234	1.560	0.169	0.060
fh123	0.562	0.275	2.045	0.262	0.080
hsdrp	−0.238	0.216	−1.103	−0.111	−0.044
coll	−0.259	0.196	−1.317	−0.120	−0.056

modeling, lowering the frequency of heavy drinking. In the two-part modeling this covariate is insignificant for the continuous part and significant only for the binary part. This implies that, holding other covariates constant, being black significantly lowers the risk of engaging in heavy drinking, but among blacks who are engaging in heavy drinking there is no difference in amount compared to other ethnic groups. These two paths of influence are confounded in the regular growth modeling.

As shown in Figure 2.11, a distal outcome can also be added to the growth model. In this example, the distal outcome is a DSM-based classification into alcohol dependence or not by age 30. The distal outcome is predicted by the age 25 random intercept using a logistic regression model part. Table 2.6 shows that the distal outcome is

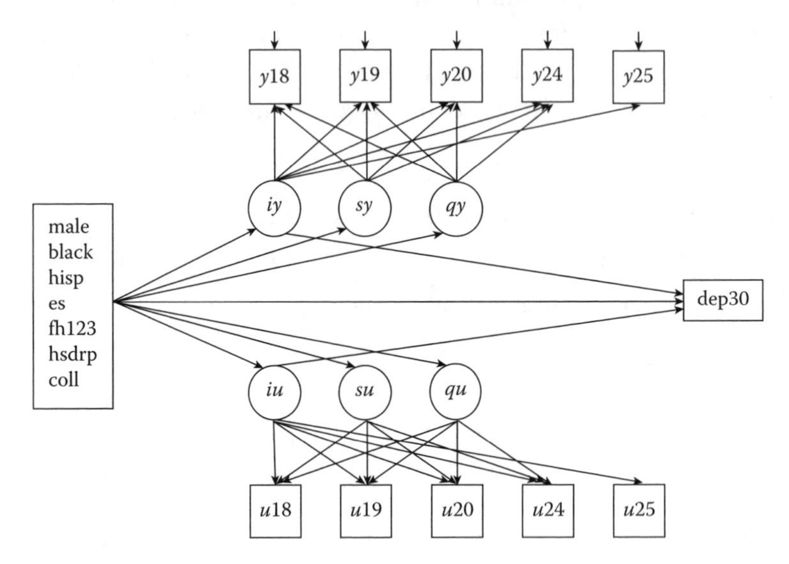

FIGURE 2.11
Two-part growth model for heavy drinking and a distal outcome.

TABLE 2.6

Two-Part Growth Modeling of Frequency of Heavy Drinking Ages 18–25 With a Distal Outcome

Parameter	Estimate	*SE*	Est./*SE*	Std	StdYX
dep30 ON					
iu	0.440	0.141	3.120	0.949	0.427
iy	0.874	0.736	1.187	0.373	0.168
dep30 ON					
male	−0.098	0.291	−0.337	−0.098	−0.022
black	0.415	0.294	1.414	0.415	0.083
hisp	0.025	0.326	0.075	0.025	0.004
es	0.237	0.286	0.830	0.237	0.038
fh123	0.498	0.325	1.532	0.498	0.069
hsdrp	0.565	0.312	1.812	0.545	0.101
coll	−0.384	0.276	−1.390	−0.384	−0.081

significantly influenced only by the age 25-defined random intercept *iu* for the binary part, not by the random intercept for the continuous part. In other words, if the probability of engaging in heavy drinking at age 25 is high, the probability of alcohol dependence by age 30 is high. But the alcohol dependence probability is not significantly influenced by the frequency of heavy drinking at age 25. The results also show that controlling for age 25 heavy drinking behavior, none of the covariates has a significant influence on the distal outcome.

2.6 GROWTH MODELING (THREE-LEVEL ANALYSIS)

This section considers growth modeling of individual- and cluster-level data. A typical example is repeated measures over grades for students nested within schools. One may again ask if there really is anything new that can be said about growth modeling in cluster data. The answer, surprisingly, is once again yes. An important extension to the conventional three-level analysis becomes clear when viewed from a general latent variable modeling perspective.

For simplicity, the notation will be chosen to coincide with that of Raudenbush and Bryk (2002). Consider the observed variables for time point t, individual i, and cluster j,

- y_{tij}: individual-level, outcome variable
- a_{1tij}: individual-level, time-related variable (age, grade)
- a_{2tij}: individual-level, time-varying covariate
- x_{ij}: individual-level, time-invariant covariate
- w_j: cluster-level, covariate

and the three-level growth model

$$\text{Level 1: } y_{tij} = \pi_{0ij} + \pi_{1ij}\, a_{1tij} + \pi_{2tij}\, a_{2tij} + e_{tij}, \tag{2.23}$$

$$\text{Level 2: } \begin{cases} \pi_{0ij} = \beta_{00j} + \beta_{01j}\, x_{ij} + r_{0ij}, \\ \pi_{1ij} = \beta_{10j} + \beta_{11j}\, x_{ij} + r_{1ij}, \\ \pi_{2tij} = \beta_{20tj} + \beta_{21tj}\, x_{ij} + r_{2tij}, \end{cases} \tag{2.24}$$

$$\text{Level 3: } \begin{cases} \beta_{00j} = \gamma_{000} + \gamma_{001}\, w_j + u_{00j}, \\ \beta_{10j} = \gamma_{100} + \gamma_{101}\, w_j + u_{10j}, \\ \beta_{20tj} = \gamma_{200t} + \gamma_{201t}\, w_j + u_{20tj}, \\ \beta_{01j} = \gamma_{010} + \gamma_{011}\, w_j + u_{01j}, \\ \beta_{11j} = \gamma_{110} + \gamma_{111}\, w_j + u_{11j}, \\ \beta_{21tj} = \gamma_{21t0} + \gamma_{21t1}\, w_j + u_{21tj}. \end{cases} \tag{2.25}$$

Here, the πs are random intercepts and slopes varying across individuals and clusters, and the βs are random intercepts and slopes varying across clusters. The residuals e, r, and u are assumed normally distributed with zero means, uncorrelated with respective right-hand side covariates, and uncorrelated across levels.

In M*plus*, growth modeling in cluster data is represented in a similar, but slightly different way that offers further modeling flexibility. As mentioned in Section 2.5 the first difference arises from the level 1 repeated measurement on y over time being represented by a multivariate outcome vector $y = (y_1, y_2, \ldots, y_T)'$ so that the number of levels is reduced from three to two. The second difference is that each variable, with the exception of variables multiplied by random slopes, is decomposed into uncorrelated within- and between-cluster components. Using subscripts w and b to represent within- and between-cluster variation, one may write the variables in Equation 2.23 as

$$y_{tij} = y_{btj} + y_{wtij}, \tag{2.26}$$

$$\pi_{0ij} = \pi_{0bj} + \pi_{0wij}, \quad (2.27)$$

$$\pi_{1ij} = \pi_{1bj} + \pi_{1wij}, \quad (2.28)$$

$$\pi_{2tij} = \pi_{2tbj} + \pi_{2twij}, \quad (2.29)$$

$$e_{tij} = e_{btj} + e_{wtij}, \quad (2.30)$$

so that the level 1 Equation 23 can be expressed as

$$y_{tij} = \pi_{0bj} + \pi_{0wij} + (\pi_{1bj} + \pi_{1wij}) a_{1tij} + (\pi_{2btj} + \pi_{2wtij})\, a_{2tij} + e_{btj} + e_{wtij}. \quad (2.31)$$

The three-level model of Equations 2.23 through 2.25) can then be rewritten as a two-level model with levels corresponding to within- and between-cluster variation,

$$\text{Within:}\begin{cases} y_{wtij} = \pi_{0wij} + \pi_{1wij}\, a_{1tij} \\ \quad\quad + \pi_{2wtij}\, a_{2tij} + e_{wtij}, \\ \pi_{0wij} = \beta_{01j}\, x_{ij} + r_{0ij}, \\ \pi_{1wij} = \beta_{11j}\, x_{ij} + r_{1ij}, \\ \pi_{2wtij} = \beta_{21tj}\, x_{ij} + r_{2tij}, \end{cases} \quad (2.32)$$

$$\text{Between:}\begin{cases} y_{btj} = \pi_{0bj} + \pi_{1bj}\, a_{1tij} \\ \quad\quad + \pi_{2btj}\, a_{2tij} + e_{btj}, \\ \pi_{0bj} = \beta_{00j} = \gamma_{000} + \gamma_{001}\, w_j + u_{00j}, \\ \pi_{1bj} = \beta_{10j} = \gamma_{100} + \gamma_{101}\, w_j + u_{10j}, \\ \pi_{2btj} = \beta_{20tj} = \gamma_{200t} + \gamma_{201t}\, w_j + u_{20tj}, \\ \beta_{01j} = \gamma_{010} + \gamma_{011}\, w_j + u_{01j}, \\ \beta_{11j} = \gamma_{110} + \gamma_{111}\, w_j + u_{11j}, \\ \beta_{21tj} = \gamma_{21t0} + \gamma_{21t1}\, w_j + u_{21tj}. \end{cases} \quad (2.33)$$

From the latent variable perspective taken in M*plus*, the first line of the within level Equation 2.32 and the first line of the between level Equation 2.33 is the measurement part of the model with growth factors π_0, π_1 measured by multiple indicators y_t. The next lines of each level contain the structural part of the model. As is highlighted in Equation 2.31, the rearrangement of the 3-level model as Equation 2.32, Equation 2.33 shows that the three-level model typically assumes that the measurement part of the model is invariant across within and between in that the same time scores a_{1tij} are used on both levels.

As seen in Equation 2.32, Equation 2.33 the decomposition into within and between components also occurs for the residual $e_{tij} = e_{wtij} + e_{btj}$. The e_{btj} term is typically fixed at zero in conventional multilevel modeling, but this is an important restriction. This restriction is not clear from the way the model is written in Equation 2.23. Time-specific, between-level variance parameters for the residuals e_{btj} are often needed to represent across-cluster variation in time-specific residuals.

Consider a simple example with no time-varying covariates and where the time scores do not vary across individuals or clusters, $a_{1tij} = a_{1t}$. To simplify notation in the actual M*plus* analyses, and dropping the *ij* and *j* subscripts, let $iw = \pi_{0w}$, $sw = \pi_{1w}$, $ib = \pi_{0b}$, and $sb = \pi_{1b}$ be the within-level and between-level growth factors, respectively. Figure 2.12 shows the model diagram for four time points using the within-level covariate x and the between-level covariate w. The model diagram may be seen as analogous to the two-level factor analysis model, adding covariates. The between-level part of the model is drawn with residual arrows pointing to the time-specific latent variables $y1 - y4$. These are the residuals e_{btj} that conventional growth analysis assumes are zero.

The model of Figure 2.12 is analyzed with and without the zero residual restriction using mathematics scores in Grades 7 through 10 from the Longitudinal Survey of American Youth (LSAY). Two

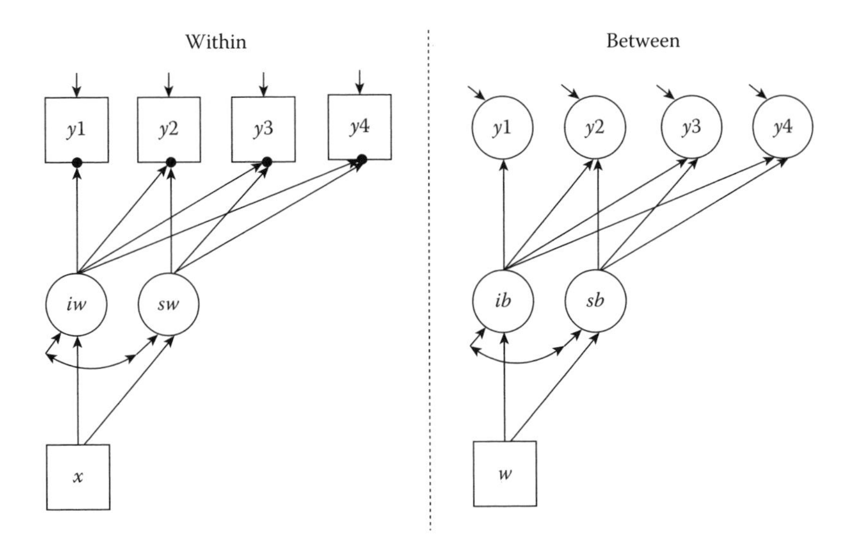

FIGURE 2.12
A two-level growth model (three-level analysis).

between-level covariates are added, *lunch* (a poverty index) and *mstrat* (math teacher workload). The between-level M*plus* ML results from the two analyses are shown in Table 2.7. The χ^2 model test of fit results show a big improvement when adding the residual variances to the model. The *sb* regression on *mstrat* also shows large differences between the two approaches with a smaller and insignificant effect in the conventional approach. Given that the *sb* residual variance estimate is larger for the conventional approach, it appears that the conventional model tries to absorb the residual variances into the slope growth factor variance. The residual variance for Grade 10 has a negative insignificant value that could be fixed at zero but does not change other results much.

2.6.1 Further Three-Level Growth Modeling Extensions

Figure 2.13 shows a student-level regression of the random slope *sw* regressed on the random intercept *iw*. With *iw* defined at the first time point, the study investigates to which extent the initial status influences the growth rate. The regression of the growth rate on the initial status has a random slope *s* that varies across clusters. For example, a researcher may be interested in how schools vary in their ability to reduce the influence of initial status on growth rate. Seltzer, Choi, and Thum (2002) studied this topic using Bayesian MCMC estimation, but ML can be used in M*plus*. Figure 2.13 shows how the school variation in *s* can be explained by a school-level covariate *w*. The rest of the school-level model is specified as in the previous section.

Figure 2.14 shows an example of a multiple-indicator, multilevel growth model. In this case the growth model simply uses a random intercept. The data have four levels in that the observations are indicators nested within time points, time points nested within individuals, and individuals nested within twin pairs. The model diagram, however, shows how this case can be expressed as a single-level model. This is accomplished using a triply multivariate

TABLE 2.7

Two-Level Growth Modeling (Three-Level Modeling) of LSAY Math Achievement, Grades 7–10

Parameter	Estimate	*SE*	Est./*SE*	Std	StdYX
Conventional growth modeling:					
Chi-square (32) = 179.58. Between-level estimates and *SE*s:					
sb ON					
lunch	−1.271	0.402	−3.160	−1.919	−0.397
mstrat	1.724	1.022	1.688	2.605	0.185
Residual variances					
math7	0.000	0.000	0.000	0.000	0.000
math8	0.000	0.000	0.000	0.000	0.000
math9	0.000	0.000	0.000	0.000	0.000
math10	0.000	0.000	0.000	0.000	0.000
ib	5.866	1.401	4.186	0.736	0.736
sb	0.354	0.138	2.564	0.809	0.809
Allowing time-specific level 3 residual variances:					
Chi-square (28) = 83.69. Between-level estimates and *SE*s:					
sb ON					
lunch	−1.312	0.367	−3.576	−2.495	−0.516
mstrat	2.281	0.771	2.957	4.338	0.308
Residual variances					
math7	1.396	0.749	1.863	1.396	0.159
math8	1.414	0.480	2.946	1.414	0.154
math9	0.382	0.381	1.002	0.382	0.042
math10	−0.121	0.518	−0.234	−0.121	−0.012
ib	5.211	1.410	3.694	0.704	0.704
sb	0.177	0.155	1.143	0.640	0.640

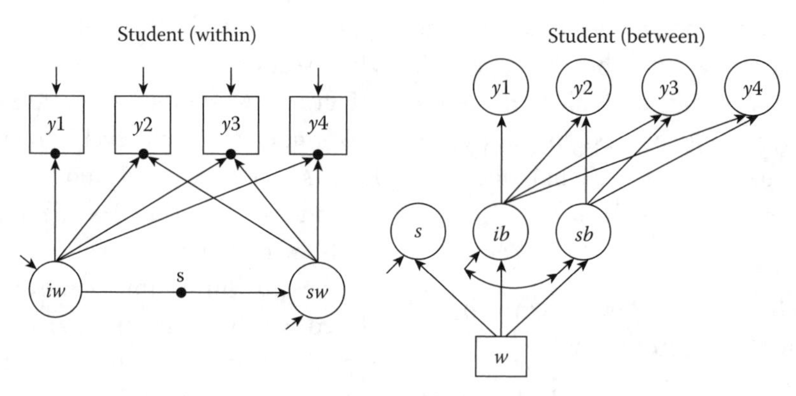

FIGURE 2.13
Multilevel modeling of a random slope regressing growth rate on initial status.

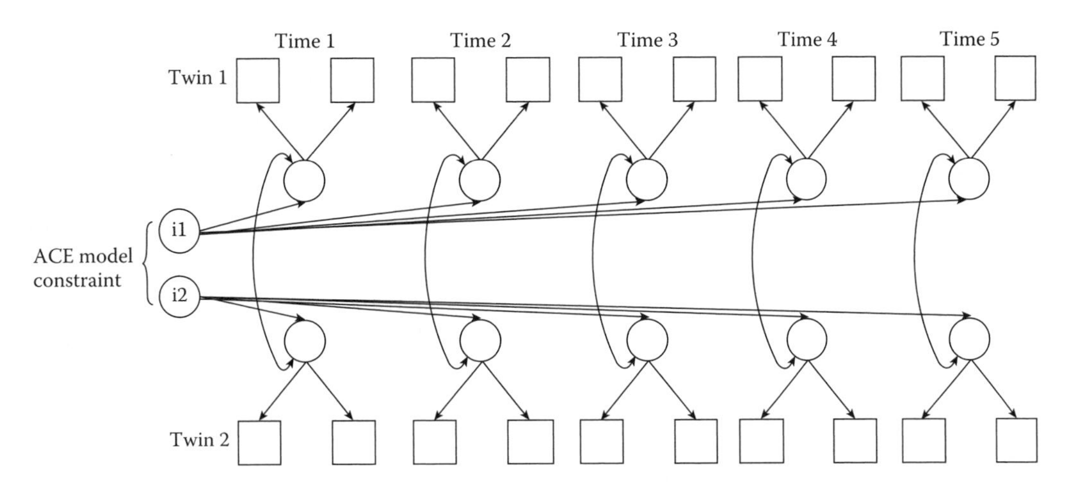

FIGURE 2.14
Multiple indicator multilevel growth.

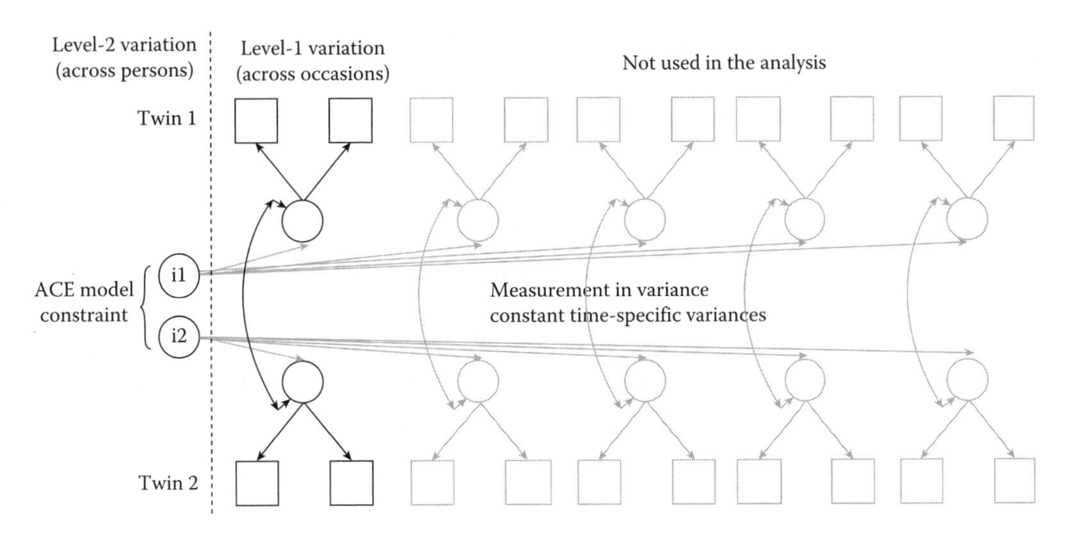

FIGURE 2.15
Multiple indicator multilevel growth in long form.

representation where the indicators (two in this case), time points (five in this case), and twins (two) create a 20-variate observation vector. With categorical outcomes, ML estimation needs numerical integration that is prohibitive given that there are 10 dimensions of integration, but weighted least squares estimation is straightforward.

Figure 2.15 shows an alternative, two-level approach. The data vector is arranged as doubly multivariate with indicators and twins creating four outcomes. The two levels are time and person. This approach assumes time-invariant measurement parameters and constant time-specific factor variances. These assumptions can be tested using the single-level approach in Figure 2.14 with weighted least squares estimation. With categorical outcomes, the two-level formulation of Figure 2.15 leads to four dimensions of integration with ML, which is possible but still quite heavy. A simple alternative is provided by the new two-level weighted least squares approach discussed for multilevel EFA in Section 2.4.

2.7 MULTILEVEL GROWTH MIXTURE MODELING

The growth model of Section 2.5 assumes that all individuals come from one and the same population. This is seen in Equation 2.22 where there is only one set of γ parameters. Similar to the two-level regression mixture example of Section 2.2, however, there may be unobserved heterogeneity in the data corresponding to different types of trajectories. This type of heterogeneity is captured by latent classes (i.e., finite mixture modeling).

Consider the following example that was briefly discussed in Muthén (2004), but is more fully presented here. Figure 2.16 shows the results of growth mixture modeling (GMM) for mathematics achievement in Grades 7 through 10 from the LSAY data. The analysis provides a sorting of the observed trajectories into three latent classes. The left-most class with poor development also shows a significantly higher percentage of students who drop out of high school, suggesting predictive validity for the classification.

Figure 2.17 shows the model diagram for the two-level GMM for the LSAY example. In the within (student-level) part of the model, the latent class variable *c* is seen to influence the growth factors *iw* and *sw*, as well as the binary distal outcome *hsdrop*. The broken arrows from *c* to the arrows from the set of covariates to the growth factors indicate that the covariate influence may also differ across the latent classes. The filled circles for the dependent variables *math7* – *math10*, *hsdrop*, and *c* indicate random intercepts. These random intercepts are continuous latent variables that are modeled in the between (school-level) part of the model. For the between part of the growth model only the intercept is random, not the slope. In other words, the slope varies only over students, not schools. Since there are three latent classes, there are two random intercepts for *c*, labeled *c*#1 and *c*#2. On between there are two covariates discussed in

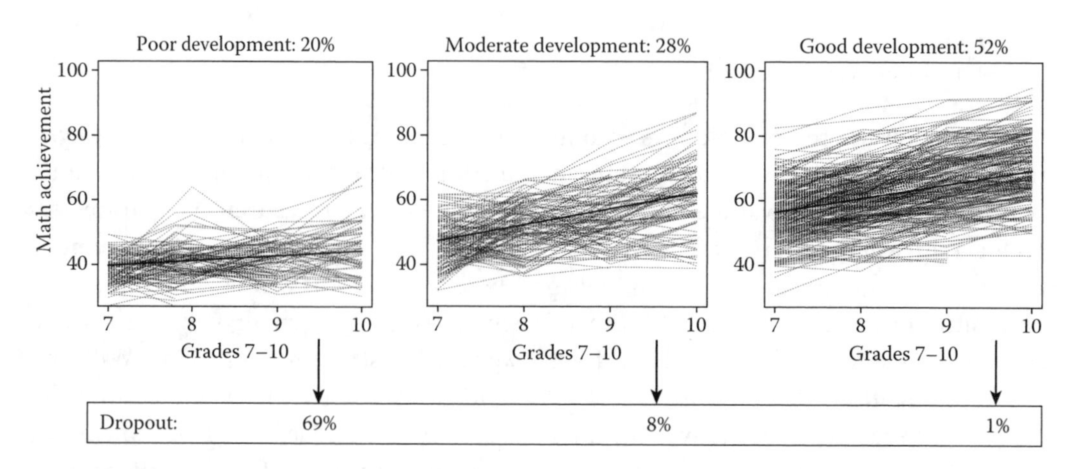

FIGURE 2.16
Growth mixture modeling with a distal outcome.

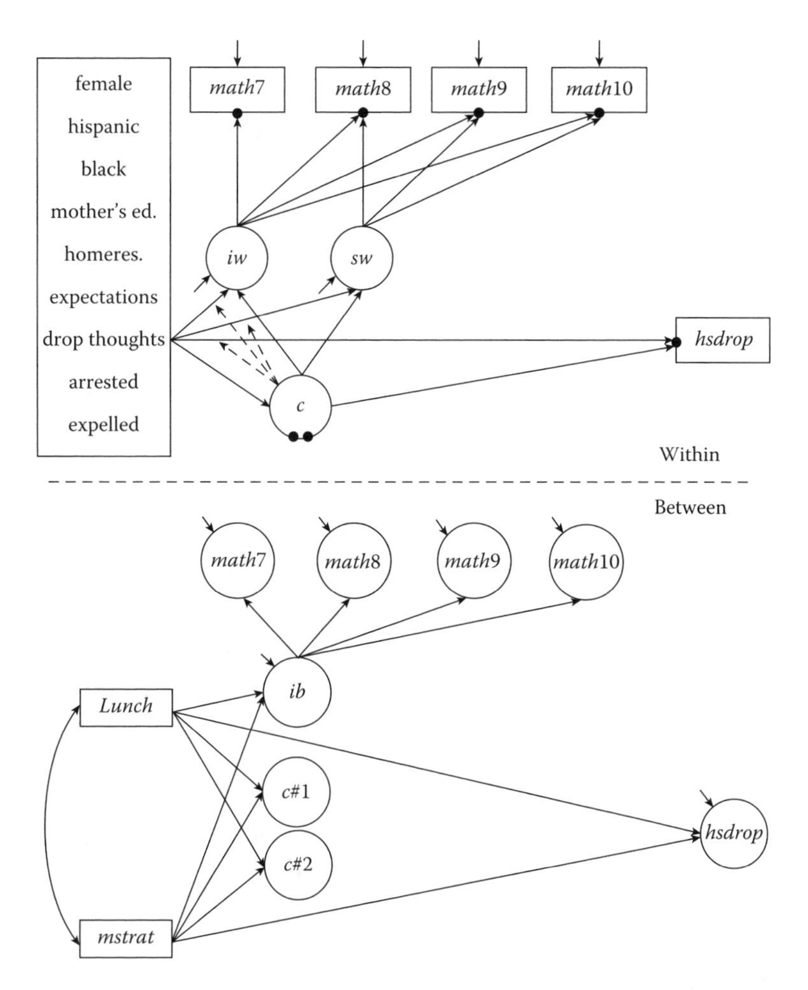

FIGURE 2.17
Two-level growth mixture modeling with a distal outcome.

earlier examples, *lunch* (a poverty index) and *mstrat* (math teacher workload).

Table 2.8 gives the estimates for the multinomial logistic regression of *c* on the covariates. On the within level (student level), the estimates are logistic regression slopes, whereas on the between level (school level), the estimates are linear regression slopes. The within level results show that the odds of membership in class 1, the poorly developing class, relative to the well-developing reference class 3 are significantly increased by being male, black, having dropout thoughts in Grade 7, and having been expelled or arrested by Grade 7. The odds are decreased by having high schooling expectations in Grade 7. The between level results pertain to how the school environment influences the student's latent class membership. The probability of membership in the poorly developing class is significantly increased by *lunch;* that is, being in the poverty category, whereas *mstrat* has no influence on this probability.

The top part of Table 2.9 shows the within-level logistic regression results for the binary distal outcome *hsdrop*. It is seen

TABLE 2.8

Two-Level GMM for LSAY Math Achievement: Latent Class Regression Results

Parameter	Estimate	*SE*	Est./*SE*
Within level			
c#1 ON			
female	−0.751	0.188	−3.998
hisp	0.094	0.705	0.133
black	0.900	0.385	2.339
mothed	−0.003	0.106	−0.028
homeres	−0.060	0.069	0.864
expect	−0.251	0.074	−3.406
droptht7	1.616	0.451	3.583
expel	0.698	0.337	2.068
arrest	1.093	0.384	2.842
Between level	1		
c#1 ON			
lunch	2.265	0.706	3.208
mstrat	−2.876	2.909	−0.988
c#2 ON			
lunch	−0.088	1.343	−0.065
mstrat	−0.608	2.324	−0.262

that the probability of dropping out of high school is significantly increased by being female, having dropout thoughts in Grade 7, and having been expelled by Grade 7. The dropout probability is significantly decreased by having high mother's education and having high schooling expectations in Grade 7.

The bottom part of Table 2.9 pertains to the between level and gives results for the random intercept *ib* of the growth model and the random intercept of the *hsdrop* logistic regression. These results concern the influence of the school environment on the level of math performance and on dropping out. For *ib* it is seen that increasing *mstrat* (math teacher workload) lowers the school average math performance. For *hsdrop* it is seen that poverty status increases the probability that a student drops out of high school. The two random intercepts are negatively correlated so that lower math performance in a school is associated with a higher dropout probability.

It is interesting to study the effects of the school level poverty index covariate *lunch*. The model says that poverty has both direct and indirect effects on dropping out of high school. The direct, school-level effect was just discussed in connection with the bottom part of Table 2.9. The indirect effect can be seen by poverty increasing the probability of being in the poorly developing math trajectory class as shown in the between-level results of Table 2.8. As seen in Figure 2.16 and also in the top part of the model diagram of Figure 2.17, the latent class variable *c* influences the probability of dropping out on the student level. In other words, poverty has an indirect, multilevel effect mediated by the within-level latent class variable. This illustrates the richness of detail that a multilevel growth mixture model can extract from the data.

2.8 CONCLUSIONS

This chapter has given an overview of latent variable techniques for multilevel modeling that are more general than those commonly described in text books. Most, if not all, of the models cannot be handled by conventional multilevel modeling or software. If space permitted, many more examples could have been given. For example, using combinations of model types, one may formulate a two-part growth model with individuals nested within clusters, or a two-part growth mixture model. Several multilevel models such as latent class analysis, latent transition analysis, and discrete- and

TABLE 2.9

Two-Level GMM for LSAY Math Achievement: Distal Outcome and School-Level Random Intercept Results

Parameter	Estimate	*SE*	Est./*SE*	Std	StdYX
Within level					
hsdrop ON					
female	0.521	0.232	2.251		
hisp	0.208	0.322	0.647		
black	−0.242	0.256	−0.944		
mothed	−0.434	0.121	−3.583		
homeres	−0.089	0.052	−1.716		
expect	−0.333	0.052	−6.417		
droptht7	0.629	0.320	1.968		
expel	1.212	0.195	6.225		
arrest	0.157	0.263	0.597		
Between level					
ib **ON**					
lunch	−1.805	1.310	−1.378	−0.851	−0.176
mstrat	−13.365	3.086	−4.331	−6.299	−0.448
hsdrop **ON**					
lunch	1.087	0.543	2.004	1.087	0.290
mstrat	−0.178	1.478	−0.120	−0.178	−0.016
ib **WITH**					
hsdrop	−0.416	0.328	−1.267	−0.196	−0.253

continuous-time survival analysis can also be combined with the models discussed. All these model types fit into the general latent variable modeling framework available in the M*plus* program.

ACKNOWLEDGMENTS

This chapter builds on a presentation by the first author at the AERA HLM SIG, San Francisco, California, April 8, 2006. The research of the first author was supported by grant R21 AA10948-01A1 from the NIAAA, by NIMH under grant No. MH40859, and by grant P30 MH066247 from the NIDA and the NIMH. We thank Kristopher Preacher for helpful comments.

REFERENCES

Asparouhov, T., Masyn, K., & Muthén, B. (2006). Continuous time survival in latent variable models. In *Proceedings of the Joint Statistical Meeting*, Seattle, August 2006. ASA section on Biometrics, 180–187.

Asparouhov, T., & Muthén, B. (2006). Constructing covariates in multilevel regression. *Mplus Web Notes: No. 11.*

Asparouhov, T., & Muthén, B. (2007). Computationally efficient estimation of multilevel high-dimensional latent variable models. In *Proceedings of the 2007 JSM meeting*, Salt Lake City, Utah, Section on Statistics in Epidemiology.

Asparouhov, T., & Muthén, B. (2009). Exploratory structural equation modeling. *Structural Equation Modeling, 16*, 397–438.

Bauer, D. J., Preacher, K. J., & Gil, K. M. (2006). Conceptualizing and testing random indirect effects and moderated mediation in multilevel models: New procedures and recommendations. *Psychological Methods, 11*, 142–163.

Little, R. J., & Rubin, D. B. (2002). *Statistical analysis with missing data* (2nd ed.). New York, NY: John Wiley & Sons.

Ludtke, O., Marsh, H. W., Robitzsch, A., Trautwein, U., Asparouhov, T., & Muthén, B. (2008). The multilevel latent covariate model: A new, more reliable approach to group-level effects in contextual studies. *Psychological Methods, 13,* 203–229.

Muthén, B. (1989). Latent variable modeling in heterogeneous populations. *Psychometrika, 54,* 557–585.

Muthén, B. (2004). Latent variable analysis: Growth mixture modeling and related techniques for longitudinal data. In D. Kaplan (Ed.), *Handbook of quantitative methodology for the social sciences* (pp. 345–368). Newbury Park, CA: Sage Publications.

Muthén, B., & Asparouhov, T. (2008). Growth mixture modeling: Analysis with non-Gaussian random effects. In G. Fitzmaurice, M. Davidian, G. Verbeke, & G. Molenberghs (Eds.), *Longitudinal data analysis* (pp. 143–165). Boca Raton, FL: Chapman & Hall/CRC Press.

Muthén, B., & Asparouhov, T. (2009). Multilevel regression mixture analysis. *Journal of the Royal Statistical Society, Series A, 172,* 639–657.

Muthén, L., & Muthén, B. (2008). *Mplus user's guide.* Los Angeles, CA: Muthén & Muthén.

Olsen, M. K., & Schafer, J. L. (2001). A two-part random effects model for semicontinuous longitudinal data. *Journal of the American Statistical Association, 96,* 730–745.

Raudenbush, S. W., & Bryk, A. S. (2002). *Hierarchical linear models: Applications and data analysis methods* (2nd ed.). Newbury Park, CA: Sage Publications.

Seltzer, M., Choi, K., & Thum, Y. M. (2002). Examining relationships between where students start and how rapidly they progress: Implications for conducting analyses that help illuminate the distribution of achievement within schools. CSE Technical Report 560. Los Angeles; CA: CRESST, University of California.

Schmidt, W. H. (1969). Covariance structure analysis of the multivariate random effects model. Unpublished doctoral dissertation. Chicago, IL: University of Chicago.

Snijders, T., & Bosker, R. (1999). *Multilevel analysis. An introduction to basic and advanced multilevel modeling.* Thousand Oaks, CA: Sage Publications.

TIMSS (Trends in Internatiopnal Mathematics and Science Study). (2003). Retrieved from: http://timss.bc.edu/timss2003.html.

3

Multilevel IRT Modeling

Akihito Kamata
Department of Educational Methodology, Policy, and Leadership, University of Oregon, Eugene, Oregon

Brandon K. Vaughn
Department of Educational Psychology, University of Texas at Austin, Austin, Texas

3.1 INTRODUCTION

In this chapter, we focus on extending the use of multilevel modeling for psychometric analyses. Such a use of multilevel modeling techniques has been referred to as multilevel measurement modeling (MMM; e.g., Beretvas & Kamata, 2005; Kamata, Bauer, & Miyazaki, 2008). When an MMM considers categorical measurement indicators, such as dichotomously and/or polytomously scored test items, we refer to such a modeling framework as multilevel item response theory (IRT) modeling. Typically, traditional IRT models do not consider a nested structure of the data, such as students nested within schools. However, data in social and behavioral science research frequently have such a nested data structure, especially when data are collected by multistage sampling. The strength of multilevel IRT modeling becomes important when we analyze psychometric data that have such a nested structure. A multilevel IRT model appropriately analyzes data by taking into account both within- and between-cluster variations of the data. Also, since multilevel modeling is essentially an extension of a regression model to multiple levels, the flexibility of multilevel IRT modeling offers the opportunity to incorporate covariates and their interaction effects.

This chapter is organized into three main sections. First, traditional IRT modeling is introduced. Then, a multilevel extension of IRT modeling is presented. In this section, three different modeling frameworks are presented. Lastly, an illustrative data analysis to estimate the variation of differential item functioning (DIF) on a statewide testing program data is presented.

3.2 ITEM RESPONSE THEORY MODELS

Item response theory modeling is a widely utilized class of traditional measurement models. For dichotomously scored test items, there are several well-recognized IRT models, such as the Rasch model, the two-parameter logistic model, and the three-parameter logistic model. For example, the two-parameter logistic model can be written as

$$p_{ip} = \frac{\exp[\alpha_i\theta_p + \delta_i]}{1+\exp[\alpha_i\theta_p + \delta_i]}, \tag{3.1}$$

where θ_p is the ability of examinee p, α_i is the discrimination power of item i, and δ_i is the threshold or location of item i. In IRT applications, the threshold is typically transformed into the difficulty parameter β_i by $\beta_i = -\delta_i/\alpha_i$, such that the exponential function has a form of $\alpha_i(\theta_p - \beta_i)$. However, in this chapter we will use the threshold parameter directly for simplicity from a modeling perspective. The metric of θ_p and $-\delta_i/\alpha_i$ are typically in a standardized scale, where 0 is the center of the distribution with a standard deviation of 1. When discrimination power is assumed to be equal for all items in the instrument and constrained to be 1, the model becomes

$$p_{ip} = \frac{\exp[\theta_p + \delta_i]}{1+\exp[\theta_p + \delta_i]}, \tag{3.2}$$

and is known as the Rasch model. The difference between θ_p and $-\delta_i = \beta_i$ is directly a logit quantity, where θ indicates a typical ability or difficulty, respectively. Furthermore, the two-parameter logistic model can be extended to the three-parameter logistic model

$$p_{ip} = \gamma_i + (1-\gamma_i)\frac{\exp[\alpha_i\theta_p + \delta_i]}{1+\exp[\alpha_i\theta_p + \delta_i]}, \tag{3.3}$$

where γ_i is the lower asymptote of the logistic curve and known as the pseudo guessing parameter. Under the three-parameter logistic model, we assume a nonzero lower asymptote, indicating a nonzero probability of endorsing an item for examinees with any ability level.

Item response modeling may be extended to polytomously scored items. One widely used model is the Graded Response Model (Samejima, 1969), which utilizes the cumulative logit principle. The model is written as

$$p^{+}_{ijp} = \frac{\exp[\alpha_i\theta_p + \delta_{ij}]}{1+\exp[\alpha_i\theta_p + \delta_{ij}]}, \tag{3.4}$$

where p^{+}_{ijp} is the probability for person p getting the scoring category j or higher on item i. In this model, δ_{ij} is the threshold parameter for the jth score boundary. As a result, the probability of getting a specific scoring category j is obtained by $p_{ijp} = p^{+}_{ijp} - p^{+}_{i(j+1)p}$. For the lowest scoring category ($j = 0$), $p_{i0p} = 1 - p^{+}_{i1p}$, while $p_{iMp} = p^{+}_{iMp}$ for the highest scoring category M. If δ_{ij} is transformed into $\beta_{ij} = -\delta_{ij}/\alpha_i$, β_{ij} is the category-boundary difficulty for the jth score boundary. By assuming the discrimination coefficients are equal across all items, it is also sensible to make a one-parameter extension from this model. Another class of IRT models for polytomously scored items is based on the adjacent logit principle. One general form is the generalized partial credit model (Muraki, 1992)

$$p_{ijp}(\theta)=\frac{\exp\left[\sum_{j=0}^{x}(\alpha_i\theta_p+\delta_{ij})\right]}{\sum_{r=0}^{m_i}\exp\left[\sum_{j=0}^{r}(\alpha_i\theta_p+\delta_{ij})\right]}, \quad (3.5)$$

where x is the target response category for the item, and m_i is the highest response category for item i ($j = 0, \ldots, r, \ldots, m_i$). Simpler variations of this model include the partial credit model with $\alpha_i = 1$ for all i (Masters, 1982), and the rating scale model with $\alpha_i = 1$ for all i and $\delta_{ij} = \eta_i + \kappa_j$, where η_i is the item location parameter and κ_j is the step parameter. In the rating scale model, step parameters κ_j are common to all items, indicating distances between step parameters are the same for all items.

3.3 MULTILEVEL ITEM RESPONSE MODELING

A multilevel IRT model extends the above mentioned IRT models, such that they consider variations of abilities between group units such as schools, as well as within group units. Accordingly, a multilevel IRT model will distinguish the individual-level abilities and group-level abilities. For example, a multilevel extended two-parameter logistic IRT model for dichotomously scored items could be expressed as

$$p_{ip}=\frac{\exp[\alpha_i(\xi_g+\zeta_{pg})+\delta_i]}{1+\exp[\alpha_i(\xi_g+\zeta_{pg})+\delta_i]}, \quad (3.6)$$

where $\theta_{pg} = \xi_g + \zeta_{pg}$. Here, θ_{pg} is the ability of person p in group g, but it is expressed with ξ_g that is the mean group ability g, and ζ_{pg} that is the amount of deviation from the group mean ability for person p in group g. This is one of the simplest forms of a multilevel IRT model. However, typical applications of multilevel IRT models involve covariates in the model.

Several different ways to formulate a multilevel IRT model have been presented in the literature. In this section, two approaches, Fox & Glas's (2001) multilevel IRT framework, and Kamata's (2001) HGLM approach to multilevel IRT will be presented. We will also describe multilevel structural equation modeling with categorical measurement indicators since both of these approaches can be viewed as special cases of the SEM.

3.3.1 Fox and Glas's Multilevel IRT Modeling

One aspect of multilevel IRT modeling traces back to the development of latent regression model (Vehelst & Eggen, 1989; Zwinderman, 1991, 1997), where observed variables are regressed on the latent variable θ. Fox and Glas (2001) extended this idea to multilevel linear modeling with two-parameter normal ogive and graded response model as the measurement model. This is a multilevel IRT model due to the nature of the multilevel model being embedded in the IRT framework. In effect, it allows modeling the relationship between observed individual and group characteristics and a latent variable represented by both dichotomous and polytomous items. In Fox and Glas's formulation, the measurement model is either a two-parameter normal ogive model or graded response model. Additionally, this model describes the structural relationship between the latent variable in the IRT model (ability) and observed covariates. Thus, the level-2 model is a structural model

$$\theta_{pg} = \beta_{0g} + \beta_{1g}x_{1pg} + \ldots + \beta_{Qg}x_{Qpg} + \zeta_{pg}^{(2)}, \quad (3.7)$$

where θ is the latent variable that represents the trait measured in the measurement (IRT) model, x are level-2 covariates, $\beta_{1g}, \ldots, \beta_{Qg}$ are corresponding coefficients, and $\zeta_{pg}^{(2)}$ is the error, where $\zeta_{ig}^{(2)} \sim N(0, \sigma^2)$. Additionally, three-level models can be written as

$$\begin{aligned}\beta_{0g} &= \gamma_{00} + \gamma_{01}w_{1g} + \ldots + \gamma_{0S}w_{Sg} + \xi_{0g}^{(3)} \\ &\vdots \\ \beta_{Qg} &= \gamma_{Q0} + \gamma_{Q1}w_{1g} + \ldots + \gamma_{QS}w_{Sg} + \xi_{Qg}^{(3)},\end{aligned} \quad (3.8)$$

where w are level-3 covariates, γ are corresponding coefficients, and $\xi_{0g}^{(3)}, \ldots, \xi_{Qg}^{(3)}$ are level-3 random effects, where $\xi_{\bullet g} \sim N(\mathbf{0}, \mathbf{\Omega})$. If there is no covariate in either levels of the structural models, the structural model is reduced to

$$\theta_{pg} = \xi_{0g}^{(3)} + \zeta_{pg}^{(2)}, \quad (3.9)$$

since β_{0g} and γ_{00} become the means of ζ_{pg} and ξ_{0g}, which are 0. This equation demonstrates its equivalency to the general multilevel IRT model equation presented in the previous section (Equation 3.6).

Fox and Glas (2001) and Fox (2005) have implemented a Markov chain Monte Carlo (MCMC) method to estimate the parameters in this model. An `R` package for the MCMC called `mlirt` has been made available for public (Fox, 2007).

3.3.2 HGLM Approach

We now focus on the use of hierarchical generalized linear models (HGLMs) for latent variable modeling. The uniqueness of the GLM over general linear models is in the dependent measure. The GLM allows response measures that follow any probability distribution in the exponential family of distributions. Generalized linear models are of great benefit in situations where the response variables follow distributions other than the normal distribution and when variances are not constant. This is of particular interest in IRT as response measures are typically dichotomous or polytomous, discrete, and nonnormal.

The analysis of the GLM incorporates the use of a link since the dependent measure in GLMs may characterize many different types of distributions and thus the relationship between the predictor and the dependent measure may not be linear in nature. Many different link functions exist, yet Table 3.1 shows the most common in research and practice.

The HGLM approach provides a flexible and efficient framework for modeling nonnormal data in situations when there may be several sources of error variation. This is accomplished by extending the familiar GLM to include additional random terms in the linear predictor. One special case of HGLMs is generalized linear mixed models (GLMMs), which constrains the additional terms to follow a normal distribution and to have an identity link. However, many HGLMs do not have such restrictions. For example, if the basic GLM is a log-linear model (Poisson

TABLE 3.1

Common Link Functions for Popular Probability Distributions

Probability Distribution	Link Function
Normal	Identity
Binomial/normal cumulative	Logit/probit
Poisson	Log
Multinomial	Logit/probit

distribution and log link), a more appropriate assumption for the additional random terms might be a gamma distribution and a log link. Thus, HGLMs bring together a wide range of models under a common approach. Each HGLM is made up of at least two levels in a multilevel model so as to incorporate several sources of error variation. This approach is especially useful in situations involving nested or clustered data. In IRT analysis, this might manifest itself in situations of students nested within schools or individuals nested within families. By considering cluster effects, innovative questions can be considered (e.g., if any differential item functioning, DIF, effects vary from cluster to cluster).

3.3.2.1 Modeling IRT as Latent HGLM

Earlier in this chapter, various IRT models were shown. An IRT model can be modeled with a two-level logistic regression where the log-odds (i.e., logit link function) of subject p providing a positive answer to an item i is represented by:

$$\eta_{ip} = \log\left(\frac{\varphi_{ip}}{1-\varphi_{ip}}\right) = \theta_p - \beta_i, \quad (3.10)$$

where φ_{ip} represents the probability that subject p gets item i correct, θ_p represents the trait level associated with subject p ($\theta_p \sim N(0, \sigma^2)$, stating that θ_p is normally distributed with 0 mean and the variance of σ^2), and β_i represents the difficulty of item i. In this model, η_{ip} represents the log-odds of subject p getting item i correct (assuming dichotomous outcomes). This simple IRT model is the Rasch model as detailed earlier. Adding one additional parameter α_i to represent the extent to which item i can discriminate between subjects of different trait levels, the model becomes:

$$\eta_{ip} = \alpha_i(\theta_p - \beta_i) = \alpha_i\theta_p - \alpha_i\beta_i. \quad (3.11)$$

Finally, if a predictor is added to this model in order to provide an explanatory approach, the formulation becomes

$$\eta_{ip} = \alpha_i\theta_p - \alpha_i\beta_i + \gamma X_p, \quad (3.12)$$

where γ is the regression coefficient for explanatory variable X_p.

For a set of r items, the logit link function can be modeled as a hierarchical two-level logistic model (e.g., Van den Noortgate & De Boeck, 2005):

$$\eta_{ip} = \log\left(\frac{\varphi_{ip}}{1-\varphi_{ip}}\right) = \beta_{1p}X_{1ip} + \ldots + \beta_{rp}X_{rip} + u_p = \sum_{q=1}^{r} \beta_q X_{qi} + u_p, \quad (3.13)$$

where $X_{qi} = 1$ if $q = i$, 0 otherwise, and $u_p \sim N(0, \sigma_u^2)$. Kamata (2001) parameterized the multilevel logistic model as:

$$\eta_{ip} = \log\left(\frac{\varphi_{ip}}{1-\varphi_{ip}}\right) = \beta_{0p} + \beta_{1p}X_{1ip} + \ldots + \beta_{(r-1)p}X_{(r-1)ip} = \beta_{0p} + \sum_{q=1}^{r-1} \beta_{qp}X_{qip}. \quad (3.14)$$

Each X_{qip} represents the qth dummy indicator variable for subject p. In order for the design matrix of the model to achieve full rank, one of the items must be dropped from the model or a no-intercept model could be fit. For the case where an item is dropped, for r set of items, only $r - 1$ items are included in the model. The coefficient, β_{0p}, is interpreted as the mean effect of the dropped item, and each β_{qp} is interpreted as the effect of the qth dummy indicator (i.e., item i, for $i = 1, \ldots, r - 1$) compared to the reference item. For a particular item i, a value of zero is assigned to X_{qip} for $q \neq i$, and a positive one when $q = i$. This gives a logit for a particular item i, for $q = i$, as:

$$\eta_{qp} = \log\left(\frac{\varphi_{qp}}{1-\varphi_{qp}}\right) = \beta_{0p} + \beta_{qp} \qquad (3.15)$$

where β_{0p} is a random effect in which $\beta_{0p} \sim N(0, \sigma_\beta^2)$.

There are a variety of methods to extend this idea to ordinal polytomous outcomes. One popular approach is the formation of a cumulative probability model. For each ordered response m ($m = 1, \ldots, M$), a probability of response y_i on item i is established for each unique response possibility:

$$\varphi_m = P(y_i = m). \qquad (3.16)$$

Defining the probability response model in this manner creates difficulty in formulating a single regression model. Thus, a cumulative probability model is incorporated:

$$\varphi_m^* = P(y_i \le m) = \varphi_1 + \varphi_2 + \cdots + \varphi_m. \qquad (3.17)$$

A cumulative logit function can be derived using the cumulative probabilities

$$\eta_m = \log\left(\frac{\varphi_m^*}{1-\varphi_m^*}\right) = \log\left(\frac{P(y_i \le m)}{P(y_i > m)}\right), \qquad (3.18)$$

for each ordinal response of $m = 1, \ldots, M - 1$. In this model, η_m represents the log-odds of responding at or below category m, versus responding above category m.

A common intercept can be introduced into this model by considering the difference (δ) between the thresholds. The general logit model now becomes

$$\eta_{mi} = \begin{cases} \beta_0 + \beta_1 X_i, & \text{for } m = 1 \\ \beta_0 + \beta_1 X_i + \delta_2 + \ldots \delta_m, & \text{for } 1 < m \le M-1 \end{cases}$$
$$= \beta_0 + \beta_1 X_i + \sum_{s=2}^{m} \delta_s. \qquad (3.19)$$

This approach can be used to model a two-level HGLM for polytomous items with the IRT perspective mentioned previously. The level-1 (item-level) model for a set of r items is represented as:

$$\eta_{mip} = \log\left(\frac{\varphi_{mip}^*}{1-\varphi_{mip}^*}\right)$$
$$= \begin{cases} \beta_{0p} + \beta_{1p} X_{1ip} + \ldots + \beta_{(r-1)p} X_{(r-1)ip}, & \text{for } m = 1 \\ \beta_{0p} + \beta_{1p} X_{1ip} + \ldots + \beta_{(r-1)p} X_{(r-1)ip} + \delta_{2p} + \ldots + \delta_{mp}, & \text{for } 1 < m \le M-1 \end{cases} \qquad (3.20)$$
$$= \beta_{0p} + \sum_{q=1}^{r-1} \beta_{qp} X_{qip} + \sum_{s=2}^{m} \delta_{sp},$$

where φ^*_{mip} is the cumulative probability as defined above. One random effect, β_{0p}, is present that represents the expected effect of the reference item for subject *p*. For a particular item *i*, a value of positive one is assigned to X_{qip} when $q = i$, and a value of zero otherwise. For a particular item *q*, this model simplifies to:

$$\eta_{mqp} = \beta_{0p} + \beta_{qp} + \sum_{s=2}^{m} \delta_{sj}. \quad (3.21)$$

One possible level-2 model (subject-level) with level-2 predictor X_p added to all effects and thresholds is expressed as:

$$\begin{cases} \beta_{0p} = \gamma_{00} + \gamma_{01}X_p + u_{0p} \\ \beta_{1p} = \gamma_{10} + \gamma_{11}X_p \\ \vdots \\ \beta_{(r-1)p} = \gamma_{(r-1)0} + \gamma_{(r-1)1}X_p \\ \delta_{2p} = \xi_{20} + \xi_{21q}X_p \\ \vdots \\ \delta_{mp} = \xi_{m0} + \xi_{m1q}X_p. \end{cases} \quad (3.22)$$

More than one predictor can be incorporated and can be a variety of variables of interest to the researcher that are subject related. For example, in DIF studies, a categorical level-2 predictor of group affiliation (reference versus focal group) can be considered. This modeling can easily be extended to a three-level model. If a third level is added, the level-2 terms can be allowed to vary among clusters of subjects and level-3 predictors (cluster related) can be added to explain the random nature of level-2 terms.

A variety of estimation procedures can be utilized for these HGLM multilevel models. With logistic regression models, estimation procedures have typically incorporated a maximum likelihood method (De Boeck & Wilson, 2004). However, use of this estimation method can prove problematic for multilevel models.

Penalized quasi-likelihood (PQL) estimation was at one time a popular approach. However, this method has been shown to produce negatively biased parameter estimates (Raudenbush, Yang, & Yosef, 2000). Raudenbush et al. (2000) and Yang (1988) suggested a sixth order Laplace (Laplace6) approximation for estimation instead. Current software, such as HLM 6 (Raudenbush, Bryk, & Congdon, 2005), allows for a Laplace6 approximation, but is limited to Bernoulli models of two and three levels. For ordinal models, however, the PQL estimation procedure is still widely used (typically because alternative methods are not widely available in some software packages).

Due to this, some suggest a Bayesian approach as a more flexible option (Johnson & Albert, 1999) and some multilevel software (e.g., MLWin) now have this estimation procedure as an option. Breslow (2003) showed that a MCMC approach is a better choice over PQL for complex problems that involve high dimensional integrals.

Many studies that do approach regular multilevel models from a Bayesian perspective use a probit link function in their formulation (Elrod, 2004; Fox, 2005; Galindo, Vermunt, & Bergsma, 2004; Hoijtink, 2000; Mwalili, Lesaffre, & Declerck, 2005; Qiu, Song, & Tan, 2002). Also popular in certain studies that have considered a Bayesian multilevel approach is the cumulative logit function (Ishwaran, 2000; Ishwaran & Gatsonis, 2000; Lahiri & Gao, 2002; Lunn, Wakefield, & Racine-Poon, 2001). Within this Bayesian framework, MCMC Gibbs sampling estimation procedures are typically used. A variety of software (e.g., WinBUGS, BRugs for R, MLwiN, etc.) allow for this Gibbs sampling estimation procedure for multilevel models. Although the use of Gibbs sampling has grown in popularity since the advent of powerful personal computers, some psychometric areas still consider Gaussian quadrature points instead for estimation.

Chaimongkol (2005) and Vaughn (2006) both incorporated this approach in estimating random DIF in multilevel models for dichotomous and polytomous items. Vague priors were used in the estimation so that the estimated values would closely mirror those using frequentist methods. In order for the model to be identified, both authors replaced the model parameters with new "adjusted" quantities that were well identified yet did not change the logit of the model.

Although the above mentioned estimation procedures are the most common in practice, there are many others available that might be considered. Goldstein and Rasbash (1992) detail a iterative generalized least squares (IGLS) method for estimation. This approach is sometimes referenced as PQL2 and is incorporated in the computer program MLWin. Also, as mentioned above, Gaussian quadrature estimation is a popular choice in other software (e.g., Sabre, Stata, and GLLAMM).

3.3.3 Multilevel SEM Approach

A more general framework for a multilevel IRT modeling is a two-level structural equation model with categorical indicators. The two-level SEM assumes that multiple individuals are sampled from each of many groups in the population (see Muthén and Asparouhov, Chapter 2 of this book).

The two-level factor model with categorical indicators can be written as

$$\mathbf{y}^*_{pg} = \mathbf{\Lambda}_W \boldsymbol{\theta}_{pg} + \boldsymbol{\varepsilon}_{pg}, \tag{3.23}$$

which represents a linear regression of the vector of I unobserved latent response variables $\mathbf{y}^*_{pg}$ on the latent variables $\boldsymbol{\theta}_{pg}$ for person p in group g. The latent response variables $\mathbf{y}^*_{pg}$ is an $I \times 1$ vector of latent response scores to I items in the test, and $\boldsymbol{\theta}_{pg}$ is a $K \times 1$ vector of factor scores (abilities) for K latent factors. As a result, $\mathbf{\Lambda}_W$ are factor loadings ($I \times K$ matrix), where the W subscript indicates "within-groups," and ε_{pg} are residuals ($I \times 1$ vector). In a unidimensional IRT application, for example, $K = 1$, and both $\mathbf{\Lambda}_W$ and ε_{pg} are $I \times 1$ vectors. Observed dichotomous response y_{ipg} is defined such that

$$\begin{aligned} y_{ipg} &= 1, \text{if } y^*_{ipg} \geq \tau_i, \text{and} \\ y_{ipg} &= 0, \text{if } y^*_{ipg} < \tau_i. \end{aligned} \tag{3.24}$$

Here, τ_i is the threshold for item i. Within groups, the latent factors are assumed to be distributed with mean vector α and covariance matrix $\boldsymbol{\psi}_W$. Similarly, for polytomously scored items with scoring categories ranging from 0 to M,

$$\begin{aligned} y_{ipg} &= M, \text{if } y^*_{ipg} \geq \tau_{iM}, \\ y_{ipg} &= M-1, \text{if } \tau_{i(M-1)} \leq y^*_{ipg} \leq \tau_{iM}, \\ &\vdots \\ y_{ipg} &= 1, \text{if } \tau_{i1} \leq y^*_{ipg} \leq \tau_{i2}, \text{and} \\ y_{ipg} &= 0, \text{if } y^*_{ipg} < \tau_{i1}. \end{aligned} \tag{3.25}$$

The residuals $\boldsymbol{\varepsilon}_{pg}$ are assumed to be distributed with means of zero and covariance matrix $\boldsymbol{\Sigma}_W$. Residuals are independent from each other according to the local independence assumption of IRT models, resulting in a diagonal $\boldsymbol{\Sigma}_W$ matrix. If errors are distributed as the logistic distribution, the model is known as the logistic model, and this will provide the basis of equivalency to logistic item response models. If we have $\boldsymbol{\theta}_{pg}$ as a 1×1 scalar (i.e., only one latent trait) and $M = 2$ (i.e., dichotomously scored items), the model is equivalent to the 2PL IRT model. On the other hand, if residuals are normally distributed, the model is known as the normal ogive model. One important assumption with this approach is that these covariance matrices are homogeneous across all groups, which will result in identical covariance structures between groups. Accordingly, for group *j*, the within-group covariance matrix is

$$V(\mathbf{y}^*)_W = \boldsymbol{\Lambda}_W \boldsymbol{\psi}_W \boldsymbol{\Lambda}'_W + \boldsymbol{\Sigma}_W, \quad (3.26)$$

which is essentially the same for the single-level SEM, except the *W* subscript for each quantity in the equation. On the other hand, the structural model of the two-level SEM can be written as

$$\boldsymbol{\theta}_{pg} = \boldsymbol{\alpha}_g + \mathbf{B}_g \boldsymbol{\theta}_{pg} + \boldsymbol{\Gamma}_g \mathbf{x}_{pg} + \boldsymbol{\zeta}_{pg}, \quad (3.27)$$

where latent factors are regressed on other latent factors and some observed covariates $\mathbf{x}$. The intercepts are given by α_g, slopes for latent predictors are $\mathbf{B}_g$, and slopes for observed covariates are Γ_g. The residuals are assumed to be normally distributed with means of zero and $K \times K$ covariance matrix ψ. If no latent variable is specified as a predictor in the model, the intercepts, α_j, are simply factor means, which are typically constrained to be 0.

Due to nested data structure, the multilevel SEM imposes an additional between-group level factor structure on the covariance matrix (e.g., Ansari, Jedidi, & Dube, 2002; Goldstein & McDonald, 1988; McDonald & Goldstein, 1989; Muthén, 1994; Muthén & Satorra, 1995). The resulting covariance structure at between-group level is

$$V(y^*)_B = \boldsymbol{\Lambda}_B \boldsymbol{\psi}_B \boldsymbol{\Lambda}'_B + \boldsymbol{\Sigma}_B, \text{ and}$$

$$V(y^* \mid \mathbf{x})_B = \boldsymbol{\Lambda}_B (\mathbf{I} - \mathbf{B}_B)^{-1} \boldsymbol{\psi}_B (\mathbf{I} - \mathbf{B}_B)'^{-1} \boldsymbol{\Lambda}'_B + \boldsymbol{\Sigma}_B. \quad (3.28)$$

Here, the structure for the within- and between-groups covariance matrices are very similar (Equations 3.25 and 3.27). However, the parameter estimates and the factor structure of the model can be different between the two parts of the model as indicated by different subscripts (*W* vs. *B*).

Traditionally, parameter estimation for this type of model has relied on the weighted least squares methods with a tetrachoric or polychoric correlation matrix, which differs from the IRT estimation tradition, where a full information maximum likelihood has been a common approach. Also, the scaling of parameters will be different from the parameter scale of IRT if the weighted least square is employed, which requires appropriate transformation of parameters (e.g., Kamata et al., 2008). More recently, a true full information maximum likelihood estimator has become available in several general SEM software programs, which is consistent with the IRT tradition. Also, the MCMC has been shown to be effective for this type of model, especially when the number of random effects becomes large (e.g., Chaimongkol, 2005; Fox, 2005; Fox & Glas, 2001; Vaughn, 2006).

3.4 ILLUSTRATIVE DATA ANALYSIS

Data used in the following illustrative analyses were sampled from the 2005 administration of mathematics assessment for eighth graders in a statewide testing program in one particular state in the United States. The mathematics test consisted of 40 items based on five subscales of skills, including: Number Sense, Measurement, Geometry, Algebraic Thinking, and Data Analysis. Here, only one subscale, Data Analysis, was used for illustrative data analysis. With the Data Analysis subscale, items measured various skills to use and interpret data through mean, median, and probability, as well as the use of a Venn diagram. Among nine items in this subscale, eight items were scored dichotomously and one item was scored polytomously with five ordered scoring categories. All nonresponded items were scored as 0. The sample of examinees included a total of 11,220 examinees from 30 schools.

By using this data set, modeling a random differential item functioning (RDIF) is demonstrated. An RDIF is a differential item functioning (DIF) that is treated as a random effect. To be more specific, we consider that the magnitude of DIF for a particular test item to vary across schools. Here, we evaluate the RDIF between English language learners (ELL) and students in standard curricula. All model parameters are estimated by full-information maximum likelihood with adaptive numerical integration. M*plus* syntax for the three analyses presented here are provided in the Appendix.

First, a DIF detection model was fitted under the multilevel CFA with a covariate. The model can be written

$$y^*_{ipg} = \lambda_i \theta^{(2)}_{pg} + \beta_i G_{pg} + \varepsilon_{ipg} \tag{3.29}$$

with structural models

$$\begin{aligned} \theta^{(2)}_{pg} &= \beta_0 + \beta_\theta G_{pj} + \zeta^{(2)}_{pg}, \text{ and} \\ \beta_0 &= \xi^{(3)}_g . \end{aligned} \tag{3.30}$$

When these measurement and structural models are combined, the model is

$$y^*_{ipg} = \lambda_i (\zeta^{(2)}_{pg} + \xi^{(3)}_g + \beta_\theta G_{pg}) + \beta_i G_{pg} + \varepsilon_{ipg}, \tag{3.31}$$

where G_{pg} is the group indicator (1 = ELL, 0 = Standard Curricula). Consequently, β_θ indicates the difference between ELL and Standard Curricula students in their ability, and β_i is the DIF magnitude for item *i*. This is essentially the same as the MIMIC approach to DIF detection (e.g., Finch, 2005) with an additional cluster level. Since this model is not identified, it was constrained with the magnitude of DIF for the first item being zero. This constraint was based on our initial data screening, including single-item DIF analysis for all items on the entire mathematics test. Therefore, we are quite confident that this constraint is not too far from reality, and that estimated DIF magnitudes for the remaining items can be interpreted as the magnitude of DIF. However, one might be more on the conservative side to interpret β_i as the difference between the first item and the *i*th item in their DIF magnitudes. Note that this limitation in scaling the DIF parameter is not unique to this current approach; it is an inherent problem in DIF analysis in general. Bolt, Hare, and Newmann (2007) and Penfield and Camilli (2007) provide more detailed discussions on this matter. Furthermore, the mean of the within-level and between-level latent factor were constrained to zero and the variance of the within-level latent factor was constrained to be one. Although it is not required for identification, item discriminations (factor loadings) were constrained to be the same for within- and between-levels. Results are summarized in Table 3.2. Values in the table without parenthesis indicate estimates of parameters and the values inside parentheses are standard errors of estimates.

The second model assumed that one of the DIF magnitudes was a random effect, where the assumption was that DIF varied across schools. This effect was modeled as another random effect $\xi_{ig}^{(3)}$ in an additional structural model

$$\beta_i = \gamma_{0i} + \xi_{ig}^{(3)}. \tag{3.32}$$

Here, the DIF for the ith item is expressed as the sum of the mean of DIF for the item across schools γ_{0i} and the random effect $\xi_{ig}^{(3)}$. The variance of this random effect is our

TABLE 3.2

Results of Data Analysis with Three-Level DIF Detection Model

a. Fixed Effects

Item	Loading	Thresholds				DIF
1	0.930	-.712				0.000
	(.029)	(.035)				
2	1.447	-2.897				-.227
	(.061)	(.086)				(.117)
3	0.849	-.188				-.158
	(.036)	(.040)				(.112)
4	1.142	-2.628:	-1.74:	1.111:	2.129	-.811
	(.045)	(.056)	(.050)	(.047)	(.040)	(.096)
5	0.889	.218				-.150
	(.040)	(.044)				(.101)
6	0.716	-.103				-.123
	(.035)	(.035)				(.093)
7	1.305	-.711				-.132
	(.063)	(.061)				(.124)
8	1.233	-.356				.009
	(.054)	(.045)				(.212)
9	0.894	-.993				-.118
	(.033)	(.041)				(.089)

b. Random Effects

	Mean	Variance
Within-level latent factor	0.000	1.000
Between-level latent factor	0.000	.091
		(.018)

main interest. The combined model can be written as

$$y^{*}_{ipg} = \lambda_i(\zeta^{(2)}_{pg} + \xi^{(3)}_{g} + \beta_\theta G_{pg}) + (\gamma_{0i} + \xi^{(3)}_{i})G_{pg} + \varepsilon_{ipg}, \quad (3.33)$$

In the data analysis, the DIF of item 7 was treated as a random effect. The results are shown in Table 3.3. Again, values in the table without parenthesis indicate estimates of parameters and the values inside parentheses are standard errors of estimates.

Note that the variance of DIF for item 7 was found to be .419, which is equivalent to $SD = .647$. Assuming a normal distribution of DIF magnitudes across schools, it can be interpreted that the range of DIF for the middle 68% of schools is nearly 1.30, which is quite large around a reasonable logit value.

Next, one covariate was entered into the model in an attempt to explain the variation of the DIF for item 7. The covariate used here is the proportion of limited English proficient (LEP) students in each school. The variable was scaled such as

TABLE 3.3

Results of Data Analysis with RDIF Model

a. Fixed Effects

Item	Loading	Thresholds				DIF
1	0.931	-.710				0.000
	(.029)	(.035)				
2	1.446	-2.893				-.225
	(.061)	(.086)				(.118)
3	0.848	-.186				-.157
	(.036)	(.040)				(.113)
4	1.142	-2.628:	-1.717:	1.113:	2.131	-.808
	(.045)	(.056)	(.050)	(.047)	(.040)	(.096)
5	0.891	.220				-.146
	(.040)	(.045)				(.100)
6	0.716	-.102				-.121
	(.035)	(.035)				(.093)
7	1.300	-.707				See the mean of random effect for item 7
	(.063)	(.061)				
8	1.237	-.354				.015
	(.054)	(.045)				(.119)
9	0.893	-.991				-.117
	(.033)	(.041)				(.089)

b. Random Effects

	Mean	Variance
Within-level latent factor	0.000	1.000
Between-level latent factor	0.000	.090
		(.018)
DIF for item 7	-.295	.419
	(.136)	(.150)

10% = 1 unit. Some descriptive statistics were as follows; N = 30, minimum = .34, maximum = 4.73, mean = 1.19, *SD* = .90. Since one part of the structural model is expanded to $\beta_i = \gamma_{0i} + \gamma_{1i}(\text{LEP\_P})_g + \xi_{ig}^{(3)}$, the model is now written

$$y^*_{ipg} = \lambda_i(\zeta_{pg}^{(2)} + \xi_g^{(3)} + \beta_\theta G_{pg}) + [\gamma_{i0} + \gamma_{i1}(\text{LEP\_P})_g + \xi_i^{(3)}]G_{pg} + \varepsilon_{ipg}. \tag{3.34}$$

The results are summarized in Table 3.4. Again, values in the table without parenthesis indicate estimates of parameters and the values inside parentheses are standard errors of estimates. The mean of the DIF of item 7 is now expressed as a linear function of LEP_P.

Note that the variance of DIF for item 7 was .419, which is equivalent to *SD* = .647 in the previous model. When a school-level covariate LEP_P is added to the model, this variance was reduced to .224 (65% reduction of the variance). The estimated coefficient for the covariate is .425 (*SE* = .086), indicating positive relationship between the proportion of ELL students and the DIF magnitude.

TABLE 3.4

Results of Data Analysis for RDIF Model with a Covariate

a. Fixed Effects

Item	Loading	Thresholds				DIF
1	0.930	-.710				0.000
	(.029)	(.035)				
2	1.446	-2.892				-.225
	(.061)	(.086)				(.117)
3	0.848	-.186				-.157
	(.036)	(.040)				(.112)
4	1.142	-2.628:	-1.717:	1.114:	2.132	-.808
	(.045)	(.056)	(.050)	(.047)	(.040)	(.095)
5	0.891	.220				-.147
	(.040)	(.045)				(.100)
6	0.716	-.102				-.122
	(.035)	(.035)				(.092)
7	1.303	-.707				See the mean of random effect for item 7
	(.063)	(.061)				
8	1.236	-.354				.014
	(.054)	(.045)				(.119)
9	0.892	-.991				-.118
	(.033)	(.041)				(.090)

b. Random Effects

	Mean	Variance
Within-level latent factor	0.000	1.000
Between-level latent factor	0.000	.090
		(.018)
DIF for item 7	-.836 + .425 (LEP_P)	.224
	(.190) (.086)	(.102)

Since the intercept is negative –.836, the interpretation is that schools with a very small proportion of ELL students had a larger disadvantage for ELL students for this item. As the proportion of ELL increases, schools had smaller disadvantages for this item. For schools with about 20% ELL students, DIF is predicted to be near zero, by substituting LEP_P = 2 into –.836 + .425(LEP_P).

3.5 CONCLUSIONS

Multilevel IRT modeling as an extension of multilevel modeling was discussed in this chapter. Several different modeling frameworks were introduced. As a practical application of the multilevel IRT modeling, estimation of random DIF (RDIF) was demonstrated with a data set sampled from a statewide testing program in the United States. As mentioned earlier, traditional IRT models typically do not consider a nested structure of the data. However, data in social and behavioral science research frequently have such a nested data structure, especially when data are collected by multistage sampling. As demonstrated in this chapter, the strength of multilevel IRT modeling becomes important when we analyze psychometric data that have such a nested structure. A multilevel IRT model appropriately analyzes data by taking into account both within- and between-cluster variations of the data. Also, since multilevel modeling is an extension of a regression model to multiple levels, the flexibility of multilevel IRT modeling offers the opportunity to incorporate previous approaches and techniques, as demonstrated in this chapter.

REFERENCES

Ansari, A., Jedidi, K., & Dube, L. (2002). Heterogeneous factor analysis models: A Bayesian approach. *Psychometrika, 67*, 49–78.

Beretvas, S. N., & Kamata, A. (2005). The multilevel measurement model: Introduction to the special issue. *Journal of Applied Measurement, 6*, 247–254.

Bolt, D. M., Hare, R. D., & Newmann, C. S. (2007). Score metric equivalence of the psychopathy checklist-revised (PCL-R) across criminal offenders in North America and the United Kingdom: A critique of Cooke, Michie, Hart, and Clark (2005) and new analyses. *Assessment, 14*, 44–56.

Breslow, N. (2003). *Whither PQL?* (Working paper 192): UW Biostatistics Working Paper Series.

Chaimongkol, S. (2005). *Modeling differential item functioning (DIF) using multilevel logistic regression models: A Bayesian perspective.* Unpublished doctoral dissertation, Florida State University, Tallahassee, FL.

De Boeck, P., & Wilson, M. (2004). *Explanatory item response models: A generalized linear and nonlinear approach.* New York, NY: Springer.

Elrod, T. (2004, December). *Bayesian modeling of consumer behavior using WinBUGS.* Paper presented at the Conference of the Japan Institute of Marketing Science, Tokyo.

Finch, H. (2005). The MIMIC model as a method for detecting DIF: Comparison with Mantel-Haenszel, SIBTEST, and the IRT likelihood ratio. *Applied Psychological Measurement, 29*, 278–295.

Fox, J.-P. (2005). Multilevel IRT model assessment. In L. A. van der Ark, M. A. Croon, & K. Sijtsma (Eds.), *New developments in categorical dat analysis for the social and behavioral sciences* (pp. 227–252). Mahwah, NJ: Lawrence Erlbaum Associates.

Fox, J.-P. (2007). Multilevel IRT modeling in practice with the package mlirt. *Journal of Statistical Software, 20*(5), 1–16.

Fox, J.-P., & Glass, C. A. W. (2001). Bayesian estimation of a multilevel IRT model using Gibbs sampling. *Psychometrika, 66*, 271–288.

Galindo, F., Vermunt, J. K., & Bergsma, W. P. (2004). Bayesian posterior estimation of logit parameters with small samples. *Sociological Methods and Research, 33*, 88–117.

Goldstein, H. I., & McDonald, R. P. (1988). A general model for the analysis of multilevel data. *Psychometrika, 53*, 455–467.

Goldstein, H. I., & Rasbash, J. (1992). Efficient computational procedures for the estimation of parameters in multilevel models based on iterative generalised least squares. *Computational Statistics and Data Analysis, 13*, 63–71.

Hoijtink, H. (2000). Posterior inference in the random intercept model based on samples obtained with Markov chain Monte Carlo methods. *Computational Statistics, 15*, 315–335.

Ishwaran, H. (2000). Univariate and multirater ordinal cumulative link regression with covariate specific cutpoints. *The Canadian Journal of Statistics, 28*(7), 1–18.

Ishwaran, H., & Gatsonis, C. A. (2000). A general class of hierarchical ordinal regression models with applications to correlated ROC analysis. *The Canadian Journal of Statistics, 28*(7), 1–23.

Johnson, V. E., & Albert, J. H. (1999). *Ordinal data modeling*. New York, NY: Springer.

Kamata, A. (2001). Item analysis by the hierarchical generalized linear model. *Journal of Educational Measurement*, 38, 79–93.

Kamata, A., Bauer, D. J., & Miyazaki, Y. (2008). Multilevel measurement model. In A. A. O'Connell & D. B. McCoach (Eds.), *Multilevel analysis of educational data* (pp. 345–388). Charlotte, NC: Information Age Publishing.

Lahiri, K., & Gao, J. (2002). Bayesian analysis of nested logit model by Markov Chain Monte Carlo. *Journal of Econometrics, 111*, 103–133.

Lunn, D. J., Wakefield, J., & Racine-Poon, A. (2001). Cumulative logit models for ordinal data: A case study involving allergic rhinitis severity scores. *Statistics in Medicine, 20*, 2261–2285.

Masters, G. N. (1982). A Rasch model for partial credit scoring. *Psychometrika, 47*, 149–174.

McDonald, R. P., & Goldstein, H. (1989). Balanced versus unbalanced designs for linear structural relations in two-level data. *British Journal of Mathematical and Statistical Psychology, 42*, 215–232.

Muraki, E. (1992). A generalized partial credit model: Application of an EM algorithm. *Applied Psychological Measurement, 17*, 59–71.

Muthén, B. O. (1994). Multilevel covariance structure analysis. *Sociological Methods & Research, 22*, 376–398.

Muthén, B. O., & Satorra, A. (1995). Complex sample data in structural equation modeling. In P. Marsden (Ed.), *Sociological methodology 1995* (pp. 216–316). Washington, DC: American Sociological Association.

Mwalili, S. M., Lesaffre, E., & Declerck, D. (2005). A Bayesian ordinal logistic regression model to correct for interobserver measurement error in a geographical oral health study. *Journal of the Royal Statistical Society, Series C, 1*, 77–93.

Penfield, R. D., & Camilli, G. (2007). Differential item functioning and item bias. In S. Sinharay & C. R. Rao (Eds.), *Handbook of statistics, Volume 27: Psychometrics*. New York, NY: Elsevier North-Holland.

Qiu, Z., Song, P. X., & Tan, M. (2002). Bayesian hierarchical models for multi-level repeated ordinal data using WinBUGS. *Journal of Biopharmaceutical Statistics, 12*(2), 121–135.

Raudenbush, S. W., Bryk, A. S., & Congdon, R. T. (2005). HLM 6: Hierarchical linear and non-linear modeling (Version 6.02) [Computer software]. Lincolnwood, IL: Scientific Software International, Inc.

Raudenbush, S. W., Yang, M. L., & Yosef, M. (2000). Maximum likelihood for hierarchical models via high order, multivariate LaPlace approximation. *Journal of Computational and Graphical Statistics, 9*(1), 141–157.

Samejima, F. (1969). Estimation of a latent ability using a response pattern of graded scores. *Psychometrika Monograph Supplement*, 17.

Van de Noortgate, W., & De Boeck, P. (2005). Assessing and explaining differential item functioning using logistic mixed models. *Journal of Educational and Behavioral Statistics*, 30, 443–464.

Vaughn, B. K. (2006). *A hierarchical generalized linear model of random differential item functioning for polytomous items: A Bayesian multilevel approach.* Unpublished doctoral dissertation, Florida State University, Tallahassee, FL.

Vehelst, N. D., & Eggen, T. J. H. M. (1989). Psychometriche en statistische aspecten van peilingsonderzoek [Psychometric and statistical aspect of measurement research.] (PPON rapport 4). Arnhem. The Netherlands: Cito.

Yang, M. L. (1988). *Increasing the efficiency in estimating multilevel Bernoulli models.* Unpublished doctoral dissertation, Michigan State University, East Lansing, MI.

Zwinderman, A. H. (1991). A generalized Rasch model for manifest predictors. *Psychometrika, 56*(4), 589–600.

Zwinderman, A. H. (1997). Response models with manifest predictors. In W. J. van der Linden & R. K. Hambleton (Eds.), *Handbook of modern item response theory* (pp. 245–256). New York, NY: Springer.

APPENDIX

Mplus Syntax for the Data Analysis Illustrations

1. DIF detections with multilevel IRT model

```
TITLE:        2PL DIF detection model
DATA:       FILE IS math_lep4.dat;
VARIABLE:     NAMES ARE sch u1-u9 lep
              lep_p;
              CATEGORICAL ARE u1-u9;
              Cluster = sch;
              within = lep;
              usevariables = sch u1-u9
              lep;
ANALYSIS:     Type = Twolevel Random;
MODEL:            %Within%
              f BY u1* (1)
                   u2 (2)
                   u3 (3)
                   u4 (4)
                   u5 (5)
                   u6 (6)
                   u7 (7)
                   u8 (8)
                   u9 (9);
              [f@0];
              f@1;
              f on lep;
              u1 on lep@0;
              u2 on lep;
              u3 on lep;
              u4 on lep;
              u5 on lep;
              u6 on lep;
              u7 on lep;
              u8 on lep;
              u9 on lep;

              %between%
              fb BY u1* (1)
                   u2 (2)
                   u3 (3)
                   u4 (4)
                   u5 (5)
                   u6 (6)
                   u7 (7)
                   u8 (8)
                   u9 (9);
              [fb@0];
```

2. RDIF detection for item 7 with multilevel 2-PL IRT model

```
TITLE:        2PL RDIF model
DATA:       FILE IS math_lep4.dat;
VARIABLE:     NAMES ARE sch u1-u9 lep
              lep_p;
              CATEGORICAL ARE u1-u9;
              Cluster = sch;
              within = lep;
              usevariables = sch u1-u9
              lep;
ANALYSIS:     Type = Twolevel Random;
MODEL:            %Within%
              f BY u1* (1)
                   u2 (2)
                   u3 (3)
                   u4 (4)
                   u5 (5)
                   u6 (6)
                   u7 (7)
                   u8 (8)
                   u9 (9);
              [f@0];
              f@1;
              f on lep;
              u1 on lep@0;
              u2 on lep;
              u3 on lep;
              u4 on lep;
              u5 on lep;
              u6 on lep;
              s7 | u7 on lep;
              u8 on lep;
              u9 on lep;

              %between%
              fb BY u1* (1)
                   u2 (2)
                   u3 (3)
                   u4 (4)
                   u5 (5)
```

```
          u6 (6)
          u7 (7)
          u8 (8)
          u9 (9);
     [fb@0];
     s7;
```

3. RDIF explanation with a school-level covariate for item 7 with multilevel 2-PL IRT model

```
TITLE:        2PL RDIF model with a
              school-level covariate
DATA:     FILE IS math_lep4.dat;
VARIABLE:    NAMES ARE sch u1-u9 lep
             lep_p;

             CATEGORICAL ARE u1-u9;
             Cluster = sch;
             within = lep;
             between = lep_p;
ANALYSIS:    Type = Twolevel Random;
MODEL:           %Within%
             f BY u1* (1)
                  u2 (2)
                  u3 (3)
                  u4 (4)
                  u5 (5)
                  u6 (6)
                  u7 (7)
                  u8 (8)
                  u9 (9);
             [f@0];
             f@1;
             f on lep;
             u1 on lep@0;
             u2 on lep;
             u3 on lep;
             u4 on lep;
             u5 on lep;
             u6 on lep;
             s7 | u7 on lep;
             u8 on lep;
             u9 on lep;

             %between%
             fb BY u1* (1)
                  u2 (2)
                  u3 (3)
                  u4 (4)
                  u5 (5)
                  u6 (6)
                  u7 (7)
                  u8 (8)
                  u9 (9);
             [fb@0];
             s7 on lep_p;
```

4

Mixture Models for Multilevel Data Sets

Jeroen K. Vermunt
Department of Methology and Statistics, Tilburg University, Tilburg, The Netherlands

4.1 INTRODUCTION

Latent class (LC) and mixture models are currently part of the standard statistical toolbox of researchers in applied fields such as sociology, psychology, marketing, biology, and medicine. In most of their applications, the aim is to cluster units into a small number of latent classes or mixture components (McLachlan & Peel, 2000). This clustering can be based on a set of categorical response variables as in the traditional LC model (Goodman, 1974), on a set of continuous items as in a latent profile model (Lazarsfeld & Henry, 1968; Vermunt & Magidson, 2002) and mixture factor analysis (FA) (McLachlan & Peel, 2000; Yung, 1997), on a set of repeated measures as is mixture growth models (Muthén, 2004; Nagin, 1999; Vermunt, 2007b) and mixture (latent) Markov models (Van de Pol & Langeheine, 1990; Vermunt, Tran, & Magidson, 2008), or on other types of two-level data sets, such as from experiments in which individuals are confronted with multiple experimental conditions (Wedel & DeSarbo, 1994) and from studies in which multiple individuals are nested within higher-level units (Aitkin, 1999; Vermunt & Van Dijk, 2001; Kalmijn & Vermunt, 2007). Whereas the main application of LC models is clustering, restricted LC models, and LC models with multiple latent variables can also be used for scaling type of applications similar to IRT and FA (Heinen, 1996; Magidson & Vermunt, 2001; Vermunt, 2001).

Recently, a multilevel extension of the LC model was proposed (Vermunt, 2003, 2008a). It can, for example, be used when individuals with multiple item responses or repeated measurements are nested within groups (Bijmolt, Paas, & Vermunt, 2004; Vermunt, 2003, 2005, 2008a), when multivariate repeated responses are nested within individuals (Vermunt et al., 2008), as well as with three-mode data sets on individuals observed with different measures in different situations (Bouwmeester, Vermunt, & Sijtsma, 2007; Vermunt, 2007a) and three-level data sets (Vermunt, 2004, 2008a). As in standard LC models, for the lower level units, the main goal will usually be to build a meaningful cluster model. One variant of this hierarchical LC model also yields a clustering of higher-level units by assuming that these belong to higher-level latent classes that differ in either lower-level responses or lower-level class membership probabilities. Another variant makes use of random effects to capture higher-level variation in LC model parameters, especially in lower-level class membership probabilities.

The aim of this chapter is threefold. The first is to explain the relationship between LC analysis and multilevel regression analysis techniques. It will be shown that LC models can be conceptualized as models for two-level data sets in which parameters vary randomly across level-2 units. Whereas in multilevel regression analysis this variation is modeled by assuming that parameters come from a particular continuous distribution (typically normal), which is equivalent to introducing one or more continuous latent variables in the model, in LC analysis variation is modeled using discrete latent variables (Aitkin, 1999; Vermunt & Van Dijk, 2001).

Our second aim is to introduce the multilevel extension of the LC model proposed by Vermunt (2003, 2008a), which is a model for univariate three-level data sets and multivariate two-level data sets. The multilevel LC model uses either continuous or discrete latent variables at the higher level.

Third, we discuss other kinds of mixture models for these types of multilevel data sets by connecting multilevel LC analysis to the general latent variable modeling framework described by Skrondal and Rabe-Hesketh (2004). This framework integrates factor analytic and random effects models, as well as models with continuous and discrete latent variables (see also Asparouhov & Muthén, 2008; Vermunt, 2008b). We present a ninefold classification of latent variable models, where eight types can be labeled multilevel mixture models. Most of these models are implemented in software packages such as gllamm (Rabe-Hesketh, Skrondal, & Pickles, 2004), M*plus* (Muthén & Muthén, 2008), and Latent GOLD (Vermunt & Magidson, 2005, 2008).

4.2 TWO-LEVEL DATA SETS: THE STANDARD LATENT CLASS MODEL

Whereas the traditional LC model was developed for the analysis of multivariate response data sets (Goodman, 1974; Lazarsfeld & Henry, 1968), LC analysis can also be used for the analysis of two-level data sets (Aitkin, 1999; Vermunt & Van Dijk, 2001; Wedel & DeSarbo, 1994), such as from students nested within schools and from longitudinal studies in which repeated measurement are nested within persons. However, when the responses are of the same type—for example, all binary or all continuous—we can also conceptualize the traditional LC model as a model for two-level data; that is, by treating the single-level multivariate responses (say on questionnaire or test items) as two-level univariate responses, as item responses nested within individuals. Note that this is similar to IRT modeling using multilevel techniques by treating multiple item responses as multiple observations nested within individuals (see De Boeck & Wilson, 2004). A similar connection has also been established between FA and multilevel linear regression analysis (Hox, 2002).

Using the typical multilevel analysis notation, let y_{ij} denote the response of level-1 unit i belonging to level-2 unit j, n_j the total the number of level-1 units within level-2 unit j, and Y_j the complete response vector of unit j. This could thus also be the n_j item responses of person j, where i refers to a particular item and where n_j takes on the same value for all persons. We refer to a particular latent class by the symbol c and to the number of latent classes by C. To stress the similarity between the discrete latent variable in a LC model and random effects in a multilevel regression model, we use the symbol u to refer to the latent class membership of unit j. More specifically, u_{jc} is one of C indicator variables, which take on the value 1 if level-2 unit j belongs to latent class c and 0 otherwise. As classes are exhaustive and mutually exclusive, exactly one of the C class indicators u_{jc} equals 1 and the others equal 0. The vector of class indicators is denoted by U_j. Using this notation, a standard LC model can be defined as follows:

$$
\begin{aligned}
f(Y_j) &= \sum_{c=1}^{C} P(u_{jc}=1) f(Y_j \mid u_{jc}=1) \\
&= \sum_{c=1}^{C} P(u_{jc}=1) \prod_{i=1}^{n_j} f(y_{ij} \mid u_{jc}=1).
\end{aligned} \tag{4.1}
$$

As can be seen, the LC model is a model for Y_j, the full response vector of level-2 unit j. The model equation shows the two basic assumptions of a LC model. The first one is that the density of Y_j, $f(Y_j)$, is a weighted sum of class specific densities $f(Y_j \mid u_{jc} = 1)$, where the class proportion $P(u_{jc} = 1)$ serve as weights. In other words, a level-2 unit has a certain prior probability of belonging to class c, and conditional on belonging to class c it has a certain probability of giving responses Y_j. The second basic assumption is that the level-1 responses are independent of one another given a level-2 unit's class membership. This assumption is typically referred to as the local independence assumption. Note that this assumption is also made in factor analytic and random effects models, in which responses are assumed to be independent conditional on a unit's latent factor scores and random effects values, respectively.

The specific form chosen for the conditional density $f(y_{ij} \mid u_{jc} = 1)$ depends on the scale type of the response variable. Examples are Bernoulli for binary responses, normal for continuous responses, and Poisson for counts. What we are typically interested in are the expected values of these conditional distributions, denoted by $E(y_{ij} \mid u_{jc} = 1)$, which can be binomial proportions, normal means, Poisson rates, and so on. These class-specific expected values can be parameterized using a generalized linear model; that is, using a linear model after applying the appropriate link function $g[\cdot]$. Let us assume that we are dealing with a traditional LC model, which means that the level-1 observations are n_j questionnaire items. A regression model for the $n_j \cdot C$ conditional means $E(y_{ij} \mid u_{jc} = 1)$ can be formulated as follows:

$$g[E(y_{ij} \mid u_{jc} = 1)] = \beta_0 + \beta_i + \lambda_{0c} + \lambda_{ic}.$$

This (unrestricted) model contains an intercept (β_0), item effects (β_i), main effects for the classes (λ_{0c}), and item-class interactions (λ_{ic}). Note that this model contains $1 + n_j + C + n_j \cdot C$ unknown parameters, which means that $1 + n_j + C$ identification restrictions should be imposed on the regression coefficients, for example, by using dummy coding where the parameters for the first item and the first class are fixed to zero, or by fixing the β_0, β_i, and λ_{0c} parameters to zero.

To show how the same regression model for the n_j item responses can be formulated in the more typical multilevel analysis notation, let X_j and Z_j be two (identical) design matrices with n_j rows and $n_j + 1$ columns:

$$X_j = Z_j = \begin{bmatrix} 1 & 1 & 0 & . & 0 \\ 1 & 0 & 1 & . & 0 \\ 1 & 0 & 0 & . & 0 \\ . & . & . & . & . \\ 1 & 0 & 0 & . & 1 \end{bmatrix}.$$

The first column of X_j and Z_j contains the 1s for the intercept term and the remaining columns contain dummies for the item effects. Using these design matrices and conditioning on the vector U_j rather than on $u_{jc} = 1$, the regression model for y_{ij} can also be written as follows:

$$g[E(y_{ij} \mid U_j, X_{ij}, Z_{ij})] = \sum_{p=0}^{P} \beta_p x_{pij} + \sum_{q=0}^{Q} \left(\sum_{c=1}^{C} \lambda_{qc} u_{jc} \right) z_{qij}. \tag{4.2}$$

Though this equation is already very similar to a two-level regression model, the similarity becomes even clearer if we define $u^*_{qj} = \sum_{c=1}^{C} \lambda_{qc} u_{jc}$; that is,

$$g[E(y_{ij} \mid U_j, X_{ij}, Z_{ij})] = \sum_{p=0}^{P} \beta_p x_{pij} + \sum_{q=0}^{Q} u^*_{qj} \; z_{qij}.$$

This equation shows that a LC model can be seen as a model with random effects u^*_{qj}. A difference compared to a standard random-effects model is that rather than assuming that the random effects come from a Q-dimensional multivariate normal distribution, here we assume that these take on only C different values, where each value λ_{qc} occurs with probability $P(u_{jc} = 1)$. Both the values of the random effects (the locations) and the associated probabilities are free parameters to be estimated. The mean of u^*_{qj} and covariance between u^*_{qj} and $u^*_{q'j}$ can be obtained as $\bar{u}^*_q = \sum_{c=1}^{C} \lambda_{qc} \; P(u_{jc} = 1)$ and $\sigma_{u_q u_{q'}} = \sum_{c=1}^{C} \lambda_{qc} \lambda_{q'c} P(u_{jc} = 1) - \bar{u}^*_q \; \bar{u}^*_{q'}$ (Aitkin, 1999; Vermunt & Van Dijk, 2001).

In the above example, we used design matrices X_j and Z_j with a very specific structure, with an intercept and a set of item effects. However, as in two-level regression models, these two matrices may also contain the values of other types of predictors. More specifically X_j may contain level-1 predictors, level-2 predictors, and cross-level interactions, and Z_j may contain level-2 predictors and cross-level interactions. This yields what is often referred to as a LC or mixture regression model (Aitkin, 1999; Vermunt & Van Dijk, 2001; Wedel & DeSarbo, 1994), which is a regression model for two-level data sets.

In LC models, level-2 predictors cannot only be used in the regression model for the response variable y_{ij}, but can also be used to predict the class membership probabilities. Typically a multinomial logit model is used for this purpose:

$$\text{logit}[P(u_{jc} = 1 \mid X_j^{(2)})] = \sum_{r=0}^{R} \gamma_{rc} x_{rj}^{(2)}, \tag{4.3}$$

where the first column of $X_j^{(2)}$ $(r = 0)$ defines the constant term. For identification, one constraint has to be imposed on γ_{rc} for each r, for example, $\gamma_{r1} = 0$.

It is important to note that also in standard two-level regression models one may define regression models for the latent variables representing the random effects, as is usually done in the hierarchical model specification of the multilevel model. However, it is always possible to substitute the random effects in the model for the response variable by their regression equations, yielding the well-known mixed model formulation of the multilevel model. Such a substitution is not possible in LC regression analysis.

Whereas the standard LC model is a latent variable model with a nominal latent variable, it is also possible to define LC models with a latent variable with ordered categories. Such an ordinal specification is obtained by using a single u_j with numeric scores instead of C class indicators u_{jc} (Heinen, 1996; Vermunt, 2001), which implies replacing the term $\sum_{c=1}^{C} \lambda_{qc} u_{jc}$ in Equation 4.2 by $\lambda_q u_j$. In a three-class model, the class locations u_j could, for example be −1, 0, and 1 or 0, 0.5, and 1.

The above model formulations can easily be adapted for LC models with more than one nominal or ordinal latent variable. For example, the LC model with multiple ordinal latent variables proposed by Magidson and Vermunt (2001) can be defined by including a term $\sum_{\ell=1}^{L} \lambda_{q\ell} u_{\ell j}$ in Equation 4.2, where $u_{\ell j}$ is one of L ordinal latent variables. Similarly, a model with L nominal latent variables can be defined by setting up a series of dummies for each latent variable and using the term $\sum_{\ell=1}^{L} \sum_{c=1}^{C_\ell} \lambda_{q\ell c} u_{\ell jc}$ in Equation 4.2. In models with multiple latent variables, one will typically restrict the joint probability density of the L latent variables, for example, using a log-linear (path) model (Hagenaars, 1993; Vermunt, 1997) or a latent Markov structure (Van de Pol & Langeheine, 1990).

4.3 THREE-LEVEL DATA SETS: THE MULTILEVEL LATENT CLASS MODEL

Let us now expand the LC model for the situation in which we have either a three-level univariate response or a two-level multivariate response data set. Note that by conceptualizing multivariate responses as nested univariate responses, the latter can also be seen as three-level data sets, which is what we will do here. The extension of the LC model discussed here yields what Vermunt (2003, 2008a) called multilevel LC analysis.

To accommodate the additional hierarchical level, two modifications of the notation introduced in the previous sections are needed: an index k is used to refer to a particular level-3 unit and, whenever necessary, a superscript (1), (2), or (3) is used to denote whether we are referring to a level-1, level-2, or level-3 quantity. For example, level-1 responses are now denoted by y_{ijk}, a level-2 response vector by Y_{jk}, level-2 class indicators by $u_{jkc}^{(2)}$, and a level-2 vector of class indicators by $U_{jk}^{(2)}$.

The main difference compared to the two-level case discussed in the previous section is that a multilevel LC model contains either continuous random effects or a discrete latent variable (=discrete random effects) at level 3. These random effects pick up variation in LC model parameters across level-3 units. Below, we first discuss the situation in which level-3 heterogeneity is modeled using discrete random effects, as well as two important special cases of this specification. Then we discuss the multilevel LC model with continuous random effects. The third part of this section introduces other types of multilevel mixture models; that is, models with discrete random effects at level 3 but that are not necessarily LC models at level 2.

4.3.1 Models with Discrete Random Effects at Level Three

Let us first look at the situation in which the heterogeneity at the highest level is modeled by assuming that level-3 units belong to one of D latent classes. The level-3 class membership is denoted using indicator variables $u_{kd}^{(3)}$, which take on the value 1 if unit k belongs to class d and 0 otherwise. The vector of level-3 class indicators is denoted by $U_k^{(3)}$. The corresponding multilevel extension of the LC model is as follows:

$$f(Y_{jk} \mid u_{kd}^{(3)}=1)=\sum_{c=1}^{C} P(u_{jkc}^{(2)}=1 \mid u_{kd}^{(3)}=1) \prod_{i=1}^{n_{jk}^{(2)}} f(y_{ijk} \mid u_{jkc}^{(2)}=1, u_{kd}^{(3)}=1). \tag{4.4}$$

As can be seen, the "only" modification compared to the standard LC model described in Equation 4.1 is that both the level-2 class proportions $P(u_{jkc}^{(2)}=1 \mid u_{kd}^{(3)}=1)$ and the class-specific densities $f(y_{ijk} \mid u_{jkc}^{(2)}=1, u_{kd}^{(3)}=1)$ may depend on $U_k^{(3)}$. It is important to note that a multilevel LC model is actually a model for Y_k, the full response vector of higher-level unit k; that is,

$$f(Y_k)=\sum_{d=1}^{D} P(u_{kd}^{(3)}=1)\prod_{j=1}^{n_k^{(3)}} f(Y_{jk} \mid u_{kd}^{(3)}=1) \tag{4.5}$$

where $f(Y_{jk} \mid u_{kd}^{(3)}=1)$ was defined in Equation 4.4. It can easily be seen that Equation 4.5 defines a LC model for the higher level units, which is very similar to Equation 4.1, the equation for a standard latent model. Equation 4.5 shows that groups are assumed to belong to one of D latent classes, as well as that level-2 observations are assumed to be independent of one another conditional on the level-3 class membership.

The fact that $P(u_{jkc}^{(2)}=1 \mid u_{kd}^{(3)}=1)$ and $f(y_{ijk} \mid u_{jkc}^{(2)}=1, u_{kd}^{(3)}=1)$ depend on $u_{kd}^{(3)}$ can also be expressed via the regression models for $u_{jkc}^{(2)}$ and y_{ijk}. These will differ from the ones in Equations 4.2 and 4.3 in that they may now contain terms for $u_{kd}^{(3)}$. In addition, a logistic regression model may be specified for $u_{kd}^{(3)}$ itself. We use (sometimes double) superscripts to distinguish the different parameters sets and design matrices, where the first index refers to the level of the dependent variable in the equation concerned and the second, if present, to the level of the random effect. The three regression equations defining the multilevel LC model are:

$$g[E(y_{ijk} \mid U_{jk}^{(2)}, U_k^{(3)}, X_{ijk}^{(1)}, Z_{ijk}^{(1)})] = \sum_{p=0}^{P} \beta_p x_{pijk}^{(1)} + \sum_{q=0}^{Q(2)} \left(\sum_{c=1}^{C} \lambda_{qc}^{(1,2)} u_{jkc}^{(2)} \right) z_{qijk}^{(1,2)} + \sum_{q=0}^{Q(3)} \left(\sum_{d=1}^{D} \lambda_{qd}^{(1,3)} u_{kd}^{(3)} \right) z_{qijk}^{(1,3)} \tag{4.6}$$

$$\text{logit}[P(u_{jkc}^{(2)}=1 \mid U_k^{(3)}, X_{jk}^{(2)}, Z_{jk}^{(2)})] = \sum_{r=0}^{R} \gamma_{rc}^{(2)} x_{rjk}^{(2)} + \sum_{s=0}^{S} \left(\sum_{d=1}^{D} \lambda_{scd}^{(2,3)} u_{kd}^{(3)} \right) z_{sjk}^{(2,3)} \tag{4.7}$$

$$\text{logit}[P(u_{kd}^{(3)}=1\,|\,X_k^{(3)})]=\sum_{t=0}^{T}\gamma_{td}^{(3)}x_{tk}^{(3)}. \quad (4.8)$$

The terms $\sum_{c=1}^{C}\lambda_{qc}^{(1,2)}u_{jkc}^{(2)}$, $\sum_{d=1}^{D}\lambda_{qd}^{(1,3)}u_{kd}^{(3)}$, and $\sum_{d=1}^{D}\lambda_{scd}^{(2,3)}u_{kd}^{(3)}$ have a similar (discrete) random effects interpretation as was explained in the previous section. As in the two-level model, the level-2 classes may capture variation in the parameters of the model for the response variable. The level-3 classes may capture variation in the parameters of the response model, as well as in the parameters of the regression model for the level-2 classes.

When the model does not contain predictors and when, as in the example used in the previous section, the design matrices are setup to yield intercepts and item parameters, the three regression equations can be written in a simpler form; that is,

$$g[E(y_{ijk}\,|\,u_{jkc}^{(2)}=1,u_{kd}^{(3)}=1)]=\beta_0+\beta_i$$
$$+\lambda_{0c}^{(1,2)}+\lambda_{ic}^{(1,2)}+\lambda_{0d}^{(1,3)}+\lambda_{id}^{(1,3)},$$

$$\text{logit}[P(u_{jkc}^{(2)}=1\,|\,u_{kd}^{(3)}=1)]=\gamma_{0c}^{(2)}+\lambda_{0cd}^{(2,3)},$$

$$\text{logit}[P(u_{kd}^{(3)}=1)]=\gamma_{0d}^{(3)},$$

which yields the more typical LC analysis notation.

Let us look at two more restricted special cases of the model defined in Equations 4.4 through 4.8. The first special case is obtained by assuming that groups differ in the lower-level class membership probabilities but have the same response variable densities. This implies that $f(y_{ijk}|u_{jkc}^{(2)}=1,u_{kd}^{(3)}=1)=f(y_{ijk}|u_{jkc}^{(2)})=1)$ or, equivalently, that the term containing the $\lambda_{qd}^{(1,3)}$ parameters is excluded from Equation 4.6. This is the special case used by Vermunt (2003, 2007a). The basic idea is that the model part linking the lower-level class memberships to the responses is the same for all groups, which is conceptually similar to saying that there is measurement equivalence across higher-level units. Groups may, however, differ with respect to the lower-level class membership probabilities of their members, as well as with respect to covariate effects on these class membership probabilities. These differences across group-level classes are defined with the second term in Equation 4.7.

The second special case is the opposite from the first. It assumes that response densities depend on the group-level class membership, but lower-level class membership not. This implies that $P(u_{jkc}^{(2)}=1\,|\,u_{kd}^{(3)}=1)=P(u_{jkc}^{(2)})=1)$ or that the term containing the $\lambda_{scd}^{(2,3)}$ parameters is omitted from Equation 4.7. This model is very similar to a standard regression model in which the variation in the responses is decomposed into independent parts (Goldstein, 2003). The model is also similar to the multilevel FA model proposed by Muthén (1994) and described by Hox (2002), in which the variation in a multivariate response vector is attributed to common latent factors at two levels of a hierarchical structure.

It should be noted that in three-level regression modeling with continuous random effects these two special cases yield equivalent models: regressing a lower-level random effect on a higher-level random effect is the same as using the terms concerned in the model for the response variables (Goldstein, 2003; Hox, 2002). In the case of a multilevel FA, the first specification is a restricted special case of the second one, which is obtained by equating the factor loadings across levels. In other words, indicating that the lower-level factor means vary randomly across higher-level units is equivalent to having a set of higher-level factors with the same loadings as the lower-level factors. Despite the conceptual similarities of these models with the multilevel LC model, these equivalences do not apply to the latter.

The full model is conceptually similar to a three-level model in which level-2 and level-3 random effects are correlated, which is a specification that is seldom used in multilevel regression analysis. Another specific aspect of the multilevel LC model is that also level-1 variances (which are free parameters in linear models with normally distributed residuals) can be allowed to differ across lower- and/or higher-level classes. In other words, level-2 and level-3 units may randomly differ with respect to their level-1 variances. There exists no equivalent specification for this in standard three-level regression analysis. A last specific feature this author would like to mention is that interactions between level-2 and level-3 class effects can easily be included in the model; that is, by adding terms containing the product $u_{jkc}^{(2)} \cdot u_{kd}^{(3)}$ to Equation 4.6. Also for such interactions there is no equivalence in standard three-level regression analysis. Though this seems to be a somewhat exotic option, this is clearly not the case. A possible application is the investigation of item bias, where not only item intercepts but also lower-level class effects on items may differ across higher-level classes. The latter would be similar to allowing that factor loadings differ randomly across groups, which is not possible in (standard) multilevel FA.

4.3.2 Models with Continuous Random Effects at Level Three

Rather than using discrete random effects, it is also possible to use a more standard specification with continuous random effects. Vermunt (2003, 2005) proposed a multilevel LC model in which the class membership probabilities of lower-level units vary randomly across level-3 units. Using as much as possible the same notation as above, but now with $u_{sk}^{(3)}$ representing the sth random effect and $U_k^{(3)}$ the vector of random effects, we can write the model as follows:

$$f(Y_{jk} | U_k^{(3)}) = \sum_{c=1}^{C} P(u_{jkc}^{(2)} = 1 | U_k^{(3)}) \prod_{i=1}^{n_{jk}^{(2)}} f(y_{ijk} | u_{jkc}^{(2)} = 1) \qquad (4.9)$$

$$f(Y_k) = \int_U f(U_k^{(3)}) \prod_{j=1}^{n_k^{(3)}} f(Y_{jk} | U_k^{(3)})\, dU_k^{(3)} \qquad (4.10)$$

with regression equations

$$g[E(y_{ijk} | U_{jk}^{(2)}, X_{ijk}^{(1)}, Z_{ijk}^{(1)})] = \sum_{p=0}^{P} \beta_p x_{pijk}^{(1)} + \sum_{q=0}^{Q(2)} \left(\sum_{c=1}^{C} \lambda_{qc}^{(1)} u_{jkc}^{(2)} \right) z_{qijk}^{(1)} \qquad (4.11)$$

$$\text{logit}[P(u_{jkc}^{(2)} = 1 | U_k^{(3)}, X_{jk}^{(2)}, Z_{jk}^{(2)})] = \sum_{r=0}^{R} \gamma_{rc}^{(2)} x_{rjk}^{(2)} + \sum_{s=0}^{S} \lambda_{sc}^{(2)} u_{sk}^{(3)} z_{sjk}^{(2)}. \qquad (4.12)$$

Note that this specification assumes measurement equivalence; that is, the parameters in the LC model for the response variable(s) do not vary across groups. Groups differ with respect of their level-2 class membership probabilities, which is specified using a random effects multinomial logistic regression model of the type proposed by Hedeker (2003) for the discrete unobserved variable $U_{jk}^{(2)}$. This model uses a lower-dimensional representation of the $S + 1$ random effects for each of the C latent classes; that is, it uses the "factor analytic" constraint $u_{sck}^{*(3)} = \lambda_{sc}^{(2)} u_{sk}^{(3)}$ which actually implies that the random effects for a particular s are perfectly correlated across latent classes (across values of c). Without covariates, Equation 4.12 reduces to a variance decomposition of the class membership logits. A more detailed description of this type of multilevel LC model is provided by Vermunt (2005).

4.3.3 Other Types of Multilevel Mixture Models

The term multilevel LC or mixture model was used so far for the situation in which we have a LC model for level-2 observations combined with an additional hierarchical level, where this third level is dealt with using either continuous or discrete random effects. However, looking at Equation 4.5 that defines a mixture model for level-3 units, one can think of other types of multilevel mixture models; that is, models with latent classes at level 3 and continuous latent variables (or random effects) at level 2. An example is a variant of the mixture FA model where mixture components are not formed by individuals but by groups (Varriale, 2008). Moreover, there is nothing that prevents applying the same logic of hierarchically structured mixture models to situations with more than three hierarchical levels.

Using a more general perspective, we get into a general latent variable modeling framework described by Skrondal and Rabe-Hesketh (2004) and expanded in certain ways by Vermunt (2008b) and Muthén (this volume). Table 4.1 shows the ninefold classification of latent variable models for three-level data sets based on the scale types of the latent variables at level 2 and level 3. At each level, one may have continuous latent variables (or random effects), discrete latent variables, or a combination of these. All models except type A1 could be called multilevel mixture models; that is, models with latent classes at either one or at two levels.

Multilevel factor and IRT models (Fox & Glas, 2001; Goldstein & Browne, 2002; Grilli & Rampichini, 2007; Muthén, 1994), as well as three-level random effects regression models belong to category A1. The multilevel latent class models described above belong, depending on whether the level-3 random effects are treated as continuous or discrete, to either the B1 or B2 type. By introduction a continuous latent variable at level 2, which is a method for dealing with

TABLE 4.1

Ninefold Classification of Latent Variable Models for Three-Level Data Sets

Level-2 Latent Variables $U_{jk}^{(2)}$	Level-3 Latent Variables $U_k^{(3)}$		
	Continuous	**Discrete**	**Combination**
Continuous	A1	A2	A3
Discrete	B1	B2	B3
Combination	C1	C2	C3

local dependencies between items in LC analysis, one would obtain a model of type C1 or C2. An example of a model of type C1 is the multilevel variant of the factor mixture model (Allua, 2007). Vermunt (2008b) and Varriale (2008) used type A2 models to define multilevel mixture IRT and factor analytic models, respectively. Palardy and Vermunt (2010) used type A2 and A3 models in the context of multilevel mixture growth modeling.

4.4 ESTIMATION, MODEL SELECTION, AND SAMPLE SIZE ISSUES

Vermunt (2003, 2004, 2005) demonstrated how to obtain maximum likelihood (ML) estimates of the parameters of multilevel LC models by means the EM algorithm. For this purpose, a (necessary) modification of the E step of the EM algorithm was developed, which was called an upward–downward algorithm. This procedure, as well as a Newton-Raphson algorithm with numerical second derivatives based on analytical first derivatives, is implemented in the Latent GOLD software package (Vermunt & Magidson, 2008). The Appendix provides various examples of Latent GOLD syntax files. Also M*plus* (Muthén & Muthén, 2008) can be used to estimate (most of) the models described in this chapter. The M*plus* manual is not very explicit about the estimation methods and algorithms that are used, but it seems to use similar procedures as Latent GOLD. The gllamm program (Rabe-Hesketh et al., 2004) can deal with the situation in which the latent variables at both levels are discrete (or both continuous) and in which only the responses depend on the higher-level class membership (special case number two described in Section 4.3.1). Optimization of the log-likelihood is performed using the Stata ML routines. Each of these three packages has options for obtaining robust standard errors as well as for dealing with missing values and complex sampling designs.

Model selection is already a rather complicated issue in standard LC analysis, but becomes even more complex in multilevel LC models, especially when the level-3 heterogeneity is modeled using level-3 latent classes. We then not only need to decide about the required number of latent classes at level 2 (the value of C), but also about the number of classes at level 3 (the value of D). In principle, in multilevel LC analysis the same types of model selection measures can be used as the ones that are used in standard LC analysis, with information criteria such as AIC, BIC, and AIC3 as the most popular ones. However, the use of BIC is somewhat problematic because it contains the sample size in its formula, and is not fully clear what sample size to use in the BIC formula in a multilevel analysis. Note that this is a problem for multilevel analysis in general, and thus not specific for multilevel LC analysis. A recent simulation study by Lukociene and Vermunt (2010) has shed some light on this issue: when deciding about the number of classes at the higher-level it is better to use the higher-level sample size in the BIC formula instead of the lower-level sample size.

Multilevel LC models have been applied with very different level-1, level-2, and level-3 sample sizes. Although little research has been done on this topic so far, some guidelines can be provided on sample size requirement. What should be understood is that in these types of models the sample size at a particular level may affect not only sampling fluctuation but also the separation of the classes at higher levels, which is similar to the reliability of the measurement of a continuous latent variable. The total level-2 sample size and the level-3 sample size affect the sampling fluctuation in the level-2 and level-3 parameters, respectively. The required level-2 and level-3 sample size depends strongly on the separation of the level-2 and level-3 classes, respectively. But this separation is itself affected by the level-1 sample size for the level-2 classes (the longer a test the more certain we can be about a person's class membership), and by the level-2 sample size for the level-3 classes (the more level-2 units, the more certain we can be about a level-3 unit's class membership). This shows that sample size requirements for one level may depend on the sample size at another level; for example, because of a better separation between level-3 classes, the required level-3 sample size is smaller with larger level-2 and level-1 sample sizes. It should be noted that there are other factors affecting the separation between classes, and thus the required sample size. One of these is how different latent classes are (the size of the λ parameters appearing in the above formulae): smaller samples are need with more dissimilar classes. Another factor is the scale type of the response variable: with continuous normal responses or Poisson counts smaller samples are needed than with the same number of dichotomous responses because the former are more informative about differences between classes.

4.5 APPLICATIONS

This section presents two applications of the multilevel LC models described in the previous section. The first is a typical three-level regression application, and concerns a data set containing repeated measurements from a longitudinal survey with individuals nested within regions. The second example uses a data set as is typically analyzed using cluster analysis or FA, with the complicating factor that individuals (children) are nested within groups (families); that is, this is an example of an analysis of a two-level data set with multiple continuous items.

For our analysis we used version 4.5 of the Latent GOLD program (Vermunt & Magidson, 2008). The Appendix presents examples of Latent GOLD syntax files. For model selection, we used two versions of BIC: BIC(2) based on the level-2 sample size and BIC(3) based on the level-3 sample size. When the two fit measures disagree with respect to the required number of level-2 classes, we select the model with the lowest BIC(2). Similarly, when disagreement concerns the number of level-3 classes, we select the model that is preferred by BIC(3).

4.5.1 Three-Level Mixture Regression Analysis

The first application uses a data set from the data library of the Centre of Multilevel Modelling, University of Bristol (http://www.cmm.bristol.ac.uk/learning-training/multilevel-m-support/datasets.shtml). The data consist of 264 participants in the 1983–1986 yearly waves from the British Social Attitudes Survey (McGrath & Waterton, 1986). It is a three-level data set: Individuals are nested within districts and time points are nested within individuals. The total number of level-3 units (districts) is 54.

The dependent variable is the number of yes responses on seven yes/no questions as to whether it is a woman's right to have an abortion under a specific circumstance. Since this variable is a count with a fixed total, it is most natural to work with a binomial error function and a logit link. Individual-level predictors in the data set are religion, political preference, gender, age, and self-assessed social class. In accordance with the results of Goldstein (2003), we found no significant effects of gender, age, self-assessed social class, and political preference. Therefore, we did not use these predictors in the further analysis. The predictors that were used are the level-1 predictor year of measurement (1983, 1984, 1985, and 1986) and the level-2 predictor religion (Roman Catholic, Protestant, Other, and No religion). As there was no evidence for a linear time effect, we used time as a categorical predictor in the regression model.

Vermunt (2004) used this data set to illustrate the similarity between three-level mixture regression models with intercept variation across latent classes and standard three-level random-intercept models. Here, the author presents a more extended analysis in which, among others, slopes are allowed to vary across level-2 and level-3 classes.

Our baseline model is a three-level mixture regression model of the form described in Equations 4.4 through 4.8. More specifically, we specified models with a fixed intercept and six fixed slopes (three for time and three for religion), a random intercept and random slopes for the time effects at level 2, and a random intercept and random slopes for the religion effects religion at level 3. This implies that $X_{ijk}^{(1)}$, $Z_{ijk}^{(1,2)}$, and $Z_{ijk}^{(1,3)}$ contain 7, 4, and 4 columns, respectively. Furthermore, we assume that the level-2 class membership does not depend on the level-3 class membership (this is the second special case discussed in Section 4.3.1) nor on covariates. This means that Equation 4.7 contains only an intercept. Also Equation 4.8 contains only an intercept since we have no level-3 predictors.

Table 4.2 reports the log-likelihood (LL) value, the number of parameters (Npar), the BIC(2) value, and the BIC(3) value for models with 1–5 level-2 classes and 1–3 level-3 classes. The fact that models with $C > 1$ (for constant D) have lower BIC values than models with $C = 1$ shows that there is evidence for level-2 variation in intercept and/or time slopes. A similar conclusion can be drawn for the level-3 variation in intercept and/or religion slopes since models with $D > 1$ perform better according to the BIC statistics than the ones with $D = 1$. BIC(2) and BIC(3) select the same model as the best one; namely, the model with $C = 4$ and $D = 2$.

As a next step, we defined five alternative models to test specific aspects of our baseline model with $C = 4$ and $D = 2$. Two more restricted models omit the random time and religion slopes, respectively. Three more extended models are estimated that add a level-3 random time effects, a direct effect of level-3 class on level-2 class, and an interaction effect between level-2 and level-3 class in the model for the response variable. The fit measures in Table 4.2 indicate that the two more restricted models perform worse than the baseline model, which means that the level-2 and level-3 variation in the time and religion slopes are significant. The models with level-3 variation in the time effects and with an association between level-2 and level-3 class membership do not perform better than the baseline model, which indicates that there is no need to include these effects. However, the last model has lower BIC values than the baseline model, which indicates that there is evidence that the level-2 class intercept differences vary across level-3 classes. This model will serve as our final model.

TABLE 4.2

Fit Measures for the Models Estimated with the Abortion Data

Model			LL	Npar	BIC(2)	BIC(3)
$C = 1$	$D = 1$	baseline	–2188	7	4416	4405
$C = 2$	$D = 1$	baseline	–1745	12	3558	3539
$C = 3$	$D = 1$	baseline	–1683	17	3460	3433
$C = 4$	$D = 1$	baseline	–1657	22	3436	3401
$C = 5$	$D = 1$	baseline	–1645	27	3441	3398
$C = 1$	$D = 2$	baseline	–2073	12	4212	4193
$C = 2$	$D = 2$	baseline	–1712	17	3519	3492
$C = 3$	$D = 2$	baseline	–1671	22	3465	3431
$C = 4$	$D = 2$	baseline	–1644	27	**3438**	**3396**
$C = 5$	$D = 2$	baseline	–1636	32	3450	3399
$C = 1$	$D = 3$	baseline	–1999	17	4092	4065
$C = 2$	$D = 3$	baseline	–1699	22	3520	3485
$C = 3$	$D = 3$	baseline	–1663	27	3477	3434
$C = 4$	$D = 3$	baseline	–1638	32	3455	3404
$C = 5$	$D = 3$	baseline	–1629	37	3465	3406
$C = 4$	$D = 2$	religion not random	–1651	24	3435	3397
$C = 4$	$D = 2$	time not random	–1683	18	3467	3439
$C = 4$	$D = 2$	time also random at level-3	–1639	30	3444	3397
$C = 4$	$D = 2$	association between U2 and U3	–1640	30	3448	3400
$C = 4$	$D = 2$	interaction between U2 and U3	–1632	30	**3432**	**3384**

Note: Bold indicates lowest value.

Figure 4.1a depicts the estimated value of the intercept for each level-2 and level-3 class combination (these are obtained by adding up the fixed and the random intercept terms, including the interaction), Figure 4.1b the time effects per level-2 class (these sum to 0 across time points and are obtained by adding up the fixed and the random time effects), and Figure 4.1c the religion effects per level-3 class (these sum to 0 across religion categories and are obtained by adding up the fixed and the random religion effects). Figure 4.1a and b show that level-2 class one contains the respondents who are most against abortion, irrespective of the level-3 (region) class, and whose opinion is most stable across the four measurement occasions. Depending on the region class, class two respondents are very much in favor or somewhat against abortion, and they become less in favor during the observation period. Class three is (moderately) against, and shows a decrease and subsequently a return to the initial position during the 4 years of the study. Class four is (moderately) in favor of abortion, but much less at the first two occasions than at the last two occasions.

As far as the level-3 classes is concerned, Figure 4.1a shows that level-3 classes of regions differ in opinion concerning abortion only among class two respondents. Moreover, the religion effects (Figure 4.1c)

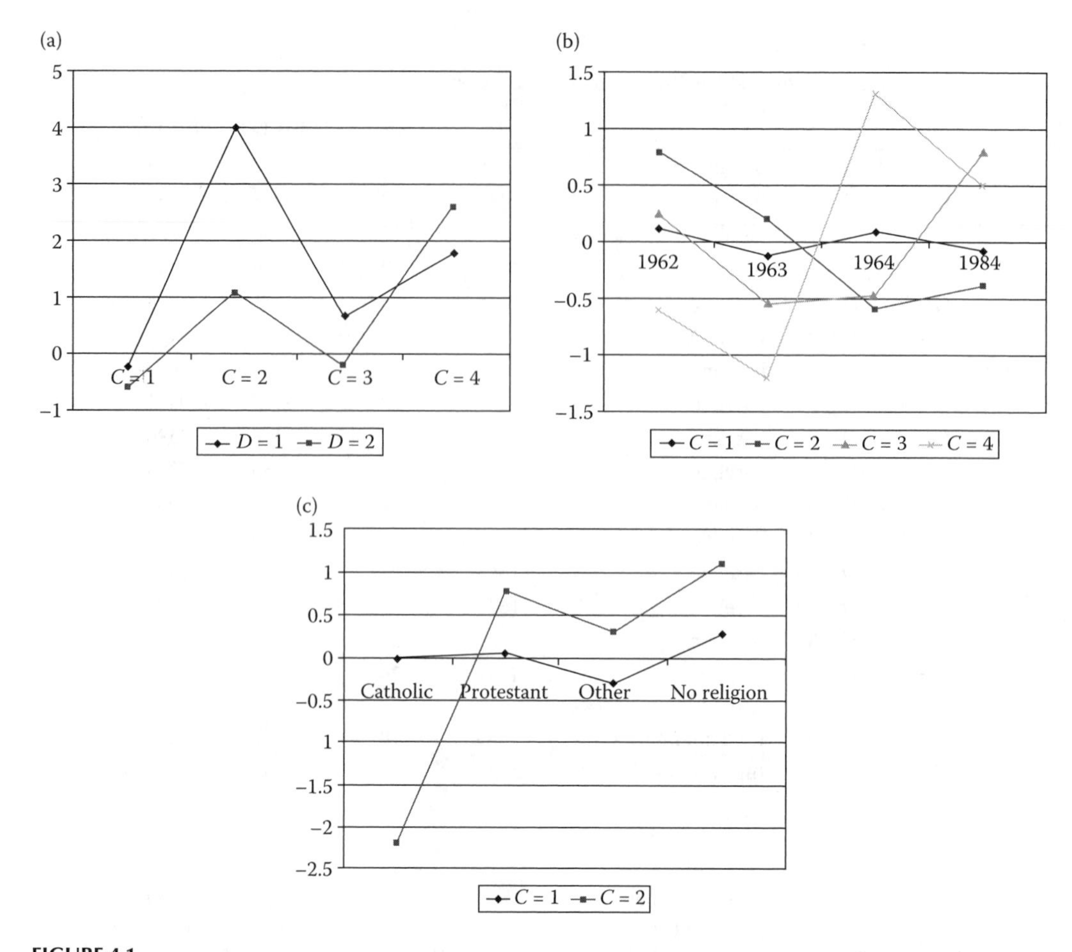

FIGURE 4.1
(a) Intercept for all combinations of level-2 and level-3 classes obtained with abortion data. (b) Time effects for level-2 classes obtained with abortion data. (c) Religion effects for level-3 classes obtained with abortion data.

are small in class two, with "other religion" slightly more against and "no religion" slightly more in favor of abortion. In class one regions, the religion effects are huge: here, Catholics are much more against abortion that the other religion categories. The latter ones show similar mutual differences as in the other latent class.

This example showed that complex but interesting level-2 and level-3 variability in intercepts and slopes can be detected using model specifications that are rather straightforward within the multilevel LC analysis framework. The most similar specification using a "standard" three-level logistic regression model would be a model with four normally distributed random effects at level 2 and four at level 3. Interpretation of the results of such an analysis would probably have been more difficult than the results presented above.

4.5.2 Multilevel Mixture Modeling with a Set of Continuous Responses

The data for this example were collected by Van Peet (1992) and used by Hox (2002) to illustrate multilevel FA. Six continuous measures supposed to be connected to intelligence—"word list," "cards," "matrices," "figures," "animals," and "occupations"—are available for 269 children belonging to 49 families. For 82 children, there is partially missing information, but these observations can be retained in the analysis using standard ML estimation methodology with missing data.

Hox (2002) analyzed this data set (excluding cases with missing values) using multilevel FA, which basically involves performing separate analyses of the within- and between-family covariance matrices. At the within level, his final model contained a "numeric" factor (loading on word list, cards, and matrices) and a "perception" factor (loading on figures, animals, and occupations), whereas at the between level a single factor sufficed. His aim was to determine whether the six measures relate to similar aspects of intelligence at the within- and between-family level, as well as to detect possible family effects, which may be explained by genetic and/or common environment factors.

To illustrate various types of multilevel mixture models, we will analyze this data set in three different ways, in each of which level-3 variation is modeled by assuming that families belong to a small number of level-3 classes. The first analysis uses a model corresponding to the first special case discussed in Section 4.3.1; that is, a model in which level-2 classes affect the item responses and level-3 classes the level-2 class memberships, but not directly the responses. Between-family differences in responses are thus explained by between-family differences in the likelihood of belonging to the child-level intelligence clusters. This is also the specification Vermunt (2008a) used in an earlier analysis of this data set.

The second analysis is conceptually similar to Hox's analysis, but with discrete instead of continuous latent variables. In this analysis, both latent variables are assumed to affect the responses directly. The role a particular item plays in the clustering of children and the clustering of families may be fully different. Moreover, the clustering of children is conditional on the family clustering, which means that it is based on within-family differences that remain after taking into account the differences between family classes.

The third analysis uses continuous latent variables at level 2. The model is a multilevel mixture factor model, a model with a mixture distribution at level 3 to capture between-family differences in the parameters of the child-level factor model. This can be seen as a kind of multiple group FA with a large number of groups. The aim is to investigate whether the factor model parameters can be assumed to be invariant across groups.

Preliminary analysis showed that simple univariate normal within-class distributions with homogeneous residual variances across lower- and higher-level classes can be assumed for the six response variables. More specifically, inspection or pairwise residuals showed that there is no need to allow for within-class correlations across responses, and comparison of models with equal and unequal variances showed that it is correct to assume that residual variances are homogeneous across lower- and higher-level classes.

4.5.2.1 Analysis 1: Indirect Effect of Family Classes

In this first analysis, the six intelligence measures were used to cluster children into intelligence classes and it was investigated whether (classes of) families differ in the distribution of (their) children over these "intelligence" clusters. Table 4.3 provides the fit measures for the estimated multilevel LC models. As can be seen, a model with four child-level classes and three family-level classes performed best according to both BIC(2) and BIC(3).

Figures 4.2a displays the estimated $\lambda_{qc}^{(1,2)}$ parameters appearing in Equation 4.6 for the model with $C = 4$ and $D = 3$. These parameters (which sum to 0 over classes) show that the means of the six intelligence indicators are nicely ordered across child-level classes one to three. These can therefore be labeled high, middle, and low. Children in class four show a somewhat mixed pattern: they perform better than the middle class on cards and figures, better than the low class on word list and matrices and worse than the low class on animals and occupations.

Figure 4.2b displays the estimated level-2 class membership probabilities for the level-3 classes. As can be seen, in family-level class three almost all children belong to the low child-level class. Children from families belonging to family-level class one are more likely to be in the high intelligence class and children from family-level class two are more often in the middle and mixed intelligence classes. These results show that there is a very strong family effect on the performance of children on these six intelligence subtests.

TABLE 4.3

Fit Measures for the Models Estimated with the Intelligence Data (First Analysis)

Model		LL	Npar	BIC(2)	BIC(3)
$C = 1$	$D = 1$	−4238	12	8543	8522
$C = 2$	$D = 1$	−4149	19	8404	8372
$C = 3$	$D = 1$	−4127	26	8400	8356
$C = 4$	$D = 1$	−4113	33	8411	8355
$C = 5$	$D = 1$	−4104	40	8432	8364
$C = 2$	$D = 2$	−4130	21	8378	8342
$C = 3$	$D = 2$	−4108	29	8379	8330
$C = 4$	$D = 2$	−4087	37	8381	8318
$C = 5$	$D = 2$	−4075	45	8402	8326
$C = 2$	$D = 3$	−4130	23	8388	8349
$C = 3$	$D = 3$	−4098	32	8374	8320
$C = 4$	$D = 3$	−4072	41	**8374**	**8304**
$C = 5$	$D = 3$	−4060	50	8400	8315
$C = 2$	$D = 4$	−4130	25	8399	8357
$C = 3$	$D = 4$	−4096	35	8388	8328
$C = 4$	$D = 4$	−4070	45	8391	8315
$C = 5$	$D = 4$	−4052	55	8412	8318
$C = 2$	$D = 5$	−4130	27	8410	8364
$C = 3$	$D = 5$	−4096	38	8405	8340
$C = 4$	$D = 5$	−4069	49	8412	8329
$C = 5$	$D = 5$	−4050	60	8436	8334

Note: Bold indicates lowest value.

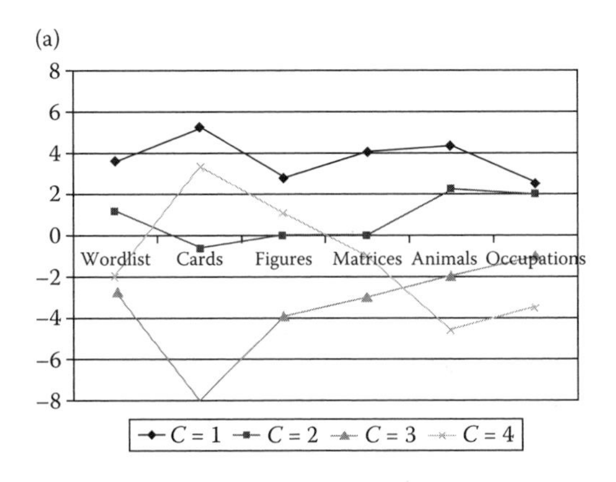

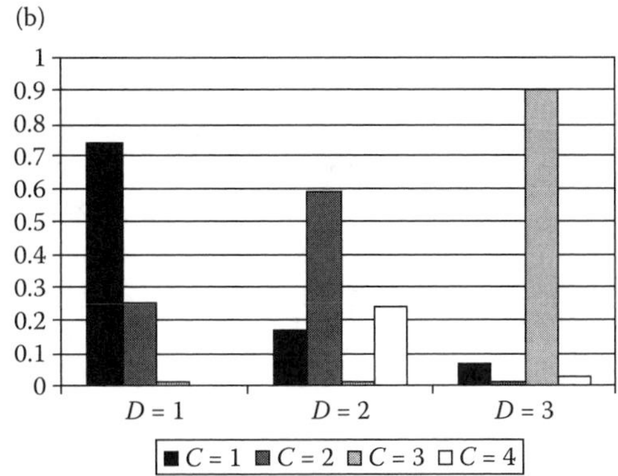

FIGURE 4.2
(a) Intercept differences between child-level classes obtained with intelligence data (first analysis). (b) Child-level class proportions for family-level classes obtained with intelligence data (first analysis).

4.5.2.2 Analysis 2: Direct Effect of Family Classes

In this second analysis, the six intelligence measures are used to simultaneously find child-level and family-level intelligence classes based on the children's responses. Table 4.4 provides the fit measures for the estimated multilevel LC models. As can be seen, the model with C=3 and D=3 performs best according to BIC(2), whereas the model with C=3 and D=4 performs best according to BIC(3). As the discrepancy is in the number of level-3 classes, we decided to keep the model selected by BIC(3) as the final model. Note that the BIC values of this model are lower than the ones found in the previous analysis, which indicates that the assumption we made earlier—that differences in responses across family classes are fully mediated by differences in child-level class membership—is not fully correct.

Figure 4.3a and b display the estimates for the $\lambda_{qc}^{(1,2)}$ and $\lambda_{qd}^{(1,3)}$ parameters from Equation 4.6, which can be used to label the level-2 and level-3 classes. The parameters for the level-2 classes show that class one scores higher on all measures than classes two and three. Class two scores higher than class three on the first four measures (with a large difference on cards), but lower than class three on animals and occupations. This pattern reveals that there is a kind of two-dimensional structure.

Although the fit measures indicated that there are significant differences between families, it is not easy to give a simple interpretation to the encountered differences between the level-3 classes. Contrary to the results by Hox, we do not find a one-dimensional structure, which would imply that classes should be (almost) ordered. Class one families score high on all measures, except for occupations on which they have a medium level score. Families belonging to class two score high on figures, medium on cards, and low on the remaining four items. Class three scores low on three items and medium on the remaining items. Class four families score extremely high on occupations, high on word list, figures and matrices, medium on cards, and somewhat low on animals.

4.5.2.3 Analysis 3: Multilevel Mixture Factor Analysis

In the third analysis, we used a factor analytic model at level 2. Similar to Hox's

TABLE 4.4

Fit Measures for the Models Estimated with the Intelligence Data (Second Analysis)

Model		LL	Npar	BIC(2)	BIC(3)
$C = 1$	$D = 1$	–4238	12	8543	8522
$C = 2$	$D = 1$	–4149	19	8404	8372
$C = 3$	$D = 1$	–4127	26	8400	8356
$C = 4$	$D = 1$	–4113	33	8411	8355
$C = 5$	$D = 1$	–4104	40	8432	8364
$C = 1$	$D = 2$	–4155	19	8417	8385
$C = 2$	$D = 2$	–4104	26	8354	8310
$C = 3$	$D = 2$	–4086	33	8356	8300
$C = 4$	$D = 2$	–4070	40	8364	8296
$C = 5$	$D = 2$	–4061	47	8385	8305
$C = 1$	$D = 3$	–4132	26	8410	8366
$C = 2$	$D = 3$	–4081	33	8346	8290
$C = 3$	$D = 3$	–4059	40	**8342**	8274
$C = 4$	$D = 3$	–4045	47	8354	8274
$C = 5$	$D = 3$	–4034	54	8370	8278
$C = 1$	$D = 4$	–4114	33	8413	8357
$C = 2$	$D = 4$	–4068	40	8361	8293
$C = 3$	$D = 4$	–4045	47	8352	**8272**
$C = 4$	$D = 4$	–4033	54	8368	8276
$C = 5$	$D = 4$	–4025	61	8391	8287
$C = 1$	$D = 5$	–4101	40	8425	8357
$C = 2$	$D = 5$	–4058	47	8379	8299
$C = 3$	$D = 5$	–4036	54	8374	8282
$C = 4$	$D = 5$	–4021	61	8384	8280
$C = 5$	$D = 5$	–4015	68	8410	8294

Note: Bold indicates lowest value.

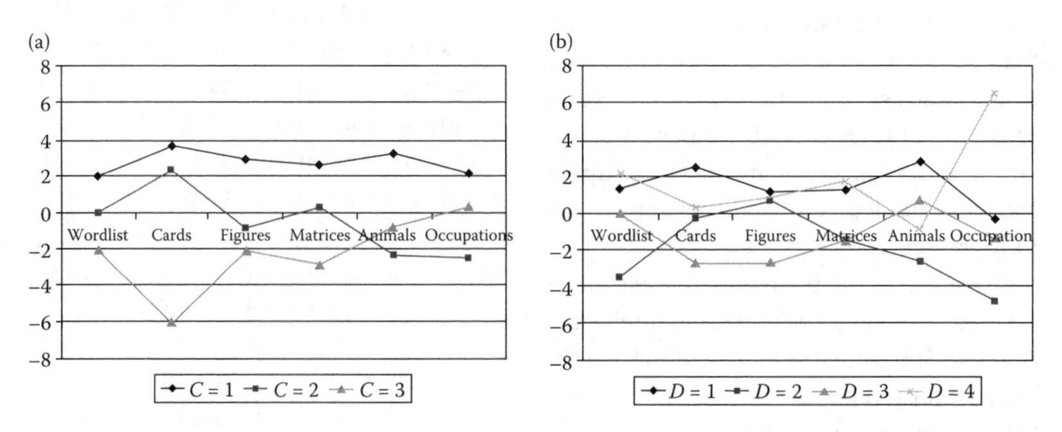

FIGURE 4.3
(a) Intercept differences between child-level classes obtained with intelligence data (second analysis). (b) Intercept differences between family-level classes obtained with intelligence data (second analysis).

analysis, it is a two factor model, but with the difference that "figure" loads on both factors. Seemingly, the factor structure changes somewhat when retaining cases with missing values in the analysis. Our baseline multilevel mixture FA model is a model in which factor (co)variances, means, and loadings, as well as item intercepts are allowed to differ across family classes. Measurement equivalence across families is achieved when factor loadings and item intercepts turn out to be the same across family clusters.

We fitted models with one to four family-level latent classes. Both BIC(2) and BIC(3) indicated that a model with three classes is the one that should be preferred. Restricting the factor (co)variances and loadings to be equal across classes did not deteriorate the fit of the model. However, assuming also that item intercepts are equal across classes yields a worse model fit. Actually, except for the two reference items (the two items for which intercepts were fixed to 0 to be able to identify the factor means), none of the item intercepts can be assumed to be equal across classes. This confirms the results we found in the models with a discrete level-2 latent variable; namely, it is not correct to assume that family effects on item responses can be assumed to be fully mediated by the child-level latent variable(s).

4.6 CONCLUSIONS

Whereas typical applications of LC analysis concern single-level multivariate response data sets, in this chapter, this author demonstrated how LC and mixture models can be used for analyzing univariate two-level data sets, univariate three-level data sets, and multivariate two-level data sets. Also discussed was how multilevel LC models fit into a more general latent variable modeling framework, which allows defining models with discrete and continuous latent variables at the multiple levels of a hierarchical structure.

The multilevel LC models were illustrated using two empirical examples. The first example showed how to use a multilevel mixture models for three-level regression analysis. Complex but interesting level-2 and level-3 variability in intercepts and slopes were detected using model specifications which are rather straightforward with the presented framework.

A second empirical data set was analyzed in three different ways. Which of the three is most appropriate depends on the exact aim of the research concerned. It is, of course, also possible that none of the three is appropriate and that another type of analysis should be used, for example, the multilevel FA used by Hox (2002). Our second analysis, as well as Hox's multilevel factor model are more suited for exploration, whereas our first and third analysis are more suited when the items can be assumed to be meaningful indicators for clustering or measuring one or more underlying factors at the lower level.

Other types of illustrations of multilevel mixture models than the ones presented here can be found in the literature. Vermunt (2003, 2005, 2007a, 2008a) gave examples of multilevel variants of standard LC models for categorical response variables. A type of model that was not illustrated with an example is the model containing continuous random effects at level 3 discussed in Section 4.3.2. Applications of this model are provided by Vermunt (2003, 2005). Other applications, which similarly to our third analysis of the intelligence data, use continuous variables at level 2 and discrete latent variables at level 3 are the multilevel mixture growth models proposed by Palardy and Vermunt (2010), the multilevel mixture IRT models used by Vermunt (2008b), and the multilevel mixture factor models by Varriale (2008).

Since multilevel mixture modeling is a rather new area of statistical methodology, it is not surprising that many issues have not yet been fully resolved. Future research should deal with issues such as sample size requirements, model selection strategies, model diagnostics, and effects of model misspecification. Possible extensions of the models presented in this chapter are multilevel LC models with ordinal latent variables and multilevel variants of latent Markov models.

REFERENCES

Aitkin, M. (1999). A general maximum likelihood analysis of variance components in generalized linear models. *Biometrics, 55*, 218–234.

Allua, S. S. (2007). *Evaluation of single- and multilevel factor mixture model estimation.* PhD thesis, The University of Texas at Austin.

Asparouhov, T., & Muthén, B. (2008). Multilevel mixture models. In G. R. Hancock & K. M. Samuelsen (Eds.), *Advances in latent variable mixture models* (pp. 27–51). Charlotte, NC: Information Age Publishing, Inc.

Bijmolt, T. H., Paas, L. J., & Vermunt, J. K. (2004). Country and consumer segmentation: Multilevel latent class analysis of financial product ownership. *International Journal of Research in Marketing, 21*, 323–340.

Bouwmeester, S., Vermunt, J. K., & Sijtsma, K. (2007). Development and individual differences in transitive reasoning: A fuzzy trace theory approach. *Developmental Review, 27*, 41–74.

De Boeck, P., & Wilson, M. (2004). *Explanatory item response models: A generalized linear and nonlinear approach.* New York, NY: Springer.

Fox, J. P., & Glas, C. A. W. (2001). Bayesian estimation of a multilevel IRT model using Gibbs sampling. *Psychometrika, 66*, 269–286.

Goldstein, H. (2003). *Multilevel statistical models* (3rd ed.). London, UK: Arnold.

Goldstein, H., & Browne, W. (2002). Multilevel factor analysis modelling using Markov chain Monte Carlo estimation. In G. A. Marcoulides & I. Moustaki (Eds.), *Latent variable and latent structure models* (pp. 225–243). Mahwah, NJ: Lawrence Erlbaum.

Goodman, L. A. (1974). Exploratory latent structure analysis using both identifiable and unidentifiable models. *Biometrika, 61*, 215–231.

Grilli, L., & Rampichini, C. (2007). Multilevel factor models for ordinal variables. *Structural Equation Modeling, 14*, 1–25.

Hagenaars, J. A. (1993). *Loglinear models with latent variables.* Newbury Park, CA: Sage.

Hedeker, D. (2003). A mixed-effects multinomial logistic regression model. *Statistics in Medicine, 22*, 1433–1446.

Heinen, T. (1996). *Latent class and discrete latent trait models: Similarities and differences.* Thousand Oaks, CA: Sage.

Hox, J. (2002). *Multilevel analysis: Techniques and applications.* Mahwah, NJ: Lawrence Erlbaum.

Kalmijn, M., & Vermunt, J. K. (2007). Homogeneity of social networks by age and marital status: A multilevel analysis of ego-centered networks. *Social Networks, 29*, 25–43.

Lazarsfeld, P. F., & Henry, N. W. (1968). *Latent structure analysis.* Boston, MA: Houghton Mill.

Lukociene, O., & Vermunt, J. K. (2010). Determining the number of components in mixture models for hierarchical data. In A. Fink, B. Lausen, W. Seidel, & A. Ultsch (Eds.), *Advances in data analysis, data handling and business intelligence* (pp. 241-250). Berlin-Heidelberg: Springer.

Magidson, J., & Vermunt, J. K. (2001). Latent class factor and cluster models, bi-plots and related graphical displays. *Sociological Methodology, 31*, 223–264.

McGrath, K., & Waterton, J. (1986). *British social attitudes, 1983–1986 panel survey.* London. UK: Social and Community Planning Research, Technical Report.

McLachlan, G. J., & Peel, D. (2000). *Finite mixture models.* New York, NY: John Wiley & Sons.

Muthén, B. O. (1994). Multilevel covariance structure analysis. *Sociological Methods & Research, 22*, 376–398.

Muthén, B. (2004). Latent variable analysis. Growth mixture modeling and related techniques for longitudinal data. In D. Kaplan (Ed.), *The Sage handbook of quantitative methodology for the social sciences* (pp. 345–368). Thousand Oaks, CA: Sage.

Muthén, B., & Muthén, L. (2008). *Mplus: User's manual*. Los Angeles, CA: Muthén & Muthén.

Nagin, D. S. (1999). Analyzing developmental trajectories: A semiparametric group-based approach. *Psychological Methods, 4*, 139–157.

Palardy, G., & Vermunt, J. K. (2010). Multilevel growth mixture models for classifying group-level observations. *Journal of Educational and Behavioral Statistics*, in press.

Rabe-Hesketh, S., Skrondal, A., & Pickles, A. (2004). Generalized multilevel structural equation modelling. *Psychometrika, 69*, 183–206.

Skrondal, A., & Rabe-Hesketh, S. (2004). *Generalized latent variable modeling: Multilevel, longitudinal and structural equation models*. London, UK: Chapman & Hall/CRC.

Van de Pol, F., & Langeheine, R. (1990). Mixed Markov latent class models. *Sociological Methodology, 20*, 213–247.

Van Peet, A. A. J. (1992). *De potentieeltheorie van intelligentie* (The potentiality theory of intelligence). PhD thesis, University of Amsterdam.

Varriale, R. (2008). *Multilevel mixture models for the analysis of the university effectiveness*. PhD thesis, University of Florence.

Vermunt, J. K. (1997). *Log-linear models for event histories*. Thousand Oaks, CA: Sage.

Vermunt, J. K. (2001). The use restricted latent class models for defining and testing nonparametric and parametric IRT models. *Applied Psychological Measurement, 25*, 283–294.

Vermunt, J. K. (2003). Multilevel latent class models. *Sociological Methodology, 33*, 213–239.

Vermunt, J. K. (2004). An EM algorithm for the estimation of parametric and nonparametric hierarchical nonlinear models. *Statistica Neerlandica, 58*, 220–233.

Vermunt J. K. (2005). Mixed-effects logistic regression models for indirectly observed outcome variables. *Multivariate Behavioral Research, 40*, 281–301.

Vermunt, J. K. (2007a). A hierarchical mixture model for clustering three-way data sets. *Computational Statistics and Data Analysis, 51*, 5368–5376.

Vermunt, J. K. (2007b). Growth models for categorical response variables: Standard, latent-class, and hybrid approaches. In K. van Montfort, H. Oud, & A. Satorra (Eds.), *Longitudinal models in the behavioral and related sciences* (pp. 139–158). Mahwah, NJ: Lawrence Erlbaum.

Vermunt, J. K. (2008a). Latent class and finite mixture models for multilevel data sets. *Statistical Methods in Medical Research, 17*, 33–51.

Vermunt, J. K. (2008b). Multilevel latent variable modeling: An application in educational testing. *Austrian Journal of Statistics, 37*, 285–299.

Vermunt, J. K., & Magidson, J. (2002). Latent class cluster analysis. In J. A. Hagenaars & A. McCutcheon (Eds.), *Applied latent class analysis* (pp. 89–106). Cambridge, UK: Cambridge University Press.

Vermunt J. K., & Magidson J. (2005). *Latent GOLD 4.0 user's guide*. Belmont, MA: Statistical Innovations Inc.

Vermunt, J. K., & Magidson, J. (2008). *LG-syntax user's guide: Manual for Latent GOLD 4.5 syntax module*. Belmont, MA: Statistical Innovations Inc.

Vermunt, J. K., Tran, B., & Magidson, J. (2008). Latent class models in longitudinal research. In S. Menard (Ed.), *Handbook of longitudinal research: Design, measurement, and analysis* (pp. 373–385). Burlington, MA: Elsevier.

Vermunt, J. K., & Van Dijk, L. (2001). A nonparametric random-coefficients approach: The latent class regression model. *Multilevel Modelling Newsletter, 13*, 6–13.

Wedel, M., & DeSarbo, W. (1994). A review of recent developments in latent class regression models. In R. P. Bagozzi (Ed.), *Advanced methods of marketing research* (pp. 352–388). Cambridge, MA: Blackwell Publishers.

Yung, Y. F. (1997). Finite mixtures in confirmatory factor-analysis models. *Psychometrika, 62*, 297–330.

APPENDIX

The models in Table 4.2 can be estimated with either the Latent GOLD Regression or Syntax module. To illustrate the Latent GOLD Syntax language, these are possible "variables" and "equations" sections of a setup for the baseline models appearing in Table 4.2:

```
variables
   groupid District_ID;
   caseid Case_ID;
   dependent Response binomial
   exposure=7;
   independent Year nominal,
   Religion nominal;
   latent U3 group nominal 2,
   U2 nominal 4;
equations
   Response <- 1 + Year +
   Religion + U2 + U2 Year
              + U3 + U3
              Religion;
   U2 <- 1;
   U3 <- 1;
```

The "variables" section defines the level-3 (group) and level-2 (case) identifiers, the dependent variable, the predictors (independent), and the latent variables in the model. Note that both latent variables are nominal, with 2 and 4 categories, respectively. Moreover, for U3, the keyword "group" indicates that it is a level-3 latent variable. The "equations" section contains, in fact, the regression Equations 4.6 through 4.8, where "1" defines an intercept term and where dummies/effects are automatically set up for nominal variables.

The two restricted models appearing in Table 4.2 can be obtained by eliminating either "+ U2 Year" or "+ U3 Religion" from the model for the response variable. The more extended models are obtained by adding "+ U3 Year" to the model for the response variable, "+ U3" to the model for the level-2 classes, and "+ U2 U3" to the model for the response variable.

The setup used for the second example differs from the one above in that the model is defined as a two-level model for multivariate responses. In other words, we have a model for six dependent variables rather than for one, but no case identifier needs to be specified since we have only one record per case. Another difference is that "equations" need to be specified for the residual variances of the response variables, because these are assumed to be normally distributed. The setup used for the models in Table 4.3 is:

```
variables
  groupid family;
  dependent wordlist
  continuous, cards
  continuous, figures
     continuous, matrices
     continuous, animals
     continuous,
     occupations continuous;
  latent U3 group nominal 4,
  U2 nominal 3;
equations
  cards <- 1 + U2;
  figures <- 1 + U2;
  matrices <- 1 + U2;
  animals <- 1 + U2;
  occupations <- 1 + U2;
  wordlist;
  cards;
  figures;
  matrices;
  animals;
  occupations;
  U2 <- 1 + U3;
  U3 <- 1;
```

A more compact specification of the equations using variable lists is

```
equations
  cards - occupations <- 1 +
  U2;
```

```
cards - occupations;
U2 <- 1 + U3;
U3 <- 1;
```

For the second analysis (models in Table 4.4), we remove "+ U3" from model for "U2" and include it in the equations for the dependent variables. The multilevel mixture factor model can be defined as follows:

```
variables
   groupid ...
   dependent ...
   latent U3 nominal group 3,
   F1 continuous, F2
   continuous;
equations
   wordlist <- (1) F1;
   cards <- 1 | U3 + F1 | U3;
   figures <- 1 | U3 + F1 | U3
   + F2 | U3;
   matrices <- 1 | U3 + F1 |
   U3;
   animals <- (1) F2;
   occupations <- 1 | U3 + F2
   | U3;
   wordlist - occupations;
   F1 <- 1 | U3;
   F2 <- 1 | U3;
   F1 | U3;
   F2 | U3;
   F1 <-> F2 | U3;
   U3 <- 1;
```

This setup illustrates various additional syntax options: two continuous latent variables are defined in "`latent`," "equations" are included for the factor means and (co) variances, the statement "`| U3`" is used to indicate that a parameter varies across level-3 clusters, and "`(1)`" is used to fix the parameter concerned to 1.

Section III

Multilevel Models for Longitudinal Data

5

Panel Modeling: Random Coefficients and Covariance Structures

Joop J. Hox
Department of Methodology and Statistics,
Utrecht University, Utrecht, The Netherlands

5.1 MULTILEVEL MODELS FOR CHANGE OVER TIME

Multilevel or mixed models are becoming standard modeling tools for longitudinal or repeated measures data. Compared to the classic MANOVA approach, they have several advantages. Firstly, they deal efficiently with panel dropout; because there is no assumption that each subject must be measured on the same number of occasions, subjects with incomplete data are simply retained in the data set. The assumption is that incomplete data are missing at random (Little, 1995), which is weaker than the assumption of missing completely at random, which is made by applying listwise deletion in MANOVA. Secondly, it is possible to include time-varying covariates in the model. Thirdly, using polynomial functions or piecewise regression the change over time can be modeled very flexibly. Finally, by allowing regression coefficients for the change model to vary across subjects, different subjects can have their own trajectory of change, which can in turn be modeled by time invariant subject characteristics.

It is useful to distinguish between repeated measures that are collected at fixed or at varying occasions. If the measurements are taken at fixed occasions, all individuals provide measurements for the same set of occasions, usually regularly spaced, such as once every year. When occasions are varying, a different number of measures is collected at different points in time for different individuals. Such data occur, for instance, in growth studies, where physical or psychological characteristics are studied for a set of individuals at different moments in their development. The data collection could be at fixed moments in the year, but the individuals would have different ages at that moment. For a multilevel analysis of the resulting data, the difference between fixed and varying occasions is not very important. For fixed occasion designs, especially when the occasions are regularly spaced and there are no missing data, repeated measures MANOVA is a viable alternative for multilevel analysis. Another possibility in such designs is latent curve analysis, also known as latent growth curve analysis. This is a structural equation (cf. Willett & Sayer, 1994) that models a repeated measures polynomial analysis of variance. This chapter focuses on fixed occasion data. In addition to the familiar multilevel model equations, it uses path diagrams to clarify the models, but the analysis concentrates on the multilevel regression approach. In fact, multilevel models and structural equation models for change over time are just different representations of the same underlying model. Bollen and Curran (2006) provide a thorough discussion of longitudinal models from a structural equation perspective, and Hedeker and Gibbons (2006) provide a comparable discussion from the multilevel regression perspective. Duncan, Duncan, and Strycker (2006) provide an introduction to longitudinal modeling from both perspectives.

The multilevel regression model for longitudinal data is a straightforward application of the standard multilevel regression mode, with measurement occasions within subjects replacing the subjects within groups structure. At the lowest, the repeated measures level, we have:

$$Y_{ti} = \pi_{0i} + \pi_{1i}T_{ti} + \pi_{2i}X_{ti} + e_{ti}. \tag{5.1}$$

In repeated measures applications, the coefficients at the lowest level are often indicated by the Greek letter π. This has the advantage that the subject level coefficients, which are in repeated measures modeling at the second level, can be represented by the usual Greek letter β, and so on. In Equation 5.1, Y_{ti} is the response variable of individual i measured at time point t, T is the time variable that indicates the time point, and X_{ti} is a time-varying covariate. Subject characteristics, such as gender, are time invariant covariates, which enter the equation at the second level:

$$\pi_{0i} = \beta_{00} + \beta_{01}Z_i + u_{0i}, \tag{5.2}$$

$$\pi_{1i} = \beta_{10} + \beta_{11}Z_i + u_{1i}, \tag{5.3}$$

$$\pi_{2i} = \beta_{20} + \beta_{21}Z_i + u_{2i}. \tag{5.4}$$

By substitution, we get the single equation model:

$$\begin{aligned} Y_{ti} = {} & \beta_{00} + \beta_{10}T_{ti} + \beta_{20}X_{ti} + \beta_{01}Z_i \\ & + \beta_{11}Z_iT_{ti} + \beta_{21}Z_iX_{ti} + u_{1i}T_{ti} \\ & + u_{2i}X_{ti} + u_{0i} + e_{ti} \end{aligned} \tag{5.5}$$

In multilevel models for subjects within groups, there is an assumed dependency between the subjects who are in the same group. Most often, there is no need to assume a specific structure for this dependency. Subjects within the same group are assumed exchangeable, and the intraclass correlation refers to the average correlation between two randomly chosen subjects from the same group. In multilevel models for occasions within subjects, measurement occasions are not freely exchangeable, because they are ordered in time. In such models, it often does make sense to assume a structure for the relationships between measurements across time. For example, an intuitively attractive assumption is that correlations between measures taken at different measurement occasions are higher when these occasions are close to each other in time.

Longitudinal designs often concern the analysis of structured change, such as growth or decline over time. The appropriate model for such research problems is a latent curve model, where change in the outcome variable is modeled as a function of time. As outlined above, the use of polynomials and varying regression coefficients makes this a very flexible analysis tool. As will be explained in more detail below, allowing coefficients for time to vary across subjects implies specific dependency structures. Since the focus is on modeling individual trajectories, the possibility of specifying specific structures across time is usually not explored.

Panel designs are longitudinal designs where the emphasis is on changes that do not follow a pattern of growth or decline across time. An example is using a panel to monitor satisfaction with the government. Satisfaction with the government is not expected to increase or decrease continuously. However, it is expected to fluctuate, and we may be able to predict these fluctuations with time-varying covariates that capture events that occur at different occasions. Here, time is not a relevant predictor variable. Still, it is logical to assume that there is a dependency structure over time, which cannot be ignored. In this case, exploring specific structures across time is very important.

Although modeling the dependencies over time can be done implicitly, by allowing random coefficients for the time variable, or explicitly, by specifying a specific structure, these approaches can also be combined. The remainder of this chapter discusses latent curve modeling, explicit modeling of dependency over time, and combining these approaches. The discussion takes up the issue when specific approaches are useful.

5.2 MULTILEVEL MODELS FOR STRUCTURED CHANGE OVER TIME

For structured change over time, we will use an example data set constructed by Patrick Curran. This data set, hereafter called the *Curran data*, was compiled from a large longitudinal data set. Supporting documentation and the original data files are available on the Internet (Curran, 1997); the following description is summarized from Curran (1997).

The Curran data are a sample of 405 children who were within the first 2 years of entry to elementary school. The data consist of four repeated measures of both the child's antisocial behavior and the child's reading recognition skills. In addition, at the first measurement occasion, measures were collected of emotional support and cognitive stimulation provided by the mother. These data are a subsample from the National Longitudinal Survey of Youth (NLSY), based on three key criteria. First, children must have been between the ages of 6 and 8 years at the first wave of measurement. Second, children must have complete data on all measures used at the first measurement occasion. Third, only one child was considered from each mother. All $N = 405$ children and mothers were interviewed at measurement occasion one; on the three following occasions the sample sizes were 374, 297, and 294. Only 221 cases were interviewed at all four occasions.

The time-varying variables are Antisocial Behavior (anti1–4) and Reading Recognition (read1–4). The time invariant variables are Emotional Support to the child (homeemo) and Cognitive Stimulation (homecog), Mother's Age (momage) and Child Age (kidage) in years at Time 1, and the child's gender (kidgen).

In this example, reading recognition is the outcome variable. A simple start model for the effect over time is to include measurement occasion as a predictor variable (coded 0,1,2,3), and allow only the intercept to vary across subjects. It turns out that the relationship between occasion and reading is nonlinear, and a quadratic term is added to the model. The results are given in the first column of Table 5.1 (REML estimates).

Table 5.1 shows in the fixed part the regression coefficients, and in the random part the residual variance at the lowest level and the (co)variances at the second level, with standard errors in parentheses. All parameter estimates are significant by the Wald test, the variances are also significant using the more accurate likelihood ratio test.

TABLE 5.1

Consecutive Models for Reading Recognition

Fixed Part	**Model 1**	**Model 2**
Predictor		
Intercept	2.54 (.05)	2.54 (.05)
Occasion	1.67 (.05)	1.67 (.05)
Occasion squared	–0.19 (.02)	–0.19 (.01)
Random Part		
Residual	0.41 (.02)	0.27 (.05)
Intercept	0.79 (.07)	0.63 (.06)
Occasion		0.09 (.01)
Intc-Occ		0.03 (.02)

The regression coefficients in the first column of Table 5.1 indicate a mean reading recognition score of 2.54 at the first measurement occasion. The linear effect indicates that the reading score goes up at each occasion, and the negative quadratic effect indicates that this effect levels off at later occasions.

The second column in Table 5.1 presents the results from the model where the regression coefficient for occasion varies across subjects. The variance of this coefficient is clearly significant. A third model with a varying coefficient for the quadratic component indicated no variance for that coefficient. To model the variance of the occasion coefficient interactions of other predictor variables with occasion must be added to the model. Cognitive stimulation and the child's age predict reading, but there are no significant interactions. To simplify the exposition, the other variables are not included in the model here. The model with varying coefficients for the linear effect of occasion is presented as a SEM diagram in Figure 5.1. Note that for complete correspondence with the multilevel regression approach, the variances of the four errors must be constrained to be the same.

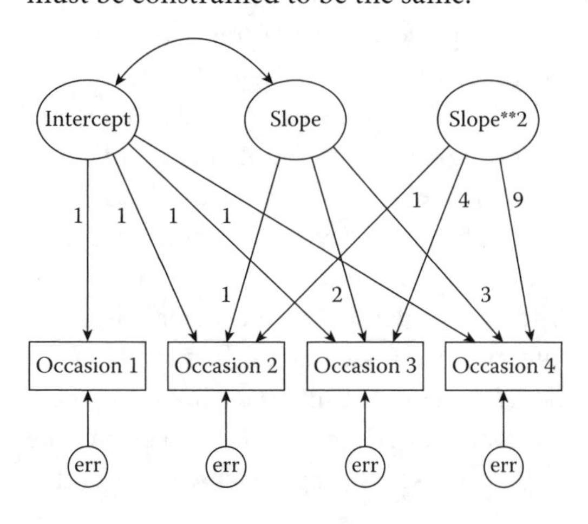

FIGURE 5.1
Path diagram corresponding to growth model with four occasions (**2 denote slope-squared).

TABLE 5.2

Observed and Modeled Means and Variances Reading

Occ.	*N*	Mean	Var. Observed	Mean Predicted	Var. *T* Fixed	Var. *T* Random
0	405	2.52	0.86	2.54	1.20	0.90
1	375	4.08	1.17	4.02	1.20	1.05
2	275	5.01	1.35	5.12	1.20	1.38
3	270	5.77	1.56	5.84	1.20	1.89

The two models that underlie Table 5.1 have different consequences for the pattern of covariances between reading measures over time. The combined model for the varying coefficient for occasion is:

$$Y_{ti} = \beta_{00} + \beta_{10}T_{ti} + u_{1i}T_{ti} + u_{0i} + e_{ti}. \quad (5.6)$$

In this model, the variance at a specific measurement occasion is given by (Goldstein, 2003; Raudenbush, 2002):

$$Var(Y_{ti}) = \sigma^2_{u0} + 2T_{ti}\sigma_{u0u1} + T^2_{ti}\sigma^2_{u1} + \sigma^2_e. \quad (5.7)$$

The covariance between two measurement occasions is given by (Goldstein, 2003; Raudenbush, 2002):

$$Cov(T_{ti,t'i}) = \sigma^2_{u0} + (T_{ti} + T_{t'i})\sigma_{u0u1} + T_{ti}T_{t'i}\sigma^2_{u1}. \quad (5.8)$$

Together, Equations 5.7 and 5.8 specify a very restricted pattern for the variances and covariances across time. The pattern for the fixed occasion model is even more specific. By removing terms that refer to T_{ti} we obtain:

$$Var(Y_{ti}) = \sigma^2_{u0} + \sigma^2_e \quad (5.9)$$

and

$$Cov(T_{ti,t'i}) = \sigma^2_{u0}. \quad (5.10)$$

The model with only a random intercept assumes that all variances are the same, and all covariances are the same. In the MANOVA context, this assumption is known as compound symmetry, and considered highly restrictive.

Table 5.2 presents the observed means and variances and the means and variances implied by the Occasion Fixed and Occasion Random model. It is clear that the random coefficient model is predicting the observed variances fairly well. It should be noted that some discrepancy is to be expected, because the observed means and variances are based on the nonmissing cases at each measurement occasion, and the model predictions are predictions for the entire sample, assuming missing at random for the missing data.

5.3 MULTILEVEL MODELS FOR UNSTRUCTURED CHANGE OVER TIME

As noted in the introduction, there are situations where it makes no sense to assume perpetual growth or decline, while it is still interesting to model change and predictors of change. The term *unstructured change* is used to indicate that there is no long-term trend to model.

The example data are simulated to reflect a diary study, in which changes are expected but no overall trend. In this hypothetical study, a sample of 60 workers who work in a stressful work environment are asked to fill in a diary for 2 weeks (only working days). The study uses a State-Trait Anxiety Inventory. Trait anxiety (TraitAnx), which is assumed to be a relatively stable individual characteristic, is measured only on the first day. State anxiety (StateAnx), which is assumed to be a transitory mood state, is measured each day. Both scales are commonly normed to T-scores that have a mean of 50 and a standard deviation of 10 in the test's norm group. In addition, the study collects daily data on perceived job demands (7-point scale) and perceived social support (7-point scale).

For such data, no large differences are expected for the average anxiety on different days. This is borne out by a repeated measures MANOVA that finds no differences across the 10 days. The linear trend over time tested by MANOVA is also not significant. Figure 5.2 presents a SEM diagram for repeated measures MANOVA. To test equality of means in a SEM context, the model in Figure 5.2 is compared to a model where the means are constrained to be equal.

Note that the path diagram explicitly shows that MANOVA estimates the variances and covariances for all measures over time. MANOVA is an unstructured model, there is no specific structure assumed for this covariance matrix. To model correlated errors in multilevel regression, we use a multivariate response model with a lowest level for the repeated measures, and a full set of dummy variables indicating the different occasions. Thus, we have 10 dummy variables, one for each day. The intercept term is removed from the model, and the variance of the lowest level residuals is constrained to zero. The dummy variables are all allowed to have random slopes at the second level. The equation for a model without further explanatory variables becomes:

$$Y_{ti} = \beta_1 D_1 \ldots \beta_{10} D_{10} + u_1 D_1 \ldots u_1 D_1. \quad (5.11)$$

Having 10 random slopes at level 2 provides us with a 10×10 covariance matrix for the 10 consecutive days. The regression slopes β_1–β_{10} are simply the estimated means at the 10 occasions. Equation 5.11 defines the multilevel model that is equivalent to the MANOVA approach. Maas and Snijders (2003) discuss this model at length, and show how the familiar *F*-ratio's can be calculated from the multilevel software output.

The model in Equation 5.11 is fully saturated; it estimates all means and all (co)variances. Both the fixed part and the random

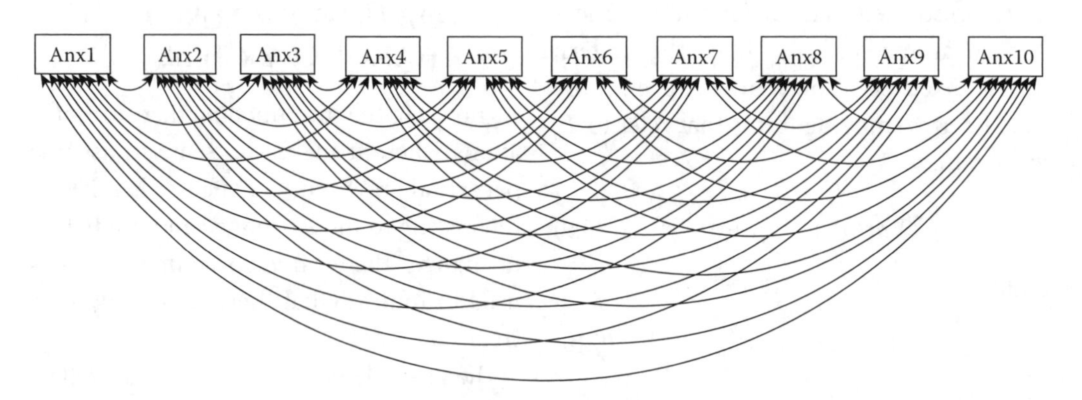

FIGURE 5.2
SEM diagram corresponding to MANOVA on 10 consecutive anxiety measures.

part can be simplified. We can replace the fixed part by a regression equation that includes predictors such as the state anxiety and the time-varying predictors job demands and social support. This models the state anxiety in a more interesting way, as the result of the combination of trait anxiety and different pressures at work. In addition we have a more parsimonious model, since we replace 10 estimated means with four estimates for the intercept and the three regression coefficients.

The covariance matrix for the 10 occasions has no restrictions. If we impose the restriction that all variances are equal, and that all covariances are equal, we have again the compound symmetry model. This shows that the model with occasion as fixed is one way to impose the compound symmetry structure on the random part of the model. Consequently, these models are nested, and we can use the overall chi-square test based on the deviance of the two models to test if the assumption of compound symmetry is tenable.

Models that assume a saturated model for the error structure are very complex. If there are k time points, the number of elements in the covariance matrix for the occasions is $k(k + 1)/2$. So, with 10 occasions, we have 55 (co)variance parameters to be estimated. If the assumption of compound symmetry is tenable, this model is preferable, because the random part contains only two parameters to be estimated. However, the compound symmetry model is very restrictive, because it assumes that there is one single value for all correlations between time points. This assumption is not very realistic, because the error term contains all omitted sources of variation, which may be correlated over time. Different assumptions about the autocorrelation over time lead to different structures of the covariance matrix across the occasions. For instance, it is reasonable to assume that occasions that are close together in time have a higher correlation than occasions that are far apart. Accordingly, the elements in the covariance matrix Σ should become smaller, the further away they are from the diagonal. Such a correlation structure is called a simplex. A more restricted version of the simplex is to assume that the autocorrelation between the occasions follow the first-order autoregressive model

$$e_t = \rho\, e_{t-1} + \varepsilon \tag{5.12}$$

where e_t is the error term at occasion t, ρ is the autocorrelation, and ε is a residual error with variance σ_ε^2. The error structure in Equation 5.15 is a first-order autoregressive process. This leads to a covariance matrix of the form:

$$\Sigma(Y) = \frac{\sigma_\varepsilon^2}{(1-\rho^2)} \begin{pmatrix} 1 & \rho & \rho^2 & \cdots & \rho^{k-1} \\ \rho & 1 & \rho & \cdots & \rho^{k-2} \\ \rho^2 & \rho & 1 & \cdots & \rho^{k-3} \\ \vdots & \vdots & \vdots & \ddots & \vdots \\ \rho^{k-1} & \rho^{k-2} & \rho^{k-3} & \cdots & 1 \end{pmatrix} \tag{5.13}$$

The first term $\sigma_\varepsilon^2/(1 - \rho^2)$ is a constant, and the autocorrelation coefficient ρ is between −1 and +1, but typically positive. It is possible to have second-order

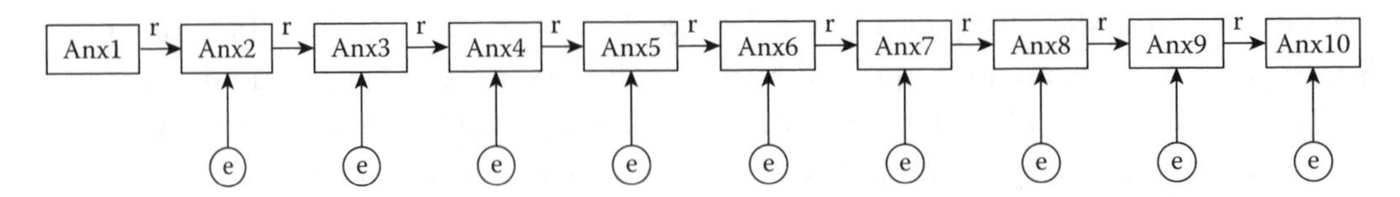

FIGURE 5.3
Path diagram for autoregressive model.

autoregressive processes, and other models for the error structure over time. The autoregressive model that produces the simplex in Equation 5.13 estimates one variance plus an autocorrelation. It is just as parsimonious as the compound symmetry model, but does not assume constant variances and covariances. The path diagram in Figure 5.3 illustrates this model.

Presenting the model as a SEM path diagram immediately suggests other structures for the dependency over time. Some often used structures are discussed by Hox (2002) and in more detail by Hedeker and Gibbons (2006). When multilevel regression is used, some structures, such as compound symmetry or the saturated model are easy to specify. Other structures are more difficult or impossible, unless the software producers have built in an option for specific structures. Many programs have these, for instance, both specialized programs like HLM and SuperMix and general packages like SPSS and SAS have a number of structures for dependency across time built in.

5.4 CHOOSING BETWEEN STRUCTURES AND COMBINED MODELS

5.4.1 Choosing Between Structures

As explained in Section 5.2, allowing time varying covariates, including indicators for the measurement occasions, to vary across subjects, implies certain covariance structures over time. In addition, some software allows direct specification of specific covariance structures, for example, an autoregressive model. As a consequence, an observed set of relationships over time can often be modeled about equally well by two different approaches. Any of such models is nested within the fully saturated model, which means that a likelihood ratio test or the equivalent chi-square deviance difference test can be used to assess their fit. However, a model allowing random slopes and a model directly specifying a covariance structure are not nested, and can only be compared using absolute fit indices such as Akaike's AIC or Schwarz's BIC. The AIC can be calculated from the deviance d and the number of estimated parameters q:

$$AIC = d + 2q, \tag{5.14}$$

and the BIC can be calculated as:

$$BIC = d + q\text{Ln}(N). \tag{5.15}$$

When the deviance goes down, indicating a better fit, both the AIC and the BIC also tend to go down. However, the AIC and the BIC include a penalty function based on the number of estimated parameters q. When the number of estimated parameters goes up, the AIC and BIC tend to go up too. For most sample sizes, the BIC places a larger penalty on complex models, which leads to a preference for smaller models compared to AIC. A problem with BIC in multilevel analysis is what the relevant sample size is: the number of groups or the total number of individuals? Most software that reports BIC uses the latter. However, in the context of longitudinal data, the number of subjects (i.e., the number of second level units) appears also reasonable. When a SEM package is used to specify the model, N is always the number of subjects. When a multilevel regression package is used, either level can supply the N, and the manual should be consulted to find out what the model does, or BIC must be calculated manually. Hedeker and Gibbons (2006), referring to Raftery (1995), advise to use the number of subjects, a practice we will follow here.

As an example, we model the state-anxiety data in several ways. The fixed part has three predictors: Trait Anxiety at the subject level and the two time-varying predictors Job Demands and Social Support. The first model in Table 5.3 models the random part using a saturated model. Models 2 and 3 use the compound symmetry and the lag-1 autocorrelation model for the covariance structure. Model 4 derives the structure for the covariances from a random intercept plus a random slope for Job Demands. Model 5 will be described in Section 5.4.2. The table reports for each model the number of parameters estimated, and the overall fit statistics deviance, AIC, BIC based on the number of subjects, and BIC based on the total number of measurements (subjects × occasions). Full Maximum Likelihood estimation is used, so both regression coefficients and (co)variances enter the likelihood function.

All models are nested within the saturated model, so they can be tested against that model using the deviance difference test. The difference between the deviances of the models is a chi-square variate with degrees of freedom equal to the difference in number of estimated parameters. The column p in Table 5.3 presents the p-value from the test of the model against the saturated model. Models 2 and 3 differ significantly from the saturated model, which means that they do not replicate the covariances well. Model 4 is significantly different from the saturated model at the 5% alpha level, but not at the 1% level. It does a better job at replicating the covariances than models 2 and 3.

It is clear that if one explores these data from a covariance structure perspective, the likely choice is for a model with all predictors fixed and a lag-1 autocorrelation. From a random slopes perspective, the likely choice is for a model with a random intercept plus a random slope for the variable Job Demands. The fit indices point toward the random slope model.

TABLE 5.3

Comparing Different Models for the State-Anxiety Data

Model	# Parameters	Deviance	p	AIC	BIC 2nd lev. N	BIC 1st lev. N
1 Saturated	59	3650.67	—	3760.67	3892.23	4002.13
2 Autocorr (1)	6	3755.88	<0.01	3767.88	3780.45	3794.26
3 Comp. Sym.	6	3777.33	<0.01	3789.33	3801.89	3815.71
4 Intercept + JobDem	7	3727.25	0.02	3741.44	3755.91	3772.02
5 Int. + JobDem + Autocorr (1)	9	3667.67	1.00	3685.67	3704.51	3725.24

5.4.2 Combined Models

The choice for a particular approach is not an either/or choice, the two approaches can be combined in a single model. The last row in Table 5.3 presents the results for a model where the intercept and the slope for Job Demands have variation, and the remaining structure over time is modeled using a lag-1 autocorrelation. All fit indices prefer this model to the simpler Intercept + Slope model. Not all combinations of regression model and covariance structure are possible. For instance, a saturated model for the covariances over time leaves no place for intercept or slope variance. Highly restricted models such as compound symmetry or lag-1 autocorrelation, which both estimate only two parameters for the (co)variances, leave much room for varying slopes.

The last row in Table 5.3 presents the fit information of a model that combines the random intercept plus slope model with a lag-1 autocorrelation structure. It fits very well, the *p*-value is 1.00, meaning that the model does not significantly differ from the saturated model. That is impressive, especially since the saturated model contains 59 parameters, and the combined model only nine. Since models 4 and 5 are also nested (model 4 is models 5 minus the autocorrelation part) they can also be compared using the formal test. The chi-square is 59.6, with two degrees of freedom, and the difference between the two models is clearly significant. Thus, model 5 is a significant improvement on model 4.

To assess the impact of various choices for the covariance structure on the fixed estimates, Table 5.4 presents the parameter estimates for the fixed part, and for the most important parameters in the random part. The values within brackets are the standard errors. These are given for the saturated model, the autocorrelation model, the random intercept plus slope model, and the combined random intercept and slope plus autocorrelation model.

Although all corresponding estimates are similar, they are clearly not identical. The combined model could be improved by removing the evidently nonsignificant

TABLE 5.4

Parameter Estimates for Four Selected Models

Model	Saturated	Diagonal + Autocorr.	Int. + Slope	Combined
Fixed Part				
Intercept	31.95 (4.77)	24.94 (4.43)	28.12 (4.17)	28.62 (4.23)
Job demands	1.46 (0.19)	1.68 (0.22)	1.61 (0.32)	1.63 (0.30)
Social support	−1.85 (0.19)	−1.66 (0.21)	−1.76 (0.22)	−1.69 (0.21)
Trait anxiety	0.41 (0.10)	0.52 (0.09)	0.46 (0.09)	0.44 (0.09)
Random Part				
Residual	n/a		18.84 (1.19)	
Intercept	n/a		12.62 (9.49)	13.53 (9.63)
Job dem slope	(all		3.12 (1.06)	2.61 (0.93)
Cov (Int-slope)	covariances)		1.61 (2.51)	2.30 (2.30)
Diagonal		92.12 (10.04)		20.16 (1.47)
Autocorrelation		0.82 (0.02)		0.25 (0.05)

covariance between the intercept and the slope. If that is done, the deviance difference test for the variance of the intercept is significant ($\chi^2 = 10.08, df = 1, p = 0.002$). The associated changes in parameter estimates are very small, so they are not reported here. The varying coefficient for Job Demands can in principle be explained by adding a cross-level interaction of Job Demands with Trait Anxiety to the model. However, the coefficient for this interaction is not significant, and the variance of the Job Demands slopes remains unexplained.

The autocorrelation in the combined model is a conditional autocorrelation (Singer & Willett, 2003), conditional on the predictors in the model and the random effects of the intercept and of Job Demands. In the autocorrelation model, without the random intercept and slopes, the autocorrelation is much higher. Since there are no random effects in this model, the structure of the covariances over time must completely be explained by the autocorrelation function. In the combined model, part of the covariance structure is explained by the random effects in the model, which leaves a lower autocorrelation.

The interpretation of the model results for the combined model is the following. The time-varying predictors Job Demands and Social Support are significant. On days that Job Demands are high, a higher state anxiety is reported. On days that social support is high, a lower state anxiety is reported. Subjects who score high on trait anxiety report in general a higher state anxiety as well. The slope variation for Job Demands shows that some subjects are more sensitive to changes in Job Demands than others. There is a medium size correlation between the residuals for state anxiety from one day to the next, which means that state anxiety has a certain amount of short-term stability over time.

5.5 CONCLUSIONS

This chapter makes a distinction between longitudinal data where indicators for time are predictors in the model, to model growth or decline over time, and models where time is not a predictor, and time-varying covariates are used to model change over time. When time or other time-varying covariates have varying regression slopes, the dependency structure in the covariances over time is modeled implicitly, as a consequence of the estimated parameters in the random part. If there are no varying coefficients, the data must be modeled otherwise.

REFERENCES

Bollen, K. A., & Curran, P. J. (2006). *Latent curve models*. Hoboken, NJ: Wiley.

Curran, P. J. (1997). *Supporting documentation for comparing three modern approaches to longitudinal data analysis: An examination of a single developmental sample*. Retrieved from http://www.unc.edu/~curran/srcd.htmlhttp://www.unc.edu/~curran/srcd.html

Duncan, T. E., Duncan, S. C., & Strycker, L. A. (2006). *An introduction to latent variable growth curve modeling*. Mahwah, NJ: Erlbaum.

Goldstein, H. (2003). *Multilevel Statistical Models*. 3rd edition. London: Edward Arnold.

Hedeker, D., & Gibbons, R. D. (2006). *Longitudinal data analysis*. Hoboken, NJ: Wiley.

Hox, J. J. (2002). *Multilevel analysis*. Mahwah, NJ: Erlbaum.

Little, R. J. A. (1995). Modeling the drop-out mechanism in repeated measures studies. *Journal of the American Statistical Association, 90*, 1112–1121.

Maas, C. J. M., & Snijders, T. A. B. (2003). The multilevel approach to repeated measures with missing data. *Quality & Quantity, 37*, 71–89.

Raftery, A. E. (1995). Bayesian model selection in social research. In P.V. Marsden (Ed.), *Sociological Methodology* (pp. 111–164). Blackwell: Oxford.

Raudenbush, S. W. (2002). Alternative covariance structures for polynomial models of individual growth and change. In D. S. Moskowitz & S. L. Hersberger (Eds), *Modeling intraindividual variability with repeated measures data* (pp. 25–57). Mahwah, NJ: Erlbaum.

Singer, J. D., & Willett, J. B. (2003). *Applied longitudinal data analysis: Modeling change and event occurrences*. New York, NY: Oxford.

Willett, J.B., & Sayer, A.G. (1994). Using covariance structure analysis to detect correlates and predictors of individual change over time. *Psychological Bulletin*, 116, 363–381.

6

Growth Curve Analysis Using Multilevel Regression and Structural Equation Modeling

Reinoud D. Stoel
Department of Digital Technology and Biometry, Netherlands Forensic Institute, The Hague, The Netherlands

Francisca Galindo Garre
TNO Quality of Life, Leiden, The Netherlands

6.1 INTRODUCTION

Data originating from a longitudinal or panel design are common in the social sciences. A wide array of statistical models is available for analyzing data from such a design. Each of these methods has specific features and the use of a particular method in a specific situation depends on such things as the type of research, the research question, and so on. The central concern of longitudinal research, however, revolves around the description of patterns of stability and change, and the explanation of how and why change does or does not take place (Kessler & Greenberg, 1981).

At the end of the last century, models that take a growth curve perspective have become increasingly used by researchers. Such growth curve models provide a parsimonious way to account for the dependency caused by the fact that the same subjects have been assessed repeatedly. The growth curve model has its roots in the models developed by Tucker (1958) and Rao (1958), the random-effects model (Laird & Ware, 1982), hierarchical model (Raudenbush & Bryk, 2002), multilevel model (Goldstein, 1995); random coefficients model (de Leeuw & Kreft, 1986), or mixed model (Longford, 1993) covering the traditions of biostatistics, education, and psychometrics. Bollen and Curran (2006) and Bollen (2007) provide a brief history of growth curve models.

One reason for the current popularity is the availability of powerful software packages for specifying and analyzing growth curve models, for example, GLLAMM: Rabe-Hesketh, Pickles, and Skrondal (2001); *Mplus*: Muthén and Muthén (2001); LISREL: Jöreskog and Sörbom (1999); Amos4.0: Arbuckle and Wothke (1999); MlwiN: Rasbash, Steele, Browne, and Prosser (2004); HLM5: Bryk, Raudenbush, and Congdon (1996), in combination with a growing amount of methodological literature (tutorials as well as specialized papers and books) dealing with growth curve analysis (e.g., Biesanz, Deeb-Sossa, Papadakis, Bollen, & Curran, 2004; Bollen & Curran, 2006; Duncan, Duncan, & Strycker, 2006; Hamagami & McArdle, 2004; Klein & Muthén, 2006; Muthén & Curran, 1997; Willet & Sayer, 1994). Probably the most important reason for the popularity of these longitudinal growth models, however, is their elegance in representing both collective and individual change as a function of time.

Although the growth curve model itself is not without debate we only touch upon that discussion here. In short, there is a fair body of research comparing the growth curve model to another important model class: the autoregressive model[1] (see Bast & Reitsma, 1997, 1998; Mandys, Dolan, & Molenaar, 1994; Rogosa & Willett, 1985). In summary, it seems difficult to empirically discriminate between the autoregressive model and the growth curve model. The autoregressive model may be a less restrictive model for change resulting as a consequence in the estimation of more parameters and a slightly more difficult interpretation. The growth curve model is a very elegant and parsimonious model, and this probably contributed in recent years, among other things, to its increased application. Of importance in this chapter is that the choice between analysis techniques is not such an issue for the autoregressive model, since in practice with panel data this model is almost always estimated using structural equation modeling (SEM) software. Growth curve models, on the other hand, are commonly estimated with either multilevel regression analysis (MLR) software or SEM software.

Standard MLR and SEM are highly similar in the case of a growth curve analysis. In MLR the individual differences in growth across time are captured by random coefficients, the SEM approach treats the individual differences as latent variables. If the basic model is used to represent the same set of longitudinal data, their models yield identical estimates of the relevant parameters. This has been repeatedly shown by several authors (among which Chou, Bentler, & Pentz, 1998; Hox, 2000; Hox & Stoel, 2005; MacCallum, Kim, Malarkey, & Kiecolt-Glaser, 1997; Mehta & West, 2000; Stoel, Van den Wittenboer, & Hox, 2003). Differences appear in estimation, in the possibilities in which the growth model can be extended, and in the ease in which such extensions can be specified in the available software.

This chapter deals with the choice between MLR and SEM once the growth curve model has been chosen. We will discuss the similarities and differences between these techniques and discuss several extensions. We will not focus on the comparison of MLR and SEM in general, the chapter by Muthén in this edited volume will have a stronger focus on this (see also Bauer, 2003; Curran, 2003; Muthén, 1997). Much has been written yet on the comparison of MLR and SEM in the analysis of growth curve models, the editors thought it useful especially for

[1] Of course, more types of longitudinal models exist, as well as combinations of both models (see Bollen & Curran, 2004). In practice, however, researchers often make a choice between these two models.

applied researchers to include a chapter on the topic in this volume. Applied researchers are often confused about the differences and similarities between the two approaches to growth curve modeling. The purpose of this chapter is to point researchers to issues that may be considered in the decision to apply one of these approaches to growth curve modeling, and to clarify the differences and similarities, so that researchers fully oversee what they may gain (or lose) from switching between the two approaches, and how the two approaches may complement each other. We will start with a discussion of the standard approach to growth curve analysis, in a later section extensions and adaptations of the standard approaches will be considered, such as the inclusion of definition variables in SEM, and the Generalized Linear Latent And Mixed Models (GLLAMM; Skrondal & Rabe-Hesketh, 2004).

In the next section we will introduce the general growth curve model, and possible extensions. Subsequently, a data set will be analyzed with both MLR and SEM, and we will show that this results in the same parameter estimates by using standard approaches.

To make matters concrete, we shall refer throughout this chapter to a hypothetical study in which data on the language acquisition of 300 children were collected during primary school at four consecutive occasions at 51 schools. Furthermore, data were collected on the children's gender and intelligence, as well as, on each occasion, a measure of their emotional well-being. Of interest is whether there is growth in language acquisition, and whether there are differences between the children concerning their growth curves. Given the interindividual differences in the growth curves, the study wants to investigate whether intelligence explains (part of) the interindividual variation in the growth curves, and whether emotional well-being can be used to explain the time specific deviations from the mean growth curve.

6.2 THE GROWTH CURVE MODEL

A general equation for a three-level linear growth curve model with two explanatory variables is:

$$
\begin{aligned}
y_{tij} &= \pi_{0ij} + \pi_{1ij}\, T_{tij} + \Sigma\pi_{2t} x_{tij} + e_{tij} \\
\pi_{0ij} &= \beta_{00j} + \beta_{01}\, z_{ij} + u_{0ij} \\
\pi_{1ij} &= \beta_{10j} + \beta_{11}\, z_{ij} + u_{1ij} \\
\beta_{00j} &= \beta_{00} + v_{0j} \\
\beta_{10j} &= \beta_{10} + v_{1j} \\
e_{tij} &\sim N(0, \sigma^2_{et}) \\
\begin{Bmatrix} u_{0ij} \\ u_{1ij} \end{Bmatrix} &\sim N(0, \Omega_2) \quad \Omega_2 = \begin{Bmatrix} \sigma^2_{u0} & \\ \sigma^2_{u01} & \sigma^2_{u1} \end{Bmatrix} \\
\begin{Bmatrix} v_{0i} \\ v_{1i} \end{Bmatrix} &\sim N(0, \Omega_3) \quad \Omega_3 = \begin{Bmatrix} \sigma^2_{v0} & \\ \sigma^2_{v01} & \sigma^2_{v1} \end{Bmatrix}
\end{aligned}
\tag{6.1}
$$

where T_{tij} is a vector denoting the time of measurement for subject i in class j. The intercept and slope (i.e., linear growth rate) for each individual subject are expressed by the coefficients π_{0ij} and π_{1ij} with random deviations u_{0ij} and u_{1ij}. The class specific intercept and slope β_{00} and β_{10} have random deviations v_{0j} and v_{1j}; π_{2t} represents the effect of the (time-varying) explanatory variable x_{tij} on time t, β_{01} and β_{11} represent, respectively, the effects of the (time-invariant) explanatory variable z_i on intercept and slope. The e_{ti} is a first-level residual, and Ω_2 and Ω_3 are the covariance matrices of intercept and slope on level 2 and level 3, respectively.

The essence of the growth curve model is that change is modeled explicitly as a function of time. Both MLR and SEM thus incorporate the factor time explicitly in the case of the growth curve model. Within the MLR framework time is modeled as an independent variable at the lowest level, the individual is defined at the second level and explanatory variables can be modeled at all existing levels. The interindividual differences in the parameters describing the growth curve are modeled as random effects. The SEM approach adopts a latent variable view, with the time dimension incorporated in the specification of the latent variables. The parameters of the individual curves are modeled as latent variables (i.e., intercept and linear growth rate). The latent variables in LGC analysis correspond to the random effects in MLR analysis. Therefore it is possible to specify exactly the same model as a LGC or MLR model, with exactly the same parameter estimates. Curran (2003, p. 540) describes an "isomorphism" between MLR and SEM because the associated matrices defining the fixed and random effects are identical (see also Bauer, 2003). This has also been noted by Bauer and Curran (2002), Neale, Boker, Xie, and Maes (1999), Rovine and Molenaar (1998, 2000, 2001), and Mehta and Neale (2005). In fact Curran (2003) proposes that *any two-level linear multilevel model can be estimated as a structural equation model given that this is essentially a data management problem.*

Estimation of growth curve models is usually done by the Maximum Likelihood (cf. Eliason, 1993). Given a sufficiently large sample size, maximum likelihood estimation is expected to produce estimates that are asymptotically efficient and consistent. Alternative estimation methods like Restricted Maximum Likelihood (RML), Markov Chain Monte Carlo (MCMC), and Bootstrapping methods (cf. Mooney & Duval, 1993). These alternative methods have their own peculiarities, and may sometimes be preferred to ML estimation. Simulation based MCMC methods, for example, do give better estimates for some problems and can be applied to more complicated models where there is, at present, no equivalent iterative procedure (Browne & Rasbash, 2002). We refer to Bollen (1989), Browne and Rasbash (2002), Goldstein (1995), and Loehlin (1987) for a description of alternative estimation methods.

6.2.1 Example

Table 6.1 shows the results of an analysis with both MLR (using MLwiN) and SEM (using M*plus*) of the data on language acquisition. The data used in this example consist of the scores on language acquisition of the 300 children, measured on four occasions (y_{ti}) at 51 schools, the repeatedly assessed measure of emotional well-being (x_{ti}), and the measure of intelligence (z_i); the covariates are mean centered. Analyzing the data with a three-level growth curve model (measurement-child-school) using both the MLR and SEM approach with Maximum Likelihood estimation leads to the parameter estimates presented in Table 6.1. The first column of Table 6.1 gives the relevant parameters; the second and third columns show the parameter estimates of, respectively, MLR and SEM. Beside parameter estimates and their standard errors of the

TABLE 6.1

Maximum Likelihood Estimates of the Parameters of Equation 6.1 using MLR and SEM

Parameter	MLR Three-Level	SEM Three-Level	MLR Two-Level	SEM Two-Level
Fixed Part				
β_{00}	9.869 (.109)	9.869 (.113)	9.807 (.056)	9.807 (.056)
β_{10}	1.925 (.053)	1.925 (.077)	1.939 (.048)	1.939 (.048)
β_{01}	.222 (.045)	.222 (.043)	.387 (.056)	.387 (.056)
β_{11}	.905 (.048)	.905 (.048)	.883 (.048)	.883 (.048)
π_{21}	.551 (.037)	.551 (.035)	.702 (.042)	.702 (.042)
π_{22}	.809 (.034)	.809 (.034)	.815 (.034)	.815 (.034)
π_{23}	.912 (.034)	.912 (.034)	.913 (.034)	.913 (.034)
π_{24}	.951 (.050)	.951 (.050)	.953 (.049)	.953 (.049)
Random Part				
σ^2_{e1}	.124 (.038)	.123 (.038)	.216 (.049)	.216 (.049)
σ^2_{e2}	.296 (.030)	.296 (.030)	.257 (.029)	.257 (.029)
σ^2_{e3}	.235 (.039)	.235 (.039)	.252 (.038)	.252 (.038)
σ^2_{e4}	.297 (.082)	.290 (.082)	.229 (.079)	.229 (.079)
σ^2_{u0}	.368 (.050)	.369 (.050)	.784 (.081)	.784 (.081)
σ^2_{u1}	.637 (.060)	.637 (.060)	.641 (.057)	.641 (.057)
σ^2_{u01}	.010 (.038)	.009 (.038)	–.004 (.048)	–.004 (.048)
σ^2_{v0}	.512 (.119)	.512 (.119)		
σ^2_{v1}	.026 (.029)	.026 (.029)		
σ^2_{v01}	–.057 (.042)	–.057 (.042)		

Note: Standard errors are given in parentheses. The chi-square test of model fit for SEM: $\chi^2(71) = 144.52$ ($p = .00$), RMSEA = .059; $\chi^2(19) = 18.55$ ($p = .49$), RMSEA = .000; for MLR: –2ll = 3233.479 for the three-level model, and –2ll = 3326.626 for the two-level model.

three-level models, Table 6.1 includes those of the two-level models.

As one can see from the second and third column of Table 6.1, presenting the results of the three-level models, the estimates of the fixed and random part of the model are the same, but their standard errors are slightly different. However, the differences are very small and as a consequence both approaches would lead to the same substantive conclusions. After controlling for the effect of the covariates, a mean growth curve emerges with an intercept, β_{01}, of 9.87 and a growth rate, β_{10}, of 1.93. The significant variation between the subjects around these mean values, σ^2_{u0} and σ^2_{u1}, respectively, implies that subjects start their growth process at different values, and grow subsequently with different speeds. The correlation between initial level and growth rate, σ^2_{u01}, is zero, implying that the initial level has no predictive value for the growth rate. Intelligence has a positive effect on both intercept and slope, β_{01} and $\beta_{11,}$ respectively. So, more intelligent children show a higher score at the first measurement occasion, and a greater increase in language acquisition than children with lower intelligence. Emotional well-being explains the time specific deviations from the mean growth curve (β_{01}, β_{11}, π_{21}, π_{21}, π_{21}, and π_{21}). That is, children with a higher emotional well-being at a specific time point show a higher score on language acquisition than is predicted by their growth curve. Furthermore, on the school-level, there are significant differences between schools in the intercepts (σ^2_{v0}), but not in slopes (σ^2_{v1}). So there appears to be some clustering with respect to the intercepts but not with respect to the slopes.

In order to show the equivalence of MLR and SEM in a two-level growth curve model, columns 4 and 5 of Table 6.1 present these results. The two-level model can be regarded to be nested within the three-level model, since we obtain the two-level model by constraining three parameters to zero (i.e., σ^2_{v0}, σ^2_{v1}, and σ^2_{v01}). The models can thus be tested against each other by means of a likelihood ratio (LR) test. The value of the LR test statistic can be directly obtained within the MLR approach and is found to be equal to 93.15 (3326.63 – 3233.48), and this value can subsequently be compared to the appropriate reference distribution for testing.[2] A direct test of these models is, however, not directly obtained in the SEM approach. If we compare the number of *df* between the models we will see the number increasing from 19 to 71, instead of decreasing by 3–16! The reason is that the SEM approach (as implemented in M*plus*) uses a different estimation procedure that decomposes the observed covariance matrix into two parts, so that effectively the number of elements in the covariance matrices doubles. With the SEM approach we would have to reestimate the three-level model with the level-3 covariance and variance constrained to zero, and subtract the χ^2-value from that of the prior model. The LR test statistic is than equal to 93,40 (237.92 – 144.52). Please note that the fit of the standard two-level growth curve model in SEM (i.e., $\chi^2(19) = 18.55$) is not equal to its three-level counterpart (i.e., $\chi^2(74) = 237.92$).

[2] Please note that this distribution is not a $\chi^2(3)$-distribution, but a mixture of a $\chi^2(1)$, a $\chi^2(2)$, and a $\chi^2(3)$-distribution due to the fact that the null-hypothesis (i.e., the two-level model) places the values of the σ^2_{v0} and σ^2_{v1} on the boundary of the parameter space. We refer to Stoel, Garre, Dolan, and Van den Wittenboer (2006) for a thorough discussion of this topic.

6.3 SEM VERSUS MLR: THE STANDARD APPROACH

A striking difference between MLR and SEM is the way time is treated in the standard growth curve model (cf. McArdle, 1986, 1988; Meredith & Tisak, 1990; Willet & Sayer, 1994). In MLR, time is introduced as an explanatory variable, whereas in SEM it is introduced via the factor loadings. So, in MLR an additional variable is added; and in SEM the factor loadings are constrained in such a way that they represent time. The consequence of this is that MLR is essentially a univariate approach to growth curve analysis, with time points treated as observations of the same variable, whereas SEM is a multivariate approach, with each time point treated as a separate variable. It therefore appears that MLR is more flexible in its treatment of time, since subject may be measured on completely different occasions, estimating a growth curve model within SEM will be a tedious exercise in this case. Random occasions may thus be more effectively modeled with MLR, whereas fixed occasions can be modeled by both MLR and SEM. In their discussion on the equivalence of SEM and MLR in general Mehta and Neale (2005) describe in detail that *univariate multilevel models are really multivariate unilevel models.* Pointing to the fact that the between-cluster and within-cluster variation in MLR parallels those of common and unique variance in measurement models from the SEM approach (Skrondal & Rabe-Hesketh, 2004). The parallel measurement model (i.e., equal factor loadings and residuals) can be shown to be mathematically equivalent to the random intercept model. An illustrating consequence of this mathematical equivalence is the following. In MLR the level-1 units are assumed to be exchangeable, resulting in the fact that the ordering of the level-1 units within the level-2 units is arbitrary. In the parallel measurement model the same is true, and changes in the ordering of the variables within a row of the data matrix will not affect the results. Although the observed covariance matrix will be different, parameter estimates are essentially the same.

That MLR analysis is essentially a univariate approach while SEM can be considered a multivariate approach is nicely illustrated by looking at the way the data are set up. Tables 6.2 and 6.3 present the data of the first two subjects from the language acquisition study for, respectively, MLR and SEM.

Table 6.2 shows the values of language acquisition (y) being treated as scores on one variable, the same holds for the time-varying covariate, emotional well-being (x). The scores of the time-invariant covariate, intelligence (z), represented by one variable with each score repeated four times for each subject. Time (t) is treated as an observed variable that will enter the model as a (fixed) time-varying covariate. The data setup for SEM in Table 6.3 shows that each variable measured at a specific occasion is treated as a separate variable. Time is not modeled as an observed variable, but instead via constraints of the factor loadings.

6.3.1 Extensions of the Measurement Model

When multiple indicators of the same construct are available SEM offers a number of possibilities for including them explicitly by means of a measurement model. Illustrative examples of a LGC model with multiple indicators can be found in Bollen and Curran (2006), Garst, Frese, and Molenaar (2000), and Hancock, Kuo, and Lawrence (2001). The multiple-indicator growth model, or curve-of-factors model (cf. McArdle, 1988), is actually a higher order factor model merging a common factor model for the multiple indicators with a growth curve model on the common

TABLE 6.2

Data Format for a Growth Curve Analysis in the MLR Approach

Variable Subject	y	X	Z	t
1	6.64	−.81	−.40	0
1	10.01	.15	−.40	1
1	12.28	−.87	−.40	2
1	15.50	−1.69	−.40	3
2	7.12	−.46	.027	0
2	9.39	.53	.027	1
2	9.92	−.33	.027	2
2	11.11	.26	.027	3

TABLE 6.3

Data Format for a Growth Curve Analysis in the SEM Approach

Variable Subject	y_0	y_1	y_2	y_3	x_0	x_1	x_2	x_3	z
1	6.64	10.01	12.28	15.50	−.81	.15	−.87	−1.69	−.40
2	7.12	9.39	9.92	11.11	−.46	.53	−.33	.26	.027

factors. In other words, the common variation in the multiple indicators is accounted for by the first-order factors, while the second-order factors serve to explain the mean and covariance structure of the first-order factors. Using similar restrictions as in the single indicator growth model, the second-order factors can be given the interpretation of intercept and slope. Thus, instead of analyzing the sum scores or item parcels, as is the standard practice within MLR, the observed variables can be included explicitly in the model. Of course, not only the dependent variables may be measured with multiple indicators, and extensions to growth curve models with a separate measurement model for the predictor variables are straightforward.

The ability to model multiple indicators under a common factor is one of the advantages of SEM, as is shown in the extensive literature dealing with issues concerning common factors and structural models in general (e.g., Bollen, 1989). Whenever multiple indicators of an outcome are available, the possibilities of MLR analysis constitute only a subset of the potentials of SEM. Because of the ability to model multiple indicators, SEM allows for an explicit test of at least one of the assumptions underlying the longitudinal factor model, the so-called assumption of measurement invariance (Meredith, 1964, 1993). This assumption ensures a comparable definition of the latent construct over time. An explicit test of this assumption is important before any further analysis can be performed. In brief, the assumption of measurement invariance implies equal factor loadings, and equal item intercepts across time for each repeatedly measured indicator. Violation of the assumption hinders the assessment of change or development of a subject over time because it will be confounded with change of the meaning of the construct over time. In a strict sense, the violation of the assumption of measurement invariance implies that analyzing the means, or any weighting of the indicators, is not permitted. However, scale scores, as is common practice in MLR, impedes the test of measurement invariance. In that sense, it adds to the discussion on measurement invariance and item-parceling (e.g., Bandalos, 2002; Byrne, Shavelson, & Muthén, 1989; Stoel, Van den Wittenboer, & Hox, 2004).

Raudenbush, Rowan, and Kang (1991) proposed a way to include multiple indicators in MLR. In this approach the indicators are modeled on a separate level, with the assumption of them being parallel measures (i.e., residual variances and factor loadings of the indicators within each factor are equal). We refer to Hox (2002) for an illustration of multilevel confirmatory factor analysis (see also Li, Duncan, Harmer, Acock, & Stoolmiller, 1998). The multiple-indicator growth model could, in principle, be analyzed using MLR using the method of Raudenbush et al. (1991). However, the restrictions on the factor structure are implicit in the model and cannot be relaxed, and the assumption of measurement invariance can, therefore, not be tested explicitly. As a consequence, the solution provided by Raudenbush et al. (1991) is only a partial (i.e., very restrictive) solution. Another way to include multiple indicators in MLR are the multilevel IRT models. An IRT model can be used to define a relation between observed categorical indicators and an underlying latent trait. Fox and Glas (2001), for instance, propose an algorith to estimate these models. Another extension to include multiple categorical indicators in MLR is Multilevel Latent Class (Vermunt,

2003). A latent class model can be used when both the indicators and the latent variables are categorical. Finally, another extension to include multiple indicators in MLR is proposed by Skrondal and Rabe-Hesketh (2004). We dedicate a subsection to this approach.

6.3.2 Estimating the Shape of the Growth Curve

The issue to be discussed in this paragraph also concerns estimation of factor loadings, and is therefore related to the issue of multiple indicators. The growth curve models discussed so far all assume linear growth, with time incorporated explicitly in the model by constraining the factor loadings (SEM) or by including it as an independent variable (MLR). As described by McArdle (1988) and Meredith and Tisak (1990), within the SEM approach it is possible to estimate a more general model in which the factor loadings for the slope are estimated. Thus, instead of constraining the corresponding factor loadings to, for instance, $[0, 1, 2, ..., T]$, it is set to $[0, 1, t_3, t_4, ..., t_T]$. In other words, the factor loadings t_3 to t_T are estimated, providing information on the shape of the growth curve. Muthén and Khoo (1998) explain this as the estimation of the time scores. The essence is captured effectively with the following citation of Garst (2000, p. 259): "Statistically, a linear model is still estimated, but the nonlinear interpretation emerges by relating the estimated time scores to the real time frame. … Therefore, a new time frame is estimated and the transformation to the real time frame gives the nonlinear interpretation." Although this model has a similar representation in MLR analysis (cf. MacCallum et al., 1997), it cannot be estimated using this approach. As a univariate approach, the repeated measurements are treated as realizations of just one dependent variable, and time is represented by another independent variable. It is required that the values of the independent variable are known. However, more flexible time functions can be used to model time, such as splines or linear models with a change-point[3] (for an illustration of splines functions see Pan & Goldstein, 1998 and for an application of linear functions with a change-point see Galindo-Garre, Zwinderman, Geskus, & Sijpkens, 2008).

6.3.3 Missing Data

One of the advantages of the MLR analysis is its ability to handle missing data (Hox, 2000; Raudenbush & Bryk, 2002; Snijders, 1996). As a univariate technique, MLR analysis does not assume "time-structured data" (cf. Bock, 1979), so that the number of measurement occasions and its spacing need not be the same for all individuals. Thus, the absence of measurements on a subject on one or more occasions poses no special problems; and/or the subjects may be measured at different occasions. In such instances, the time variable, as a fixed independent variable, will just have a different number of scores for the subjects. In an extreme case, there may be many measurement occasions, but there may be just one observation at each occasion. This can be seen as an advantage of the MLR model. As a result, MLR analysis easily models

[3] Spline functions in general are piecewise polynomials of degree n whose function values and first n–1 derivatives agree at the points where they join. The abscissae of these joined points are called knots. A linear function with a change-point is a function composed of two lines that are connected by an estimated change point. This point allows a different slope for both linear functions.

longitudinal data with a less extreme pattern of observations, such as, for example, panel dropout. The ability of MLR analysis to easily handle cases with missing values on the outcome variable (i.e., missing occasions) is the natural result of the fact that MLR does not assume balanced data to begin with. The MLR has no special provisions for incomplete data otherwise, so it cannot deal with missing values on the covariates.

On the contrary, SEM as such assumes time-structured data (fixed occasions). Special procedures are needed, if the number and spacing of the measurement occasions varies. One possibility is to estimate the growth model using all available data of all cases using full-information maximum likelihood estimation (Muthén, Kaplan, & Hollis, 1987; Wothke, 2000). This approach actually constitutes a principled way to deal with incomplete data in any part of the data matrix, and is now commonly applied in the case of missing data.

Similar to MLR, full-information maximum likelihood does not make the restrictive assumption of Missing Completely At Random (MCAR), but instead it is based on the less restrictive assumption that the missing values are Missing At Random (MAR; Little & Rubin, 1987). The MAR assumes that the missing values can be predicted from the available data. It is not our purpose here to give a broad explanation of this procedure and of the different estimation techniques. However, the advantage of MLR analysis concerning missing observations only holds for missing observations on the dependent variable. Whenever explanatory variables are missing, the growth model as such, cannot be estimated by MLR analysis. If an explanatory variable is missing, the usual treatment within MLR analysis is to remove the subject from the analysis by listwise deletion, whereas the SEM can still be estimated using the full-information maximum likelihood approach.

6.3.4 Extensions of the Structural Model

Frequently, research hypotheses are not restricted to the pattern of change in a single process, but they focus, instead, on simultaneously modeling the change in several outcome variables (cf. cross-domain analysis, MacCallum et al., 1997; Willett & Sayer, 1996), or on modeling relationships between growth parameters and variables serving as outcomes of those parameters (e.g., Garst et al., 2000). In the language acquisition study, for example, the interest may be in whether growth in language acquisition (of their mother tongue) in primary education (w), and intelligence can be used to predict the achievement of foreign language acquisition at the first year of secondary education.

A distinction between MLR and SEM relates to this kind of extensions of the "structural part" of the model. Compared to SEM, the multilevel approach is severely limited in modeling extensions of this kind. The typical MLR software (i.e., MlwiN and HLM) merely allows for the inclusion of measured predictors on all existing levels, and the estimation of the covariances between growth parameters in a multivariate model (e.g., MacCallum et al., 1997). Also, posterior Bayes estimates (see Raudenbush & Bryk, 2002) of the intercept and slope parameters can be computed, and subsequently be included as predictors in a separate multiple regression analysis. The SEM, on the other hand, is more flexible. It is possible to estimate all means and covariances associated with the latent growth parameters, or they can be modeled explicitly. As a consequence, SEM is better prepared to estimate (1) interrelationships (directional) among several growth processes, (2) simultaneous and joint associations of these growth processes and covariates, and (3) mediated effects of covariates (see also Willett & Sayer, 1994). In other words, the growth curve model as modeled in SEM, can be part of a larger structural model, in which, for example, the effects of the latent initial level and linear shape factor on other variables can be modeled simultaneously.

6.3.5 GLLAMM

Skrondal and Rabe-Hesketh (2004) proposed a general framework to estimate multilevel, structural equation models and longitudinal models. The essence of their model formulation is that it allows for the specification of hierarchical conditional relationships and that the latent variables in the structural part may be regressed on other latent and observed variables. Their model includes three parts: (1) The response model or measurement model that is a model for the observed responses conditional on the latent variables and covariates. Observed responses do not need to be continuous. (2) The structural model for the latent variables in which the latent variables may be regressed on other latent variables or on observed covariables. (3) The distributions of the disturbances.

Let $\mathbf{y}_j$ be the vector of responses for individual j, $\mathbf{X}_j\beta$ be the fixed part; Δ_j be the structure matrix, which may refer to both the design matrix $\mathbf{Z}_j$ in the multilevel formulation or the factor loading matrix $\boldsymbol{\Delta}$ in SEM formulation; $\boldsymbol{\eta}_j$ be latent variables, which may be random effects or common factors; and ε_j be the vector of errors. The response model can be written as:

$$\mathbf{y}_j = \mathbf{X}_j\beta + \Delta_j\eta_j + \varepsilon_j \qquad (6.2)$$

The response model allows for the formulation of a latent growth model as a random coefficient model or as a factor model. The fixed part is written in the same way. The main difference between multilevel and SEM formulations is that the design matrix $\mathbf{Z}_j$ for the random-effects model is a known matrix of covariates and constants whereas the factor loading matrix $\boldsymbol{\Delta}$ for factor models may be an unknown parameter matrix. Factor models and random coefficient models can be unified by transforming the design (factor loadings) matrix in a product of two matrices $\mathbf{Z}_j\lambda$ where $\mathbf{Z}_j$ is a design matrix and $\boldsymbol{\lambda}$ a parameter vector. In the latent growth model the factor loadings matrix is also fixed, and therefore the parameter vector will be a unit vector. In the same way as in structural equation models the structural model is used to define relationship between latent variables. The advantage of this formulation is that it easily allows relationships between latent variables of different levels.

Contrary to standard SEM, the distribution of the disturbances may be continuous (parametric and nonparametric), discrete, or mixed discrete and continuous.

6.3.6 Software

Besides the well-known specialized multilevel and SEM programs, standard statistical packages also include functions, packages of macros to estimate both multilevel and/or SEM models. The SAS contains a library to estimate factor models, called PROC FACTOR, and a library to estimate SEM models, called PRO CALIS. More information can be found at http://www.als.ucla.edu/stat/sas/library and http://support.sas.com/onlinedoc/913/docMainpag.jsp.

The STATA has a function to estimate multilevel models, called xtmixed, but does not have a specific function to estimate SEM models. The program GLLAMM, which has been programmed in STATA, can be used to estimate both models. The GLLAMM is a very flexible program, but it may work very slowly if the number of random effects or the number of records is very large.

The program R has several packages to estimate multilevel models and a package to estimate SEM models. These packages can be downloaded from http://www.r-project.org. The most popular package to estimate multilevel models is lmre. An excellent description of multilevel (hierarchical) models and their estimation with R can be found in Gelman and Hill (2007). A description of the SEM package can be found in Fox (2006). The complexity of the models that can be estimated with this package is limited compared to dedicated multilevel and SEM software like MLwiN and M*plus*.

Finally, WinBUGS is a very flexible Bayesian package that uses a Gibb sampling algorithm. In principle every model can be estimated with this program. The only problem is that it may work very slow if the model is very complex or the number of records is very large. WinBUGS can be downloaded from http://www.mrc-bsu.cam.ace.uk/bugs. Gelman and Hill (2007) also describe how to estimate different multilevel models with WinBUGS and examples of WinBUGS code for SEM models can be found in Congdon (2001).

REFERENCES

Arbuckle, J. L., & Wothke, W. (1999). AMOS 4.0 User's Guide. Chicago, IL: SPSS Inc.

Bandalos, D. L. (2002). The effects of item parceling on goodness-of-fit and parameter estimate bias. *Structural Equation Modeling, 9,* 78–102.

Bast, J., & Reitsma, P. (1997). Matthew effects in reading: A comparison of latent growth curve models and simplex models with structured means. *Multivariate Behavioral Research, 32*(2), 135–167.

Bast, J., & Reitsma, P. (1998). The simple view of reading: A developmental perspective. In P. Reitsma & L. Verhoeven (Eds.), *Problems and interventions in literacy development* (pp. 95–109). Dordrecht, The Netherlands: Kluwer.

Bauer, D. J. (2003). Estimating multilevel linear models as structural equation models. *Journal of Educational and Behavioral Statistics, 28*(2), 135–167.

Bauer, D. J., & Curran, P. J. (2002). *Estimating multilevel linear models as structural equations models.* Paper presented at the annual meeting of the Psychometric Society, Chapel Hill, NC.

Biesanz, J., Deeb-Sossa, N. Papadakis, A. Bollen, K. A., & Curran, P. (2004). The role of coding time in estimating and interpreting growth curve models. *Psychological Methods*, 9, 30–52.

Bock, R. D. (1979). Univariate and multivariate analysis of variance of time-structured data. In J. R. Nesselroade & P. B. Baltes (Eds.), *Longitudinal research in the study of behavior and development* (pp. 199–231). New York, NY: Academic Press.

Bollen, K. A. (1989). *Structural equations with latent variables.* New York, NY: Wiley.

Bollen, K. A. (2007). On the origins of latent curve models. In R. Cudeck & R. MacCallum (Eds.), *Factor Analysis at 100* (pp. 79–98). Mahwah, NJ: Lawrence Erlbaum.

Bollen, K. A., & Curran, P. J. (2004). Autoregressive latent trajectory (ALT) models: A synthesis of two traditions. Sociological Methods as Reasearch, 32, 336–338.

Bollen, K. A., & Curran, P. J. (2006). Latent curve models: A structural equation approach. *Wiley series on probability and mathematical statistics* (pp. 9–14). Hoboken, NJ: John Wiley & Sons.

Browne, W. J., & Rasbash, J. (2002). Multilevel modelling. In A. Bryman & M. Hardy (Eds.), *Handbook of data analysis* (pp. 459–479). London: Sage.

Bryk, A. S., Raudenbush, S. W., & Congdon, R. T. (1996). *HLM. Hierarchical linear and nonlinear modelling with the HLM/2L and HLM/3L programs.* Chicago, IL: Scientific Software International.

Byrne, B. M., Shavelson, R. J., & Muthén, B. (1989). Testing for the equivalence of factor covariance and mean structures: The issue of partial measurement invariance. *Psychological Bulletin, 105*, 456–466.

Chou, C. P., Bentler, P. M., & Pentz, M. A. (1998). Comparison of two statistical approaches to study growth curves. *Structural Equation Modeling, 5*, 247–266.

Congdon, P. (2001). *Bayesian statistical modelling.* Chichester, UK: Wiley.

Curran, P. J. (2003). Have multilevel models been structural equation models all along? *Multivariate Behavioral Research, 38*, 529–569.

De Leeuw, J., & Kreft, I. G. G. (1986). Random coefficient models for multilevel analysis. *Journal of Educational Statistics, 11*, 57–86.

Duncan, S. C., Duncan, T. E., & Strycker, L. A. (2006). Alcohol use from ages 9–16: A cohort-sequential latent growth model. *Drug and Alcohol Dependence, 81*, 71–81.

Eliason, S. R. (1993). *Maximum likelihood estimation: Logic and practice.* Thousand Oaks, CA: Sage.

Fox, J. (2006). Structural equation modeling with the SEM package in R. *Structural Equation Modeling, 13*(3), 465–486.

Fox, J.-P., & Glas, C. A. W. (2001). Bayesian estimation of a multilevel IRT model using Gibbs sampling. *Psychometrika, 66*, 271–288.

Galindo-Garre, F., Zwinderman, A. H., Geskus, R. B., & Sijpkens, Y. W. J. (2008). A joint latent class changepoint model to improve the prediction of time to graft failure. *Journal of the Royal Statistics Society, 171*, 299–308.

Garst, H. (2000). *Longitudinal research using structural equation modeling applied in studies of determinants of psychological well-being and personal initiative in East Germany after the unification.* Unpublished doctoral dissertation, University of Amsterdam.

Garst, H., Frese, M., & Molenaar, P. C. M. (2000). The temporal factor of change in stressor-strain relationships: A growth curve model on a longitudinal study in East Germany. *Journal of Applied Psychology, 85*, 417–438.

Gelman, A., & Hill, J. (2007). Data analysis using regression and mutilevel/hierarchical models. New York, NY: Cambridge University Press.

Goldstein, H. (1995). *Multilevel statistical models.* London, UK: Edward Arnold.

Hamagami, F., & McArdle, J. J. (2004). Modeling latent growth curves with incomplete data using different types of structural equation modeling and multilevel software. *Structural Equation Modeling, 11*, 452–483.

Hancock, G. R., Kuo, W. L., & Lawrence, F. R. (2001). An illustration of second-order latent growth models. *Structural Equation Modeling, 8*, 470–489.

Hox, J. J. (2000). Multilevel analysis of grouped and longitudinal data. In T. D. Little, K. U. Schnabel, & J. Baumert (Eds.), *Modeling longitudinal and multilevel data* (pp. 15–32). Mahwah, NJ: Lawrence Erlbaum.

Hox, J. J. (2002). *Multilevel analysis: Techniques and applications.* Mahwah, NJ: Lawrence Erlbaum.

Hox, J. J., & Stoel, R. D. (2005). Multilevel and SEM approaches to growth curve modeling. In Brian S. Everitt & David C. Howell (Eds.), *Encyclopedia of statistics in behavioral science* (pp. 1296–1305). Chichester, UK: John Wiley & Sons.

Jöreskog, K. G., & Sörbom, D. (1999). LISREL 8 *user's reference guide.* Lincolnwood, IL: Scientific Software International.

Kessler, R. C., & Greenberg, D. F. (1981). *Linear panel analysis.* New York, NY: Academic Press.

Klein, A. G., & Muthén, B. O. (2006). Modeling heterogeneity of latent growth depending on initial status. *Journal of Educational and Behavioral Statistics,* 31(4), 357–375.

Laird, N. M., & Ware, J. H. (1982). Random effects models for longitudinal data. *Biometrics, 38*, 963–974.

Li, F., Duncan, T. E., Harmer, P., Acock, A., & Stoolmiller, M. (1998). Analyzing measurement models of latent variables through multilevel confirmatory factor analysis and hierarchical modeling approaches. *Structural Equation Modeling, 5*, 294–306.

Little, R. J. A., & Rubin, D. B. (1987). *Statistical analysis with missing data.* New York, NY: John Wiley & Sons.

Loehlin, J. C. (1987). *Latent variable models.* Baltimore, MD: Lawrence Erlbaum.

Longford, N. T. (1993). *Random coefficients models.* Oxford, UK: Clarendon.

MacCallum, R. C., Kim, C., Malarkey, W. B., & Kiecolt-Glaser, J. K. (1997). Studying multivariate change using multilevel models and latent growth curve models. *Multivariate Behavioral Research, 32*, 215–253.

Mandys, F., Dolan, C. V., & Molenaar, P. C. M. (1994). Two aspects of the simplex model: Goodness of fit to linear growth curve structures and the analysis of mean trends. *Journal of Educational and Behavioral Statistics, 19*, 201–215.

McArdle, J. J. (1986). Latent variable growth within behavior genetic models. *Behavior Genetics, 16*(1), 163–200.

McArdle, J. J. (1988). Dynamic but structural equation modeling of repeated measures data. In R. B. Cattel & J. Nesselroade (Eds.), *Handbook of multivariate experimental psychology* (2nd ed., pp. 561–614). New York, NY: Plenum Press.

Mehta, P. D., & Neale, M. C. (2005). People are variables too: Multilevel structural equations modeling. *Psychological Methods, 10*, 259–284.

Mehta, P. D., & West, S. G. (2000). Putting the individual back into growth curves. *Psychological Methods, 5*, 23–41.

Meredith, W. (1964). Notes on factorial invariance. *Psychometrika, 29*, 177–185.

Meredith, W. (1993). Measurement invariance, factor analysis, and factorial invariance. *Psychometrika, 58*, 525–543.

Meredith, W. M., & Tisak, J. (1990). Latent curve analysis. *Psychometrika, 55*, 107–122.

Mooney, C. Z., & Duval, R. D. (1993). *Bootstrapping: A nonparametric approach to statistical inference.* Newbury Park, CA: Sage.

Muthén, B. (1997). Latent variable modeling of longitudinal and multilevel data. In A. Raftery (Ed.), *Sociological methodology* (pp. 453–480). Boston, MA: Blackwell Publishers.

Muthén, B., & Curran, P. J. (1997). General longitudinal modeling of individual differences in experimental designs: A latent variable framework for analysis and power estimation. *Psychological Methods, 2*, 371–402.

Muthén, B., Kaplan, D., & Hollis, M. (1987). On structural equation modeling with data that are not missing completely at random. *Psychometrika, 52*, 431–462.

Muthén, B., & Khoo, S. (1998). Longitudinal studies of achievement growth using latent variable modeling. *Learning and Individual Differences, 10*, 73–102.

Muthén, B., & Muthén, L. (2001). *Mplus user's guide.* Los Angeles, CA: Muthén & Muthén.

Neale, M., Boker, S., Xie, G., & Maes, H. (1999). *Mx: Statistical modeling.* Richmond, VA: Medical College of Virginia, Department of Psychiatry.

Pan, H., & Goldstein, H. (1998). Multi-level repeated measures growth modelling using extended splines functions. *Statistics in Medicine, 17*, 2755–2770.

Rabe-Hesketh, S., Pickles, A., & Skrondal, A. (2001). *GLLAMM manual.* Technical Report 2001, Department of Biostatistics and Computing, Institute of Psychiatry, King's College, London.

Rao, C. R. (1958). Some statistical methods for comparison of growth curves. *Biometrics, 14*, 1–17.

Rasbash, J., Steele, F., Browne, W., & Prosser, B. (2004). *A user's guide to MLwiN version 2.0.* London, UK: Institute of Education.

Raudenbush, S. W., & Bryk, A. S. (2002). *Hierarchical linear models: Applications and data analysis methods.* (2nd ed.). Thousand Oaks, CA: Sage.

Raudenbush S. W., Rowan, B., & Kang, S. J. (1991). A multilevel, multivariate model for studying school climate with estimation via the EM algorithm and application to U.S. high-school data. *Journal of Educational Statistics, 16*, 295–330.

Rogosa, D., & Willett, J. B. (1985). Understanding correlates of change by modeling individual differences in growth. *Psychometrika, 50*, 203–228.

Rovine, M. J., & Molenaar, P. C. M. (1998). A nonstandard method for estimating a linear growth model in LISREL. *International Journal of Behavioral Development, 22*, 453–473.

Rovine, M. J., & Molenaar, P. C. M. (2000). A structural modeling approach to a multilevel random coefficients model. *Multivariate Behavioral Research, 35*, 51–88.

Rovine, M. J., & Molenaar, P. C. M. (2001). A structural equation modeling approach to the general linear mixed model. In L. Collins & A. Sayer (Eds.), *New methods for the analysis of change* (pp. 65–96). Washington, DC: American Psychological Association.

Skrondal, A. & Rabe-Hesketh, S. (2004). *Generalized latent variable modeling: Multilevel, longitudinal and structural equation models.* Boca Raton, FL: Chapman & Hall/CRC.

Snijders, T. A. B. (1996). Analysis of longitudinal data using the hierarchical linear model. *Quality and Quantity, 30*, 405–426.

Stoel, R. D., Garre, F. G., Dolan, C., & Van den Wittenboer, G. (2006). On the likelihood ratio test in structural equation modeling when parameters are subject to boundary constraints. *Psychological Methods, 11*, 439–455.

Stoel, R. D., Van den Wittenboer, G., & Hox J. J. (2003). Analyzing longitudinal data using multilevel regression and latent growth curve analysis. *Metodologia de las Ciencas del comportamiento, 5*, 21–42.

Stoel, R. D., Van den Wittenboer, G., & Hox, J. J. (2004). Methodological issues in the application of the latent growth curve model. In K. van Montfort, H. Oud, & A. Satorra (Eds.), *Recent developments in structural equation modeling: Theory and applications.* (pp. 241–246). Amsterdam: Kluwer Academic Press.

Tucker, L. R. (1958). Determination of parameters of a functional relation by factor analysis. *Psychometrika, 23*, 19–23.

Vermunt, J. K. (2003). Multilevel latent class models. *Sociological Methodology*, *33*, 213–239.

Willett, J. B., & Sayer, A. G. (1994). Using covariance structure analysis to detect correlates and predictors of individual change over time. *Psychological Bulletin*, *116*, 363–381.

Willett, J. B., & Sayer, A. G. (1996). Cross-domain analyses of change over time: Combining growth modeling and covariance structure analysis. In G. A. Marcoulides & R. E. Schumacker (Eds.), *Advanced structural equation modeling: Issues and techniques* (pp. 125–157). Mahwah, NJ: Lawrence Erlbaum.

Wothke, W. (2000). Longitudinal and multi-group modeling with missing data. In T. D. Little, K. U. Schnabel, & J. Baumert (Eds.), *Modeling longitudinal and multilevel data* (pp. 219–240). Mahwah, NJ: Lawrence Erlbaum.

Section IV

Special Estimation Problems

7

Multilevel Analysis of Ordinal Outcomes Related to Survival Data

Donald Hedeker
Division of Epidemiology & Biostatistics
University of Illinois, Chicago, Illinois

Robin J. Mermelstein
Institute for Health Research and Policy,
University of Illinois, Chicago, Illinois

7.1 INTRODUCTION

Models for grouped-time survival data are useful for analysis of failure-time data when subjects are measured repeatedly at fixed intervals in terms of the occurrence of some event, or when determination of the exact time of the event is only known within grouped intervals of time. For example, in many school-based studies of substance use, students are typically measured annually regarding their smoking, alcohol, and other substance use during the past year. An important question is then to determine the degree to which covariates are related to substance use initiation. In these studies it is often of interest to model the student outcomes while controlling for the nesting of students in classrooms and/or schools. In analysis of such grouped-time initiation (or survival) data, use of grouped-time regression models that assume independence of observations (Allison, 1982; Prentice & Gloeckler, 1978; Thompson, 1977) is therefore problematic because of this clustering of students. More generally, this same issue arises for other types of clustered data sets in which subjects are observed nested within various types of clusters (e.g., hospitals, firms, clinics, counties), and thus cannot be assumed to be independent. To account for the data clustering, multilevel models (also called hierarchical linear or mixed models) provide a useful approach for simultaneously estimating the parameters of the regression model and the variance components that account for the data clustering (Goldstein, 1995; Raudenbush & Bryk, 2002).

For continuous-time survival data that are clustered, several authors (Clayton & Cuzick, 1985; Guo & Rodriquez, 1992; Lancaster, 1979; Paik, Tsai,

& Ottman, 1994; Self & Prentice, 1986; Shih & Louis, 1995; Vaupel, Manton, & Stallard, 1979) have developed mixed-effects survival models. These models are often termed frailty models or survival models including heterogeneity, and review articles describe many of these models (Hougaard, 1995; Pickles & Crouchley, 1995). An alternative approach for dealing with correlated data uses the generalized estimating equations (GEE) method to estimate model parameters. In this regard, Lee, Wei, and Amato (1992) and Wei, Lin, and Weissfeld (1989) have developed continuous-time survival models.

Application of these continuous-time models to grouped or discrete-time survival data is generally not recommended because of the large number of ties that result. Instead, models specifically developed for grouped or discrete-time survival data have been proposed. Both Han and Hausman (1990) and Scheike and Jensen (1997) have described proportional hazards models incorporating a log-gamma distribution specification of heterogeneity. Also, Ten Have (1996) developed a discrete-time proportional hazards survival model incorporating a log-gamma random effects distribution, additionally allowing for ordinal survival and failure categories. Ten Have and Uttal (1994) used Gibbs sampling to fit continuation ratio logit models with multiple normally distributed random effects. In terms of a GEE approach, Guo and Lin (1994) have developed a multivariate model for grouped-time survival data.

Several authors have noted the relationship between ordinal regression models (using complementary log-log and logit link functions) and survival analysis models for grouped and discrete time (Han & Hausman, 1990; McCullagh, 1980; Teachman, Call, & Carver, 1994). Similarly, others (Allison, 1982; D'Agostino et al., 1990; Singer & Willett, 1993) have described how dichotomous regression models can be used to model grouped and discrete time survival data. The ordinal approach simply treats survival time as an ordered outcome that is either right-censored or not. Alternatively, in the dichotomous approach each survival time is represented as a set of indicators of whether or not an individual failed in each time unit (until a person either experiences the event or is censored). As a result, the dichotomous approach is more useful for inclusion of time-dependent covariates and relaxing of the proportional hazards assumption.

Several authors have generalized these fixed-effects regression models for categorical responses to the multilevel setting (Barber, Murphy, Axinn, & Maples, 2000; Grilli, 2005; Hedeker, Siddiqui, & Hu, 2000; Muthén & Masyn, 2005; Rabe-Hesketh, Yang, & Pickles, 2001; Reardon, Brennan, & Buka, 2002; Scheike & Jensen, 1997; Ten Have, 1996; Ten Have & Uttal, 1994). The resulting models are generally based on dichotomous and ordinal mixed-effects regression models, albeit with the extension of the ordinal model to allow for right-censoring of the response. Typically, these models allow multiple random effects and a general form for model covariates. In many cases, proportional or partial proportional hazards or odds models are considered. In this chapter, this class of two-level models will be described where the random effects are assumed to be normally distributed. Because we assume the normal distribution for the random effects, standard software (e.g., SAS PROC NLMIXED) can be used to estimate these models, and therefore broaden the potential application of this approach. Syntax examples will be provided to facilitate this.

7.2 MULTILEVEL GROUPED-TIME SURVIVAL ANALYSIS MODEL

Let i denote the level-2 units ($i = 1, \ldots, N$) and let j denote the level-1 units ($j = 1, \ldots, n_i$). Note that this use of i for level-2 units and j for level-1 units is consistent with usage often found in statistics (Verbeke & Molenberghs, 2000) and biostatistics (Diggle, Heagerty, Liang, & Zeger, 2002), but is opposite from the typical usage in the multilevel literature (de Leeuw & Meijer, 2008; Goldstein, 1995). If subjects are nested within clusters, the subjects and clusters represent the level-1 and level-2 units, respectively. Alternatively, if there are multiple failure times per subject, then the level-2 units are the subjects and the level-1 units are the repeated failure times. Suppose that there is a continuous random variable for the uncensored time of event occurrence (which may not be observed), however assume that time (of assessment) can take on only discrete positive values $t = 1, 2, \ldots, m$. For each level-1 unit, observation continues until time t_{ij} at which point either an event occurs or the observation is censored, where censoring indicates being observed at t_{ij} but not at $t_{ij} + 1$. Define P_{ijt} to be the probability of failure, up to and including time interval t; that is,

$$P_{ijt} = \Pr[t_{ij} \leq t] \tag{7.1}$$

and so the probability of survival beyond time interval t is simply $1 - P_{ijt}$.

Because $1 - P_{ijt}$ represents the survivor function, McCullagh (1980) proposed the following grouped-time version of the continuous-time proportional hazards model:

$$\log[-\log(1-P_{ijt})] = \alpha_{0t} + \boldsymbol{x}'_{ij}\boldsymbol{\beta}. \tag{7.2}$$

This is the so-called complementary log-log function, which can be re-expressed in terms of the cumulative failure probability, $P_{ijt} = 1 - \exp(-\exp(\alpha_{0t} + \boldsymbol{x}'_{ij}\boldsymbol{\beta}))$. In this model, x_{ij} is a $p \times 1$ vector including covariates that vary either at level 1 or 2, however they do not vary with time (i.e., they do not vary across the ordered response categories). They may, however, represent the average of a variable across time or the value of the covariate at the time of the event.

Since the integrated hazard function equals $-\log(1 - P_{ijt})$, this model represents the covariate effects (β) on the log of the integrated hazard function. The covariate effects are identical to those in the grouped-time version of the proportional hazards model described by Prentice and Gloeckler (1978). As such, the β coefficients are also identical to the coefficients in the underlying continuous-time proportional hazards model. Furthermore, as noted by Allison (1982), the regression coefficients of the model are invariant to interval length. Augmenting the coefficients β, the intercept terms α_{0t} are a set of m constants that represent the logarithm of the integrated baseline hazard (i.e., when $\boldsymbol{x} = \boldsymbol{0}$). As such, these terms represent cut points on the integrated baseline hazard function; these parameters are often referred to as threshold parameters in descriptions of ordinal regression models. While the above model is the same as that described in McCullagh (1980), it is written so that the covariate effects are of the same sign as the Cox proportional hazards model. A positive coefficient for a regressor then reflects increasing hazard with greater values of the regressor.

Adding random effects to this model, we get

$$\log[-\log(1-P_{ijt})] = \alpha_{0t} + \boldsymbol{x}'_{ij}\boldsymbol{\beta} + \boldsymbol{w}'_{ij}\boldsymbol{\upsilon}_i, \tag{7.3}$$

or

$$P_{ijt} = 1 - \exp(-\exp(\alpha_{0t} + \boldsymbol{x}'_{ij}\boldsymbol{\beta} + \boldsymbol{w}'_{ij}\boldsymbol{\upsilon}_i)) \\ = 1 - \exp(-\exp z_{iit}), \tag{7.4}$$

where $\boldsymbol{\upsilon}_i$ is the $r \times 1$ vector of unknown random effects for the level-2 unit i, and $\boldsymbol{w}_{ij}$ is the design vector for the r random effects. The distribution of the r random effects $\boldsymbol{\upsilon}_i$ is assumed to be multivariate with mean vector $\mathbf{0}$ and covariance matrix $\boldsymbol{\Sigma}_\upsilon$. An important special case is when the distribution is assumed to be multivariate normal. For convenience, the random effects are often expressed in standardized form. Specifically, let $\boldsymbol{\upsilon} = \boldsymbol{S\theta}$, where $\boldsymbol{SS}' = \boldsymbol{\Sigma}_\upsilon$ is the Cholesky decomposition of $\boldsymbol{\Sigma}_\upsilon$. The model for z_{ijt} then is written as:

$$z_{ijt} = \alpha_{0t} + \boldsymbol{x}'_{ij}\boldsymbol{\beta} + \boldsymbol{w}'_{ij}\boldsymbol{S}\boldsymbol{\theta}_i. \tag{7.5}$$

As a result of the transformation, the Cholesky factor $\boldsymbol{S}$ is estimated instead of the covariance matrix $\boldsymbol{\Sigma}_\upsilon$. As the Cholesky factor is essentially the square-root of the covariance matrix, this allows more stable estimation of near-zero variance terms.

7.2.1 Proportional Odds Model

As applied to continuous-time (cross-sectional) survival data, the proportional odds model is described by Bennett (1983). For grouped-time, the multilevel proportional odds model is written in terms of the logit link function as

$$\log[P_{ijt}/(1 - P_{ijt})] = z_{ijt} \tag{7.6}$$

or alternatively as $P_{ijt} = 1/(1 + \exp(-z_{ijt}))$. The choice of which link function to use is not always clear-cut. Bennett (1983) noted that the proportional odds model is useful for survival data when the hazards of groups of subjects are thought to converge with time. This contrasts to the proportional hazards model where the hazard rates for separate groups of subjects are assumed proportional at all time points. However, this type of nonproportional hazards effect can often be accommodated in the complementary log-log link model by including interactions of covariates with the baseline hazard cut points (Collett, 1994). Also as Doksum and Gasko (1990) note, large amounts of high quality data are often necessary for link function selection to be relevant. Since these two link functions often provide similar fits, Ten Have (1996) suggests that the choice of which to use depends upon whether inference should be in terms of odds ratios or discrete hazard ratios. Similarly, McCullagh (1980) notes that link function choice should be based primarily on ease of interpretation.

7.2.2 Pooling of Repeated Observations and Nonproportional Hazards

Thus far, survival time has been represented as an ordered outcome t_{ij} that is designated as censored or not. An alternative approach for grouped-time survival data, described by several authors (Allison, 1982; D'Agostino et al., 1990; Singer & Willett, 1993, and others) treats each individual's survival time as a set of dichotomous observations indicating whether or not an individual failed in each time unit until a person either experiences the event or is censored. Thus, each survival time is represented as a $t_{ij} \times 1$ vector of zeros for censored individuals, while for individuals experiencing the event the last element of this $t_{ij} \times 1$ vector of zeros is changed to a one. These multiple person-time indicators are then treated as distinct observations in a dichotomous regression model. In the case of clustered data, a multilevel dichotomous regression model is used. This method has been called the pooling of repeated observations method by Cupples, D'Agostino, Anderson, and Kannel (1985) and is extensively described in Singer and Willett (2003). For multilevel models, Reardon et al. (2002) provide a useful illustration of this approach. The dichotomous treatment is particularly useful for handling time-dependent covariates and fitting nonproportional hazards models because the covariate values can change across each individual's t_{ij} time points.

For the dichotomous approach, define p_{ijt} to be the probability of failure in time interval t, conditional on survival prior to t:

$$p_{ijt} = \Pr[t_{ij} = t \mid t_{ij} \geq t]. \quad (7.7)$$

Similarly, $1 - p_{ijt}$ is the probability of survival beyond time interval t, conditional on survival prior to t. The proportional hazards model is then written as

$$\log[-\log(1 - p_{ijt})] = \alpha_{0t} + \boldsymbol{x}'_{ijt}\boldsymbol{\beta} + \boldsymbol{w}'_{ij}\boldsymbol{S}\boldsymbol{\theta}_i, \quad (7.8)$$

and the corresponding proportional odds model is

$$\log[p_{ijt}/(1 - p_{ijt})] = \alpha_{0t} + \boldsymbol{x}'_{ijt}\boldsymbol{\beta} + \boldsymbol{w}'_{ij}\boldsymbol{S}\boldsymbol{\theta}_i, \quad (7.9)$$

where now the covariates $\boldsymbol{x}$ can vary across time and so are denoted as $\boldsymbol{x}_{ijt}$. Augmenting the model intercept α_{01}, the remaining intercept terms α_{0t} ($t = 2, \ldots m$) are obtained by including as regressors $m - 1$ dummy codes representing deviations from the first time point. Because the covariate vector $\boldsymbol{x}$ now varies with t, this approach automatically allows for time-dependent covariates, and relaxing the proportional hazards assumption only involves including interactions of covariates with the $m - 1$ time-point dummy codes.

Under the complementary log-log link function, the two approaches characterized by (7.3) and (7.8) yield identical results for the parameters that do not depend on t, namely the regression coefficients of time-independent covariates and the Cholesky factor (Engel, 1993; Läärä & Matthews, 1985). For the logit link, similar, but not identical, results are obtained for these parameters. Comparing these two approaches, notice that for the ordinal approach each observation consists of only two pieces of data: the (ordinal) time of the event and whether it was censored or not. Alternatively, in the dichotomous approach each survival time is represented as a vector of dichotomous indicators, where the size of the vector depends upon the timing of the event or censoring. Thus, the ordinal approach can be easier to implement and offers savings in terms of the data set size, especially as the number of time points gets large, while the dichotomous approach is superior in its treatment of time-dependent covariates and relaxing of the proportional hazards or odds assumption.

7.2.3 Proportional Hazards/Odds Assumption

Relaxing the proportional hazards or odds assumption in the ordinal model is possible; Hedeker and Mermelstein (1998) have described a multilevel partial proportional odds model. For this, the model can be rewritten as:

$$z_{ijt} = \alpha_{0t} + (\boldsymbol{u}^*_{ij})'\boldsymbol{\alpha}^*_t + \boldsymbol{x}'_{ij}\boldsymbol{\beta} + \boldsymbol{w}'_{ij}\boldsymbol{S}\boldsymbol{\theta}_i, \quad (7.10)$$

or absorbing α_{0t} and $\boldsymbol{\alpha}_t^*$ into $\boldsymbol{\alpha}_t$,

$$z_{ijt} = \boldsymbol{u}_{ij}'\boldsymbol{\alpha}_t + \boldsymbol{x}_{ij}'\boldsymbol{\beta} + \boldsymbol{w}_{ij}'\boldsymbol{S}\boldsymbol{\theta}_i, \qquad (7.11)$$

where, $\boldsymbol{u}_{ij}$ is a $(l+1) \times 1$ vector containing (in addition to a 1 for α_{0t}) the values of observation ij on the set of l covariates for which interactions with the cut points of the integrated baseline hazard are desired. Here, $\boldsymbol{\alpha}_t$ is a $(l+1) \times 1$ vector of regression coefficients associated with the l variables (and the intercept) in $\boldsymbol{u}_{ij}$. Under the complementary log-log link this provides a partial proportional hazards model, while under the logistic link it would be a partial proportional odds model. Of course, all covariates could be in $\boldsymbol{u}$, and none in $\boldsymbol{x}$, to provide purely nonproportional hazards/odds models.

Tests of the proportional hazards or odds assumption for a set of covariates can then be performed by running and comparing models: (a) requiring proportional odds/hazards (i.e., covariates are in $\boldsymbol{x}$), and (b) relaxing proportional odds/hazards assumption (i.e., covariates are in $\boldsymbol{u}$). Comparing the model deviances from these two using a likelihood ratio test then provides a test of the proportional odds/hazards assumption for the set of covariates under consideration.

Note that because the dichotomous and ordinal approaches only yield identical results under the proportional hazards model (i.e., the complementary log-log link and covariates with effects that do not vary across time), differences in interpretation emerge for covariates allowed to have varying effects across time under these two approaches. For covariates of this type, the dichotomous approach is generally preferred because it models the covariate's influence in terms of the conditional probability of failure given prior survival (i.e., the hazard function), rather than the cumulative probability of failure (i.e., the integrated or cumulative hazard function) as in the ordinal representation of the model.

7.3 MAXIMUM LIKELIHOOD ESTIMATION

For the dichotomous approach, standard methods and software for multilevel analysis of dichotomous outcomes can be used. This is well-described in Barber et al. (2000). For the ordinal treatment of survival times, the solution must be extended to accommodate right-censoring of the ordinal outcome. For this, let $\delta_{ij} = 0$ if level-1 unit ij is a censored observation and equal to 1 if the event occurs (fails). Thus, t_{ij} denotes the value of time ($t = 1, \ldots, m$) when either the ijth unit failed or was censored. It is assumed that the censoring and failure mechanisms are independent. In the multilevel model the probability of a failure at time t for a given level-2 unit i, conditional on $\boldsymbol{\theta}$ (and given $\boldsymbol{\alpha}_t$, $\boldsymbol{\beta}$, and $\boldsymbol{S}$) is:

$$\Pr(t_{ij} = t \cap \delta_j = 1 \mid \boldsymbol{\theta}; \boldsymbol{\alpha}_t, \boldsymbol{\beta}, \boldsymbol{S}) = P_{ijt} - P_{ij,t-1} \qquad (7.12)$$

where $P_{ij0} = 0$ and $P_{ij,m+1} = 1$. The corresponding probability of being right censored at time t equals the cumulative probability of not failing at that time, $1 - P_{ijt}$.

Let $\boldsymbol{t}_i$ denote the vector pattern of failure times from level-2 unit i for the n_i level-1 units nested within. Similarly, let δ_i denote the vector pattern of event indicators. The joint probability of patterns $\boldsymbol{t}_i$ and δ_i, given $\boldsymbol{\theta}$, assuming independent censoring is equal to the product of the probabilities of the level-1 responses:

$$\ell(\mathbf{t}_i,\boldsymbol{\delta}_i \mid \boldsymbol{\theta};\boldsymbol{\alpha}_t,\boldsymbol{\beta},S)=\prod_{j=1}^{n_i}\prod_{t=1}^{m}\left[(P_{ijt}-P_{ij,t-1})^{\delta_{ij}} \times (1-P_{ijt})^{1-\delta_{ij}}\right]^{d_{ijt}} \tag{7.13}$$

where $d_{ijt} = 1$ if $t_{ijt} = t$ (and $= 0$ if $t_{it} \neq t$).

The marginal density of $\mathbf{t}_i$ and δ_i in the population is expressed as the following integral of the conditional likelihood, $\ell(\cdot)$, weighted by the prior density $g(\cdot)$:

$$h(\mathbf{t}_i,\boldsymbol{\delta}_i)=\int_{\theta}\ell(\mathbf{t}_i,\boldsymbol{\delta}_i \mid \boldsymbol{\theta};\boldsymbol{\alpha}_t,\boldsymbol{\beta},S)\,g(\boldsymbol{\theta})\,d\boldsymbol{\theta} \tag{7.14}$$

where $g(\boldsymbol{\theta})$ represents the multivariate distribution of the standardized random effects vector $\boldsymbol{\theta}$ in the population. The marginal log-likelihood for the patterns from the N level-2 units is then written as $\log L = \sum_i^N \log h(\mathbf{t}_i,\boldsymbol{\delta}_i)$. Maximizing this log-likelihood yields maximum likelihood (ML) estimates, which are sometimes referred to as maximum marginal likelihood estimates (Bock, 1989) because integrating the joint likelihood of random effects and responses over the distribution of random effects translates to marginalization of the data distribution. Specific details of the solution, utilizing numerical quadrature to integrate over the random effects distribution, are provided in Hedeker et al. (2000). As mentioned, SAS PROC NLMIXED can be used to obtain the ML estimates; several syntax examples are provided in the Appendix.

7.4 EXAMPLES

Three examples will be presented to illustrate the ordinal representation of the grouped-time survival analysis multilevel model. The first two examples are from school-based studies and treat students nested within schools. In the first example, there is right-censoring only at the last time point; these censored observations then form an additional category of the ordered time to event outcome. Thus, the model is akin to an ordinary multilevel ordinal regression using a complementary log-log link function to yield a proportional hazards model. The second example has intermittent right-censoring (i.e., observations can be right-censored at any time point) and so the likelihood function must be adapted for the censored observations, as described in the section on Estimation. This is perhaps the more usual situation in survival or time-to-event data. The final example is a joint longitudinal and survival model that allows the two processes to be correlated. In this example, time until study dropout is the survival outcome that is related to the longitudinal outcomes via the random effects of the latter.

7.4.1 Example 1: EMA Study

The data for this example are drawn from a natural history or Ecological Momentary Assessment (EMA; Smyth & Stone, 2003; Stone & Shiffman, 1994) study of adolescent smoking. Participants included in this study were in either 9th or 10th grade at baseline, and reported on a screening questionnaire 6–8 weeks prior to baseline that they had smoked at least one cigarette in their lifetimes. The majority (57.6%) had smoked at least one cigarette in the past month at baseline. A total of 461 students completed the baseline measurement wave. Baseline measurement was coordinated in

TABLE 7.1

Onset of Smoking Event Across Days Frequencies (and Percentages), $N = 461$

Sunday	Monday	Tuesday	Wednesday	Thursday	Friday	Saturday	Never
68	56	39	21	18	24	8	227
(14.75)	(12.15)	(8.46)	(4.56)	(3.90)	(5.21)	(1.74)	(49.24)

the schools of these students. In all, there were 16 schools in this study.

The study utilized a multi-method approach to assess adolescents including a week-long time/event sampling method via hand-held computers (EMA). Adolescents carried the hand-held computers with them at all times during a data collection period of seven consecutive days and were trained both to respond to random prompts from the computers and to event record (initiate a data collection interview) in conjunction with smoking episodes. Random prompts and the self-initiated smoking records were mutually exclusive; no smoking occurred during random prompts. Questions concerned place, activity, companionship, mood, and other subjective variables. The hand-held computers date and time-stamped each entry.

In a previous paper based on an earlier EMA data set (Hedeker, Mermelstein, & Flay, 2006), our group observed that adolescent smoking reports were most commonly observed on Fridays and Saturdays (i.e., weekend smoking), but that mid-week smoking was more informative in determining the level of an adolescent's smoking behavior. Here, our interest is in modeling time until the first smoking report following the weekend.[1] The idea being that earlier post-weekend smoking is a potentially greater indicator of dependency among adolescents. Thus, for each subject, we recorded the first day in which a smoking report was made, ordering the days as Sunday, Monday, …, Friday, Saturday (this variable will be denoted as Smk). Table 7.1 lists the frequencies of responses across these seven days, and also the number of students who did not provide a smoking report during the week. Smk is thus an ordinal outcome with eight response categories.

In terms of predictor variables, for simplicity, we only considered a subject's level of social isolation (denoted as SocIso), which has previously been shown to be related to smoking in adolescents (Ennett & Bauman, 1993; Johnson & Hoffmann, 2000). This variable was based on responses from the random prompts and consisted of a subject's average, across all prompts, on several individual mood items, each rated from 1 (not at all) to 10 (very much), which were identified via factor analysis. Specifically, SocIso consisted of the average of the following items that reflected a subject's assessment of their social isolation before the prompt signal: I felt lonely, I felt left out, and I felt ignored. Over all prompts, and ignoring the clustering of the data, the marginal mean of SocIso was 2.709 (sd = 1.329) reflecting a relatively low level of social isolation on average.

A proportional hazards model was fit to these data using Smk as the time to event variable and SocIso as the independent variable. First, a model was run that ignored the clustering of students in schools. The effect

[1] Here, we consider Friday and Saturday to be the weekend days, and Sunday to be the first post-weekend day. This, of course, is technically incorrect, but gets at the notion that for adolescents most social events involving peers occur on Friday and Saturday.

of social isolation was observed to be significant in this analysis ($\hat{\beta} = .1085$, se = .04650, $p < .02$). Thus, students with higher average social isolation scores had increased hazards for earlier post-weekend smoking. However, when the clustering of students in schools was taken into account, by including a random school effect in the model, the effect was no longer significant at the .05 level ($\hat{\beta} = .0986$, se = .04813, $p < .059$). While not a dramatic change between these two, nonetheless, the conclusion based on the multilevel model would be that the effect of social isolation was only marginally significant.

The random effect variation, expressed as a standard deviation ($\hat{\sigma}_\upsilon$), was estimated to be .2926 (se = .09882). This estimated school variability can be expressed as an intraclass correlation, $\hat{\sigma}_\upsilon^2 / (\hat{\sigma}_\upsilon^2 + \sigma^2)$, where σ^2 represents the variance of the latent continuous event time variable. For the complementary log-log link, the standard variance $\sigma^2 = \pi^2/6$, while for the logit link, $\sigma^2 = \pi^2/3$ (Agresti, 2002). Applying this formula, the estimated intraclass correlation equals .04947 under the proportional hazards model (i.e., complementary log-log link). This value is certainly in the range reported by Siddiqui, Hedeker, Flay, and Hu (1996), who examined school-based ICCs for a number of similar outcomes, and suggests a fair degree of similarity in terms of time to smoking within schools.

To test the proportional hazards assumption, a model was also fit allowing the effect of SocIso to vary across the ordinal Smk outcome. With eight categories of Smk, seven coefficients for SocIso were estimated, one for each cumulative comparison of these categories. The deviance for this extended model was 1450.8, while the deviance for the above proportional hazards model was 1453.2. Based on these values, the likelihood ratio chi-square statistic is 2.4, which on six degrees of freedom is not significant. Thus, the proportional hazards assumption is reasonable. Appendix 1 provides the SAS PROC NLMIXED syntax for both of these analyses.

7.4.2 Example 2: TVSFP Study

This example is taken from the Hedeker et al. (2000) article that used MIXOR (Hedeker & Gibbons, 1996) to estimate model parameters. Here, we replicate the results using SAS NLMIXED code. Relative to the previous example, this one will have right-censored observations at all time points, so the program code must reflect this, as noted just below Equation 7.12. Namely, the probability of being right censored at time t equals the cumulative probability of not failing at that time, $1 - P_{ijt}$.

The Television School and Family Smoking Prevention and Cessation Project (TVSFP) study (Flay et al., 1988) was designed to test independent and combined effects of a school-based social-resistance curriculum and a television-based program in terms of tobacco use prevention and cessation. The sample consisted of seventh-grade students who were assessed at pretest (Wave A), immediate post-intervention (Wave B), 1-year follow-up (Wave C), and 2-year follow-up (Wave D). A cluster randomization design was used to assign schools to the design conditions, while the primary outcome variables were at the student level. Schools were randomized to one of four study conditions: (a) a social-resistance classroom curriculum (CC); (b) a media (television) intervention (TV); (c) a combination of CC and TV conditions; and (d) a no-treatment control group.

An outcome of interest from the study is the onset of cigarette experimentation. At each of the four time points, students answered the question: "Have you ever smoked a cigarette?" In analyzing the data below, because the intervention was implemented following the pretest, we focused on the three post-intervention time points and included only those students who had not answered yes to this question at pretest. In all, there were 1556 students included in the analysis of smoking initiation. Of these students, approximately 40% ($n = 634$) answered yes to the smoking question at one of the three post-intervention time points, while the other 60% ($n = 922$) either answered no at the last time point or were censored prior to the last time point. The breakdown of cigarette onset for gender and condition subgroups is provided in Table 7.2. In terms of the clustering, these 1,556 students were from 28 schools with between 13 to 151 students per school ($\bar{n} = 56$, $sd = 38$).

A proportional hazards model was fit to these data to examine the effects of gender and intervention group on time to smoking initiation. Specifically, gender was included as a dummy variable expressing the male versus female difference. For the condition terms, because the CC by TV interaction was observed to be nonsignificant, a main effects model was considered. The maximum likelihood estimates (standard errors) were: Male = 0.05736 (0.07983), CC = 0.04461 (0.08418), and TV = 0.02141 (0.08311). Thus, none of the regressors are close to being significant, though the direction of the effects are increased hazards of smoking initiation for males and those exposed to the CC and TV interventions. The random effect variation, expressed as a standard deviation ($\hat{\sigma}_\upsilon$), was estimated to be 0.05119 (0.1242). So the clustering effect attributable to schools is not large, and is much smaller than the previous example. Expressed as in intraclass correlation, it equals .00159, which reflects a rather low

TABLE 7.2

Onset of Cigarette Experimentation Across Three Waves Frequencies (and Percentages) for Gender and Condition Subgroups

	Wave B			Wave C			Wave D		
	Event	Censored	Total	Event	Censored	Total	Event	Censored	Total
Males	156	83	742	89	134	503	63	217	280
	(21.0)	(11.2)		(17.7)	(26.6)		(22.5)	(77.5)	
Females	130	105	814	117	154	579	79	229	308
	(16.0)	(12.9)		(20.2)	(26.6)		(25.6)	(74.4)	
Control	66	60	401	53	69	275	34	119	153
	(16.5)	(15.0)		(19.3)	(25.1)		(22.2)	(77.8)	
CC only	75	27	392	53	61	290	49	127	176
	(19.1)	(6.9)		(18.3)	(21.0)		(27.8)	(72.2)	
TV only	71	54	410	60	79	285	38	108	146
	(17.3)	(13.2)		(21.1)	(27.7)		(26.0)	(74.0)	
CC and TV	74	47	353	40	79	232	21	92	113
	(21.0)	(13.3)		(17.2)	(34.1)		(18.6)	(81.4)	

degree of similarity in smoking initiation times within schools. The SAS PROC NLMIXED syntax for this example is listed in Appendix 2.

7.4.3 Example 3: Joint Longitudinal and Survival Model

The data for this example come from the National Institute of Mental Health Schizophrenia Collaborative Study. In terms of the longitudinal outcome, we will examine Item 79 of the Inpatient Multidimensional Psychiatric Scale (IMPS; Lorr & Klett, 1996). Item 79, "Severity of Illness," was originally scored on a 7-point scale ranging from *normal, not at all ill* (0) to *among the most extremely ill* (7). Here, as in Hedeker and Gibbons (1994), we will analyze the outcome as an ordinal variable, specifically recoding the original seven ordered categories of the IMPS 79 severity score into four: (1) normal or borderline mentally ill, (2) mildly or moderately ill, (3) markedly ill, and (4) severely or among the most extremely ill.

In this study, patients were randomly assigned to receive one of four medications: placebo, chlorpromazine, fluphenazine, or thioridazine. Since our previous analyses revealed similar effects for the three antipsychotic drug groups, they were combined in the present analysis. The experimental design and corresponding sample sizes are presented in Table 7.3. In this study, the protocol called for subjects to be measured at weeks 0, 1, 3, and 6; however, a few subjects were additionally measured at weeks 2, 4, and 5. There was some intermittent missingness in this study, however, dropout was a much more common pattern of missingness. In all, 102 of 437 subjects did not complete the trial.

The main question of interest is addressing whether there is differential change across time for the drug groups, here combined, relative to the control group. We have previously addressed this question by analyzing these data using multilevel ordinal regression using both a probit (Hedeker & Gibbons, 1994) and logistic link (Hedeker & Gibbons, 2006). A potential issue with these previous analyses is that the assumption of missing at random (MAR), inherent in the full maximum likelihood estimation of model parameters of the ordinal multilevel models, may not be plausible. That is, it could be that there was an association between missingness and the value of the dependent variable (i.e., severity of illness) that would have been measured. Though the observed data cannot confirm or refute this possibility, one can fit missing not at random (MNAR) models as a means of doing a sensitivity analysis.

One class of MNAR models augment the usual multilevel model for longitudinal data with a model of dropout (or missingness), in which dropout depends on the random effects of the multilevel model. These

TABLE 7.3

Experimental Design and Weekly Sample Sizes Across Time

	Sample Size at Week							
Group	**0**	**1**	**2**	**3**	**4**	**5**	**6**	**Total**
Placebo	107	105	5	87	2	2	70	108
Drug	327	321	9	287	9	7	265	329

Note: Drug = Chlorpromazine, fluphenazine, or thioridazine.

models have been called random-coefficient selection models (Little, 1995), random-effects-dependent models (Hogan & Laird, 1997), and shared parameter models (De Gruttola & Tu, 1994; Schluchter, 1992; Ten Have, Kunselman, Pulkstenis, & Landis, 1998; Wu & Bailey, 1989; Wu & Carroll, 1988) in the literature. Appealing aspects of this class of models is that they can be used for nonignorable missingness and can be fit using some standard software. Guo and Carlin (2004) present an excellent description for longitudinal outcomes and continuous time until event data, and provide a Web page with `PROC NLMIXED` syntax to carry out the analysis. Here, we will modify their approach and syntax for a longitudinal ordinal outcome and a grouped-time survival event (i.e., time until study dropout).

In terms of the ordinal severity of illness outcome (denoted `imps79`), as in Hedeker and Gibbons (2006), we will use a multilevel ordinal logistic regression model including random subject intercepts and time trends. Here, if i represents subjects, j time points, and c ordinal response categories, the longitudinal outcome model is

$$\log\left[\frac{P_{ijc}}{1-P_{ijc}}\right]=\gamma_c-[\beta_0+\beta_1\sqrt{W_j}+\beta_2\mathrm{Tx}_i+\beta_3(\mathrm{Tx}_i\times\sqrt{W_j})+\upsilon_{0i}+\upsilon_{1i}\sqrt{W_j}\,], \tag{7.15}$$

where, Tx denotes treatment group (0 = placebo and 1 = drug) and W denotes week (the square root of week is used to linearize the relationship between the cumulative logits and week). This is a cumulative logit model where $P_{ijc} = \Pr(\texttt{imps79}_{ij} \le c)$, and the threshold parameters γ_c represent the cumulative logit values when the covariates and random subject effects equal zero. As parameterized above, a positive regression coefficient β indicates that as the regressor increases so does the probability of a higher value of imps79.

For time to dropout, let the variable $D_i = j$ if subject i drops out after the jth time point; namely, $\texttt{imps}_{ij}$ is observed, but $\texttt{imps}_{i,j+1}, \ldots, \texttt{imps}_{i,n}$ are all missing (here n represents the last possible time point, week 6). Note that we are ignoring intermittent missingness and focusing on time until a subject is no longer measured. Because there were no subjects who were only measured at week 0 in this study, D_i will take on values of 1 to 5 for subjects dropping out prior to the end of the study, and a value of 6 for subjects completing the study (i.e., was measured at week 6). Table 7.4 lists the frequencies of time to dropout (D) by treatment group.

For analysis of time to dropout, consider the following proportional hazards survival model (i.e., ordinal regression model with complementary log-log link):

$$\log(-\log(1-P(D_i\le j)))=\alpha_{0j}+\alpha_1\mathrm{Tx}_i+\alpha_2(1-\mathrm{Tx}_i)\upsilon_{0i}+\alpha_3(1-\mathrm{Tx}_i)\upsilon_{1i}+\alpha_4\mathrm{Tx}_i\upsilon_{0i}+\alpha_5\mathrm{Tx}_i\upsilon_{1i}. \tag{7.16}$$

This model specifies that the time of dropout is influenced by a person's treatment group (α_1) and their intercept and trend in `imps79`. For the latter, α_2 and α_3 represent the effect of these on dropout among the placebo group (i.e., when $\mathrm{Tx}_i = 0$), whereas

TABLE 7.4

Crosstabulation of Treatment Group by Time to Dropout—Frequencies (Percentages)

	Time to Dropout						
Treatment Group	**1**	**2**	**3**	**4**	**5**	**6**	**Total**
Placebo	13	5	16	2	2	70	108
	(.12)	(.05)	(.15)	(.02)	(.02)	(.65)	
Drug	24	5	26	3	6	265	329
	(.07)	(.02)	(.08)	(.01)	(.02)	(.81)	

TABLE 7.5

Separate and Shared Parameter Models

	Separate			Shared		
Parameter	**Estimate**	**SE**	***P*** <	**Estimate**	**SE**	***P*** <
Outcome						
Tx β_1	.048	.392	.90	.149	.382	.70
$\sqrt{W}\ \beta_2$	−.887	.218	.0001	−.708	.221	.002
Tx × $\sqrt{W}\ \beta_3$	−1.692	.252	.0001	−1.909	.257	.0001
Dropout						
Tx α_1	−.693	.205	.0008	−.719	.281	.02
Placebo random intercept α_2				.242	.094	.02
Placebo random slope α_3				.570	.289	.05
Drug random intercept α_4				−.150	.071	.04
Drug random slope α_5				−.553	.177	.002
Deviance		4056.7			4038.5	

α_4 and α_5 are the analogous effects for the drug group (i.e., when $\text{Tx}_i = 1$).

Table 7.5 lists the regression coefficient estimates from separate and shared parameter modeling of these data. The separate parameter model sets $\alpha_2 = \alpha_3 = \alpha_4 = \alpha_5 = 0$. The separate parameter model yields identical parameter estimates and standard errors as running these two models, one for the longitudinal outcome and one for time to dropout, separately (not shown). Thus, the results for the longitudinal component represent the usual ordinal model with random subject intercepts and time-trends assuming MAR. As can be seen, both the $\sqrt{W}$ and Tx × $\sqrt{W}$ terms are significant. This indicates that, respectively, the placebo group is improving across time and that the drug group is improving at an even faster rate across time. Additionally, because the Tx term is not significant, these groups are not different when week = 0 (i.e., at baseline). The dropout component indicates that the drug group has a significantly diminished hazard. Exponentiating the estimate of −.693 yields approximately .5, indicating that the hazard for dropout is double in the placebo group, relative to the drug group.

The shared parameter model fits these data significantly better, as evidenced by the likelihood ratio test, $X_4^2 = 18.2, p < .002$. This is not necessarily a rejection of MAR, but it is a rejection of this particular MAR model in favor of this particular MNAR shared parameter model. In terms of the longitudinal component, we see that the conclusions are the same as in the MAR model. If anything, the results are slightly stronger for the drug group in that the drug by time interaction is somewhat larger in the MNAR shared parameter model. In terms of the dropout component, all of the terms are significant. The significant Tx effect indicates that the drug group has a significantly diminished hazard of dropping out. For the terms involving the random effects, higher (i.e., more positive) intercepts and slopes are associated with dropout for the placebo group, whereas for the drug group it is lower (i.e., more negative) intercepts and slopes that are associated with dropout. In other words, among placebo subjects, those who start off worse (i.e., higher severity scores) and who are not improving, or improving at a slower rate, are more likely to drop out. Conversely, for the drug subjects, those who start off relatively better and who have more negative slopes (i.e., greater improvement) are more likely to drop out. `SAS PROC NLMIXED` code for the shared parameter model is provided in Appendix 3.

7.5 DISCUSSION

Multilevel categorical regression models have been described for analysis of clustered grouped-time survival data, using either a proportional or partial proportional hazards or odds assumption. For models without time-dependent covariates, and assuming proportional hazards or odds, the data are analyzed utilizing an ordinal mixed-effects regression model. In this approach, survival times are represented as ordinal outcomes that are right-censored or not. Alternatively, in the dichotomous representation of the model, survival times are represented as sets of binary indicators of survival and analyzed using multilevel methods for dichotomous outcomes. In this chapter we have focused on the ordinal representation of the model; for extensive information on the dichotomous version see Barber et al. (2000) and Singer and Willett (2003).

Three examples were presented to illustrate the flexibility of this approach. The first did not have any intermittent right-censoring, and so the model was akin to a standard ordinal multilevel model using a complementary log-log link. Intermittent right-censoring was present for the second example, and the resulting likelihood function was modified to account for this. The final example illustrated how the ordinal representation of the survival analysis model can readily be used in longitudinal trials where there is interest in jointly modeling the longitudinal process and time to study dropout. For all examples, `SAS PROC NLMIXED` code was provided to allow data analysts to apply these methods.

ACKNOWLEDGMENTS

Thanks are due to Siu Chi Wong for assisting with data analysis. This work was supported by National Cancer Institute grant 5PO1 CA98262.

REFERENCES

Agresti, A. (2002). *Categorical data analysis* (2nd ed.). New York, NY: Wiley.

Allison, P. D. (1982). Discrete-time methods for the analysis of event histories. In S. Leinhardt (Ed.), *Sociological methodology 1982* (pp. 61–98). San Francisco, CA: Jossey-Bass.

Barber, J. S., Murphy, S. A., Axinn, W. G., & Maples, J. (2000). Discrete-time multilevel hazard analysis. *Sociological Methodology, 30,* 201–235.

Bennett, S. (1983). Analysis of survival data by the proportional odds model. *Statistics in Medicine, 2,* 273–277.

Bock, R. D. (1989). Measurement of human variation: A two stage model. In R. D. Bock (Ed.), *Multilevel analysis of educational data.* New York, NY: Academic Press.

Clayton, D., & Cuzick, J. (1985). Multivariate generalizations of the proportional hazards model (with discussion). *Journal of the Royal Statistical Society, Series A, 148,* 82–117.

Collett, D. (1994). *Modelling survival data in medical research.* New York, NY: Chapman and Hall.

Cupples, L. A., D'Agostino, R. B., Anderson, K., & Kannel, W. B. (1985). Comparison of baseline and repeated measure covariate techniques in the Framingham heart study. *Statistics in Medicine, 7,* 205–218.

D'Agostino, R. B., Lee, M.-L., Belanger, A. J., Cupples, L. A., Anderson, K., & Kannel, W. B. (1990). Relation of pooled logistic regression to time dependent Cox regression analysis: The Framingham heart study. *Statistics in Medicine, 9,* 1501–1515.

De Gruttola, V., & Tu, X. M. (1994). Modelling progression of CD4-lymphocyte count and its relationship to survival time. *Biometrics, 50,* 1003–1014.

de Leeuw, J., & Meijer, E. (Eds.). (2008). *Handbook of multilevel analysis.* New York, NY: Springer.

Diggle, P. J., Heagerty, P., Liang, K.-Y., & Zeger, S. L. (2002). *Analysis of longitudinal data* (2nd ed.). New York, NY: Oxford University Press.

Doksum, K. A., & Gasko, M. (1990). On a correspondence between models in binary regression analysis and in survival analysis. *International Statistical Review, 58,* 243–252.

Engel, J. (1993). On the analysis of grouped extreme-value data with GLIM. *Applied Statistics, 42,* 633–640.

Ennett, S. T., & Bauman, K. E. (1993). Peer group structure and adolescent cigarette smoking: A social network analysis. *Journal of Health and Social Behavior, 34,* 226–236.

Flay, B. R., Brannon, B. R., Johnson, C. A., Hansen, W. B., Ulene, A., Whitney-Saltiel, D. A. … Spiegel, D. C. (1988). The television, school and family smoking prevention/cessation project: I. Theoretical basis and program development. *Preventive Medicine, 17*(5), 585–607.

Goldstein, H. (1995). *Multilevel statistical models* (2nd ed.). New York, NY: Halstead Press.

Grilli, L. (2005). The random-effects proportional hazards model with grouped survival data: A comparison between the grouped continuous and continuation ratio versions. *Journal of the Royal Statistical Society: Series A, 168,* 8394.

Guo, G., & Rodriquez, G. (1992). Estimating a multivariate proportional hazards model for clustered data using the em algorithm with an application to child survival in Guatemala. *Journal of the American Statistical Association, 87,* 969–976.

Guo, S. W., & Lin, D. Y. (1994). Regression analysis of multivariate grouped survival data. *Biometrics, 50,* 632–639.

Guo, X., & Carlin, B. P. (2004). Separate and joint modeling of longitudinal and event time data using standard computer packages. *The American Statistician, 58,* 16–24.

Han, A., & Hausman, J. A. (1990). Flexible parametric estimation of duration and competing risk models. *Journal of Applied Econometrics, 5,* 1–28.

Hedeker, D., & Gibbons, R. D. (1994). A random effects ordinal regression model for multilevel analysis. *Biometrics, 50,* 933–944.

Hedeker, D., & Gibbons, R. D. (1996). MIXOR: A computer program for mixed-effects ordinal probit and logistic regression analysis. *Computer Methods and Programs in Biomedicine, 49,* 157–176.

Hedeker, D., & Gibbons, R. D. (2006). *Longitudinal data analysis.* New York, NY: Wiley.

Hedeker, D., & Mermelstein, R. J. (1998). A multilevel thresholds of change model for analysis of stages of change data. *Multivariate Behavioral Research, 33,* 427–455.

Hedeker, D., Mermelstein, R. J., & Flay, B. R. (2006). Application of item response theory models for intensive longitudinal data. In T. A. Walls & J. F. Schafer (Eds.), *Models for intensive longitudinal data* (pp. 84–108). New York, NY: Oxford.

Hedeker, D., Siddiqui, O., & Hu, F. B. (2000). Random-effects regression analysis of correlated grouped-time survival data. *Statistical Methods in Medical Research, 9,* 161–179.

Hogan, J. W., & Laird, N. M. (1997). Model-based approaches to analysing incomplete longitudinal and failure time data. *Statistics in Medicine, 16,* 259–272.

Hougaard, P. (1995). Frailty models for survival data. *Lifetime Data Analysis, 1,* 255–273.

Johnson, R. A., & Hoffmann, J. P. (2000). Adolescent cigarette smoking in U.S. racial/ethnic subgroups: Findings from the national education longitudinal study. *Journal of Health and Social Behavior, 41,* 392–407.

Läärä, E., & Matthews, J. N. S. (1985). The equivalence of two models for ordinal data. *Biometrika, 72,* 206–207.

Lancaster, T. (1979). Econometric methods for the duration of unemployment. *Econometrica, 47,* 939–956.

Lee, E. W., Wei, L. J., & Amato, D. A. (1992). Cox-type regression analysis for large numbers of small groups of correlated failure time observations. In J. P. Klein & P. K. Goel (Eds.), *Survival analysis: State of the art* (pp. 237–247). Dordrecht, The Netherlands: Kluwer.

Little, R. J. A. (1995). Modeling the drop-out mechanism in repeated-measures studies. *Journal of the American Statistical Association, 90,* 1112–1121.

Lorr, M., & Klett, C. J. (1966). *Inpatient multidimensional psychiatric scale: Manual.* Palo Alto, CA: Consulting Psychologists Press.

McCullagh, P. (1980). Regression models for ordinal data (with discussion). *Journal of the Royal Statistical Society, Series B, 42,* 109–142.

Muthén, B., & Masyn, K. (2005). Discrete-time survival mixture analysis. *Journal of Educational and Behavioral Statistics, 30,* 27–58.

Paik, M. C., Tsai, W.-Y., & Ottman, R. (1994). Multivariate survival analysis using piecewise gamma frailty. *Biometrics, 50,* 975–988.

Pickles, A., & Crouchley, R. (1995). A comparison of frailty models for multivariate survival data. *Statistics in Medicine, 14,* 1447–1461.

Prentice, R. L., & Gloeckler, L. A. (1978). Regression analysis of grouped survival data with application to breast cancer data. *Biometrics, 34,* 57–67.

Rabe-Hesketh, S., Yang, S., & Pickles, A. (2001). Multilevel models for censored and latent responses. *Statistical Methods in Medical Research, 10,* 409–427.

Raudenbush, S. W., & Bryk, A. S. (2002). *Hierarchical linear models* (2nd ed.). Thousand Oaks, CA: Sage.

Reardon, S. F., Brennan, R., & Buka, S. L. (2002). Estimating multi-level discrete-time hazard models using cross-sectional data: Neighborhood effects on the onset of adolescent cigarette use. *Multivariate Behavioral Research, 37,* 297–330.

Scheike, T. H., & Jensen, T. K. (1997). A discrete survival model with random effects: An application to time to pregnancy. *Biometrics, 53,* 318–329.

Schluchter, M. D. (1992). Methods for the analysis of informatively censored longitudinal data. *Statistics in Medicine, 11,* 1861–1870.

Self, S. G., & Prentice, R. L. (1986). Incorporating random effects into multivariate relative risk regression models. In S. H. Moolgavkar & R. L. Prentice (Eds.), *Modern statistical methods in chronic disease epidemiology* (pp. 167–178). New York, NY: Wiley.

Shih, J. H., & Louis, T. A. (1995). Assessing gamma frailty models for clustered failure time data. *Lifetime Data Analysis, 1,* 205–220.

Siddiqui, O., Hedeker, D., Flay, B. R., & Hu, F. B. (1996). Intraclass correlation estimates in a school-based smoking prevention study: Outcome and mediating variables, by gender and ethnicity. *American Journal of Epidemiology, 144,* 425–433.

Singer, J. D., & Willett, J. B. (1993). It's about time: Using discrete-time survival analysis to study duration and the timing of events. *Journal of Educational and Behavioral Statistics, 18,* 155–195.

Singer, J. D., & Willett, J. B. (2003). *Applied longitudinal data analysis.* New York, NY: Oxford University Press.

Smyth, J. M., & Stone, A. A. (2003). Ecological momentary assessment research in behavioral medicine. *Journal of Happiness Studies, 4,* 35–52.

Stone, A., & Shiffman, S. (1994). Ecological momentary assessment (EMA) in behavioral medicine. *Annals of Behavioral Medicine, 16,* 199–202.

Teachman, J. D., Call, V. R. A., & Carver, K. P. (1994). Marital status and the duration of joblessness among white men. *Journal of Marriage and the Family, 56,* 415–428.

Ten Have, T. R. (1996). A mixed effects model for multivariate ordinal response data including correlated discrete failure times with ordinal responses. *Biometrics, 52,* 473–491.

Ten Have, T. R., Kunselman, A. R., Pulkstenis, E. P., & Landis, J. R. (1998). Mixed effects logistic regression models for longitudinal binary response data with informative drop-out. *Biometrics, 54,* 367–383.

Ten Have, T. R., & Uttal, D. H. (1994). Subject-specific and population-averaged continuation ratio logit models for multiple discrete time survival profiles. *Applied Statistics, 43,* 371–384.

Thompson, W. A. (1977). On the treatment of grouped observations in life studies. *Biometrics, 33,* 463–470.

Vaupel, J. W., Manton, K. G., & Stallard, E. (1979). The impact of heterogeneity in individual frailty on the dynamics of mortality. *Demography, 16,* 439–454.

Verbeke, G., & Molenberghs, G. (2000). *Linear mixed models for longitudinal data.* New York, NY: Springer.

Wei, L. J., Lin, D. Y., & Weissfeld, L. (1989). Regression analysis of multivariate incomplete failure time data by modeling marginal distributions. *Journal of the American Statistical Association, 84,* 1065–1073.
Wu, M. C., & Bailey, K. R. (1989). Estimation and comparison of changes in the presence of informative right censoring: Conditional linear model. *Biometrics, 45,* 939–955.
Wu, M. C., & Carroll, R. J. (1988). Estimation and comparison of changes in the presence of right censoring by modeling the censoring process. *Biometrics, 44,* 175–188.

APPENDIX 1

In this listing for Example 1, and in subsequent syntax listings, expressions with all uppercase letters are used for SAS-specific syntax, while expressions including lowercase letters are used for user-defined entities. In this example, `SocIso` is the regressor and `bSocIso` is its regression coefficient. Note that `PROC NLMIXED` requires the user to name all model parameters in the syntax. The variable Smk indicates the day of the first smoking event with Sunday = 1, Monday = 2, …, Saturday = 7, and Never Smoked = 8. The final category of Never Smoked represents all right-censored observations (i.e., there is no intermittent right-censoring). Under the complementary log-log link, the cumulative probability of an event occurring up to a particular time point is given by Equation 7.4. Because this is a cumulative probability, the actual probability for a given time point is obtained by subtraction of these cumulative probabilities, except for the first and last categories. The last category (i.e., Never Smoked) is obtained as 1 minus the cumulative probability of smoking up to and including Saturday (i.e., the last day). The parameters `a1,a2,…,a7` represent the baseline hazard (i.e., hazard when all covariates equal 0); there are seven of these in this example because the total number of `Smk` categories is eight. These are akin to the threshold parameters in ordinal regression models and the values of these parameters should be increasing to reflect increased hazard across time. Finally, the variable `Schoolid` is the cluster (level-2) id, which indicates the students that belong to what schools. The random effect variance attributable to schools is estimated as a standard deviation and named `sd`. For clustered data, where the cluster variance is thought to be small, it is usually better to estimate the standard deviation than the variance because the latter will be much smaller and close to zero. Also, the random effect, named `theta`, is multiplied by its standard deviation in the model, as in Equation 7.5, and so it is in standardized form (i.e., the variance of `theta` equals 1 on the `RANDOM` statement).

```
PROC NLMIXED;
PARMS a1 = -1.9 a2 = -1.7 a3 = -1.4 a4 = -1.1 a5 = -.9 a6 = -.8 a7 = -.6
         bSocIso = .1 sd = .2;
      z = bSocIso*SocIso + sd*theta;
IF (Smk = 1) THEN
      p = 1 - EXP( - EXP(a1 + z));
ELSE IF (Smk = 2) THEN
      p = (1 - EXP( - EXP(a2 + z))) - (1 - EXP( - EXP(a1 + z)));
ELSE IF (Smk = 3) THEN
      p = (1 - EXP( - EXP(a3 + z))) - (1 - EXP( - EXP(a2 + z)));
```

```
ELSE IF (Smk = 4) THEN
      p = (1 - EXP( - EXP(a4 + z))) - (1 - EXP( - EXP(a3 + z)));
ELSE IF (Smk = 5) THEN
      p = (1 - EXP( - EXP(a5 + z))) - (1 - EXP( - EXP(a4 + z)));
ELSE IF (Smk = 6) THEN
      p = (1 - EXP( - EXP(a6 + z))) - (1 - EXP( - EXP(a5 + z)));
ELSE IF (Smk = 7) THEN
      p = (1 - EXP( - EXP(a7 + z))) - (1 - EXP( - EXP(a6 + z)));
ELSE IF (Smk = 8) THEN
      p = EXP( - EXP(a7 + z));
logl = LOG(p);
MODEL Smk ~ GENERAL(logl);
RANDOM theta ~ NORMAL(0,1) SUBJECT = Schoolid;
```

Users must provide starting values for all parameters on the PARMS statement. To do so, it is beneficial to run the model in stages using estimates from a prior stage as starting values and setting the additional parameters to zero or some small value. For example, one can start by estimating a fixed-effects model to provide starting values for the regression coefficients using SAS PROC PHREG.

In order to test the proportional hazards assumption, one can compare the above model to one in which the effect of `SocIso` is allowed to vary across the cumulative comparisons of the ordinal outcome (i.e., a nonproportional hazards model). For this, seven response models (named `z1`, `z2`, ..., `z7`) with varying effects of `SocIso` (named `bSocIso1`, `bSocIso2`, ..., `bSocIso7`) are defined. The appropriate response models are then indicated in the calculations for the category probabilities.

```
PROC NLMIXED;
PARMS a1 = -1.9 a2 = -1.7 a3 = -1.4 a4 = -1.1 a5 = -.9 a6 = -.8 a7 = -.6 sd = .2
          bSocIso1 = .1 bSocIso2 = .1 bSocIso3 = .1 bSocIso4 = .1 bSocIso5 = .1
          bSocIso6 = .1 bSocIso7 = .1;
      z1 = bSocIso1*SocIso + sd*theta;
      z2 = bSocIso2*SocIso + sd*theta;
      z3 = bSocIso3*SocIso + sd*theta;
      z4 = bSocIso4*SocIso + sd*theta;
      z5 = bSocIso5*SocIso + sd*theta;
      z6 = bSocIso6*SocIso + sd*theta;
      z7 = bSocIso7*SocIso + sd*theta;
IF (Smk = 1) THEN
      p = 1 - EXP( - EXP(a1 + z1));
ELSE IF (Smk = 2) THEN
      p = (1 - EXP( - EXP(a2 + z2))) - (1 - EXP( - EXP(a1 + z1)));
ELSE IF (Smk = 3) THEN
      p = (1 - EXP( - EXP(a3 + z3))) - (1 - EXP( - EXP(a2 + z2)));
ELSE IF (Smk = 4) THEN
      p = (1 - EXP( - EXP(a4 + z4))) - (1 - EXP( - EXP(a3 + z3)));
ELSE IF (Smk = 5) THEN
      p = (1 - EXP( - EXP(a5 + z5))) - (1 - EXP( - EXP(a4 + z4)));
ELSE IF (Smk = 6) THEN
      p = (1 - EXP( - EXP(a6 + z6))) - (1 - EXP( - EXP(a5 + z5)));
```

```
ELSE IF (Smk = 7) THEN
      p = (1 - EXP( - EXP(a7 + z7))) - (1 - EXP( - EXP(a6 + z6)));
ELSE IF (Smk = 8) THEN
      p = EXP( - EXP(a7 + z7));
logl = LOG(p);
MODEL Smk ~ GENERAL(logl);
RANDOM theta ~ NORMAL(0,1) SUBJECT = Schoolid;
```

APPENDIX 2

For Example 2, `Male`, `Cc`, and `Tv` are indicator variables of male, CC intervention, and TV intervention, respectively. The regression coefficients for these variables are named `bMale`, `bCc`, and `bTv`. The variable `Onset` indicates whether the student either answered yes to the smoking question or was censored at immediate post-intervention (`Onset` = 2), the 1-year follow-up (`Onset` = 3), or the second year follow-up (`Onset` = 4). The indicator `statuse` distinguishes censored observations (`Statuse` = 0) from smoking observations (`Statuse` = 1). Notice that if the event occurs, then the cumulative probability, under the complementary log-log link, is given by Equation 7.4, whereas if the event is censored then the probability equals 1 minus this expression. The parameters `a1`, `a2`, and `a3` represent the baseline `hazard`; there are three in this example because there are three time points and there is right-censoring. As in Example 1, the variable `Schoolid` is the cluster (level-2) id, and the random effect variance attributable to schools is estimated as a standard deviation and named `sd`.

```
PROC NLMIXED;
PARMS a1 = -3.5 a2 = -2.9 a3 = -2.4 bMale = .2 bCc = .1 bTv = .1 sd = .2;
      z = bMale*Male+bCc*Cc + bTv*Tv + sd*theta;
IF (Onset = 2 AND Statuse = 1) THEN
      p = 1 - EXP( - EXP(a1 + z));
ELSE IF (Onset = 2 AND Statuse = 0) THEN
      p = EXP( - EXP(a1 + z));
ELSE IF (Onset = 3 AND Statuse = 1) THEN
      p = (1 - EXP( - EXP(a2 + z))) - (1 - EXP( - EXP(a1 + z)));
ELSE IF (Onset = 3 AND Statuse = 0) THEN
      p = EXP( - EXP(a2 + z));
ELSE IF (Onset = 4 AND Statuse = 1) THEN
      p = (1 - EXP( - EXP(a3 + z))) - (1 - EXP( - EXP(a2 + z)));
ELSE IF (Onset = 4 AND Statuse = 0) THEN
      p = EXP( - EXP(a3 + z));
logl = LOG(p);
MODEL Onset ~ GENERAL(logl);
RANDOM theta ~ NORMAL(0,1) SUBJECT = Schoolid;
```

APPENDIX 3

To estimate the shared parameter model in Example 3 with `NLMIXED`, first, one must do a bit of data managing to produce a data set with both the longitudinal outcomes and the time to dropout. The code for this is listed below.

```
DATA one; INFILE 'c:\schzrepo.dat'; INPUT id imps79 week tx ;

/* The coding for the variables is as follows:
id = subject id number
imps79 = overall severity (ordinal version from 1 to 4)
week = 0,1,2,3,4,5,6 (most of the obs. are at weeks 0,1,3, and 6)
tx 0 = placebo 1 = drug (chlorpromazine, fluphenazine, or thioridazine)

/* compute the square root of week to linearize relationship */
sweek = SQRT(week);

/* calculate the maximum value of WEEK for each subject */
PROC MEANS NOPRINT; CLASS id; VAR week;
OUTPUT OUT = two MAX(week) = d;

/* adding the max of week (d) to the original data set */
DATA all; MERGE one two; BY id; IF id NE.;

/* create an indicator of the last observation of a subject (last) */
PROC SORT DATA = all; BY id week;
DATA all; SET all; BY id;
last = LAST.id;
RUN;
```

Each subject contributes n_i records in the input data file, namely, the file `schzrepo.dat` only contains datalines for which a subject has a valid measurement at a given time point. Thus, the maximum value of the time variable `week` for a given subject is the last time point that a subject is measured on. This variable, named d, then serves as the time to dropout variable in the analysis.

Below is the `NLMIXED code` for the shared parameter model, which, as mentioned, is based on the code of the Guo and Carlin (2004) article. The first part of the code is for the time to dropout model, and the latter part is for the ordinal longitudinal outcome.

For time to dropout, `Zsurv` represents the response model and `Psurv` is the probability for a given observation based on the complementary log–log link function. As time to dropout is ordinal, with values of 1 to 6 indicating the final week of measurement for the individual, the cumulative probabilities represent the probability of response in a given category and below. Individual category probabilities, the `Psurv's`, are therefore obtained by subtraction. As noted, the `i1` to `i5` parameters represent the cumulative baseline hazard, and the parameters `aTx` to `aDslp` are the effects on time to dropout. In particular, the latter four, `aPint` to `aDslp`, indicate the effect of the random subject intercepts and time-trends of the longitudinal model on time to dropout.

In terms of the longitudinal model (i.e., the ordinal `imps79` scores), the `b` terms are for the regression coefficients (i.e., the β's), the u terms are for the random subject effects, and the v terms are for the random-effects variance–covariance parameters. The Cholesky factorization of the variance–covariance matrix of the random effects is used in the code below. The ordinal outcome `imps79` can take on values from 1 to 4 and the cumulative probabilities (i.e., the `Plong's`) are calculated using the logistic response function. As in the ordinal survival model, individual category probabilities are obtained by subtraction. The t parameters represent the threshold parameters. Because an intercept is in the model (b0), the number of thresholds equals the number of ordinal categories minus two.

```
PROC NLMIXED DATA=all;
PARMS b0=7.3 bTx=0 bSwk=-.88 bTxSwk=-1.7 t2=4 t3=6.5 s11=1 s12=0 s22=.5
      aTx=0 aPint=0 aPslp=0 aDint=0 aDslp=0 i1=-1 i2=-.7 i3=-.5 i4=0
          i5=.2;

/* Compute log likelihood contribution of the survival data part */
/* when the last observation of a subject is reached */
IF (last) THEN DO;
     Zsurv = aTx*Tx + aPint*(1-Tx)*u0 + aPslp*(1-Tx)*u1 + aDint*Tx*u0 +
             aDslp*Tx*u1;
     IF (d = 1) THEN
        Psurv = 1 - EXP( - EXP(i1+Zsurv));
ELSE IF (d = 2) THEN
        Psurv = (1 - EXP( - EXP(i2+Zsurv))) - (1 - EXP( - EXP(i1+Zsurv)));
ELSE IF (d = 3) THEN
        Psurv = (1 - EXP( - EXP(i3+Zsurv))) - (1 - EXP( - EXP(i2+Zsurv)));
ELSE IF (d = 4) THEN
        Psurv =(1 - EXP( - EXP(i4+Zsurv))) - (1 - EXP( - EXP(i3+Zsurv)));
ELSE IF (d = 5) THEN
        Psurv =(1 - EXP( - EXP(i5+Zsurv))) - (1 - EXP( - EXP(i4+Zsurv)));
ELSE IF (d = 6) THEN
        Psurv = EXP( - EXP(i5+Zsurv));
IF (Psurv > 1e-8) THEN Lsurv = LOG(Psurv);
ELSE Lsurv = -1e100;
END; ELSE Lsurv = 0;

/* Cholesky parameterization of the random effects var-covar matrix */
/* This ensures that the matrix is nonnegative definite */
v11 = s11*s11;
v12 = s11*s12;
v22 = s12*s12 + s22*s22;

/* Compute the contribution of the longitudinal part */
/* Every observation in the data set makes a contribution */
Zlong = b0 + bTx*tx + bSwk*sweek + bTxSwk*tx*sweek + u0 + u1*sweek;
IF (imps79 = 1) THEN
      Plong = 1 / (1 + EXP(-(-Zlong)));
ELSE IF (imps79 = 2) THEN
      Plong = (1/(1 + EXP(-(t2-Zlong)))) - (1/(1 + EXP(-(-Zlong))));
ELSE IF (imps79 = 3) THEN
      Plong = (1/(1 + EXP(-(t3-Zlong)))) - (1/(1 + EXP(-(t2-Zlong))));
ELSE IF (imps79 = 4) THEN
      Plong = 1 - (1 / (1 + EXP(-(t3-Zlong))));
IF (Plong > 1e-8) THEN Llong = LOG(Plong);
ELSE Llong = -1e100;

/* Any numeric variable can be used as the response in the MODEL
   statement */
```

```
/* It has no bearing on the results */
MODEL last ~ GENERAL(Llong + Lsurv);
RANDOM u0 u1 ~ NORMAL([0, 0],[v11,v12,v22]) SUBJECT=id;
/* Compute the variances and covariance of the random effects */
ESTIMATE 'Var[u0]' v11;
ESTIMATE 'Cov[u0,u1]' v12;
ESTIMATE 'Var[u1]' v22;
RUN;
```

8

Bayesian Estimation of Multilevel Models

Ellen L. Hamaker and Irene Klugkist
Department of Methods and Statistics, Faculty of Social Sciences,
Utrecht University Utrecht, The Netherlands

8.1 INTRODUCTION

There is a growing interest among both statisticians and empirical researchers for Bayesian statistics (Poirier, 2006). Two key factors that have contributed to this development are the rapid increase in computational power of PCs, and the development of Markov Chain Monte Carlo (MCMC) methods in the nineties (Gelfand & Smith, 1990; Gilks, Richardson, & Spiegelhalter, 1996). The latter has revolutionized Bayesian statistics, as it allows for the empirical estimation of probability distributions that are difficult (or impossible) to derive analytically (Gill, 2008; Poirier, 2006). Together, the PC and MCMC methods have made Bayesian statistics flexible and computationally feasible, and it is to be expected that in the next decade we will witness a steady increase in Bayesian applications in the social sciences. In fact, several popular statistical packages already include the option of Bayesian analysis (e.g., MLwiN, SAS), and recently several books have appeared that encourage the social scientist to explore Bayesian analysis (Gelman & Hill, 2007; Gill, 2008; Lynch, 2007).

The fundamental difference between the dominant approach in statistics—which is referred to as frequentistic or classic—and Bayesian statistics, is the way in which the concept *probability* is handled. Frequentists interpret probability as "the limit that the relative frequency approaches when the same experiment is repeated infinitely many times" (hence, the name frequentists). Bayesians on the other hand use the term probability to refer to "uncertainty" or "degree of believe." This difference has far reaching consequences for the way in which model estimation and model selection are handled by frequentists and Bayesians. For instance, Bayesians place a probability distribution on a model parameter because they are uncertain about the true value. However, to frequentists this is unacceptable, since model parameters are considered to be fixed and the frequentists' interpretation of probability only allows for probability distributions to be placed on random phenomena. Similarly, when comparing a number of competing models, the Bayesian can determine the probability that a particular model generated the data. Again, to the frequentist this is unacceptable, because a model (i.e., a hypothesis) is either true or not, and it is not a random quantity for which a probability distribution can be defined.[1]

This chapter is meant as a first introduction into Bayesian estimation for readers already familiar with multilevel modeling. We begin with discussing a number of general arguments for considering Bayesian estimation. This is followed by a section on the basics of Bayesian estimation, and a section on Bayesian estimation of multilevel models. To illustrate some of the features and details of Bayesian estimation, we make use of an empirical illustration, which is included as a running example throughout these two sections. All the analyses in this chapter are performed using WinBUGS (Spiegelhalter, Thomas, Best, & Lunn, 2003), which was called from R using the function bugs() from the package R2WinBUGS (Sturtz, Ligges, & Gelman, 2007; for a thorough description of how to use bugs(), see Gelman & Hill, 2007). Both WinBUGS and R can be downloaded freely. We end this chapter with a discussion of the specific merits of Bayesian estimation for multilevel modeling.

[1] Although the frequentists' and Bayesian viewpoints are incompatible, it is important to note that both interpretations of probability are compatible with the axioms of probability; that is, mathematically they behave like probabilities (Gill, 2008).

8.2 WHY CONSIDER BAYESIAN ESTIMATION?

In a nutshell Bayesian estimation consists of placing probability distributions on model parameters to represent prior uncertainty, and to update this in the light of current data to obtain a posterior distribution for the model parameters that contains less uncertainty (Lynch, 2007). Hence, the posterior distribution is a compromise between the prior distribution and the data, and it represents the status of knowledge about the model parameters.

Having to place a prior distribution on the model parameters is seen as both a virtue and a weakness of Bayesian statistics. On a positive note, if one has prior knowledge from previous research, the prior distribution forms a means by which this knowledge can be incorporated in the current study. As such it offers the opportunity to build on previous research results, rather than having to start from scratch with each new study. However, it can be difficult to formulate reasonable prior distributions, especially if one has no prior knowledge, but also if the current study differs from previous studies in that the sample comes from a (slightly) different population, or that it includes some additional (or different) variables. In such cases researchers may decide to specify vague prior distributions that represent a lot of uncertainty (i.e., distributions with a large variance), such that the prior distribution plays a negligible role and the posterior distribution is dominated by the data.[2]

[2] Note that this only applies to estimation. In case of Bayesian model selection, prior distributions on the parameters do not become negligible when they are vague.

This leads to the question: Why should one be interested in Bayesian estimation, if informative priors are hard to get by, while vague priors are negligible and lead to parameter estimates that are (almost) the same as the parameter estimates obtained within the frequentists' approach? This is a legitimate question that deserves a convincing answer. In fact, there are several advantages of Bayesian estimation over the frequentists' approach, besides the option of building on prior knowledge.

First, the Bayesian approach is not based on normality assumptions or on asymptotic results (Gill, 2008; Lynch, 2007). That is, Bayesian estimation results in a posterior distribution from which the mean, median, or mode can be used as the parameter estimate, and in addition a 100 $(1 - \alpha)$% interval around this estimate can be derived for inferential purposes. The construction of such *credibility intervals* is discussed in more detail in the following section. Here it suffices to state that these credibility intervals do not depend on normality assumptions and/or asymptotic results, making this a potentially advantageous approach for small sample sizes.

Second, the posterior distributions of parameters always lie in the sample space—given the prior was defined in the sample space. This implies that where maximum likelihood estimation may lead to negative estimates of variance terms, this problem does not arise in Bayesian estimation. Similarly, while the classical approach may lead to the estimation of a correlation larger than one, this will not occur in Bayesian estimation.

Third, a Bayesian credibility interval around a parameter estimate has the intuitive interpretation that is often erroneously given to frequentistic confidence intervals, namely, there is a $1 - \alpha$ chance that the true value lies in this interval. In contrast, confidence intervals require the interpretation that 100 $(1 - \alpha)$% of such intervals contain the true parameter value. Hence, Bayesian results are easier to communicate (Gill, 2008).

Fourth, it is extremely easy to obtain parameter estimates and credibility intervals for quantities that are functions of parameters in the model (e.g., for some probability p and for the logit (p)). This also implies that one can easily construct hypotheses concerning parameters that are not directly estimated in the model, which can then be investigated by determining whether a particular value lies inside the credibility interval of the quantity of interest.

Fifth, Bayesian estimation algorithms can easily handle missing data, an issue we will elaborate on in the following section. The Bayesian package WinBUGS treats missing data in the dependent variable automatically.

Finally, Bayesian model estimation is still likely to work even for complex models, when the classical approach breaks down (Berger, 2006). Especially with MCMC methods, Bayesian estimation becomes extremely flexible, as will become clear below.

8.3 ABC OF BAYESIAN ESTIMATION

Bayesian estimation consists of finding the posterior distribution of unknown parameters based on the observed data and a prior distribution—the latter being defined by the researcher. In this section, we begin with explaining how the combination of a prior distribution and the data lead to the posterior distribution. Then we discuss the Gibbs sampler—an MCMC method—which is a convenient method for obtaining the posterior distribution. This is followed by an example in which the Gibbs sampler is used to estimate means of different groups. In addition, it is shown that it is easy to include an extra step in the Gibbs sampler to handle missing data. At the end, posterior estimates for parameters and for functions of parameters are provided and discussed.

8.3.1 Bayes' Theorem and How To Get From the Prior To the Posterior

In a Bayesian approach uncertainty about parameters is represented by a probability distribution: the prior distribution represents the uncertainty or knowledge before observing any data, and the posterior distribution expresses the knowledge after taking the observed data into account. The posterior distribution is obtained through Bayes' theorem, which, in general notation, states

$$p(\theta \mid y) = \frac{p(y \mid \theta)p(\theta)}{p(y)},$$

where $p(\theta|y)$ is the posterior distribution for a parameter vector θ given observed data y, $p(y|\theta)$ is the sampling density for the data, which is identical to the likelihood $L(\theta|y)$ with all (normalizing) constants included, $p(\theta)$ is the prior distribution for the model parameters, and $p(y)$ is the marginal probability of the data. Note that the denominator $p(y)$ is a normalizing constant and is required to make the posterior a proper density—meaning it integrates to one. Often, the posterior distribution is required up to proportionality only; that is,

$$p(\theta \mid y) \propto p(y \mid \theta)p(\theta). \qquad (8.1)$$

Hence, the posterior distribution is proportional to a combination of the information contained in data and prior.

The role of prior distributions is a point of discussion, and often seen as the bottleneck of Bayesian analyses. Since a prior distribution represents (subjective) prior knowledge, different researchers may come up with different priors and subsequently obtain different results. In many applications, however, vague or uninformative priors can be used, leading to results that are determined by the observed data only. As an example, consider the uniform distribution: $p(\theta) \propto c$. This prior states that any value for the parameter θ between $-\infty$ and $+\infty$ is a priori equally likely. Using this uniform or constant prior results in a posterior distribution that is proportional to the likelihood function, as all values of the likelihood are weighted by the same value c to obtain the posterior in Equation 8.1.

However, note that the unbounded uniform prior is *improper*, in that it is not a density because the integral is not finite. While for relatively simple models it may be straightforward to show that an improper prior still leads to a proper posterior, in more complex models, like the multilevel model with several regression parameters and variance terms, improper priors may lead to posteriors that are not proper densities. To avoid this risk, often proper prior distributions are preferred. WinBUGS, for instance, does not allow for improper priors.

To make proper priors vague or uninformative, one can choose the hyperparameters, that is, the parameters that define the prior distribution, in such a way that the prior becomes very wide and flat. For instance, a normal prior distribution $N(a, b^2)$ approaches the uniform prior if b^2 is specified large enough, but it remains a proper density. A popular choice for proper prior distributions are *conjugate* priors, which means that the functional form of the prior distribution is such that, combined with a specific density for the data, it leads to a posterior distribution of the same functional form. This is an intuitively appealing choice, and furthermore it has many computational advantages.

In models with multiple parameters, instead of using a conjugate prior for all model parameters, one can also use semi-conjugate priors (Gelman, Carlin, Stern, & Rubin, 2004), which are sometimes referred to as conditional conjugate priors (Gelman, 2006). Such priors are conjugate priors for some parameter(s) in the model, if the other parameter(s) are known. Because Gibbs sampling, discussed in detail below, is based on sampling from conditional posterior distributions, choosing conditional conjugate priors such that the conditional posterior is of a known functional form is advantageous when using this algorithm. Throughout this chapter we will use vague conditional conjugate priors for the illustrations.

8.3.2 Gibbs Sampler

Bayesian estimation is based on deriving the posterior and then summarizing the entire distribution. In general, summarizing a distribution with quantities like, for instance, means and variances involves solving (complex) integrals and is, at best, analytically challenging and more often unsolvable (for instance, for complex, multivariate distributions or when nonconjugate priors are used). Alternatively, one can make use of sampling approaches to approximate the integrals of interest. The MCMC methods are a class of algorithms for sampling from complex probability distributions using iterative methods and certain, less complex, proposal distributions. A Markov chain is constructed by using the previous sample values to randomly generate the next sample value. The state of the chain after a large number of steps is then used as a sample from the desired distribution.

Perhaps the most popular MCMC sampling method is the Gibbs sampler (Gelfand & Smith, 1990; Geman & Geman, 1984; Smith & Roberts, 1993), which is a special case of the Metropolis-Hastings algorithm (Chib & Greenberg, 1995; Hastings, 1970). Gibbs sampling can be used when sampling from a multivariate posterior is not feasible, but sampling from the univariate conditional distributions is. For the illustrations in this chapter, the Gibbs sampler can be applied to obtain a sample from the posterior distribution and, therefore, other MCMC sampling methods are not discussed (but see Chen, Shao, & Ibrahim, 2000; Chib & Greenberg, 1995; Hastings, 1970; Robert & Casella, 1999).

The basic idea of the Gibbs sampler is that sampling from a multidimensional posterior distribution can be done via repeatedly sampling from the univariate distribution of each parameter, *conditional on the current values of the other parameters.* For a parameter vector $\theta = \{\theta_1, \ldots, \theta_Q\}$ and observed data y, a sample from the joint posterior $p(\theta|y)$ is obtained by a Gibbs sampler consisting of the following steps:

1. Set $h = 0$ and assign arbitrary starting values to $\theta^{(h)}$
2. Set $h = h + 1$
3. Sample $\theta_1^{(h)}$ from $p(\theta_1|\theta_2^{(h-1)},\ldots,\theta_Q^{(h-1)}, y)$
4. Sample $\theta_2^{(h)}$ from $p(\theta_2\,|\,\theta_1^{(h)},\ldots,\theta_Q^{(h-1)}, y)$
5. …
6. Sample $\theta_Q^{(h)}$ from $p(\theta_Q\,|\,\theta_1^{(h)},\ldots,\theta_{Q-1}^{(h)}, y)$
7. If $h < H$ go to Step 2

The conditional posterior distributions generally have simpler structures than the joint posterior, so that they are easier to sample from. A key issue in the successful implementation of any MCMC sampler is the number of runs (iterations) until the chain converges. Typically a first set of iterations is discarded (the burn-in period) since the initial samples are affected by the arbitrary starting values. The required length of burn-in can be shortened by choosing sensible initial values. For the remaining iterations one of various checks is used to assess whether convergence has indeed been reached. A good impression of length of required burn-in and convergence of the sampling can be obtained by visual inspection of plots of the sampled values, especially if several chains with different starting values are compared. Also, one can compute formal diagnostics as, for instance, the $\hat{R}$ (Gelman & Rubin, 1992) that evaluates the between and within chains variation to get information on convergence of the MCMC sample. For a more elaborate presentation and discussion of convergence diagnostics, see Cowles and Carlin (1996).

The remaining iterations after burn-in constitute a sample from the intended posterior distribution. Discrete formulas can be applied to these samples to summarize the (posterior) knowledge about parameters. For instance, the mean of the distribution of a parameter θ_q is estimated by the average of the sampled values of $\theta_q^{(h)}$. In a similar way, the posterior standard deviation and various quantiles can be computed (e.g., 95% credibility interval).

8.3.3 Example

To illustrate some of the issues presented above, we make use of an empirical example. The data we use in this chapter come from the Panel Study of Income Dynamics (PSID), and consist of four biannual measurements of body mass index (BMI) from 3355 males between 1999 and 2005.[3] In the example here the object is to estimate the mean BMI at the first measurement occasion for four (independent) groups based on the self-reported importance to maintain a healthy weight: the first group indicated it was "not at all important"; the second group indicated it was "not too important"; the third group indicated it was "somewhat important"; and the fourth group indicated it was "very important." In Table 8.1, the sample BMI scores at the first measurement occasion are summarized for each of the four groups. Subsequently, the sample mean (*M*), standard deviation (*SD*), sample size (*n*), and number of missing BMI-scores (n_{miss}) are presented for each of the four groups.

TABLE 8.1

Sample Descriptives on BMI at the First Measurement

Group	*M*	*SD*	*n*	n_{miss}
1	25.90	4.46	110	1
2	27.95	5.67	208	2
3	27.88	4.48	1013	6
4	26.92	4.05	1992	23

Note: Group 1 indicates maintaining a healthy weight is "not at all important." Group 2 indicates it is "not too important." Group 3 indicates it is "somewhat important." and Group 4 indicates it is "very important."

8.3.3.1 Bayesian Estimation of Group Means and Residual Variance

Let y_{ji} denote the BMI-score of the ith individual in the jth group (for $i = 1, \ldots, n_j$ and $j = 1, \ldots, 4$), and let μ_j denote the mean BMI

[3] See http://psidonline.isr.umich.edu/. Note that the PSID is not a random sample from the U.S. population, but it includes sample weights such that proper inferences can be made. Since this is a subsample from the PSID sample—depending on whether the information was given to compute the BMI (i.e., weight and height)—the current subsample can not be used to make inferences to the population.

in group j. Then the observed BMI score can be expressed as

$$y_{ji} = \mu_j + e_i, \tag{8.2}$$

where the residual e_i is assumed to be normally distributed with mean zero and variance σ^2.

From the expression in Equation 8.1 it is clear that in order to obtain the posterior distribution we need both the density of the data, and a prior distribution for the parameters. For the current illustration the density of the data can be expressed as

$$p(y \mid \mu, \sigma^2) = (2\pi\sigma^2)^{-N/2} \exp \left\{ -\frac{1}{2\sigma^2} \sum_{j=1}^{4} \sum_{i=1}^{n_j} (y_{ji} - \mu_j)^2 \right\}, \tag{8.3}$$

where $y = \{y_{11}, \ldots, y_{1n_1}, \ldots, y_{41}, \ldots, y_{4n_4}\}$, $\mu = \{\mu_1, \mu_2, \mu_3, \mu_4\}$, and $N = \Sigma_j n_j$.

To specify a prior for all five model parameters (i.e., the four μ's and the variance σ^2), we begin by assuming the means and the variance are independent, such that the joint prior can be written as $p(\mu, \sigma^2) = p(\mu)p(\sigma^2)$. Using conditional conjugate priors for the four μ's and for σ^2, we get

$$p(\mu, \sigma^2) = \prod_{j=1}^{4} N(a_j, b_j^2) \times IG(c, d), \tag{8.4}$$

where $N(a_j, b_j^2)$ denotes a normal distribution with mean a_j and variance b_j^2 (for $j = 1, \ldots, 4$) and $IG(c, d)$ denotes an inverse gamma distribution with hyperparameters c and d. The prior in Equation 8.4 is vague if b_j^2 is chosen large, and $c = d$ approaches zero.

Through combining the density of the data and the joint prior we obtain the posterior distribution

$$p(\mu, \sigma^2 \mid y) \propto p(y \mid \mu, \sigma^2) p(\mu, \sigma^2)$$
$$= (2\pi\sigma^2)^{-N/2} \exp \left\{ -\frac{1}{2\sigma^2} \sum_{j=1}^{4} \sum_{i=1}^{n_j} (y_{ji} - \mu_j)^2 \right\}$$
$$\times \prod_{j=1}^{4} N(a_j, b_j^2) \times IG(c, d). \tag{8.5}$$

To make inferences for a specific parameter based on this joint posterior distribution, we need to integrate the above expression over the other parameters. For instance, if we want to determine a credibility interval for μ_1, we have to integrate μ_2, μ_3, μ_4, and σ^2 out first to obtain the marginal posterior distribution of μ_1. While this may be fairly easy for simple models, it can become rather difficult or even impossible for more complex models.

Alternatively, instead of evaluating the expression in Equation 8.5 directly, we can obtain an empirical approximation of it using the Gibbs sampler. To this end we need the conditional posterior distributions of each μ_j and of σ^2. Since the means are assumed to be independent in the model defined in Equation 8.3, the conditional posterior distribution of each mean is also independent of the other means. Using the normal density Equation 8.3 and a normal prior for μ_j, and assuming σ^2 is known, results in a normal posterior (see for instance, Lynch, 2007, pp. 62–64; Gelman et al. 2004, p. 49); that is

$$p(\mu_j \mid \sigma^2, y) = N\left(\frac{a_j / b_j^2 + n_j \bar{y}_j / \sigma^2}{1/b_j^2 + n_j / \sigma^2}, \frac{1}{1/b_j^2 + n_j / \sigma^2} \right),$$

where $\bar{y}_j$ denotes the average BMI-score in group j. From the parameters of the normal distribution it can be seen that the posterior is a compromise of the information in data and prior. For instance, the mean of the conditional posterior for μ_j is the average of prior mean a_j, weighted with corresponding precision $1/b_j^2$, and data mean $\bar{y}_j$ weighted with precision n_j/σ^2.

The conditional posterior distribution of the variance needed for the Gibbs sampler is obtained by combining the normal likelihood Equation 8.3 with the inverse gamma prior Equation 8.4 and assuming the means are known. This results in a posterior distribution that is also an inverse gamma distribution (see, for instance, Gill, 2008, pp. 77–78)

$$p(\sigma^2 \mid \mu, y) = IG\left(c + N/2, d + \frac{1}{2}\sum_{j=1}^{4}\sum_{i=1}^{n_j}(y_{ji} - \mu_j)^2\right).$$

Note that, as with the conditional posterior for the means, the parameters of the posterior inverse gamma distribution are a compromise between prior (i.e., c and d) and data information (i.e., $N/2$ and y_{ji}).

Using these conditional posterior distributions, the Gibbs sampler for estimating the means and residual variance can be written as

1. Set $h = 0$ and assign an arbitrary (positive) starting value to $\sigma^{2(h)}$
2. Set $h = h + 1$
3. For $j = 1, \ldots, 4$, sample $\mu_j^{(h)}$ from $p(\mu_j|\sigma^{2(h-1)}, y)$
4. Sample $\sigma^{2(h)}$ from $p(\sigma^2|\mu^{(h)}, y)$
5. If $h <$ H go to Step 2

8.3.3.2 *Missing Data*

The Gibbs sampler discussed above can be easily extended to handle missing data on the dependent variable. At the first measurement, 32 respondents failed to report the information needed to compute their BMI. Let $y = \{y_{obs}, y_{mis}\}$, where y_{obs} are the observed cases, and y_{mis} are the cases for which BMI is missing. Note that for all respondents group membership is known; that is, respondents did indicate how important it was to them to maintain a healthy weight. The missing data assumption, in the algorithm presented hereafter, is that—conditional on all information in the model (i.e., here just group membership)—the missing observations on BMI are random. Under this missing at random (MAR) assumption, missing data are easily dealt with using MCMC sampling and is implemented in WinBUGS. Compared with the Gibbs sampler presented before, initial values for y_{mis} are specified and in each iteration h, new values for y_{mis} are sampled along with the samples for the model parameters. Sampling of the missing values is conditional on the current values of the parameters; that is

1. Set $h = 0$ and assign an arbitrary (positive) starting value to $\sigma^{2(h)}$ and arbitrary starting values to $y_{mis}^{(h)}$
2. Set $h = h + 1$
3. For $j = 1, \ldots, 4$, sample $\mu_j^{(h)}$ from $p(\mu_j \mid \sigma^{2(h-1)}, y_{mis}^{(h-1)}, y_{obs})$
4. Sample $\sigma^{2(h)}$ from $p(\sigma^2 \mid \mu^{(h)}, y_{mis}^{(h-1)}, y_{obs})$
5. Sample $y_{mis}^{(h)}$ from $p(y_{mis}|\mu^{(h)}, \sigma^{2(h)}, y_{obs})$
6. If $h < H$ go to Step 2

Sampling $y_{mis}^{(h)}$ in Step 5 is straightforward, since within each iteration h, the parameters

μ and σ^2 have known current values. If a person's score is missing, this score y_{ji} is sampled from $N(\mu_j^{(h)}, \sigma^{2(h)})$. Again, choosing sensible starting values for $y_{mis}^{(0)}$ shortens the required burn-in of the Gibbs sampler. WinBUGS generates such starting values automatically. Note that the algorithm described above is a form of multiple imputation (Rubin, 1987). This means that the uncertainty in the model parameters is taken into account by multiple imputing values for the missing scores conditional on the current values of the parameters.

8.3.3.3 Posterior Estimates

We use WinBUGS to obtain a sample from the posterior distribution based on the prior in Equation 8.4 with $a_j = 25$ and $b_j^2 = 10^6$ (for $j = 1, \ldots, 4$), and $c = d = .001$. To examine the required burn-in and total number of iterations, two samples are obtained using different starting values. In Figure 8.1, the first 50 iterations of two chains are plotted for each of the parameters. It can be seen that, in this example, the effect of the arbitrary starting values disappears within just a couple of iterations.

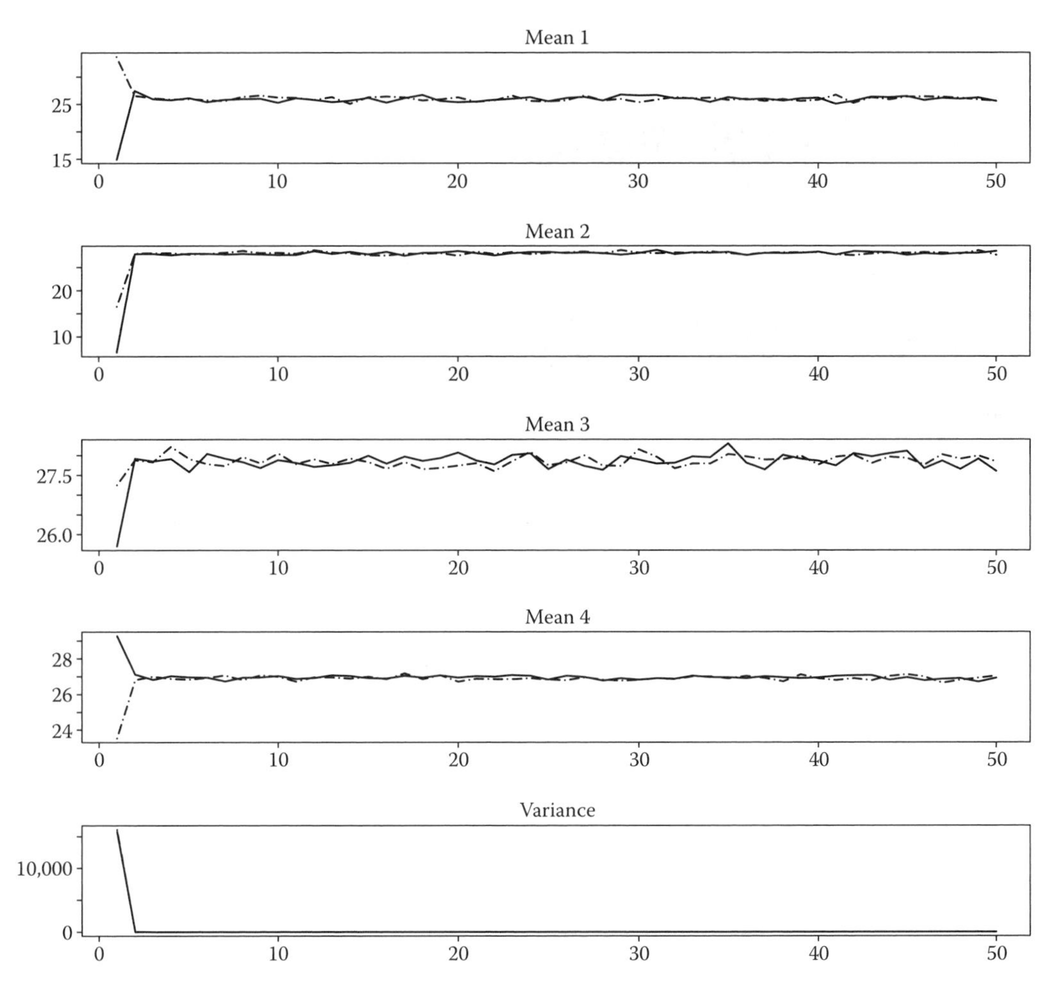

FIGURE 8.1
Comparison of the first 50 iterations of two chains with different starting values.

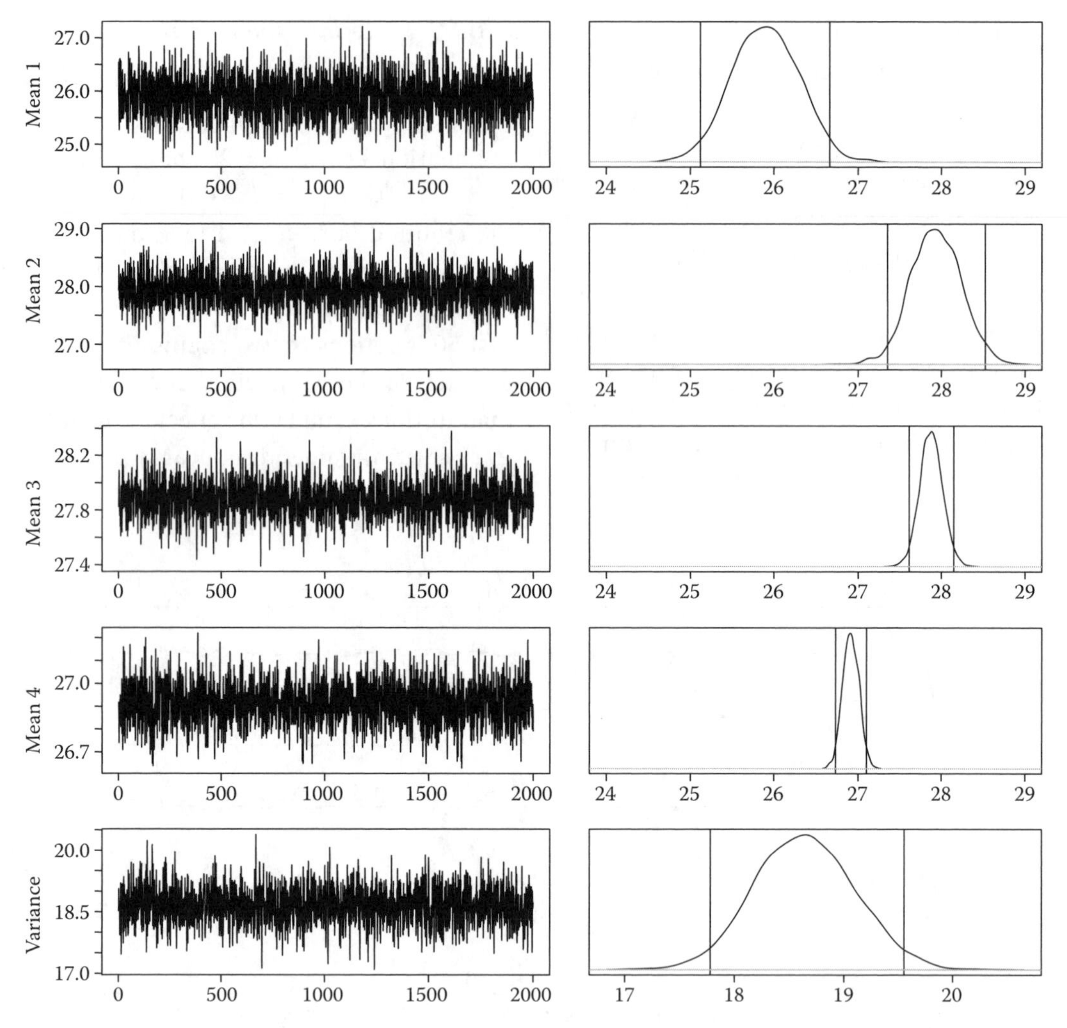

FIGURE 8.2
Chains of 2,000 iterations after burn-in (left) and summary of posterior density (right) for all parameters. Vertical lines mark the 95% credibility interval.

All estimates in this section are based on 4000 iterations after discarding the first 2000. Figure 8.2 shows the iterations plots on the left and the posterior density plots on the right for all five parameters. For the means, the same scale on the x-axis is used to show the difference in (mean) location as well as in the variance of the resulting posterior distributions. The latter is caused by the large differences in sample size, ranging from 110 in the first group to 1992 in the fourth group. This can also be seen in Table 8.2 where the posterior summaries are provided. The posterior mean is obtained by computing the average of the 2000 draws for the corresponding parameter. Likewise, the posterior SD is the standard deviation across all iterations. Quantiles are obtained by sorting the sampled values and reporting the parameter value corresponding with the desired quantiles. The 2.5th and the 97.5th percentile, for instance, provide the lower and upper bound of the 95%

TABLE 8.2

Posterior Distributions for Four Means and the Variance

	PM	PSD	.025	.25	.50	.75	.975
μ_1	25.90	0.40	25.12	25.62	25.90	26.18	26.66
μ_2	27.93	0.30	27.36	27.73	27.93	28.14	28.53
μ_3	27.88	0.14	27.61	27.79	27.88	27.97	28.15
μ_4	26.91	0.09	26.73	26.85	26.91	26.98	27.10
σ^2	18.64	0.45	17.79	18.33	18.64	18.95	19.55

Note: Posterior mean (PM), posterior standard deviation (PSD), and several percentiles of the posterior distribution (.025, .25, .50, .75, .975).

central credibility interval for each parameter. Note that to obtain such an interval no normality assumption is made, so for skewed distributions the interval is, correctly, not symmetric around the mean. Skewed posterior distributions are not uncommon, especially for variance parameters in small samples. In this illustration however, all marginal posteriors look quite normally distributed, as can be seen both in Figure 8.2, and when comparing, for instance, the posterior mean and median (they are equivalent for all parameters) in Table 8.2. This is to be expected given the large sample sizes.

The power of the Gibbs sampler, or any MCMC method, is that by obtaining a sequence of univariate conditional random variables we can compute any feature of all marginal distributions (as shown above), as well as any feature of the marginal distributions of functions of the parameters. We will illustrate the latter with two examples. First, suppose we are interested in differences between groups 2 and 3. It may be expected that differences in BMI will be observed for the extreme groups "not at all important" and "very important" compared to the in-between group "not too important" or "somewhat important," while there is no clear distinction between the latter two. Since μ_2 and μ_3 have values in each iteration, we can also compute a value for the difference $Q_1^{(h)} = \mu_3^{(h)} - \mu_2^{(h)}$ in each iteration. This provides 2000 values for the difference scores and thus a discrete summary of the posterior distribution of Q_1. The results are plotted in Figure 8.3 (top). The resulting posterior of Q_1 indicates that the difference between the mean BMI scores of groups 2 and 3 is very likely to be (close to) zero.

The two plots at the bottom of Figure 8.3 represent another question of interest: Can we observe a difference in average BMI scores for the fourth group compared to the two in-between groups 2 and 3? Within each iteration, $Q_2^{(h)} = \mu_4^{(h)} - (\mu_2^{(h)} + \mu_3^{(h)})/2$ is computed. The posterior distribution of Q_2 is plotted and shows that the entire posterior is smaller than zero. This indicates that men in the group that indicates maintaining a healthy weight is very important to them, have, on average, a lower BMI than the two groups that indicate that it is not too important and somewhat important. A possible explanation for this finding is that those who find maintaining a healthy weight very important will put in more effort to actually maintain a healthy weight than those who do not find it that important.

In this section we have introduced several basic concepts concerning Bayesian

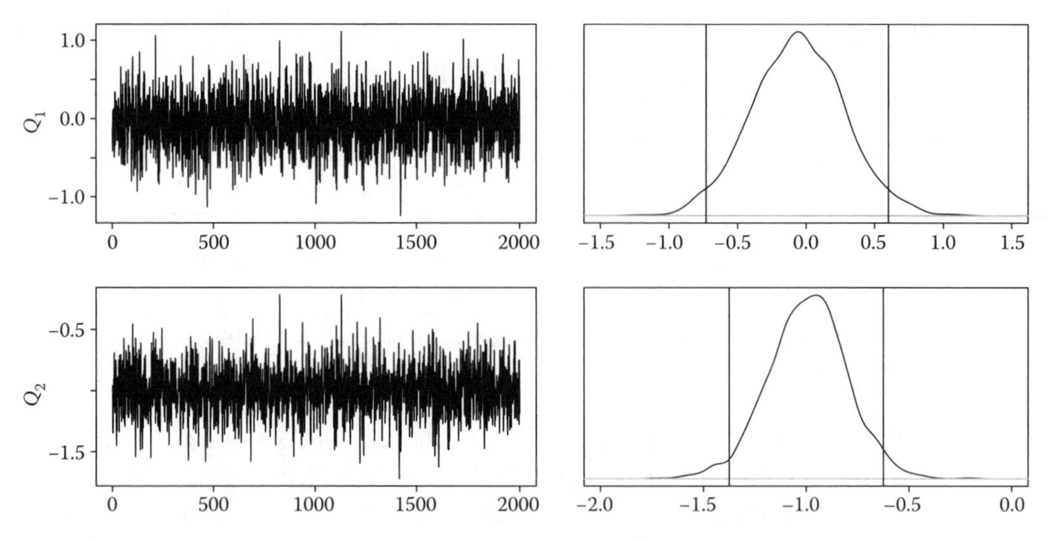

FIGURE 8.3
Chains of 2000 iterations after burn-in (left) and summary of posterior density (right) for Q_1 and Q_2. Vertical lines mark the 95% credibility interval.

estimation, and provided illustrations in the context of the estimation of independent group means. In the remainder of this chapter, we will extend these ideas and illustrate them for multilevel modeling.

8.4 BAYESIAN ESTIMATION IN MULTILEVEL MODELING

Multilevel modeling fits so naturally into a Bayesian framework that virtually any textbook on Bayesian analysis includes a chapter on multilevel modeling (e.g., Gelman, 2006; Gill, 2008; Lynch, 2007). In contrast to those presentations, which present multilevel modeling as the novel issue, here the Bayesian aspects are considered the novel part. We discuss three multilevel models; that is, the well-known random intercept model, the growth curve model, and a less typical multilevel model with time-varying covariates. The latter illustrates the flexibility of formulating alternative models in WinBUGS.

8.4.1 The Basics of Bayesian Estimation of a Multilevel Model

A typical approach to multilevel modeling consists of specifying the model at each level. Hence in a multilevel model with two levels, one begins with specifying the model at level 1, and then proceeds with specifying the model at level 2. When using Bayesian estimation for multilevel modeling, an additional step is required, in which the hyperprior distributions are specified for the parameters in the model at level 1 and 2.

Let y_{km} be the m-th observation in cluster k, where $k = 1, \ldots, K$, and $m = 1, \ldots, M_k$. Let's assume that within each cluster the data are distributed according to a particular distribution, Q with parameter θ, such that we can define the model at level 1 as

$$y_{km} \sim Q(\theta_k). \tag{8.6}$$

Next, we assume that the parameters θ_k come from a distribution R that is characterized by parameter γ, such that we can define the model at level 2 as

$$\theta_k \sim R(\gamma). \tag{8.7}$$

Thus far, this does not differ from defining a multilevel model in the classical approach. The next step however is unique to the Bayesian approach: We define a prior distribution for the hyperparameter γ; that is,

$$\gamma \sim S(a, b). \tag{8.8}$$

The last step is sometimes referred to as the third level (e.g., Press, 2003, p. 342), because the model in Equation 8.6 through Equation 8.8 can be thought of as a multilevel model with 3 levels, for which we have observed just one unit at level 3. To avoid confusion however, we refer to the distribution in Equation 8.8 as the hyperprior distribution, because this distribution is the prior for the hyperparameters in Equation 8.7.

Now let $y=\{y_{11},\ldots,y_{1M_1},\ldots,y_{K1},\ldots,y_{KM_K}\}$ denote all observed data. Then the joint posterior distribution of *all unknowns* in the multilevel model, that is, the population parameter γ *and* the cluster parameters $\theta = \{\theta_1, \ldots, \theta_k\}$, can be obtained through employing Bayes' theorem; that is,

$$p(\gamma,\theta \mid y) \propto p(y \mid \theta,\gamma)p(\theta \mid \gamma)p(\gamma), \tag{8.9}$$

where the density of the data is obtained through

$$p(y \mid \theta,\gamma) = \prod_{k=1}^{K}\prod_{m=1}^{M_k} p(y_{km} \mid \theta_k,\gamma)$$

and the prior for the random effects is

$$p(\theta \mid \gamma) = \prod_{k=1}^{K} p(\theta_k \mid \gamma).$$

From the joint posterior distribution in Equation 8.9, one can derive the conditional posterior distributions through selecting only those terms that contain the parameter of interest, but this can be very tedious (cf. Lynch, 2007, pp. 242–245). When using WinBUGS, it suffices to specify the model and the hyperprior distributions.

8.4.2 Model 1: Random Intercept Model

For the multilevel models considered here, we make use of all four measurement occasions between 1999 and 2005. The sample consists of 3355 males, for whom we have at least one BMI measurement over a 6-year period. In Table 8.3 it can be seen that the

TABLE 8.3

Frequencies of BMI Categories Over a 6-Year Period for 3355 Males

	BMI Category					
Year	**< 18.5**	**[18.5–25)**	**[25–30)**	**[30–35)**	**≥ 35**	**Miss**
99	10	1066	1518	545	184	32
01	15	983	1534	589	211	23
03	11	940	1521	619	239	25
05	13	883	1548	628	258	25

Note: Four measurement occasions between 1999 and 2005. A BMI below 18.5 is considered underweight; between 18.5 and 25 is considered healthy; between 25 and 30 is considered overweight; between 30 and 35 is considered obese; and above 35 is considered severely obese. Last column contains the number of missing observations per measurement occasion.

number of men in the obese and severely obese categories increases over the span of the study, while the number of men in the healthy category steadily decreases. As a result the number of men in the overweight category remains more or less in equilibrium. We interpret this as an indication that BMI of most individuals in the study increases over the 6 years between 1999 and 2005. We will investigate this with growth curve models below.

We start with a simple multilevel model known as the random intercept model. The measurements are denoted as y_{it}, where i refers to individual, and t refers to measurement occasion (i.e., $t = \{1, 2, 3, 4\}$). Each individual is allowed its own intercept denoted as α_i. Hence at level 1 we have $y_{it} = \alpha_i + e_{it}$, with $e_{it} \sim N(0, \sigma^2)$, and at level 2 we have $\alpha_i = \alpha_{(0)} + \varepsilon_i$, with $\varepsilon_i \sim N(0, \tau^2_\alpha)$. Using probability notation (Lynch, 2007), we can write this model as

$$\begin{aligned} \text{Level 1:} \quad & y_{it} \sim N(\alpha_i, \sigma^2) \\ \text{Level 2:} \quad & \alpha_i \sim N(\alpha_{(0)}, \tau^2_\alpha) \\ \text{Hyperpriors:} \quad & \alpha_{(0)} \sim N(a, b^2) \\ & \tau^2_\alpha \sim IG(g, h) \\ & \sigma^2 \sim IG(q, r) \end{aligned}$$

To ensure the priors are uninformative, we set $b^2 = 10^6$ with $a = 0$, and $g = h = q = r = .001$. We used 2000 iterations as burn-in and another 2000 iterations for estimation. The results are given in Table 8.4.

In Bayesian estimation of a multilevel model, the random effects are typically not integrated out of the model. As a result, we also obtain estimates of the actual random

TABLE 8.4

Results for Four Multilevel Models on Repeated Measures of BMI

Parameter	Model 1	Model 2a	Model 2b	Model 3
$\alpha_{(0)}$	27.64 (.077)	27.28(.075)	27.29(.075)	27.29(.078)
	[27.50, 27.81]	[27.14, 27.43]	[27.13, 27.43]	[27,13, 27.44]
$\beta_{(0)}$		.12(.008)	.12(.008)	.11(.007)
		[.10, .13]	[.10, .13]	[.10, .12]
τ^2_α	18.75(.459)	18.08(.464)	18.03(.483)	18.10(.465)
	[17.85, 19.64]	[17.19, 18.99]	[17.06, 18.99]	[17.24, 19.07]
τ^2_β		.09(.005)	.09(.005)	.09(.005)
		[.08, .10]	[.08, .10]	[.08, .10]
$\tau_{\sigma\beta}$			.02(.034)	
			[−.05, .09]	
$\alpha_{(1)}$				.67(.105)
				[.46, .88]
$\alpha_{(2)}$				−.22(.11)
				[−.44, .01]
σ^2	2.62(.036)	1.94(.03)	1.94(.034)	1.94(.032)
	[2.55, 2.69]	[1.88, 2.01]	[1.88, 2.01]	[1.88, 2.00]

Note: For each parameter in the model the posterior mean is given, with the posterior standard deviation in parentheses. Underneath the 95% credibility interval is given between square brackets. Parameters: $\alpha_{(0)}$ is the average intercept; $\beta_{(0)}$ is the average slope; τ^2_α is the variance of the random intercept; τ^2_β is the variance of the random slope; $\tau_{\sigma\beta}$ is the covariance between the random intercept and random slope; $\alpha_{(1)}$ is the effect of transitioning into a relationship; $\alpha_{(2)}$ is the effect of transitioning out of a relationship; and σ^2 is the residual variance.

effects for each individual. In Figure 8.4 we have plotted the intercept estimates α_i (i.e., the posterior distributions) of all 3355 individuals (panel a), and the density of these (panel b). From this it can be seen that the distribution of the intercepts is skewed upwards. In addition we have plotted the 2000 draws from the conditional posteriors of the mean $\alpha_{(0)}$ and the variance of the random intercept τ_{α}^2. The random draws over 2000 iterations in panels c and e show that the Gibbs sampler converged. The density plots given in panels d and f show that these distributions are quite symmetric.

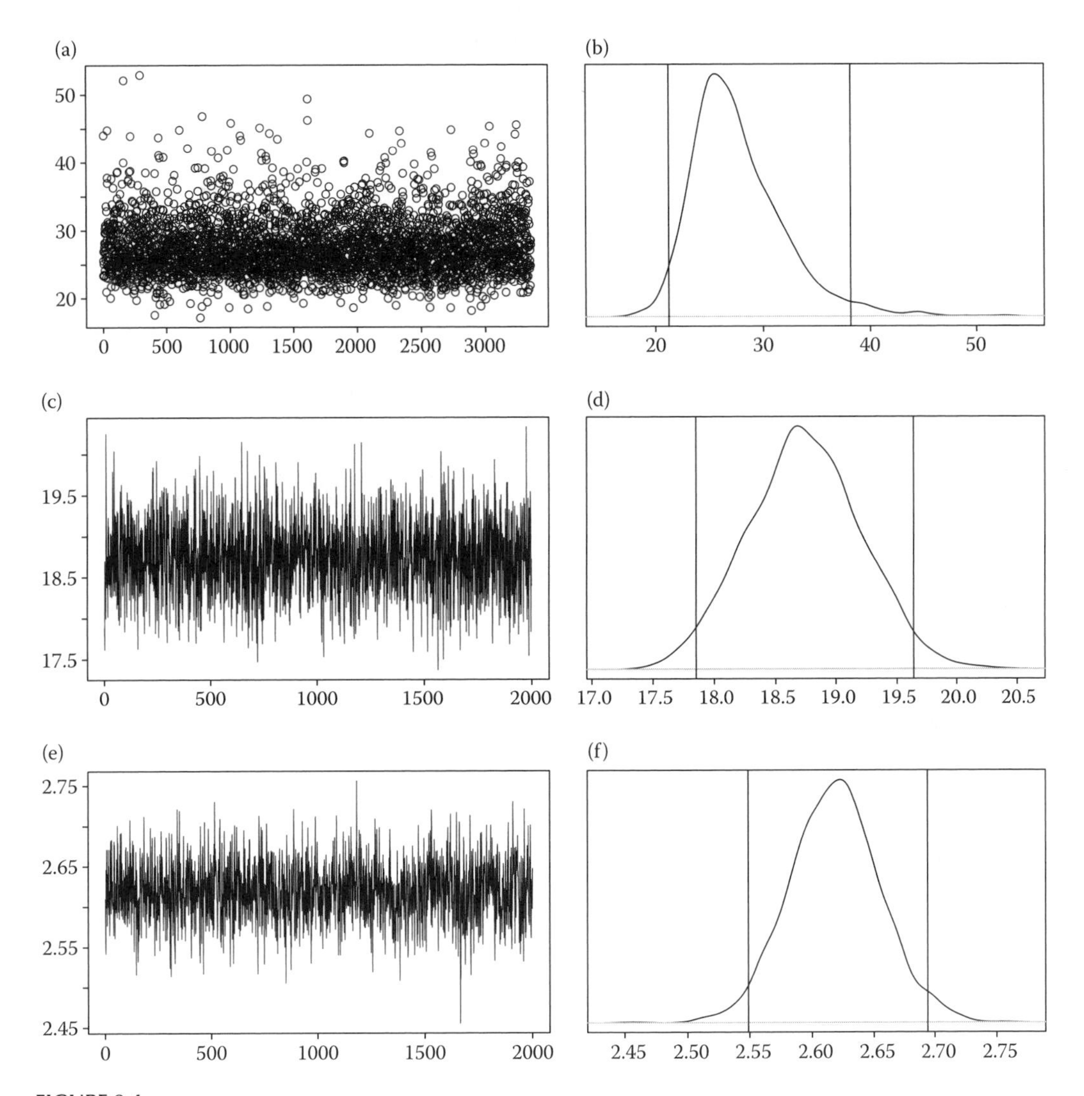

FIGURE 8.4
Panel a represents the individual intercepts plotted against participant number. These estimates were obtained by taking the mean of each individual's posterior distributions of the intercept. Panel b contains the density of these individual intercepts with the 95% (central) interval indicated by the vertical lines. Panel c are the Gibbs samples for the mean intercept (plotted against iteration), and panel d is the posterior density of the mean intercept with the 95% credibility interval. Panel e consists of the Gibbs samples of the variance of the intercept (plotted against iteration), and panel f is the posterior distribution of the variance with the 95% credibility interval.

Moreover, the Gibbs sampler can also be used to determine a credibility interval for a specific individual's intercept. In Figure 8.5 we plotted the posterior densities of the intercepts of the first nine individuals in the sample. Based on this density we have also determined the 95% credibility intervals for each of these individual intercepts. Hence, we can determine for each individual in the sample whether his average BMI over the 6-year period covered by the current study differs from, for instance, 25 (the upper bound of a healthy BMI), or 30 (the upper bound of the category "overweight").

8.4.3 Model 2: Growth Curve Model

The random intercept model can be extended to include a random slope β_i for the predictor time that we denote as T_{it}. Because we

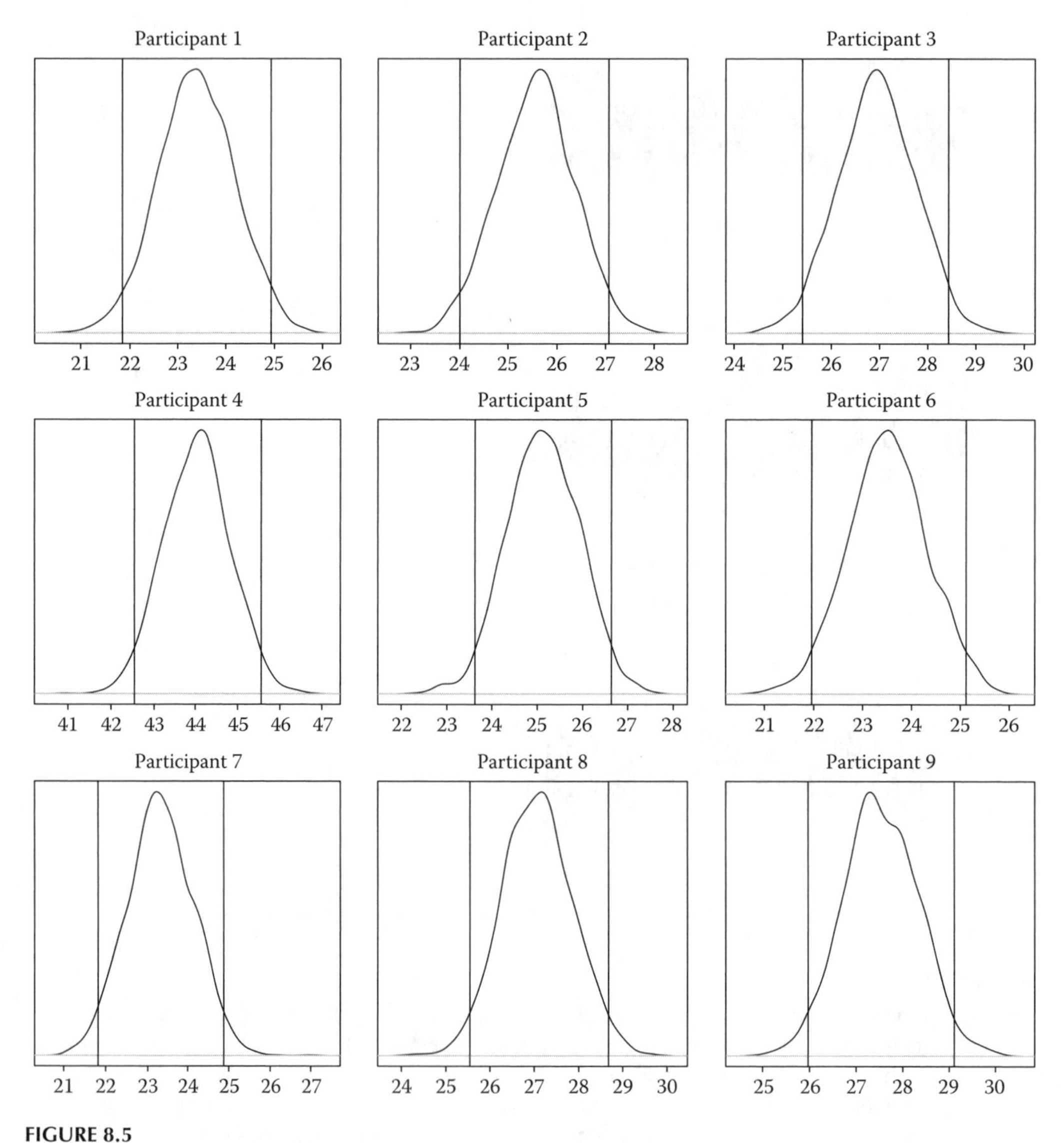

FIGURE 8.5
Posterior distributions of the intercepts of the first nine individuals in the sample. Vertical lines indicate the bounds of the 95% credibility interval.

have biannual measurements, we use $T_{i1} = 0$, $T_{i2} = 2$, $T_{i3} = 4$, and $T_{i4} = 6$. This implies that α_i is the (predicted) BMI value of individual i at the first measurement occasion, while β_i is the annual change of individual i over the 6 years.

We begin with a growth model in which the random intercept and random slope are uncorrelated. This model, which we refer to as model 2a, can be written as

$$
\begin{aligned}
\text{Level 1:} \quad & y_{it} \sim N(\alpha_i + \beta_i T_{it}, \sigma^2) \\
\text{Level 2:} \quad & \alpha_i \sim N(\alpha_{(0)}, \tau^2_\alpha) \\
& \beta_i \sim N(\beta_{(0)}, \tau^2_\beta) \\
\text{Hyperpriors:} \quad & \alpha_{(0)} \sim N(a, b^2) \\
& \tau^2_\alpha \sim IG(g, h) \\
& \beta_{(0)} \sim N(c, d^2) \\
& \tau^2_\beta \sim IG(u, v) \\
& \sigma^2 \sim IG(q, r)
\end{aligned}
$$

To ensure the priors are vague, we set $a = c = 0$, $b^2 = d^2 = 10^6$, and $g = h = q = r = .001$. The results are presented in Table 8.4. From this we can see that there is an average increase in BMI of .12 per year.

Alternatively, one can include a covariance between the random intercept α_i and the random slope β_i. To this end, the random parameters need to be drawn from a bivariate normal distribution. Let $\boldsymbol{\theta}_i = [\alpha_i \; \beta_i]'$ come from a multivariate normal distribution with mean $\boldsymbol{\theta}_{(0)} = [\alpha_{(0)} \; \beta_{(0)}]'$, and covariance matrix Σ, which contains τ^2_α and τ^2_β on the diagonal, and the covariance $\tau_{\alpha\beta}$ as the off-diagonal element. Using this setup requires us to specify hyperprior distributions for the parameters in $\boldsymbol{\theta}_{(0)}$, for instance, a multivariate normal distribution, and $\boldsymbol{\Sigma}_{(0)}$.

To ensure the covariance matrix $\boldsymbol{\Sigma}$ is symmetric and positive definite, we use an inverse-Wishart distribution as its prior distribution. This distribution is defined by degrees of freedom k, which can be interpreted as the prior effective sample size (i.e., it is the weight we wish to give to the prior), and a matrix $\mathbf{R}$, which can be interpreted as the prior estimate for the covariance matrix. While the effect of specific choices for k and $\mathbf{R}$ have been investigated in a simulation study (Browne & Draper, 2000), defining a vague inverse Wishart prior remains a difficult task. Here we consider two options suggested by Browne and Draper (2000): an inverse-Wishart distribution with $\mathbf{R}$ equal to the covariance matrix obtained with restricted maximum likelihood estimation and $k = 4$; and $\mathbf{R} = \mathbf{I}_2$, and $k = 2$.

The growth curve model with correlated intercept and slope, which we refer to as model 2b, can be written as

$$
\begin{aligned}
\text{Level 1:} \quad & y_{it} \sim N(\alpha_i + \beta_i T_{it}, \sigma^2) \\
\text{Level 2:} \quad & \boldsymbol{\theta}_i \sim MN(\boldsymbol{\theta}_{(0)}, \boldsymbol{\Sigma}) \\
\text{Hyperpriors:} \quad & \boldsymbol{\theta}_{(0)} \sim MN(\boldsymbol{f}, \mathbf{G}) \\
& \boldsymbol{\Sigma}^{-1} \sim Wish(\mathbf{R}, k) \\
& \sigma^2 \sim IG(q, r)
\end{aligned}
$$

By setting the diagonal elements of $\mathbf{G}$ large, we ensure that the prior distribution of $\boldsymbol{\theta}_{(0)}$ is vague too.

Given the large data set, the different hyperparameters for the inverse-Wishart distribution had virtually no effect on the results. The results are presented in Table 8.4. Note that the 95% credibility interval of the covariance includes zero, such that we can conclude that the covariance is probably not (very) different from zero. Since

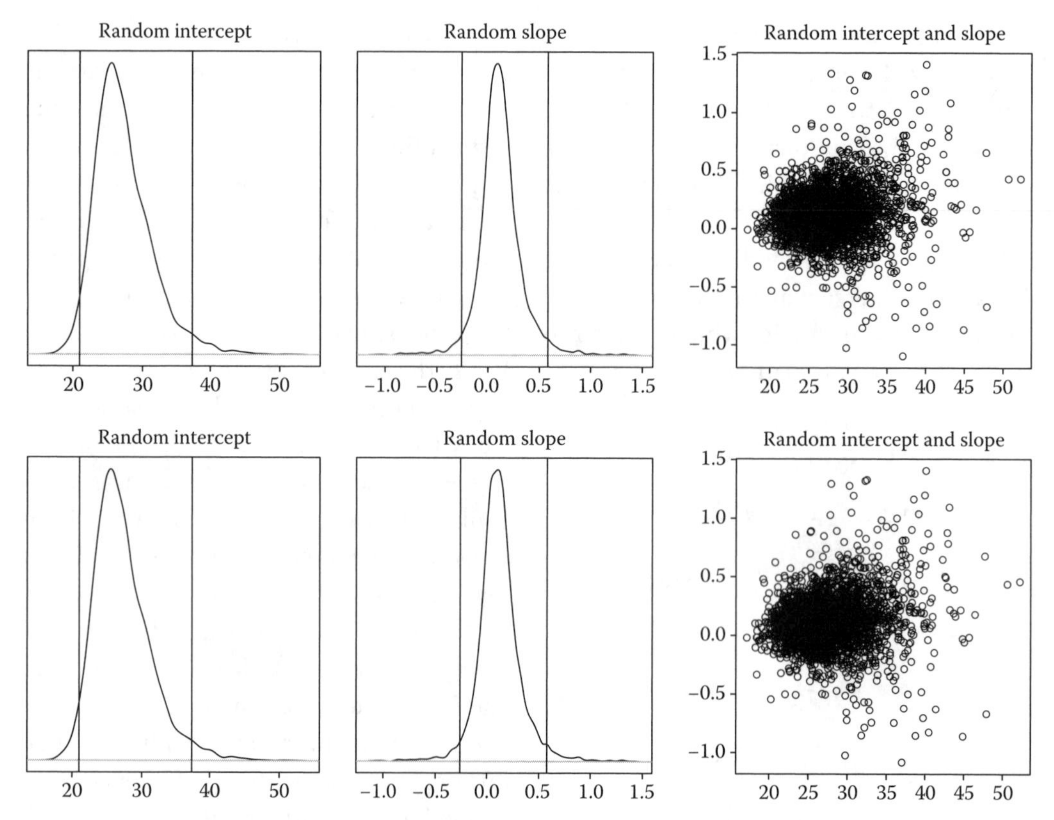

FIGURE 8.6
Results for the random intercept and random slope of model 2a (top) and model 2b (bottom).

we find correlations easier to interpret, we also let WinBUGS compute the correlation between the intercept and slope at each iteration of the Gibbs sampler, based on the latest samples of the variances and the covariance. This way, we also obtain an approximation of the marginal posterior distribution of the correlation, from which we can make inferences. The posterior mean for the correlation is .013 (posterior SD = .027), and its 95% credibility interval lies between −.037 and .067.

In Figure 8.6 the random intercepts and random slopes (i.e., the posterior means per individual) are presented for the two models: the top contains the results for the uncorrelated intercept and slope (model 2a), while the bottom contains the results for the correlated intercept and slope (model 2b). This also shows clearly that adding the correlation to the model has virtually no influence on the estimation of the random intercepts and slopes.

8.4.4 Model 3: Growth Model With Time-Varying Predictors

The growth curve model used above can be easily extended with predictors that are either time-invariant—meaning they are included in the level-2 equation(s)—or time-varying predictors—meaning they are included in the level 1 equation. Here we consider an atypical example of the latter, for which we use information obtained at each measurement occasion on whether a participant was cohabiting with a spouse, or living without a spouse.

We are interested in whether the transition from singlehood to a relationship, or vice versa, affects the change in BMI. To this end, we make six dummy variables, two for each of the occasions 2, 3, and 4, such that x_{1it} is a dummy that identifies the individuals who *transition into a relationship* between $t-1$ and t, and x_{2it} is a dummy that identifies individuals that *transition out of a relationship* between $t-1$ and t. Note that since we have no information on relationship status before the observations started, we do not have dummies for the first occasion.

We assume each individual has a single growth parameter for all time points, denoted as β_i. If a transition into or out of a relationship is associated with an increase or decrease in BMI, this effect is assumed to continue to exist at the subsequent occasions. For instance, suppose that a particular individual $i = g$ transitioned into a relationship between occasions 2 and 3, and did not experience transitions out of a relationship. Then, for the four occasions in the study, the scores of this individual g can be expressed as

$$
\begin{aligned}
y_{g1} &= \alpha_g + e_{g1} \\
y_{g2} &= \alpha_g + 2\beta_g + e_{g2} \\
y_{g3} &= \alpha_g + 4\beta_g + \alpha_{(1)} + e_{g3} \\
y_{g4} &= \alpha_g + 6\beta_g + \alpha_{(1)} + e_{g4}.
\end{aligned}
$$

We interpret $\alpha_{(1)}$ as a shift in the individual's intercept, such that it persists after the transition has taken place. Hence, we obtain an intercept that is time-dependent, such that for this individual we can write $\alpha_{g1} = \alpha_{g2} = \alpha_g$, and $\alpha_{g3} = \alpha_{g4} = \alpha_g + \alpha_{(1)}$.

More generally, we can express the time-dependent intercepts as: $\alpha_{i1} = \alpha_i$, and for $t = 2, 3, 4$ we can write $\alpha_{it} = \alpha_{i,t-1} + \alpha_{(1)} x_{1it} + \alpha_{(2)} x_{2it}$. Note that by including the individual's intercept at the previous occasion (i.e., $\alpha_{i,t-1}$), changes that happened in the past are maintained in the model.

This model, which we refer to as model 3, can be expressed as

$$
\begin{aligned}
\text{Level 1:}\quad & y_{i1} \sim N(\alpha_{i1}, \sigma^2) \\
& y_{it} \sim N(\alpha_{it} + \beta_i T_{it}, \sigma^2) \\
& \quad \textit{for}\ t = 2, 3, 4 \\
& \alpha_{it} = \alpha_{i,t-1} + \alpha_{(1)} x_{1it} + \alpha_{(2)} x_{2it} \\
& \quad \textit{for}\ t = 2, 3, 4 \\
\text{Level 2:}\quad & \alpha_{i1} \sim N(\alpha_{(0)}, \tau_\alpha^2) \\
& \beta_i \sim N(\beta_{(0)}, \tau_\beta^2) \\
\text{Hyperpriors:}\quad & \alpha_{(0)} \sim N(a_0, b_0^2) \\
& \alpha_{(1)} \sim N(a_1, b_1^2) \\
& \alpha_{(2)} \sim N(a_2, b_2^2) \\
& \tau_\alpha^2 \sim IG(g, h) \\
& \beta_{(0)} \sim N(c_0, d_0^2) \\
& \tau_\beta^2 \sim IG(u, v) \\
& \sigma^2 \sim IG(q, r)
\end{aligned}
$$

such that $\alpha_{(1)}$ denotes the average additional change in BMI for men transitioning into a relationship, while $\alpha_{(2)}$ denotes the average additional change in BMI for men transitioning out of a relationship.

The results for this model are presented in the last column of Table 8.4. Note that in comparison to model 2a, the estimates of the fixed effects $\alpha_{(0)}$ and $\beta_{(0)}$, and the

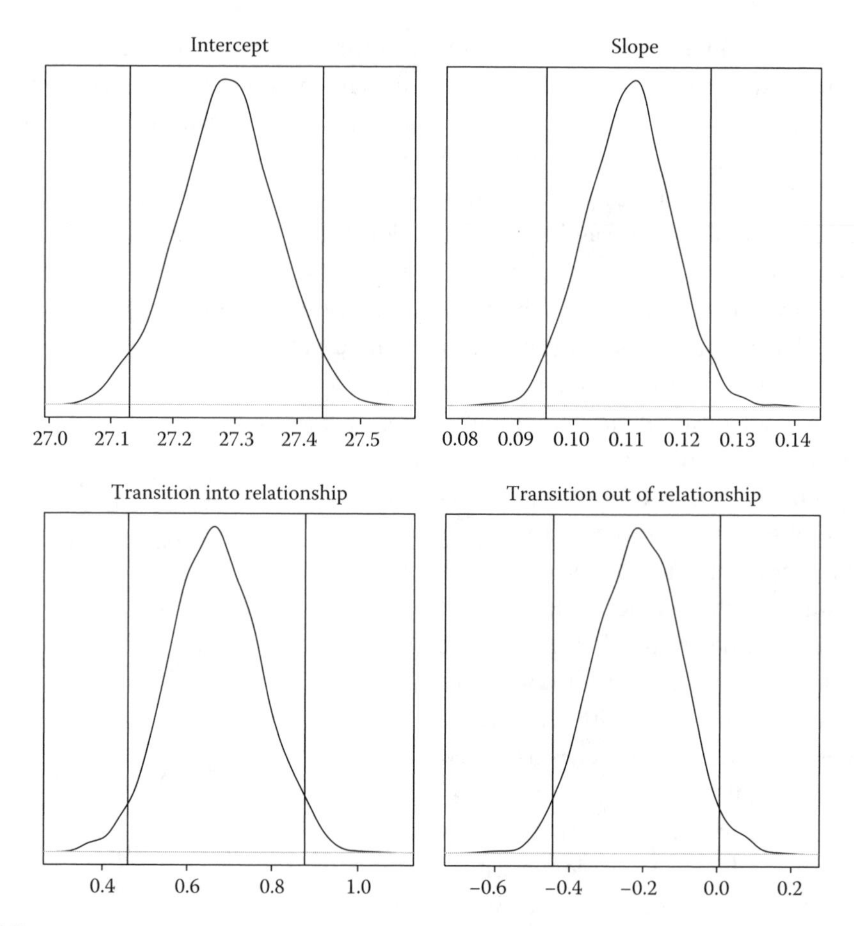

FIGURE 8.7
Posterior distributions of the intercept, slope and the two regression parameters for the dummy variables indicating a transition into or out of a relationship; 95% credibility intervals are given for each parameter.

variance components τ^2_α, τ^2_β, and σ^2 have not changed much. In Figure 8.7 the posterior distributions are plotted of the average intercept $\alpha\beta_{(0)}$, the average slope $\beta_{(0)}$, the shift in intercept due to a transition into a relationship $\alpha_{(1)}$, and the shift in intercept due to a transition out of a relationship $\alpha_{(2)}$. From the 95% credibility intervals it can be concluded that the transition into a relationship implies an additional increase in BMI, while a transition out of a relationship is not associated with any additional change in BMI. The posterior mean for $\alpha_{(1)}$ is .67. Comparing this to the average annual change in BMI (i.e., $\beta_{(0)} = .11$), it can be stated that transitioning into a relationship is associated with an additional increase in BMI that is equal to the average increase in BMI over a 6-year period!

8.4.5 Some Additional Remarks on Priors for Variance Components in Multilevel Models

For the priors of the random effects in the illustration above we have used either inverse-gamma(ε, ε) distributions, or an inverse-Wishart($\mathbf{R}$, k) distribution (to allow for a correlated random intercept and random slope). These priors are popular choices in Bayesian multilevel analysis, because they are conditionally conjugate, which makes them easy to combine with Gibbs sampling (Gelman, 2006, p. 517).

However, there are some indications that these prior distribution may not be appropriate in multilevel modeling if one wishes to use vague priors. Regarding the inverse gamma distribution, Gelman (2006) has shown that this can lead to large shrinkage of the random effects (i.e., they are shrunk to the mean), such that the posterior distribution of the standard deviation of the random effect peaks close to zero. Moreover, if the standard deviation of the random effect is estimated close to zero, the resulting inferences will be sensitive to the choice of ε. Therefore, Gelman (2006) advises using a noninformative uniform prior on the standard deviation (instead of the variance) of the random effect.

We investigated whether the choice of ε had an effect on the results in the random intercept model (model 1), by comparing the results obtained with $\varepsilon = .001$, with the results obtained with $\varepsilon = 1$. This only had a very minor effect on the posterior distribution of the variance of the random intercept: the posterior mean changed from 18.75 to 18.73; the posterior SD remained the same; and the credibility interval changed from [17.85,19.64] to [17.84,19.63].

Regarding the inverse-Wishart distribution, we have already indicated that it is difficult to choose **R** and k such that it is a vague prior. Gelman & Hill (2007) indicate that choosing k equal to the dimension of $\theta_{(0)}$ plus 1, results in a uniform prior on the correlations associated with Σ, but it has the undesirable effect that the variances in Σ are quite constrained. Using a different value for k results in a less informative prior for the variances, but constrains the correlations (Gelman & Hill, p. 289). To overcome these restrictions, Gelman and Hill (2007) suggest using a scaled inverse Wishart prior, which is obtained by pre- and post-multiplying the matrix from the inverse Wishart distribution by a vector with scale parameters (see p. 377, Gelman & Hill, 2007 for an example using bugs()). Note however that in our example (model 2b), the different choices of **R** and k had no effect on the results, due to the large sample size involved.

8.5 DISCUSSION

The current chapter presented an introduction into Bayesian estimation of multilevel models. At the beginning of this chapter we gave a number of general arguments for preferring a Bayesian approach over the classical approach. Here we emphasize several of these arguments specifically in the context of multilevel modeling, and discuss an additional argument that is (more) exclusively associated with multilevel modeling.

As indicated before, Bayesian estimation is not based on normality assumptions and asymptotic results. In the context of multilevel modeling this is an important vantage over the classical approach, because (higher level) variances play a key role in these models, and it is well-known that the sampling distribution of a variance is only approximately normal if sample size is large enough. Maas and Hox (2004) performed a simulation study in which they showed that standard errors of level-2 variances may be seriously underestimated when the number of level-2 units is small (say less than 30 clusters), or when the level 2 residuals are nonnormal (e.g., the random intercept in model 1, see panel b in Figure 8.4). Moreover, even robust standard errors (based on the sandwich estimator) are too small for level 2 variance components (Maas & Hox, 2004).

Browne (1998) and Browne and Draper (2000) performed a number of simulation studies to compare diverse likelihood based estimation procedures with Bayesian results obtained with MCMC methods in multilevel models. Overall the conclusion from these simulation studies is that while Bayesian estimates are biased, the coverage rate of their $100(1-\alpha)\%$ credibility intervals are in general much closer to $1-\alpha$ than the rates obtained with the likelihood based $100(1-\alpha)\%$ confidence intervals.[4] Because the normal distribution is a bad approximation for level 2 variances in small samples, Browne (1998) also constructed $100(1-\alpha)\%$ confidence intervals based on the inverse Gamma distribution, but this did not solve the problem. In sum it can be concluded that if the focus is not just on the point estimates, but also on making inferences based on intervals around point estimates, Bayesian estimation outperforms the maximum likelihood approaches for small (level 2) samples in multilevel modeling.

Another issue that comes up in multilevel modeling—particularly with small sample sizes at level 2—is that one can obtain a negative estimate for the variance of a random effect (cf. Browne, 1998; Maas & Hox, 2004). Clearly, from a statistical point of view, this is impossible, and the typical way to handle this is by fixing the variance to zero. In Bayesian estimation however, the entire posterior distribution will lie in the parameter space, and therefore the parameter estimate obtained will always lie in the sample space.

A third advantage of Bayesian estimation is the ease with which missing data are handled. We have already shown how missing observations in the dependent variable are dealt with in the Gibbs sampler, and indicated that this is done automatically in WinBUGS. Moreover, Carrigan, Barnett, Dobson, and Mishra (2007) show how missingness in a covariate in longitudinal data can be handled using WinBUGS. They also indicate that WinBUGS can easily deal with missing categorical data, whereas most other packages rely on an assumption of normality such that categorical data cannot be imputed (Carrigan et al., 2007).

Another advantage of Bayesian estimation is that it can handle complicated problems that are difficult to tackle within the classical approach. For instance, Wang and McArdle (2008) showed in a simulation study that Bayesian estimation of multilevel change-point models leads to better results than two classical estimation procedures. In particular, the Bayesian method was not affected by the initial values, while the two classical methods only gave unbiased results if the initial values were close to the true values. Another example can be found in Oravecz, Tuerlinckx, and Vandekerckhove (2009), who based their multilevel model on differential equations to allow for unequal intervals between the repeated measurements of individuals. Such unequal intervals are characteristic of many diary studies. They indicate that the classical approach would require high-dimensional integration over the numerous random effects distributions, which is not necessary in the Bayesian approach. Finally, in Bayesian estimation it is easy to incorporate inequality constraints on model parameters, including

[4] For multilevel logistic regression Browne (1998) found that the MCMC methods outperformed the likelihood based methods both in terms of bias and coverage rates.

multilevel models (Kato & Hoijtink, 2006; Kato & Peeters, 2008).

A fifth advantage that was mentioned and illustrated in this chapter, is that with MCMC methods it is easy to obtain additional quantities that may be of interest. Another example of this in multilevel modeling can be found in Hoijtink (2000), who considered a traditional random intercept model for a data set consisting of test results from children who are nested in schools. Within each iteration of the Gibbs sampler, the rank order of the schools was determined, based on the random intercepts. By saving the rank order numbers, and plotting these afterwards with their 95% credibility intervals, it became easy to see that for instance the first 10 schools were not different with respect to their ranking. Such information would be extremely difficult, if not impossible, to obtain within the classical approach.

Besides these general advantages of Bayesian estimation, there is an additional advantage that specifically applies to multilevel modeling. In the previous section, we have shown that Bayesian analysis results in the simultaneous estimation of model parameters (i.e., both fixed effects and the variance components), and the actual random effects. The latter are of interest for model evaluation, for instance, to check for outliers (Hox, 2002, pp. 22–30). Moreover, estimates of the actual random effects are essential if one wishes to make inferences at the cluster level, for instance, to rank schools (Hoijtink, 2000), or to select individuals who need treatment (Candel, 2004).

However, in the classical approach to multilevel modeling, the cluster effects are integrated out of the likelihood function, such that only the population parameters (i.e., fixed effects and variance components) are estimated. In order to obtain estimates of the random effects, one can use what is known as the empirical Bayes estimates (EBE). However, Candel (2004) has shown in a simulation study that the quality of the EBEs depends on whether one uses full maximum likelihood or restricted maximum likelihood, and whether negative second level variances are fixed to zero or not. In addition, the performance may depend on sample sizes at level 1 and 2, and on the intraclass correlation (Candel, 2004).

In comparison, the advantage of taking a truly Bayesian approach as discussed in this chapter is that the random effects α_i are not integrated out of the model. Hence, rather than having to estimate the random effects using other parameter estimates (for the variance components and the fixed effects), the random effects are sampled along with all the other unknown model parameters. In the random effects model we illustrated that for each level-2 unit we obtain a posterior distribution, from which we can use the mean or median as the parameter estimate, and construct a credibility interval for which we do not have to rely on large sample theory.

In conclusion, Bayesian estimation helps to tackle some of the problems that are frequently encountered in multilevel modeling. The availability of software packages such as WinBUGS and MLwiN, which allow for Bayesian estimation of multilevel models, makes it an easy to use and attractive alternative to the classical approach. Moreover, Bayesian estimation opens up possibilities that are difficult or impossible within the classical context, but may be valuable to the researcher using multilevel modeling.

In this chapter we have focused exclusively on Bayesian estimation, as it is the first step in Bayesian analysis. Additional steps in Bayesian analysis consist of hypothesis testing using posterior predictive p-values, and model selection using posterior model probabilities or the Bayes factor. While these topics are beyond the scope of the current chapter, it is important to note that, besides the advantages of Bayesian estimation brought forward in this chapter, there are additional reasons for choosing a Bayesian approach, which are associated with Bayesian hypothesis testing and Bayesian model selection (cf. Kass & Raftery, 1995; Meng, 1994).

ACKNOWLEDGMENT

This work was supported by the Netherlands Organization for Scientific Research (NWO), VENI grant 451-05-012 awarded to E. L. Hamaker.

REFERENCES

Berger, J. (2006). The case for objective Bayesian analysis. *Bayesian Analysis, 1,* 385–402.

Browne, W. J. (1998). *Applying MCMC methods to multilevel models.* PhD dissertation, Department of Mathematical Sciences, University of Bath, UK.

Browne, W. J., & Draper, D. (2000). Implementation and performance issues in the Bayesian single and likelihood fitting on multilevel models. *Computational Statistics, 15,* 391–420.

Candel, M. J. J. M. (2004). Performance of empirical Bayes estimators of random coefficients in multilevel analysis: Some results for the random-intercept model. *Statistica Neerlandica, 58,* 197–219.

Carrigan, G., Barnett, A., Dobson, A. J., & Mishra, G. (2007). Compensation for missing data from longitudinal studies using WinBUGS. *Journal of Statistical Software, 19,* 1–17.

Chen, M. H., Shao, Q. M., & Ibrahim, J. G. (2000). *Monte Carlo methods in Bayesian computation.* New York, NY: Springer-Verlag.

Chib, S., & Greenberg, E. (1995). Understanding the Metropolis Hastings algorithm. *The American Statistician, 49,* 327–335.

Cowles, M. K., & Carlin, B. P. (1996). Markov chain Monte Carlo convergence diagnostics: A comparative review. *Journal of the American Statistical Association, 91,* 883–904.

Gelfand, A. E., & Smith, A. F. M. (1990). Sampling-based approaches to calculating marginal densities. *Journal of the American Statistical Association, 85,* 398–409.

Gelman, A. (2006). Prior distribution for variance parameters in hierarchical models. *Bayesian Analysis, 1,* 515–533.

Gelman, A., Carlin, J. B., Stern, H. S., & Rubin, D. (2004). *Bayesian data analysis* (2nd ed.). Boca Raton, FL: Chapman and Hall/CRC.

Gelman, A., & Hill, J. (2007). *Data analysis using regression and multilevel/hierarchical models.* Cambridge, NY: Cambridge University Press.

Gelman, A., & Rubin, D. B. (1992). Inference from iterative simulation using multiple sequences. *Statistical Science, 7,* 457–511.

Geman, S., & Geman, D. (1984). Stochastic relaxation, Gibbs distributions, and the Bayesian restoration of images. *IEEE Transactions on Pattern Analysis and Machine Intelligence, 6,* 721–741.

Gilks, W. R., Richardson, S., & Spiegelhalter, D. J. (1996). *Markov Chain Monte Carlo in practice.* London, UK: Chapman and Hall.

Gill, J. (2008). *Bayesian methods: A social and behavioral sciences approach* (2nd ed.). Boca Raton, FL: Chapman and Hall/CRC.

Hastings, W. K. (1970). Monte Carlo sampling methods using Markov chains and their applications. *Biometrika, 57,* 97–109.

Hoijtink, H. (2000). Posterior inference in the random intercept model based on samples obtained with Markov chain Monte Carlo methods. *Computational Statistics, 15,* 315–336.

Hox, J. (2002). *Multilevel analysis: Techniques and application.* Mahwah, NJ: Lawrence Erlbaum Associates.

Kass, R., & Raftery, A. (1995). Bayes factors. *Journal of the American Statistical Association, 90,* 773–795.

Kato, B. S., & Hoijtink, H. (2006). A Bayesian approach to inequality constrained linear mixed models: Estimation and model selection. *Statistical Modeling, 6,* 231–249.

Kato, B. S., & Peeters, C. F. W. (2008). Inequality constrained multilevel models. In H. Hoijtink, I. Klugkist, & P. A. Boelen (Eds.), *Bayesian evaluation of informative hypotheses* (pp. 273–295). New York, NY: Springer.

Lynch, S. M. (2007). *Introduction to applied Bayesian statistics and estimation for social scientists.* New York, NY: Springer Science + Bussiness Media.

Maas, C. J. M., & Hox, J. J. (2004). Robustness issues in multilevel regression analysis. *Statistica Neerlandica, 58,* 127–137.

Meng, X.-L. (1994). Posterior predictive p-values. *The Annals of Statistics, 33,* 1142–1160.

Oravecz, Z., Tuerlinckx, F., & Vandekerckhove, J. (2009). A hierarchical Ornstein-Uhlenbeck model for continuous repeated measurement data. *Psychometrika, 74,* 395–418.

Poirier, D. J. (2006). The growth of Bayesian methods in statistics and economics since 1970. *Bayesian Analysis, 1,* 969–980.

Press, S. J. (2003). *Subjective and objective Bayesian analysis: Principles, models and applications* (2nd ed.). Hoboken, NJ: John Wiley and Sons, Inc.

Robert, C. P., & Casella, G. (1999). *Monte Carlo statistical methods.* New York, NY: Springer-Verlag.

Rubia, D. B. (1987). Multiple imputation for nourespouse in surveys. New York, NY: John Wiley & Sons.

Smith, A. F. M., & Roberts, G. O. (1993). Bayesian computation via the Gibbs sampler and related Markov chain Monte Carlo methods. *Journal of the Royal Statistical Society, Series B, 55,* 3–23.

Spiegelhalter, D., Thomas, D., Best, N., & Lunn, D. (2003). *Win-BUGS version 1.4 user manual.* Cambridge, UK: MRC Biostatistics Unit. Retrieved from: http://www.mrc-bsu.cam.ac.uk/bugs/

Sturtz, S., Ligges, U., & Gelman, A. (2007). R2winbugs version 2.1-6: A package for running winbugs from r. *Journal of Statistical Software*: Retrieved from: http://ideas.repec.org/a/jss/jstsof / 12i03. html.

Wang, L., & McArdle, J. J. (2008). A simulation study comparison of Bayesian estimate with conventional methods for estimating unknown change points. *Structural Equation Modeling, 15,* 52–74.

9

Bootstrapping in Multilevel Models

Harvey Goldstein
Centre for Multilevel Modelling, Graduate School of Education, University of Bristol, Bristol, United Kingdom

9.1 WHAT IS A BOOTSTRAP SAMPLE?

The bootstrap was devised for situations where the usual estimators may be biased, or where exact inference is required, for example, because distributional assumptions are not satisfied or only large sample properties hold (Davison & Hinckley, 1997).

Consider a simple random sample of n observations $x_1, \ldots x_n$, from which we wish to estimate a population quantity, say a mean or median. We choose an estimator, say $\bar{x}$, and we wish to estimate features of the distribution of this estimator, say its standard error or population quantiles. The simplest *nonparametric* bootstrap is obtained as described below.

We sample randomly (i.e., according to the assumed mechanism that generated the observations), with replacement, n observations from the original sample. Denote this by $X^* = \{x_1^*, \ldots, x_n^*\}$. Then we can obtain B of these bootstrap samples, $X^{*1}, X^{*2}, \ldots, X^{*B}$. For each of these we calculate our estimate, say of the mean, and each of these is referred to as a bootstrap *replicate*. It is such replicates that are used for inference. In this chapter we shall first review some standard uses for the bootstrap and then discuss extensions to multilevel data.

9.2 STANDARD ERROR AND QUANTILE ESTIMATES FROM THE BOOTSTRAP

For a set of nonparametric bootstrap replicates we can calculate the standard deviation of, say, a mean, or a set of distribution quantiles, to make our inferences. Thus, we can get an *estimate* of the standard error of the mean simply by calculating the standard deviation of the bootstrap replicates for the sample mean. Note, however, that this is only an estimate: it is based on a finite sample of bootstraps. As the sample size tends to infinity, this becomes more accurate and approximates the *ideal bootstrap estimate.* If θ^* is a bootstrap estimator then the ideal bootstrap estimate for the standard error is the square root of $E_F(\theta^* - \theta)^2$, where F is the distribution function for the data.

We note, however, that even as the number of bootstrap replications tends to infinity, the estimate of the population density function that is used to generate the bootstrap samples is the empirical "plug in" one derived from the actual sampled observations by placing mass points (e.g., equal probabilities) at each one. In other words the sample is assumed to be a reasonable representation of the population. Thus, with nonparametric bootstrapping, we do not have exact inference. This does not carry over to the parametric case that we describe below, where the model-based (assumed) population distribution is used for sampling: we shall return to this case later. In fact, in some situations the nonparametric bootstrap can perform very badly, for example, in small or moderate samples where the statistic of interest is the smallest or largest value, say of a set of higher level residuals in a multilevel model.

In practice, we would normally wish to stop generating bootstrap replicates by inspecting the running estimate of the quantity of interest. This estimate is simply the value computed after a particular chosen number of replications. When this "settles down" to a value with a predetermined accuracy—for example in terms of the coefficient of variation—we do not sample further replicates. In fact the coefficient of variation depends on the underlying distribution so will often not be useful when that is unknown. Clearly we require a general practical stopping rule. What will be important also is visual inspection of the updated histogram of values and a smoothed density function.

A further consideration, as with all statistical analysis, is the detection of rogue values or "outliers," in this case individual replicates. Density displays and box and whisker plots are useful diagnostic tools here. We should be careful about discarding extreme values, since these will naturally occur given enough bootstrap replicates, and as an alternative we might possibly use robust estimators of the standard error of replicates. One such would be

$$\frac{\hat{\theta}^{*(\alpha)} - \hat{\theta}^{*(1-\alpha)}}{2z^{(\alpha)}} \tag{9.1}$$

where $z^{(\alpha)}$ is the 100 αth percentile of the standard Normal distribution and $\theta^{*(\alpha)}$ is the 100 α th percentile of the bootstrap estimator from the observed bootstrap replications, with α typically being taken as 0.90 or 0.95. Unless the distribution of the bootstrap estimator is Normal, this is biased and inspection of the bootstrap density function and the use of Normal plots will show how good the Normal approximation is in any particular case.

In some cases, for example, estimating a mean or a set of regression coefficients, the standard error of a bootstrap sample can be obtained analytically, depending only on functions of covariates (e.g., a cross product matrix) and the residual variance of the observations that is obtained from the original analysis. This does not carry over directly to the multilevel case, where the standard errors are functions of the parameters, but we can study the accuracy of these estimated standard errors via the bootstrap replications.

9.3 BOOTSTRAPS FOR COMPLEX DATA STRUCTURES

We shall use as an illustration a two-level variance components model and the JSP data set containing test scores for primary school pupils (Goldstein, 2003, Chapter 2).

$$y_{ij} = (X\beta)_{ij} + u_j + e_{ij} \tag{9.2}$$

where the explanatory variables are 8 year math scores and gender and the response is 11 year math scores. We have simulated the response from the model results given in Goldstein (2003, Chapter 2) to ensure an approximately Normal distribution. Based on 887 pupils in 48 schools the maximum likelihood estimates are given in Table 9.1.

We now consider drawing a bootstrap sample. We first consider nonparametric versions. To do this we need to decide whether we are going to sample complete units or just residuals. In general it seems that we would wish to use the latter, since mostly we are concerned with conditional inference; that is, fixing the explanatory variables. In some situations, however, such as survey samples, it is more natural to think of all the variables as generated randomly so that complete unit selection is to be preferred.

TABLE 9.1

JSP 2 Level Variance Components Model Parameter Estimates

Fixed	
Intercept	16.06 (0.93)
Gender	−0.17 (0.37)
8 year math	0.58 (0.033)
Random	
Level-2	4.61 (1.32)
Level-1	29.32 (1.43)

The process of selecting a bootstrap sample corresponds to the supposed probabilistic mechanism that generated the data. This can be modeled as the selection of a simple random sample of school residuals according to their estimated distribution and within each school a sample of students according to the estimated distribution of the pupil residuals. As we shall see, this is appropriate for the parametric bootstrap but raises difficulties in the nonparametric case.

9.4 A NONPARAMETRIC MULTILEVEL BOOTSTRAP

In the nonparametric complete level-2 unit bootstrap suppose we sample, with replacement, a random sample of schools. This will in general lead to variable total numbers (N) of students across bootstrap samples. This procedure, however, does retain the data structure. The variability of N will add some noise to our estimates, but for moderate or large sample sizes this will be negligible and the procedure will be consistent. For each bootstrap sample we then fit our model.

Another possibility is to sample level-1 units directly. Having selected a random sample of the required size we then sort into their actual level-2 units. This leads to a variable number of level-2 units and a variable number of level-1 units per level-2 unit, but retains the overall number of level-1 units. This procedure, however, pays no attention to the sample structure since each level-1 unit is sampled independently so that the within-unit correlation structure is not preserved. Likewise, if we sample level-2 units and then sample level-1 units from within each level-2 unit, the *joint* probability of selection for two level-1 units within a level-2 unit is the product of their separate selection probabilities. This is also the case for two level-1 units from different level-2 units—in the balanced case these joint probabilities are $1/n^2$, where n is the size of each level-2 unit. Thus, again the within-unit correlation structure is not preserved since this should be a function also of the between-unit variation. In both these cases the independent selection of level-1 units will tend to add precision to the estimation of level-2 effects and so overestimate the level-2 variation in the bootstrap samples. We note that the same considerations apply if the level-1 units are selected with replacement, sorted into their level-2 units and then these level-2 units selected with replacement.

9.5 SIMPLE RESIDUALS BOOTSTRAP PROCEDURES

We now describe some different ways of sampling residuals to preserve the multilevel structure, in a bootstrap where we work with the estimated (posterior) residuals (possibly after centering them to ensure they have zero means). We show that these procedures, although intuitively attractive, do not provide a satisfactory solution, and in a later section we shall look at extensions.

We first sample with replacement the level-2 residuals, one for each level-2 unit. For each level-2 unit, sample with replacement the required number of level-1 residuals associated with that same bootstrap sampled level-2 residual. The required number is the number in the original data set for that level-2 unit. Note that in some cases this will mean sampling more level-1 residuals (with replacement) than there actually are associated with the chosen level-2 residual. The reason for this is that the level-1 and level-2 residual estimates are correlated and we need to preserve this correlation structure in our bootstrap sampling. Note that both level-1 and level-2 residuals are "shrunken": the variance of each is less than the population variances, but the correlation between them ensures that the variance of the sum is equal to the total residual variance. In the variance components case this is equivalent to using the raw residuals for each chosen level-2 unit and then sampling with replacement from these raw residuals to achieve the required number.

For each bootstrap sample we then carry out the estimation of the parameters of the model. The results of doing this, sampling residuals, for the JSP data is given in Table 9.2. We have chosen a sample of 500 after inspection of the running estimates of the parameters. A difficulty with this procedure, however, is that the amount of shrinkage is correlated with the school size. Thus, the larger level-2 residuals will also tend to have the largest number of level-1 residuals so that these will be given greater weight in the estimation. This violates the assumption that the random errors provided for the bootstrap should be independent of the unit sizes. This will tend to lead to an upward bias of the level-2 variance and this is confirmed in Table 9.2 (linked level-1 residuals). The alternative procedure of selecting level-1 residuals from the overall set of level-1 residuals will tend to reduce both level-2 and level-1 variances as is also shown in Table 9.2 (unlinked level-1 residuals).

A final possibility is to select, for each school, a set of linked level-1 + level-2 residuals and attaching these to the same number of sets of fixed variables by selecting these with replacement from each school. This,

TABLE 9.2

Results of 500 Bootstrap Replications for Four Bootstrap Procedures for the JSP Data in Table 9.1. Mean of Bootstraps (s.d. in brackets—estimating model s.e.)

	Sampling Complete Level-2 Units	Random Sample of Level-1 Units	Posterior Residuals—Linked Level-1	Posterior Residuals—Unlinked Level-1
Fixed coefficient				
Intercept	16.11 (0.91)	16.03 (1.09)	15.97 (1.40)	16.09 (0.95)
Gender	−0.14 (0.38)	−0.18 (0.38)	−0.18 (0.36)	−0.18 (0.36)
8 year math	0.55 (0.033)	0.58 (0.034)	0.58 (0.032)	0.58 (0.032)
Random				
Level-2 variance	4.46 (1.00)	6.34 (1.09)	6.62 (1.40)	3.11 (0.95)
Level-1 variance	29.27 (1.46)	27.75 (1.38)	26.69 (1.96)	28.12 (1.35)

however, destroys the sample structure and again leads to overestimation of the level-2 variation. In a later section we show how a restandardizing procedure can improve the residuals bootstrap.

We notice in Table 9.2 that only in the complete level-2 unit case do we obtain satisfactory estimates of the random parameters, for the reasons we have discussed.

9.6 THE PARAMETRIC BOOTSTRAP

In the parametric case, we sample first the level-2 residuals from the (estimated) level-2 distribution, in this case a simple Normal distribution. Then we sample level-1 residuals from the (estimated) level-1 distribution. The structure is preserved since, according to the model assumptions, the distributions are independent across levels. The procedure extends naturally to the random coefficient case.

Table 9.3 shows the results of a fully parametric bootstrap obtained by simulating from the estimated random parameters of the model.

We see now that the bootstrap estimates are very close to those from the fitted model and similar to those from the complete level-2 nonparametric bootstrap in Table 9.2, although the standard deviation for the level-2 variance in the latter case appears to be an underestimate. The restricted maximum likelihood level-2 estimate that corrects for the maximum likelihood bias is higher by an amount, which is the difference between the fitted estimate and the bootstrap one, implying that the bootstrap accurately corrects for the bias in the ML estimate of this parameter. This leads us to the topic of bias correction.

9.7 BOOTSTRAP BIAS CORRECTION

If the bootstrap estimate of a parameter (or other function of the data) is θ^* then the bias in the estimate $\hat{\theta}$ is $\theta^* - \hat{\theta}$. Thus the bias corrected estimate is $\hat{\theta} - (\theta^* - \hat{\theta}) = 2\hat{\theta} - \theta^*$. In some models the bias of the estimation procedure is a function of the parameter values, so that a simple bias correction will be an approximation only and an iterative

TABLE 9.3

JSP 2 Level Variance Components Model Parameter Estimates (Maximum Likelihood) and 500 Parametric Bootstraps

Fixed	Fitted Model (s.e.)	Bootstrap (s.d.)
Intercept	16.06 (0.93)	16.06 (0.93)
8 year math	–0.17 (0.37)	–0.17 (0.37)
Gender	0.58 (0.033)	0.58 (0.033)
Random		
Level-2	4.61 (1.32)	4.46 (1.29)
Level-1	29.32 (1.43)	29.20 (1.36)

Note: The restricted maximum likelihood estimates for the level-2 and level-1 variances are 4.76 and 29.30, respectively.

procedure will be necessary and this is the case for generalized linear multilevel models. We shall not give details here but a discussion is given by Goldstein (2003). Because a bias corrected estimate of a parameter may have greater variability, it is also routinely useful to check the *accuracy* of a bias corrected estimate by using it (or a set of these) as the basis for another set of bootstrap replications. As a rule of thumb, 500 bootstrap replications should be used for bias correction.

9.8 THE RESIDUALS BOOTSTRAP

One potential drawback to the fully parametric bootstrap is that it relies upon the (Normal) distribution assumption for the residuals. In single level linear models a residuals bootstrap can be implemented by fitting the model, calculating the empirical residuals by subtracting the predicted response from the observed, and then for each bootstrap iteration sampling from these residuals with replacement. In a multilevel model, however, the situation is more complicated. Consider first a "crude" residuals bootstrap as follows for the Equation 9.2:

1. Estimate residuals $(\hat{u}_j, \hat{e}_{ij})$. These are the standard posterior shrunken estimates.
2. Sample with replacement m level-2 residuals and N level-1 residuals, add these to the fixed part estimate to generate a new set of Y values.
3. Fit the model to these new data and obtain the parameter estimates.
4. Repeat steps 2 and 3, say 1000 times.

Such a procedure, however, leads to biases because the residuals are shrunken and the estimates across levels are correlated, negatively in the present case, so independent resampling is inappropriate. Therefore, we require both to reflate residuals and create independence before resampling.

Consider first just reflating the residuals separately at each level. We illustrate with the general random coefficient two-level model

$$y_{ij} = (X\beta)_{ij} + (ZU)_j + e_{ij}$$
$$U^T = \{U_0, U_1 ....\} \tag{9.3}$$

Having fitted the model we estimate the residuals for each level-2 unit j

$$\{\hat{u}_{0j}, \hat{u}_{1j} ....\}, \hat{e}.$$

Analogous operations can be carried out at all levels. Write the empirical covariance matrix of the estimated residuals at level-2 in Equation 9.3 as

$$S = \frac{\hat{U}\hat{U}^T}{m}, \quad (9.4)$$

and denote the corresponding model estimated covariance matrix of the random coefficients at level-2 as R. The empirical covariance matrix is estimated using the number of level-2 units, m, as divisor rather than m–1. We assume that the estimated residuals have been centered, although centering will only affect the overall intercept value. We also note that no account is taken of the relative sizes of the level-2 units and we could consider a weighted form of Equation 9.5

$$S = \frac{\hat{U}W\hat{U}^T}{N} \quad (9.5)$$

where W is the (m × m) diagonal matrix with unit sizes on the diagonal. This is equivalent to defining new residual variables

$$U'_{hj} = U_{hj}\sqrt{mn_j / N} \quad (9.6)$$

and using these with Equation 9.4.

We now seek a transformation of the residuals of the form

$$\hat{U}^* = \hat{U}A$$

where A is an upper triangular matrix of order equal to the number of random coefficients at level-2, and such that

$$\hat{U}^{*T}\hat{U}^* / m = A\hat{U}^T\hat{U}A = A^TSA = R, \quad (9.7)$$

so that these new residuals have the required model covariance matrix, and we now can sample sets of residuals with replacement from $\hat{U}^*$. This will be done at every level of the model, with sampling being independent across levels, thus retaining the independence assumption of the model. Having sampled a set of these residuals we back transform these using $\hat{U} = \hat{U}^* A^{-1}$ and add to the fixed part of the model, along with the corresponding level-1 residuals to obtain the new set of responses.

We form A as follows. Write the Cholesky decomposition of S, in terms of a lower triangular matrix as $S = L_S L_S^T$ and the Cholesky decomposition of R as $R = L_R L_R^T$. We have

$$L_R L_S^{-1}\hat{U}^T\hat{U}(L_R L_S^{-1})^T = L_R L_S^{-1} S (L_S^{-1})^T (L_R)^T$$
$$= L_R (L_R)^T = R$$

and the required matrix is therefore $A = (L_R L_S^{-1})^T$.

Carpenter, Goldstein, and Rasbash (1999) use this procedure, unweighted, to demonstrate the improved (confidence interval based) coverage probabilities compared to the parametric bootstrap when the level-1 residuals have a chi-squared distribution rather than a Normal. Further work confirms this but also suggests that the procedure may underestimate coverage for certain departures from an assumed Normal distribution. (Carpenter, Goldstein, & Rasbash, 2003). In the present case running an unweighted residuals bootstrap for 500 samples gives estimates that are virtually indistinguishable from those in Table 9.1.

The above reflating procedure takes no account of dependencies across levels. If we now write

$$Q^T = \{U_0, U_1, \ldots, e\}, \; Q^T \text{ is } (N \times [p+1]) \quad (9.8)$$

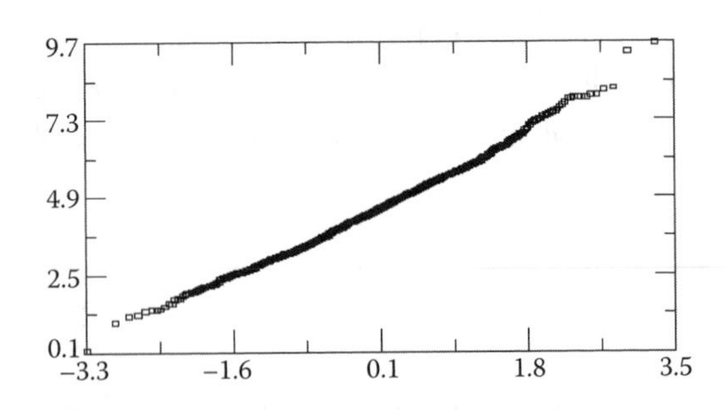

FIGURE 9.1
Normal score plot for level-2 variance bootstrap replications of Table 9.3.

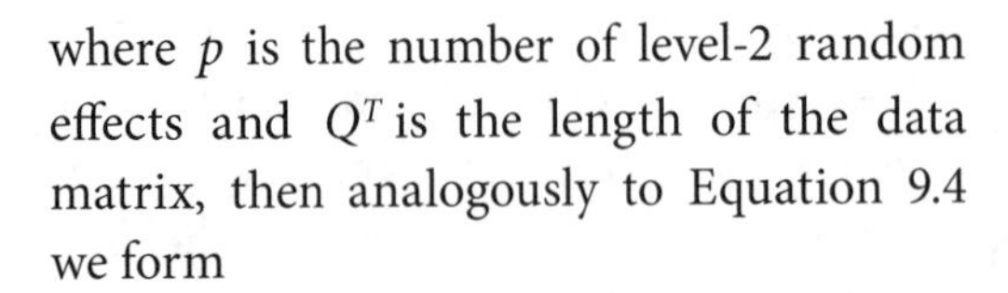

where p is the number of level-2 random effects and Q^T is the length of the data matrix, then analogously to Equation 9.4 we form

$$S = \frac{\hat{Q}\hat{Q}^T}{N},$$

and proceed as before with computing transformed residual sets for the resampling bootstrap. The R matrix has the form

$$\begin{pmatrix} \Omega_u & 0 \\ 0 & \sigma_e^2 \end{pmatrix}$$

so that the Cholesky decomposition is formed in the same way as the separate ones above. The S matrix, however, does not have this form since there are cross product terms for the level-2 and level-1 residuals of the form

$$\frac{1}{N}\sum_j \hat{u}_{hj} \sum_i \hat{e}_{ij}. \qquad (9.9)$$

We note that the Q vectors are still uncorrelated across levels so that we can sample them separately at each level and maintain the model independence assumption as before. We also note that if we ignore the terms in Equation 9.9 we obtain the weighted procedure given by Equation 9.5.

9.9 CONFIDENCE INTERVALS

For many purposes the Normal approximation for the bootstrap replications is adequate and we can use the estimated standard errors for constructing confidence intervals (and significance tests). For example Figure 9.1 is a Normal plot based on 1000 bootstrap replications for the level-2 variance from Table 9.3.

In general, however, we may not be able to rely upon the Normal approximation (although studying plots such as Figure 9.1 should help in making a decision in any particular case). In this case the simplest procedure is to use the empirical bootstrap distribution by simply reading off the 100 α – percentile points, interpolating where necessary. Call these confidence limits $\hat{\theta}^{*(\alpha_1)}, \hat{\theta}^{*(\alpha_2)}$ where in the standard symmetrical case $\alpha_2 = 1 - \alpha_1$ and the coverage is 2α. This does, however, require a large number of replicates, as a rule of thumb 2000 can be used for a 95% interval.

Where there may be biases, a better interval is the bias corrected one computed as follows. Define

$$\alpha_1 = \Phi(2\hat{z}_0 + z^{(\alpha)})$$

$$\alpha_2 = \Phi(2\hat{z}_0 + z^{(1-\alpha)})$$

$$\hat{z}_0 = \Phi^{-1}\left(\frac{\#\{\hat{\theta}^* < \hat{\theta}\}}{B}\right)$$

where Φ is the standard Normal cumulative distribution function and B is the number of bootstrap replicates.

9.10 BOOTSTRAP LIKELIHOOD

The likelihood, considered as a function of a parameter θ, is proportional to

$$L(\theta) = \prod_i p(y_i \mid \theta)$$

where i indexes the data units, assumed conditionally independent. The *partial likelihood* based on a parameter estimate $\hat{\theta}$ rather than the data $\{y_i\}$ can be approximated by a bootstrap as follows. We consider the parametric bootstrap.

The first stage is to generate B_1 bootstrap replications to produce bootstrap parameter estimates: label the set of the parameter of interest $S_1 = \{\hat{\theta}_1^*, \ldots, \hat{\theta}_b^*, \ldots, \hat{\theta}_{B_1}^*\}$. For each replication (i.e., from the parameter estimates associated with each replication) we generate a second stage bootstrap set of replicates giving the set of interest $S_{2b} = \{\hat{\theta}_{b1}^{**}, \ldots, \hat{\theta}_{bB_1}^{**}\}$. For S_{2b} we estimate the (Normal) kernel density $\hat{p}(t \mid \hat{\theta}_b^*)$ as a function of t and evaluate it at $t = \hat{\theta}$. Because the set S_{2b} was generated from a population with parameter $\hat{\theta}_b^*$, $\hat{p}(\hat{\theta} \mid \hat{\theta}_b^*)$ is an estimate of the partial likelihood of θ at $\theta = \hat{\theta}$. We thus have estimates of the likelihood for all the values in S_1 and we can use a suitable smoother (such as LOESS) to plot the likelihood function. In fact, for the region of interest a simple polynomial or fractional polynomial function may be adequate. From this function we can obtain the maximum and by plotting - *2log(likelihood)* we can obtain confidence intervals using the asymptotic chi-squared approximation.

This can be extended to more than one parameter (the estimates for all the parameters are available from the bootstrap replications), but this will then involve smoothing in more than one dimension, although again we may be able to achieve a satisfactory smoothing via an additive function of polynomials.

9.11 CONCLUSIONS

The bootstrap is a very general procedure for bias correction and especially for providing accurate estimates where standard assumptions do not hold. In the multilevel case the fully parametric bootstrap, if the model assumptions are accepted, is usually preferred, especially for complex models such as those with several random coefficients or cross classifications. Where model assumptions may not be acceptable, the modified residuals bootstrap will often perform adequately. One problem with the bootstrap, however, is the computing load associated with it. An alternative, especially for accurate quantile estimates, is a Markov Chain Monte Carlo (MCMC) estimation that is generally faster and allows a fully Bayesian approach.

REFERENCES

Carpenter, J., Goldstein, H., & Rasbash, J. (1999). A non-parametric bootstrap for multilevel models. *Multilevel Modelling Newsletter, 11,* 2–5.

Carpenter, J. R., Goldstein, H., & Rasbash, J. (2003). A novel bootstrap procedure for assessing the relationship between class size and achievement. *Journal of the Royal Statistical Society, Series C, 52,* 431–443.

Davison, A. C., & Hinckley, D. V. (1997). *Bootstrap methods and their application.* Cambridge, UK: Cambridge University Press.

Goldstein, H. (2003). *Multilevel statistical models* (3rd ed.). London, UK: Edward Arnold.

10

Multiple Imputation of Multilevel Data

Stef van Buuren
TNO Quality of Life, Leiden and University of Utrecht, The Netherlands

10.1 INTRODUCTION

In the early days of multilevel analysis, Goldstein (1987, p. 8) wrote: "We shall require and assume that all the necessary data at each level are available" (Goldstein, 1987). Despite the many conceptual and computational advances that have been made over the last two decennia, Goldstein's requirement is still dominant today. To illustrate this, consider how modern software for fitting multilevel models deals with missing data. Dedicated packages like MLwiN (Rasbash, Steel, Browne, & Prosser, 2005) and HLM (Raudenbush, Bryk, & Congdon, 2008) remove all level-1 units with missing values on any level-1 variable. If level-2 explanatory variables have missing values, the associated level-2 units are deleted, including all level-1 data. Thus, if the age of the teacher is unknown, all data of all children within the class are removed prior to analysis. Multilevel procedures in general purpose statistical software, like SAS PROC MIXED (Littell, Milliken, Stroup, & Wolfinger, 1996), SPSS MIXED (SPSS Inc., 2008), STATA xtmixed (StataCorp LP, 2008), S-PLUS library `nlme3` and the R package `nlme` (Pinheiro & Bates, 2000), and the R package `arm` (Gelman & Hill, 2007) use a similar approach. Deletion is not only wasteful of costly collected data, but it may also bias the estimates of interest (Little, 1992; Little & Rubin, 2002).

Alternative approaches have been tried. In older versions of HLM it was possible to perform pairwise deletion, a method to calculate the covariance matrix where each element is based on the full number of complete cases for that pair of variables. However, this approach causes estimation problems due to the possibility of nonpositive definite covariance matrices. Also, model comparisons in terms of the log-likelihood are debatable since there is no clear-cut way to calculate the degrees of freedom. Version 6 of HLM therefore dropped this feature.

M*plus* (Muthén & Muthén, 2007) uses full information maximum likelihood. This approach specifically deals with the case of multiple outcome variables. If one or more outcomes are missing, the values of the remaining dependent variables are still used. In this way, there is no need to delete the whole level-1 unit. When there are missing data in any covariates however, M*plus* resorts to listwise deletion.

Some general purpose programs offer modules to impute missing data (e.g., SAS PROC MI and the new Multiple Imputation procedure in SPSS V17.0). These approaches generally ignore the clustering structure in hierarchical data. Not much is known how imputation by such procedures affects the complete data analysis.

This chapter discusses critical issues associated with imputation of multilevel data. Section 10.2 introduces the notation used and outlines how two formulations of the same model are related. Section 10.3 dissects the multilevel missing data problem into five main questions that need to be addressed. Section 10.4 outlines six different strategies for dealing with the missing data problem. Section 10.5 describes a multilevel imputation method for univariate data, and discusses its properties. Section 10.6 describes a method to apply the univariate method iteratively to multivariate missing data. Finally, Section 10.7 sums up the major points and provides directions for future research.

10.2 TWO FORMULATIONS OF THE LINEAR MULTILEVEL MODEL

Let y_j denote the $n_j \times 1$ vector containing observed outcomes on units i ($i = 1,\ldots,n_j$) within class j ($j = 1, \ldots, J$). The univariate linear mixed-effects model (Laird & Ware, 1982) is written as

$$y_j = X_j\beta + Z_j u_j + e_j \qquad (10.1)$$

where X_j is a known $n_j \times p$ design matrix in class j associated with the common $p \times 1$ fixed effects vector β, and where Z_j is a known $n_j \times q$ design matrix in class j associated with the $q \times 1$ random effect vectors u_j. The random effects u_j are independently and interchangeably normally distributed as $u_j \sim N(0, \Omega)$. The number of random effects q is typically smaller than the number of fixed effects p. Symbol e_j denotes the $n_j \times 1$ vector of residuals, which are independently normally distributed as $e_j \sim N(0, \sigma_j^2 I(n_j))$ for $j = 1, \ldots, J$. It is often assumed that the residual variance is equal for all classes: $\sigma_j^2 = \sigma^2$. In addition, e_j and u_j are uncorrelated so $\text{cov}(e_j, u_j) = {}_{n_j}0_q$, an $n_j \times q$ matrix of zeroes. Model formulation of Equation 10.1 clearly separates fixed from random effects.

It is also convenient to conceptualize Equation 10.1 as constructed from a set of different levels. To see how this works, write the two-level linear model as

$$y_j = Z_j\beta_j + e_j \quad \text{level-1 equation} \qquad (10.2a)$$

where β_j is a $q \times 1$ vector of regression coefficients that vary between the J classes. At level-2, we model β_j by the linear regression model

$$\beta_j = W_j\beta + u_j \quad \text{level-2 equation} \qquad (10.2b)$$

where W_j is a $q \times p$ matrix of a special structure (see below), and where u_j can be interpreted as the $q \times 1$ vector of level-2 residuals. Equations 10.2a and b are sometimes collectively called the slopes-as-outcome model (Bryk & Raudenbush, 1992). Note that the regression coefficient β is identical in all level-2 classes. Substituting Equation 2b into Equation 2a yields

$$y_j = Z_j W_j\beta + Z_j u_j + e_j, \qquad (10.3)$$

which is a special case of the linear mixed model (Equation 10.1) with $X_j = Z_j W_j$.

Matrix W_j has a special structure for the linear multilevel model. Suppose the model contains $q = 2$ random effects (an intercept and a slope) and a level-2 predictor whose values are denoted by w_j ($j = 1, \ldots, J$). The structure of W_j is then

$$W_j = \begin{bmatrix} 1 & 0 & w_j & 0 \\ 0 & 1 & 0 & w_j \end{bmatrix}. \quad (10.4)$$

The first two columns of W_j correspond to the random intercept and random slope terms, respectively. In the expression $X_j = Z_j W_j$, this part effectively copies Z_j into X_j. Multiplication of Z_j by the third column W_j replicates w_j as n_j elements in class j, thus forming a covariate associated with the main (fixed) effect in matrix X_j. Multiplication by the fourth column adds the interaction between the random slope and the fixed level-2 predictor, also known as the cross-level interaction term. In applications where this term is not needed, one may simply drop the fourth column of W_j. It is easy to extend Equation 10.4 to multiple level-2 predictors by padding additional columns with the same structure. Note that Equation 10.2 implicitly assumes that all level-1 variables are treated as random effects. It is straightforward to exclude the random part for the lth ($l = 1, \ldots, q$) variable by requiring $u_{1l} = \ldots = u_{jl} = \ldots = u_{Jl} = 0$, or equivalently, by setting the corresponding diagonal element in Ω to zero. In the sequel, we assume that all level-1 data are collected into Z_j.

Equation 10.1 separates the fixed and random effects, but the same covariates may appear in both X_j and Z_j. This complicates imputation of those covariates. To make matters more complex, X_j can also contain interactions between covariates at level-1 and level-2. Equation 10.2 distinguishes the level-1 from the level-2 predictors. There is no overlap between W_j and Z_j. This is a convenient parameterization if we are trying to understand the missing data processes that operate on different levels of the data collection.

10.3 CLASSIFICATION OF MULTILEVEL INCOMPLETE DATA PROBLEMS

This section provides a typology of incomplete data problems that can appear in a multilevel context. There are five major factors to consider: the role of the variables in the model, the pattern of the missingness, the missing data mechanism, the distribution of the variable, the design of the study. In order to be able to provide an adequate treatment to the missing data we need answers on the following questions:

- Role: In which variables do the missing data occur?
- Pattern: Do the missing data form a pattern in the data?
- Mechanism: How is the probability to be missing related to the data?
- Scale: What is the scale of the incomplete variables?
- Design: What is the design of the study (e.g., random, clustered, longitudinal)?

This section classifies problems in incomplete multilevel data into five subproblems: role, pattern, mechanism, scale, and design. We briefly indicate the major difficulties and consequences of missing data in each case. The typology can be used to characterize particular data analytic problems. In addition, the typology provides insight into what fields are well covered in the literature and those less covered. Different combinations of the five factors correspond to different analytic situations and may thus require specialized approaches.

10.3.1 Role of the Variable In the Model

Missing data can occur in y_j, Z_j, W_j, and j. The consequences of incompleteness of a variable depend on the role the variable plays in the multilevel model.

10.3.1.1 Missing Data in y_j

Many classical statistical techniques for experimental designs require balanced data with equal group sizes (Cochran & Cox, 1957). The experimental factors are under control of the experimenter and the missing data typically occur in y_j. The problem of missing data in y_j is that they may destroy the balance present in the original design. In the days of Fisher, this used to be a major setback since the calculations required for the analysis of unbalanced data are much more demanding than those for the balanced case. In a similar vein, the classic approach to analyzing change relies on repeated measurements of the same subject on a fixed number of occasions (de Leeuw & Meijer, 2008). Missing data that occur in repeated measures result in incompleteness of the subject's response vector, which leads to severe complications in MANOVA. Many techniques have been proposed to circumvent and deal with problems of missing outcomes in experiments (Dodge, 1985).

The advent of multilevel modeling opened up new ways of analyzing data with missing y_j. Modern likelihood-based methods have been developed in which missing data in y_j no longer present a problem. Snijders and Bosker (1999, p. 52) write that the model can be applied "even if some groups have sample size $n_j = 1$, as long as other groups have greater sizes." We add that this statement will only go as far as the assumptions of the model are met: data in y_j are missing at random and the model is correctly specified. Section 10.4.5 discusses the likelihood-based approach in more detail.

The problem of missing data in y_j has received vast attention. There is an extensive literature, which often concentrates on the longitudinal case (Daniels & Hogan, 2008; Molenberghs & Verbeke, 2005; Verbeke & Molenberghs, 2000). For more details, see the overview of the state-of-the-art including direct likelihood approaches, Generalized Estimating Equations (GEE), Weighted GEE, and others (Beunckens, Molenberghs, Thijs, & Verbeke, 2007).

10.3.1.2 Missing Data in Z_j

Missing data can also occur in the level-1 predictors Z_j. In applications where pupils are nested within classes, missing data in Z_j occur at the child level: age of the pupil, occupational status of the father, ethnic background, and so on. In longitudinal applications where time is nested within persons, missing data in Z_j may occur on time-varying covariates. Examples include breast-feeding status and stage of pubertal development at a particular age.

The effect of missing data in Z_j is that the estimators become undefined. The usual solution is simply to remove the incomplete cases before analysis. This is not only wasteful, but may also bias estimates of the regression weights (Little, 1992). Some authors suggest that data missing at the micro units may not need to be replaced or imputed if the data are to be aggregated and the analysis is to be done at the macro level (McKnight, McKnight, Sidani, & Figueredo, 2007). While easy to perform, this advice is only sound under the restrictive assumption that the process that caused the missing data is missing completely at random.

Several solutions for handling missing data in Z_j have been offered. Goldstein proposed to extend the multilevel model with one extra level that contains a dummy variable for each incomplete variable (1987). Petrin implemented this suggestion, and noted that the procedure is "susceptible to producing biased parameters estimates." The procedure requires reorganization of the data and, according to Petrin, is "very tedious" (2006). Schafer noted that missing values in Z_j are problematic since they require a probability model on the covariates (1997). Handling this in general "would require us to incorporate random effects into the imputation model, which remains an open problem." Longford observed that drawing imputations using random effects models is hard because the relevant parameter distributions depend on the within–between classes variance ratio, which is often not estimated with high precision (Longford, 2005).

Schafer and Yucel (2002) suggested transferring incomplete variables in Z_j to the other side of the equation, and impute the missing data in the multivariate outcomes under a joint multivariate model (Yucel, 2008). This approach has been implemented in their PAN package. There is a macro for MLwiN that implements this approach (Carpenter & Goldstein, 2004). Multiple imputation of multilevel data is possible using the chained equations approach (Jacobusse, 2005). This method is implemented in the WinMICE computer program, which can be downloaded from www.multiple-imputation.com. Similar research was done by Yucel, Schenker, and Raghunathan (2006), who called their approach SHRIMP. Longford (2008) proposed an Expectation–Maximization (EM) algorithm to estimate the parameters in the multilevel model in case of missing Z_j. In its generality, this approach requires substantial programming effort and, according to Longford, would only be practical if few missing data patterns arise.

10.3.1.3 Missing Data in W_j

The problem of missing data in W_j has received little attention. Missing data in the level-2 predictors W_j occur if, for example, it is not known whether a school is public or private. In a longitudinal setting, missing data in fixed person characteristics, like sex or education, lead to incomplete W_j.

Missing entries in W_j complicate the estimation of group-level effects. The typical fix is to delete all records in the class. For example, suppose that the model contains the professional qualification of the teacher (e.g., teacher school, university, PhD). If the qualification is missing, the data of all pupils in the class are removed before the analysis. Again, this strategy is not only wasteful, but may also lead to selection effects at level 2.

Some have studied the use of (inappropriate) flat-file imputation methods that ignore the hierarchical group structure in multilevel data. Standard errors are underestimated, leading to confidence intervals that are too short (Cheung, 2007; Gibson & Olejnik, 2003; Roudsari, Field, & Caetano, 2008). Zhang (2005) reports however that flat multiple imputation worked well with multilevel data, and advises that future researchers should feel confident applying the procedure with a missing data level up to 30%. There is no consensus yet on this issue, and some more work is needed to clear things up.

Imputation methods for level-2 predictors should assign the same imputed value to all members within the same class. Some authors suggest creating two data sets, one with only individual-level data, and one with group-level data, and do separate imputations within each data set while using the results from one in the other (Gelman & Hill, 2007; Petrin, 2006). Note that the steps can also be iterated.

10.3.1.4 Missing Data in j

It is also possible that the group identification is unknown. For example, some pupils may have failed to fill in their class number on the form. The result is that the investigator cannot allocate the pupil to a group. Though one might envisage applications of imputing class memberships, we will not deal with the case of missing data in j. For now, the only action one could do is to eliminate the record from the data.

10.3.2 Missing Data Pattern

For both theoretical and practical reasons, it is useful to distinguish between monotone and nonmonotone missing data patterns, and between univariate and multivariate missing data patterns. A pattern is monotone if the variables can be ordered such that, for each person, all earlier variables are observed if all subsequent variables are observed. Monotone patterns often occur as a result of *drop out* in a longitudinal study. It is often useful to sort variables and cases to approach a monotone pattern.

Little and Rubin (2002) graphically demonstrate the univariate/multivariate and the monotone/nonmonotone distinctions for flat files. Things become more complicated in the context of multilevel data. Figure 10.1 demonstrates several possibilities. Figure 10.1a is the case where all missing data are confined to the outcome y_j, and where a person is lost once dropped out. Figure 10.1b depicts the situation where the person only misses one or more visits, but does not completely drop out. This leads to missing data that are *intermittent*. Note that the difference between 10.1a and b only makes sense for longitudinal data (i.e., when Z_j can be interpreted as time).

If Z_j attains identical values in each group (i.e., if the data are repeated measures at fixed time points), we can reorder the file into a broad matrix where each cluster occupies one record, and where a set of columns represent the time points. It is then easy to see that drop out leads to a monotone missing data problem, whereas intermittent missing data result in a nonmonotone pattern. The practical usefulness of a monotone pattern is that it opens up the possibility to solve the missing data problem by a sequence of simple steps without the need to iterate (Little & Rubin, 2002).

Figure 10.1c represents the situation where there are also missing data in level-1 predictors Z_j. For example, Z_j could contain body height and y_j could be body weight. Multilevel multivariate missing data usually correspond to a missing data pattern that is nonmonotone. Figure 10.1d depicts the

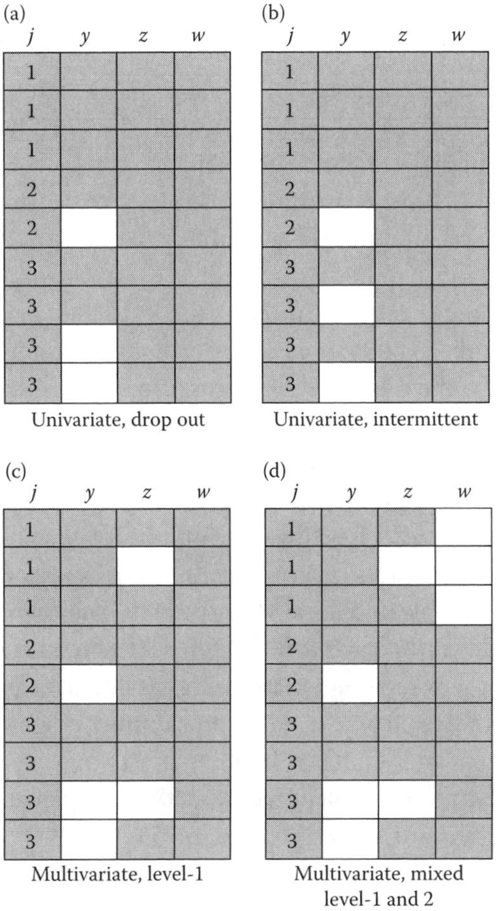

(a)

j	y	z	w
1			
1			
1			
2			
2			
3			
3			
3			
3			

Univariate, drop out

(b)

j	y	z	w
1			
1			
1			
2			
2			
3			
3			
3			
3			

Univariate, intermittent

(c)

j	y	z	w
1			
1			
1			
2			
2			
3			
3			
3			
3			

Multivariate, level-1

(d)

j	y	z	w
1			
1			
1			
2			
2			
3			
3			
3			
3			

Multivariate, mixed level-1 and 2

FIGURE 10.1
Four typical missing data patterns in the multilevel data with two levels and three groups. The grey parts represent observed data, whereas the transparent cells indicate the missing data.

one most general situation where missing data occur in level-2 predictors W_j, level-1 predictors Z_j and level-1 outcomes y_j. Note that all level-1 units have missing level-2 predictors if W_j is missing. This is perhaps the most complex case, but also a case that occurs frequently.

10.3.3 Missing Data Mechanism

The process that created the missing data influences the way the data should be analyzed. Except in artificial cases, the precise form of the missingness process is generally unknown, so one has to make assumptions. If the probability to be missing is independent of both unobserved and observed data, then the data are said to be Missing Completely at Random (MCAR; Rubin, 1976). If, conditional on the observed data, the probability to be missing does not depend on the unobserved data, then the data are said to be Missing at Random (MAR). Note that MCAR is a special case of MAR. A mechanism that is neither MCAR nor MAR is called Missing Not at Random (MNAR).

It is possible to test between MCAR and MAR. For data missing due to drop out, Diggle (1988) proposed a test for the hypothesis that the probability a unit drops out at time t_j is independent of the measurement on that unit up to time t_{j-1}. An alternative for general monotone data was developed by

Little (1988). It is not possible to test MNAR versus MAR since the data needed for such a test are, by definition, missing.

A closely related concept is ignorability of the missing data process. If the data are MAR and if the parameters of the complete data model are independent of those of the missing data mechanism, then likelihood inference of the observed data can ignore the missing data process. Suppose that the random variable $R = 1$ indicates that Y is observed, whereas $R = 0$ for missing Y. The information about Y that is present in X, Z, and R is summarized by the conditional distribution $P(Y|X, Z, R)$. Cases with missing Y; that is, with $R = 0$, do not provide any information about $P(Y|X, Z, R)$, and so we have only information to fit models for $P(Y|X, Z, R = 1)$. However, we need the distribution $P(Y|X, Z, R = 0)$ to model the missing Ys. Assuming that the missing data mechanism is ignorable corresponds to equating $P(Y|X, Z, R = 0) = P(Y|X, Z, R = 1)$ (Rubin, 1987).

The assumption of ignorability generally provides a natural starting point for analysis. If the assumption is clearly not reasonable (e.g., when data are censored), we may use other forms for $P(Y|X, Z, R = 0)$. The fact that $R = 0$ allows for the possibility that the $P(Y|X, Z, R = 1) \neq P(Y|X, Z, R = 0$; cf. Rubin, 1987, p. 205), so nonignorable nonresponse can be modeled by specifying $P(Y|X, Z, R = 0)$ different from $P(Y|X, Z, R = 1)$. The difference can be just a simple shift in the mean of the distribution (Van Buuren, Boshuizen, & Knook, 1999), but it may also consist of highly customized (selection, pattern mixture, shared parameter) models that mimic the nonresponse mechanism (Daniels & Hogan, 2008; Demirtas & Schafer, 2003; Little & Rubin, 2002). Daniels and Hogan (2008) suggest viewing the effects of alternative missing data assumption in terms of departures from MAR. A key requirement is that the assumed nonignorable model should be more reasonable and sensible than the model implied by the assumption of ignorability.

A somewhat different strategy to bypass the assumption of ignorability is to construct double robust estimators. An estimator is double robust if it remains consistent when either (but not necessarily both) a model for the missing data mechanism or a model for the distribution of the complete data is correctly specified (Bang & Robins, 2005; Scharfstein, Rotnitzky, & Robins, 1999). The approach uses inverse probability weighting, and its pros and cons with respect to multiple imputation have been the subject of debate (Kang & Schafer, 2007). The literature is now moving toward using the best of both worlds from inverse probability weighting and multiple imputation (Beunckens, Sotto, & Molenberghs, 2008; Carpenter, Kenward, & Vansteelandt, 2006).

10.3.4 Scale

Data can be measured on many types of scales: continuous (but are usually rounded to whole units), ordered categorical, unordered categorical, binary, semicontinuous (i.e., a mixture of a binary and a continuous variable), counts, censored (with known or unknown censoring points), truncated (with known or unknown truncation points), below the detection limit, bracketed response (e.g., obtained by a format that zooms in by posing successively more detailed questions), constrained by other data (e.g., a sum score or interaction term), and so on. In addition the data can take almost any distribution, including bimodal, skewed, and kurtotic shapes. Moreover, the relations can be highly nonlinear.

All these factors can occur in conjunction with multilevel data. The most advanced methods for dealing with missing data in a multilevel context invariably assume that variables follow a multivariate normal distribution. Though multiple imputation is generally robust to violations of the multivariate normality assumption (Schafer, 1997), advances could be made that respect the scale, the distribution, and nonlinear relations of the data.

10.3.5 Study Design

The study design determines the class of incomplete data models that can be usefully applied to the data. Popular designs that lead to hierarchical data include:

Multistage sample: A design where sampling progresses in a number of stages, for example, first sample from school, then sample classes within schools, and then sample pupils within classes. Missing data can occur at any stage of sampling, but usually only missing data in the level-1 outcomes are explicitly considered as missing data. This is a common design in the social sciences.

Longitudinal study with fixed occasions: A design where data are collected according to a number of planned visits. Missing data may result from missed visits (intermittent missing data) or panel attrition (drop out). This design is common in the biomedical field.

Longitudinal study, varying occasions: A design where the data are ordered according to time and nested within individuals. There is no such thing as a complete data vector. The number of observations per individuals may vary widely, can be as low as one, and can occur anywhere in time (Snijders & Bosker, 1999).

Planned missing data: A design where intentional missing data occur in the data as a consequence of the administration procedures. For example, the investigator could use matrix-sampling to minimize the number of questions posed to a student (Thomas & Gan, 1997). Missing data are an automatic part of the data. The percentage of missing data is typically large, sometimes over 75%.

File matching: A post-hoc procedure for combining two or more data sets measured on the same units. Missing data occur in the rows and in the columns since different data sources can measure different units on different attributes (Rässler, 2002; Rubin, 1986).

Relational databases: A common way for storing information on different types of units (e.g., customers, products, stores) as a set of linked tables. Missing data result from partial tables and imperfect links.

10.4 STRATEGIES TO DEAL WITH INCOMPLETE DATA

10.4.1 Prevention

The best solution to the missing data problem is not to have any. Consequently, the best strategy is to deal with unintentional missing data and to minimize their number. There are many factors that influence the response rate in social and medical studies: design of the study (number of variables collected, number and spacing of time repeated measures, follow-up time, missing data retrieval strategy), data collection method (mode of collection, intrusive measures, sensitivity of information collected, incentives, match of the interviewer and the respondent), measures (clarity, layout), treatment burden (intensity of the intervention) and data entry coding errors. For more information, we refer to the appropriate literature (De Leeuw, Hox, & Dillman, 2008; McKnight et al., 2007; Stoop, 2005). When carefully planned and executed, prevention of missing data may substantially increase the completeness of the information.

10.4.2 Listwise Deletion

Listwise deletion (or complete case analysis) is the simplest and most popular way of dealing with missing data. Listwise deletion simply eliminates any incomplete record from the analysis. This is potentially a very wasteful strategy because valuable data are thrown away, especially when variables at the higher levels have missing data. If the missing data are confined to y_j and if the missing data mechanism is MAR, then listwise deletion followed by the appropriate likelihood-based analysis is unbiased. Note that any covariates that predict the missingness in y_j should be included into the model, even if they are of no scientific interest to the researcher. For missing data in W_j or Z_j, analysis of the complete cases will generally bias parameter estimates, even under MCAR (Little, 1992).

10.4.3 Last Observation Carried Forward

Last Observation Carries Forward (LOCF) is a technique applicable only to longitudinal data with drop out. The LOCF substitutes any missing y_j after drop out by the last observation. LOCF is popular for clinical trials in order to be able to perform an "intention to treat" analysis; that is, an analysis of the subject as randomized, irrespective of treatment compliance. However, LOCF makes the strong and often unrealistic assumptions that the response profile of the subject remains constant after dropping out of the study. The LOCF does not even work under MCAR (Molenberghs & Kenward, 2007). The magnitude and direction of this bias depend on the true but unknown treatment effects. In contrast to the widespread belief that LOCF leads to conservative tests, it is entirely possible that LOCF induces effects where none exist. Furthermore, because there is no distinction between the observed and the imputed data, LOCF artificially increases the amount of information in the data. This results in confidence intervals that are too short. All in all, the use of LOCF is discouraged (Lavori, 1992; Little & Yau, 1996).

10.4.4 Class Mean Imputation

Class mean imputation replaces each missing value with the class or cluster mean. The method is applicable to both longitudinal and nonlongitudinal data. Thus, class mean imputation substitutes the missing grade of a pupil by the average of the known grades of all pupils in the class. Just like LOCF, the method is unconditional on any other information from the pupil, so the method may distort relations between variables. Unless special methods are used to analyze the imputed data, the variability may be severely underestimated (Little & Rubin, 2002; Schafer & Schenker, 2000). All in all, class mean imputation can be as damaging as LOCF and should generally not be used.

10.4.5 Likelihood-Based Methods

Likelihood-based methods attempt to analyze the entire data without systematically biasing the conclusions of the subject matter question. The method maximizes the likelihood function derived from the underlying model. If there are missing data, the likelihood function is restricted to the observed data only. If the missing data mechanism is ignorable, we may write the likelihood of the observed data $L(\theta|Y_{obs})$ as

$$L(\theta|Y_{obs}) = \int L(\theta|Y_{obs}, Y_{mis})dY_{mis} \quad (10.5)$$

where θ are the parameters of interest, and where $L(\theta|Y_{obs}, Y_{mis})$ is the likelihood of the hypothetically complete data. The observed data likelihood averages over the distribution of the missing data. The EM algorithm (Dempster, Laird, & Rubin, 1977) maximizes $L(\theta|Y_{obs})$ by filling in the complete data sufficient statistics.

The linear mixed-effects model (Equation 10.1) subsumes repeated-measures ANOVA and growth curve models for longitudinal data as special cases. The model parameters can be estimated efficiently via likelihood-based methods. Laird and Ware developed an EM algorithm that can be used to fit the mixed linear model to longitudinal data (1982). Jennrich and Schluchter (1986) improved the speed of the method by Fisher scoring and Newton–Raphson. Currently, full-information maximum likelihood (FIML) is widely used to estimate the model parameters. Restricted maximum likelihood estimation (REML) is a closely related alternative that is less sensitive to small-sample bias of maximum likelihood (Fitzmaurice, Laird, & Ware, 2004; Verbeke & Molenberghs, 2000).

Software for fitting mixed models has the ability to handle unbalanced longitudinal data, where the response data y_j are observed at arbitrary time points for each subject. Missing data in y_j are ignored by the maximum likelihood and REML methods along with their values on W_j and Z_j. An advantage of the multilevel model for the analysis of longitudinal data is its ability to handle arbitrary time points. Missing values in W_j and Z_j are however problematic (Longford, 2008; Schafer, 1997). No generally applicable likelihood-based approach has yet been developed for the case of missing values in W_j and Z_j.

Despite the attractive properties of the multilevel model, likelihood-based methods should be used with some care when data are incomplete. First, the standard multilevel model implicitly assumes is that the missing data in the outcomes are MAR. This assumption can be suspect in some settings. For example, patients who drop out early from a trial often have slopes that differ from patients who stay in the trial. Another assumption is that the individual patient slopes have a common normal distribution. This assumption may not be realistic if drop out occurs. There is an active statistical literature on the problem of estimating the linear mixed model under MNAR situations (Daniels & Hogan, 2008).

In the case that the MAR assumption is correct, the factors that govern the probability of the missing data must be included into the multilevel model, for example, as covariates. Failing to do so may introduce biases in the estimate of the treatment effect. Note that this requirement complicates the interpretation of the complete-data model, and may lead to models that are impossible to estimate and more complex to interpret. Also, missing data problems may actually worsen if the additional covariate(s) contain missing values themselves.

Third, the missing data may increase the sensitivity of inferences to misspecification of the model for the complete data. Incorrectly assuming a linear relationship between an outcome and a covariate may lead to more serious bias when missingness depends on the value of the covariate than when it does not (Little, 2008). Zaidman-Zait and Zumbo (2005) performed simulations where the missing data mechanism depended on a person factor. Theoretically, including the person factor into the model should adequately deal with the missing data. However, they found bias in the MAR case and attribute that to the incorrect specification of the level-1 model.

Fourth, it is generally more difficult to derive appropriate standard errors if there are missing data. For example, the occurrence of missing data may destroy the block-diagonal structure of the information matrix in many repeated measure designs. Hence, the full matrix needs to be inverted, which can be time consuming (Little, 2008).

In summary, likelihood-based methods are the preferred approach to missing data if all of the following hold:

1. The missing data are confined to y_j,
2. The MAR assumption is plausible,
3. Any factors in the MAR mechanism are included into the multilevel model,
4. The multilevel model for the complete data is correctly specific.

If one or more of these conditions are not met, using likelihood methods for incomplete data could be problematic. Not much is yet known about the relative importance of each factor.

10.4.6 Multiple Imputation

The likelihood-based approach attempts to solve both the missing data and complete data problems in one step. An alternative strategy is to attack the problem in two steps: First solve the missing data problem by imputing the missing data, and then fit the complete data analysis on the imputed data. Such a modular approach breaks down the model complexity in each step. It is well known that the precision of the complete-data estimates is overestimated if no distinction is made between observed and imputed data. The solution to this problem is to use multiple imputation (MI), which can produce correct estimates of the sampling variance of the estimates of interest (Rubin, 1987, 1996).

10.5 IMPUTATION OF UNIVARIATE MISSING DATA IN y_j

10.5.1 Multilevel Imputation Algorithm

The linear mixed model formulation of the multilevel model is given by Equation 10.1: $y_j = X_j\beta + Z_ju_j + e_j$ with $u_j \sim N(0, \Omega)$ and $e_j \sim N(0, \sigma^2 I(n_j))$. In order to derive imputations under this model, we adopt a Bayesian approach. For complete data, the distribution of the parameters can be simulated by Markov chain Monte Carlo (MCMC) methods (Schafer & Yucel, 2002; Zeger & Karim, 1991). The main steps are:

1. Sample β from $p(\beta \mid y, u, \sigma^2)$
2. Sample u_j from $p(u \mid y, \beta, \Omega, \sigma^2)$
3. Sample Ω from $p(\Omega \mid u)$ (10.6)
4. Sample σ^2 from $p(\sigma^2 \mid y, \beta, u)$
5. Repeat step 1–4 until convergence

The rate of convergence of this Gibbs sampler depends on the magnitude of the correlation between the steps. Many variations on the above scheme have been proposed (Chib & Carlin, 1999; Cowles, 2002; Gelman, Carlin, Stern, & Rubin, 2004; Gelman, Van Dyk, Huang, & Boscardin, 2008).

Let us first consider the case where y contains missing data. Let y^{obs} represent the observed data and let y^{mis} be the missing data, so that $y = [y^{obs}, y^{mis}]$. If the MAR assumption is plausible, we can replace y by y^{obs} in the above steps, and simulate the parameter distribution using only the complete records. At the end, we append an additional step to generate imputations for the missing data:

6. Sample y^{mis} from $p(y^{mis} \mid y^{obs}, \beta, u, \Omega, \sigma^2)$. (10.7)

Under model Equation 10.1, we calculate imputations by drawing

$$e_j^* \sim N(0, \sigma^2) \quad (10.8)$$

$$y_j^* = X_j\beta + Z_j u_j + e_j^* \quad (10.9)$$

where all parameters that appear on the right are replaced by their values drawn under the Gibbs sampler.

The classic algorithm outlined above will not produce good imputations for incomplete predictors. A considerable advance in imputation quality is possible by using a slightly more general version of model Equation 10.1, where the within-cluster variance σ_j^2 is allowed to vary over the clusters. Kasim and Raudenbush (1998) proposed a Gibbs sampler for this heterogeneous model. They specify

$$p(\sigma_j^2 \mid \sigma_0^2, \phi) \sim \frac{\sigma_0^2 \chi_1^2}{\phi} \quad (10.10)$$

where σ_0^2 and ϕ are hyperparameters. The hyperparameter σ_0^2 describes the location of prior belief about residual variance σ_j^2 in the conjugate prior distribution for σ_j^2. The hyperparameter ϕ is a measure of variability of the variances σ_j^2. Both σ_0^2 and ϕ are also updated within the Gibbs sampler. The algorithm was implemented in R by Roel de Jong, where $\sigma_j^2 = 1$ and $\phi = 1$ are used as starting parameters. Below, we will refer to this method as multilevel imputation (ML).

10.5.2 Simulation Study

Data with a multilevel structure were generated according to the model $y_{ij} = 0.5 z_{ij} + u_j + e_{ij}$ with $e_j \sim N(0, \sigma^2)$ and $u_j \sim N(0, \Omega)$. This model is a special case of Equation 10.1 and 10.2), where $X_j = Z_j = (1, z_{ij})$ with $i = 1,\ldots,n_j$ is the $n_j \times 2$ data matrix of class j, where $\Omega = \text{diag}(\omega^2,0)$, where $\beta = (0,0.5)^T$ is a 2×1 vector of fixed parameters, and where W_j is the identity matrix. We varied the variance parameters (σ^2, ω^2) in pairs as {(0.75,0.00), (0.65,0.10), (0.45,0.30), (0.25, 0.50)}. Since variable z_{ij} was drawn as $z_{ij} \sim N(0,1)$, the intraclass correlation coefficient (ICC) under the stated model equals ω^2, so the ICC effectively varies between 0.0 and 0.5. We fixed the total number of respondents to 1,200. The number of classes was chosen 12, 24, and 60, yielding 100, 50, and 20 respondents per class, respectively.

Two missing data mechanisms were specified: Y and Z. Mechanism Y generates 50% missing data in y_{ij} under MAR. For values of $z_{ij} < 0$, the nonresponse probability in y_{ij} is 10%. For $z_{ij} \geq 0$, this probability is 90%. Vice versa, mechanism Z generates 50% missing data in z_{ij} under MAR given y_{ij}. For values of $y_{ij} < 0$, the nonresponse probability is 10%. For $y_{ij} \geq 0$, the probability is 90%.

The following methods for handling the missing data were used:

- Complete Case Analysis (CC). This method removes any incomplete records before analysis, also known as listwise deletion.
- Multiple Imputation Flat File (FF). This method multiple imputes missing data while ignoring any clustering structure in the data by standard linear regression imputation.
- Multiple Imputation Separate Classes (SC). This method multiple imputes missing data by treating the cluster allocation as a fixed factor, so that differences in intercepts between classes are modeled.
- Multiple Imputation Multilevel Imputation (ML). This method applies the Gibbs sampler as described above to generate multiple imputations from posterior of the missing data given the observed data.

The number of multiple imputation was fixed to 5. Parameter estimates are pooled using Rubin's rules (Rubin, 1987; Rubin, 1996). The complete-data model was fitted by the `lmer()` function in R package `lme4` (Pinheiro & Bates, 2000).

10.5.3 Results

Table 10.1 contains results of the simulations. When missing data are confined to y_{ij}, then CC is unbiased for both the fixed and random parameters, as expected. Method FF is unbiased in the fixed parameters, but severely biased in the random parameters for clustered data (i.e., when $\omega^2 > 0$). Method SC produces unbiased estimates of both the fixed and random parameters. Note that this is related to the fact that the model that generated the data included only random intercepts and no random slopes. Also, method ML is unbiased in both the fixed and random parameters.

If missing data occur in z_{ij}, the results are drastically different. The estimates under CC are severely biased, both for the fixed and random parameters. Thus even under MAR, the standard practice of eliminating incomplete records can produce estimates that are plainly wrong. Of the three imputation methods, SC and ML yield estimates that are close to population values, FF is generally less successful. Method SC had computational problems for small cluster sizes ($n_j = 20$) because the number of observations in the cluster that remain after missing data were created could become too low (≤ 3). The FF and ML methods are insensitive to this problem since they combine information across clusters.

Table 10.2 contains estimates of the coverage of the 95% confidence interval for the fixed parameters. The number of replications used is equal to 100, so the simulation standard error is $\sqrt{(0.95(1-0.95)/100)} = 2.2\%$. For missing data in y_{ij}, CC has appropriate coverage. However, coverage for missing data in z_{ij} is dismal, so statistical inferences are unwarranted under incomplete z_{ij}. The FF is generally not well calibrated, and may achieve both under- or overcoverage depending on the amount of clustering. The SC has appropriate coverage of β_0, but coverage is suboptimal for β_x. The ML has appropriate coverage for larger cluster sizes for both β_0 and β_x. Coverage for small cluster sizes is however less than ideal, though still reasonable.

This section addressed the properties of four methods for dealing with univariate

TABLE 10.1

MAR Missing Data in Either y_{ij} or z_{ij}

	J	n_j	β_0	CC	FF	SC	ML	β_x	CC	FF	SC	ML	σ^2	CC	FF	SC	ML	ω^2	CC	FF	SC	ML
Y																						
A	12	100	0.00	0.00	0.00	0.00	0.01	0.50	0.51	0.50	0.50	0.50	0.75	0.75	0.75	0.75	0.76	0.00	0.00	0.00	0.02	0.02
B	12	100	0.00	0.00	0.01	−0.02	0.01	0.50	0.50	0.49	0.50	0.50	0.65	0.65	0.71	0.65	0.65	0.10	0.10	0.03	0.12	0.11
C	12	100	0.00	−0.01	0.00	0.01	0.00	0.50	0.50	0.50	0.50	0.50	0.45	0.45	0.63	0.45	0.45	0.30	0.30	0.08	0.33	0.31
D	12	100	0.00	0.03	−0.01	0.00	0.02	0.50	0.50	0.49	0.50	0.50	0.25	0.25	0.55	0.25	0.25	0.50	0.49	0.13	0.51	0.51
E	24	50	0.00	0.00	0.00	0.00	0.00	0.50	0.49	0.50	0.50	0.50	0.75	0.74	0.74	0.75	0.75	0.00	0.01	0.00	0.03	0.02
F	24	50	0.00	0.02	0.00	0.00	0.00	0.50	0.51	0.50	0.50	0.50	0.65	0.65	0.71	0.65	0.66	0.10	0.11	0.03	0.12	0.12
G	24	50	0.00	0.01	0.00	0.00	−0.01	0.50	0.50	0.51	0.50	0.50	0.45	0.44	0.62	0.45	0.45	0.30	0.30	0.07	0.32	0.31
H	24	50	0.00	−0.02	0.00	−0.01	−0.02	0.50	0.50	0.50	0.51	0.51	0.25	0.25	0.57	0.25	0.25	0.50	0.48	0.13	0.48	0.50
I	60	20	0.00	0.00	−0.01	0.00	−0.01	0.50	0.49	0.50	0.50	0.50	0.75	0.74	0.74	0.74	0.74	0.00	0.01	0.00	0.08	0.03
J	60	20	0.00	−0.01	0.01	0.00	0.00	0.50	0.50	0.51	0.50	0.50	0.65	0.65	0.71	0.65	0.65	0.10	0.10	0.03	0.17	0.12
K	60	20	0.00	0.00	−0.01	0.00	0.02	0.50	0.50	0.50	0.50	0.50	0.45	0.45	0.64	0.45	0.45	0.30	0.29	0.07	0.36	0.31
L	60	20	0.00	−0.01	0.01	0.00	−0.01	0.50	0.50	0.50	0.49	0.49	0.25	0.25	0.57	0.25	0.25	0.50	0.49	0.13	0.53	0.49
Z																						
A	12	100	0.00	−0.53	0.00	0.00	0.00	0.50	0.32	0.49	0.49	0.48	0.75	0.49	0.75	0.75	0.74	0.00	0.00	0.00	0.00	0.00
B	12	100	0.00	−0.49	0.00	0.00	−0.01	0.50	0.34	0.48	0.49	0.48	0.65	0.44	0.66	0.65	0.66	0.10	0.05	0.08	0.11	0.10
C	12	100	0.00	−0.36	0.01	0.01	0.01	0.50	0.40	0.45	0.50	0.49	0.45	0.34	0.50	0.45	0.46	0.30	0.20	0.23	0.31	0.30
D	12	100	0.00	−0.22	−0.01	−0.01	−0.01	0.50	0.43	0.40	0.50	0.50	0.25	0.21	0.34	0.25	0.25	0.50	0.39	0.42	0.48	0.52
E	24	50	0.00	−0.53	0.00	0.00	0.00	0.50	0.33	0.50	0.48	0.48	0.75	0.49	0.75	0.75	0.74	0.00	0.00	0.00	0.01	0.01
F	24	50	0.00	−0.49	0.00	0.00	−0.01	0.50	0.35	0.48	0.50	0.47	0.65	0.45	0.67	0.65	0.66	0.10	0.06	0.07	0.10	0.10
G	24	50	0.00	−0.39	−0.01	−0.01	0.01	0.50	0.39	0.44	0.50	0.49	0.45	0.33	0.51	0.45	0.46	0.30	0.20	0.23	0.30	0.29
H	24	50	0.00	−0.23	−0.02	0.00	0.00	0.50	0.43	0.40	0.50	0.49	0.25	0.21	0.35	0.25	0.25	0.50	0.41	0.39	0.50	0.50

(Continued)

TABLE 10.1

MAR Missing Data in Either y_{ij} or z_{ij} (*Continued*)

	J	n_j	β_0	CC	FF	SC	ML	β_x	CC	FF	SC	ML	σ^2	CC	FF	SC	ML	ω^2	CC	FF	SC	ML
I	60	20	0.00	−0.53	0.00	−0.01	−0.01	0.50	0.33	0.50	0.47	0.48	0.75	0.49	0.74	0.74	0.73	0.00	0.00	0.00	0.02	0.01
J	60	20	0.00	−0.50	0.00	0.00	−0.01	0.50	0.34	0.49	0.49	0.48	0.65	0.44	0.66	0.65	0.65	0.10	0.05	0.08	0.12	0.09
K	60	20	0.00	−0.41	−0.01	#	−0.01	0.50	0.38	0.44	#	0.47	0.45	0.33	0.52	#	0.47	0.30	0.18	0.25	#	0.27
L	60	20	0.00	−0.26	−0.01	#	0.00	0.50	0.42	0.40	#	0.49	0.25	0.20	0.35	#	0.27	0.50	0.39	0.41	#	0.46

Notes: Average estimates of fixed (β_0, β_x) and random variance (σ^2, ω^2) parameters in four methods for handling missing data (CC = complete case analysis, FF = MI flat file, SC = MI separate group, ML = MI multilevel).

solution could not be calculated due to almost empty classes.

TABLE 10.2

Coverage (in Percentage) of the True Values by the 95% Confidence Interval for Fixed Parameter Estimates Under Four Methods for Treating Missing Data in *Y* or *Z*, Respectively

	J	n_j	β_0	CC	FF	SC	ML	β_x	CC	FF	SC	ML
Y												
A	12	100	95	96	72	90	90	95	96	73	72	90
B	12	100	95	89	69	96	87	95	96	82	76	91
C	12	100	95	94	71	94	91	95	97	98	70	93
D	12	100	95	94	68	94	97	95	94	100	78	91
E	24	50	95	95	71	91	87	95	97	66	68	88
F	24	50	95	96	73	90	89	95	97	76	63	87
G	24	50	95	92	63	93	88	95	96	90	66	94
H	24	50	95	91	73	94	95	95	96	95	72	87
I	60	20	95	98	66	92	84	95	98	73	69	90
J	60	20	95	99	64	88	88	95	93	71	68	89
K	60	20	95	97	67	88	98	95	97	79	76	86
L	60	20	95	92	66	96	88	95	97	89	73	87
Z												
A	12	100	95	0	88	92	95	95	0	84	84	93
B	12	100	95	0	84	94	87	95	2	83	85	94
C	12	100	95	25	82	90	94	95	23	49	86	94
D	12	100	95	75	91	91	92	95	39	5	87	95
E	24	50	95	0	88	93	90	95	0	94	80	87
F	24	50	95	0	88	99	95	95	1	78	84	87
G	24	50	95	5	96	96	95	95	11	25	94	91
H	24	50	95	54	91	94	94	95	29	1	94	94
I	60	20	95	0	91	92	89	95	0	77	78	85
J	60	20	95	0	87	95	98	95	1	83	86	83
K	60	20	95	0	90	#	96	95	2	35	#	79
L	60	20	95	17	88	#	91	95	16	1	#	85

Notes: CC = complete case analysis, FF = MI flat file, SC = MI separate group, ML = MI multilevel.
solution could not be calculated due to almost empty classes.

missing data within a multilevel context. The CC method is easy and works well under MAR when missing data are restricted to y_{ij}. However, the performance CC with z_{ij} missing at random is bad. We therefore recommend against CC if many z_{ij} are missing. An alternative is to apply multiple imputation. Three such methods were studied. The overall best performance was obtained by the ML Gibbs sampling method.

10.6 MULTIVARIATE MISSING DATA IN y_j AND z_j

10.6.1 General Approach

Missing data may also occur in y_{ij} and z_{ij} simultaneously. The present section deals with the case where both y_{ij} and z_{ij} are incomplete. There are two general approaches to impute multivariate missing data: Joint

Modeling (JM) and Fully Conditional Specification (FCS).

Joint modeling partitions the observations into groups of identical missing data patterns, and imputes the missing entries within each pattern according to a joint model for all variables. The first such model was developed for the multivariate normal model (Rubin & Schafer, 1990). Schafer (1997) extended this line and developed sophisticated JM methods for generating multivariate imputations under the multivariate normal, the log-linear, and the general location model. This work was extended to include multilevel models (Schafer & Yucel, 2002; Yucel, 2008).

The fully conditional specification imputes data on a variable-by-variable basis by specifying an imputation model per variable. The FCS is an attempt to specify the full multivariate distribution of the variables by a set of conditional densities for each incomplete variable. This set of densities is used to impute each variable by iteration, where we start from simple initial guesses. Though convergence can only be proved in some special cases, the method has been found to work well in practice (Raghunathan, Lepkowski, van Hoewyk, & Solenberger, 2001; Van Buuren et al., 1999; Van Buuren, Brand, Groothuis-Oudshoorn, & Rubin, 2006). The R `mice` package (Van Buuren & Groothuis-Oudshoorn, 2000) enjoys a growing popularity. Van Buuren (2007) provides an overview of the similarities and contrasts of JM and FCS.

10.6.2 Simulation Study

Using the same complete-data model as before, we created missing data in both x_{ij} and y_{ij} by applying mechanisms Y and Z each to a random split of the data. For missing z_{ij} the procedure is identical to that given before. For missing y_{ij}, the procedure is reversed. For values of $z_{ij} < 0$, the nonresponse probability in y_{ij} is 90%. For $z_{ij} \geq 0$, this probability is 10%. Thus, many high z_{ij} and many low y_{ij} will be missing.

We created five multiple imputed data sets with `mice` using the three imputation methods. The number of iterations in `mice` was fixed to 20.

10.6.3 Results

Table 10.3 contains the parameter estimates averaged over 100 simulations. Complete case (CC) analysis severely biases the estimates of the intercept term β_0 and the within-group variance σ^2, especially when the clustering is weak. Methods FF and SC have a somewhat better performance for the fixed effects, and behave differently for the variance estimates. The best overall method is ML, but note that ML is not yet ideal since β_0 is biased slightly upward while β_x is biased slightly downward. No systematic bias appears to be present in the variance estimates, so ML seems to recover the multilevel structure present in the original data quite well.

Table 10.4 contains the accompanying coverage percentages. The best method is ML, but none of the methods is really satisfactory. Trouble cases include A, E, and I, where $\omega^2 = 0$. The Gibbs sampler can get stuck if there is no between-cluster variation (Gelman et al., 2008), so this might be a reason for the low coverage. It also appears to be difficult to get appropriate coverage for small cluster sizes.

The simulations suggest that FCS is a promising option for imputing incomplete

TABLE 10.3

MAR Missing Data in both y_{ij} and z_{ij}. Average Estimates of Fixed (β_0, β_x) and Random Variance (σ^2, ω^2) Parameters in Four Methods for Handling Missing Data

	J	n_j	β_0	CC	FF	SC	ML	β_x	CC	FF	SC	ML	σ^2	CC	FF	SC	ML	ω^2	CC	FF	SC	ML
YZ																						
A	12	100	0.00	−0.46	−0.16	−0.16	0.09	0.50	0.40	0.39	0.38	0.46	0.75	0.55	0.77	0.77	0.75	0.00	0.01	0.00	0.05	0.01
B	12	100	0.00	−0.42	−0.16	−0.18	0.09	0.50	0.41	0.39	0.38	0.44	0.65	0.49	0.75	0.69	0.68	0.10	0.06	0.01	0.16	0.10
C	12	100	0.00	−0.30	−0.15	−0.15	0.08	0.50	0.44	0.41	0.41	0.46	0.45	0.36	0.69	0.47	0.48	0.30	0.24	0.02	0.34	0.28
D	12	100	0.00	−0.14	−0.13	−0.17	0.04	0.50	0.48	0.40	0.41	0.47	0.25	0.22	0.70	0.28	0.27	0.50	0.47	0.03	0.54	0.49
E	24	50	0.00	−0.46	−0.16	−0.16	0.08	0.50	0.40	0.39	0.38	0.44	0.75	0.55	0.77	0.76	0.76	0.00	0.01	0.00	0.11	0.01
F	24	50	0.00	−0.42	−0.17	−0.16	0.08	0.50	0.41	0.40	0.39	0.45	0.65	0.48	0.77	0.67	0.68	0.10	0.07	0.01	0.22	0.10
G	24	50	0.00	−0.32	−0.18	−0.15	0.07	0.50	0.45	0.38	0.39	0.44	0.45	0.35	0.73	0.48	0.49	0.30	0.20	0.02	0.38	0.29
H	24	50	0.00	−0.20	−0.15	−0.15	0.09	0.50	0.47	0.37	0.39	0.47	0.25	0.21	0.72	0.28	0.28	0.50	0.47	0.03	0.59	0.52
I	60	20	0.00	−0.47	−0.17	#	0.08	0.50	0.39	0.37	#	0.44	0.75	0.52	0.78	#	0.73	0.00	0.01	0.00	#	0.02
J	60	20	0.00	−0.44	−0.14	#	0.09	0.50	0.41	0.39	#	0.44	0.65	0.49	0.76	#	0.67	0.10	0.07	0.01	#	0.09
K	60	20	0.00	−0.36	−0.17	#	0.09	0.50	0.42	0.38	#	0.43	0.45	0.35	0.75	#	0.50	0.30	0.19	0.02	#	0.28
L	60	20	0.00	−0.23	−0.17	#	0.08	0.50	0.46	0.40	#	0.44	0.25	0.21	0.70	#	0.30	0.50	0.40	0.03	#	0.47

Notes: CC = complete case analysis, FF = MI flat file, SC = MI separate group, ML = MI multilevel.
solution could not be calculated due to almost empty classes.

TABLE 10.4

Coverage (in Percentage) of the True Values by the 95% Confidence Interval for Fixed Parameter Estimates Under Four Methods for Treating Missing Data in Both *Y* and *X*

	J	n_j	β_0	CC	FF	SC	ML	β_x	CC	FF	SC	ML
YZ												
A	12	100	95	0	5	42	37	95	46	29	27	85
B	12	100	95	2	18	64	81	95	55	23	22	77
C	12	100	95	45	25	83	89	95	71	32	26	76
D	12	100	95	83	38	85	90	95	88	29	17	82
E	24	50	95	0	6	39	37	95	48	28	30	64
F	24	50	95	0	9	56	79	95	56	30	27	67
G	24	50	95	16	21	76	84	95	75	25	15	55
H	24	50	95	69	28	81	87	95	82	28	13	72
I	60	20	95	0	1	#	34	95	42	19	#	55
J	60	20	95	0	13	#	50	95	53	24	#	57
K	60	20	95	1	12	#	73	95	52	22	#	42
L	60	20	95	28	17	#	82	95	76	27	#	43

Notes: CC = complete case analysis, FF = MI flat file, SC = MI separate group, ML = MI multilevel.
solution could not be calculated due to almost empty classes.

multilevel data. The FCS used in conjunction with multiple multilevel imputation is a considerable improvement over standard practice. The methodology is not yet ideal however, and further optimization and tuning is needed.

10.7 CONCLUSIONS

Multilevel data can be missing at different levels. Variables in which missing data occur can have different roles in the analysis. The optimal way to deal with missing data depends on both the level and the role of the variable in the analysis.

Multilevel models are often presented in the form of the linear mixed model Equation 10.1. This formulation complicates conceptualization of the missing data problem because the same variable can appear at different places. It is useful to write the multilevel model as a slopes-as-outcomes model Equation 10.2, which clearly separates the variables at the different levels. Section 10.2 describes how Equations 10.1 and 10.2 are related.

Missing data can occur in y_j (level-1 outcomes), Z_j (level-1 predictors) are W_j (level-2 predictors) and j (class variable). The problem of missing data in y_j has received considerable attention. The linear multilevel model provides an efficient solution to this problem if the data are missing at random and if the model fits the data. There is a large literature on what can be done if the MAR assumption is suspect, or when models for other outcome distributions are needed. By comparison, the problem of missing data in Z_j, W_j, and j received only scant attention. The usual solution is to remove any incomplete records, which is wasteful and could bias the estimates of interest. Several fixes have been proposed, but none of these have yet gained wide use.

Other questions that need to be addressed are less particular to the multilevel setting: the missing data pattern, the missing data mechanism, the measurement scales used, and the study design. A successful attack on a given incomplete data problem depends on our capability to address these factors for the application at hand.

Section 10.3 outlines six strategies. Quick fixes like listwise deletion, last observation carried forward and class mean imputation will only work in a limited set of circumstances and are generally discouraged. Prevention, likelihood-based methods, and multiple imputation are methodologically sound approaches based on explicit assumptions about the missing data process.

Multiple imputation is a general statistical technique for handling incomplete data problems. Some work on MI in multilevel setting has been done, but many open issues remain. We performed a simulation study with missing data in y_{ij} or z_{ij}, and compared complete case analysis with three MI techniques: flat file (FF) imputation that ignores the multilevel structure, separate clusters (SC) imputation that includes a group factor, and multilevel (ML) imputation by means of the Gibbs sampler. Complete case analysis was found to be a bad strategy with missing data in z_{ij}. The best imputation technique was ML. A second simulation addressed the question of how the methods behave when missing data occur simultaneously in y_{ij} or z_{ij}. Though its performance is not yet ideal, multiple imputation by ML within the FCS framework considerably improves upon standard practice.

Simulation is not reality. The missing data mechanisms we have used in the simulation have a considerable amount of missing information, and are probably more extreme than those encountered in practice. The simulations are still useful though. Differences between methods in absolute terms may be smaller in practice, but the best methods will continue to dominate others in less extreme situations. All other things being equal, we therefore prefer to use imputation methods that performs best "asymptotically" in extreme situations.

Since ML requires more work than complete case analysis it would be useful to have clear-cut rules that say when doing ML is not worth the trouble. No such rules have yet been devised. This would be a useful area of further research. Another area for research would be to further optimize and tune the ML imputation method to the multivariate missing data problem. For example, taking alternative distributions for within-cluster residual variance σ_j^2 could improve performance. The current implementation of the method uses a full Gibbs sampler. Though the algorithm is robust, it is not particularly fast. Adding parameter expansion (Gelman et al., 2008) could be useful to prevent the Gibbs sampler from getting stuck at the border of the parameter space at $\omega^2 = 0$. Computations could be speeded up, for example by obtaining marginal maximum likelihood estimates of β and Ω using numerical integration via Gauss–Hermite (Pinheiro & Bates, 2000). Extensions toward higher level models are also possible (Yucel, 2008). Finally, we can classify missing data problems by combining the answers on the five questions posed in Section 10.3. Classification of the combinations opens up a whole research agenda with many white spots.

REFERENCES

Bang, K., & Robins, J. M. (2005). Doubly robust estimation in missing data and causal inference models. *Biometrics, 61,* 962–972.

Beunckens, C., Molenberghs, G., Thijs, H., & Verbeke, G. (2007). Incomplete hierarchical data. *Statistical Methods in Medical Research, 16,* 457–492.

Beunckens, C., Sotto, C., & Molenberghs, G. (2008). A simulation study comparing weighted estimating equations with multiple imputation based estimating equations for longitudinal binary data. *Computational Statistics and Data Analysis, 52,* 1533–1548.

Bryk, A. S., & Raudenbush, S. W. (1992). *Hierarchical linear models.* Newbury Park, CA: Sage Publications, Inc.

Carpenter, J., & Goldstein, H. (2004). Multiple imputation in MLwiN. MLwin newsletter [On-line]. Retrieved from http://www.lshtm.ac.uk/msu/missing data/papers/newsletterdec04.pdf

Carpenter, J., Kenward, M. G., & Vansteelandt, S. (2006). A comparison of multiple imputation and doubly robust estimation for analyses with missing data. *Journal of the Royal Statistical Society. Series A: Statistics in Society, 169,* 571–584.

Cheung, M. W. L. (2007). Comparison of methods of handling missing time-invariant covariates in latent growth models under the assumption of missing completely at random. *Organizational Research Methods, 10,* 609–634.

Chib, S., & Carlin, B. P. (1999). On MCMC sampling in hierarchical longitudinal models. *Statistics and Computing, 9,* 26.

Cochran, W. G., & Cox, G. M. (1957). *Experimental designs.* New York, NY: John Wiley & Sons, Ltd.

Cowles, M. K. (2002). MCMC sampler convergence rates for hierarchical normal linear models: A simulation approach. *Statistics and Computing, 12,* 377–389.

Daniels, M. J., & Hogan, J. W. (2008). *Missing data in longitudinal studies. Strategies for Bayesian modeling and sensitivity analysis.* Boca Raton, FL: Chapman & Hall/CRC.

De Leeuw, E. D., Hox, J. J., & Dillman, D. A. (2008). *International handbook of survey methodology.* New York, NY: Lawrence Erlbaum Associates.

de Leeuw, J., & Meijer, E. (2008). *Handbook of multilevel analysis.* New York, NY: Springer.

Demirtas, H., & Schafer, J. L. (2003). On the performance of random-coefficient pattern-mixture models for non-ignorable drop-out. *Statistics in Medicine, 22,* 2553–2575.

Dempster, A. P., Laird, N. M., & Rubin, D. B. (1977). Maximum likelihood estimation from incomplete data via the EM algorithm (with discussion). *Journal of the Royal Statistical Society. Series B: Statistical Methodology,* 1–38.

Diggle, P. J. (1989). Testing for random dropouts in repeated measurement data. *Biometrics, 45,* 1258.

Dodge, Y. (1985). *Analysis of experiments with missing data.* New York, NY: John Wiley & Sons, Ltd.

Fitzmaurice, G. M., Laird, N. M., & Ware, J. H. (2004). *Applied longitudinal analysis.* New York, NY: John Wiley & Sons, Ltd.

Gelman, A., Carlin, J. B., Stern, H. S., & Rubin, D. B. (2004). *Bayesian data analysis* (2nd ed.). London, UK: Chapman and Hall.

Gelman, A., & Hill, J. (2007). *Data analysis using regression and multilevel and hierarchical models.* New York, NY: Cambridge University Press.

Gelman, A., Van Dyk, D. A., Huang, Z., & Boscardin, W. J. (2008). Using redundant parameterizations to fit hierarchical models. *Journal of Computational and Graphical Statistics, 17,* 95–122.

Gibson, N. M., & Olejnik, S. (2003). Treatment of missing data at the second level of hierarchical linear models. *Educational and Psychological Measurement, 63,* 204–238.

Goldstein, H. (1987). *Multilevel models in educational and social research.* London, UK: Charles Griffin & Company Ltd.

Jacobusse, G. W. (2005). WinMICE user's manual [Computer software]. Leiden, The Netherlands: TNO Quality of Life.

Jennrich, R. I., & Schluchter, M. D. (1986). Unbalanced repeated-measures models with structured covariance matrices. *Biometrics, 42,* 805–820.

Kang, J. D. Y., & Schafer, J. L. (2007). Demystifying double robustness: A comparison of alternative strategies for estimating a population mean from incomplete data. *Statistical Science, 22,* 523–539.

Kasim, R. M., & Raudenbush, S. W. (1998). Application of Gibbs sampling to nested variance components models with heterogeneous within-group variance. *Journal of Educational and Behavioral Statistics, 23,* 93–116.

Laird, N. M., & Ware, J. H. (1982). Random-effects models for longitudinal data. *Biometrics, 38,* 963–974.

Lavori, P. W. (1992). Clinical trials in psychiatry: Should protocol deviation censor patient data? (with discussion). *Neuropsychopharmacology, 6,* 39–63.

Littell, R. C., Milliken, G. A., Stroup, W. W., & Wolfinger, R. D. (1996). *SAS system for mixed models.* Cary, NC: SAS Institute.

Little, R. J. A. (1988). A test of missing completely at random for multivariate data with missing values. *Journal of the American Statistical Association, 83,* 1198–1202.

Little, R. J. A. (1992). Regression with missing X's: A review. *Journal of the American Statistical Association, 87,* 1227–1237.

Little, R. J. A. (2008). Selection and pattern-mixture models. In G. M. Fitzmaurice, M. Davidian, G. Verbeke, & G. Molenberghs (Eds.), *Longitudinal data analysis: A handbook of modern statistical methods* (pp. 409–431). New York, NY: Wiley.

Little, R. J. A., & Rubin, D. B. (2002). *Statistical analysis with missing data.* (2nd ed.). New York: Wiley.

Little, R. J. A., & Yau, L. (1996). Intent-to-treat analysis for longitudinal studies with drop-outs. *Biometrics, 52,* 1324–1333.

Longford, N. T. (2005). *Missing data and small-area estimation.* New York, NY: Springer.

Longford, N. T. (2008). Missing data. In J. De Leeuw & E. Meijer (Eds.), *Handbook of multilevel analysis* (pp. 377–399). New York, NY: Springer.

McKnight, P. E., McKnight, K. M., Sidani, S., & Figueredo, A. J. (2007). *Missing data. A gentle introduction.* New York, NY: Guilford Press.

Molenberghs, G., & Kenward, M. G. (2007). *Missing data in clinical studies.* Chichester, UK: John Wiley & Sons, Ltd.

Molenberghs, G., & Verbeke, G. (2005). *Models for discrete longitudinal data.* New York, NY: Springer.

Muthén, L. K., & Muthén, B. O. (2007). *Mplus user's guide* (Version V5.1) [Computer software]. Los Angeles, CA: Muthén & Muthén.

Petrin, R. A. (2006). *Item nonresponse and multiple imputation for hierarchical linear models.* Paper presented at the annual meeting of the American Sociological Association, Montreal Convention Center, Montreal, Quebec, Canada Retrieved from http://www.allacademic.com/meta/p102126_index.html

Pinheiro, J. C., & Bates, D. M. (2000). *Mixed-effects models in S and S-PLUS.* New York, NY: Springer.

Raghunathan, T. E., Lepkowski, J. M., van Hoewyk, J., & Solenberger, P. (2001). A multivariate technique for multiply imputing missing values using a sequence of regression models. *Survey Methodology, 27,* 85–95.

Rasbash, J., Steel, F., Browne, W., & Prosser, B. (2005). A user's guide to MLwiN Version 2.0 [Computer software]. Bristol, UK: Centre for Multilevel Modelling, University of Bristol.

Rässler, S. (2002). *Statistical matching. A frequentist theory, practical applications, and alternative Bayesian approaches.* New York, NY: Springer.

Raudenbush, S. W., Bryk, A. S., & Congdon, R. (2008). *HLM 6* [Computer software]. Chicago, IL: SSI Software International.

Roudsari, B., Field, C., & Caetano, R. (2008). Clustered and missing data in the US National Trauma Data Bank: Implications for analysis. *Injury Prevention, 14,* 96–100.

Rubin, D. B. (1976). Inference and missing data. *Biometrika, 63,* 581–590.

Rubin, D. B. (1986). Statistical matching using file concatenation with adjusted weights and multiple imputations. *Journal of Business Economics and Statistics, 4,* 87–94.

Rubin, D. B. (1987). *Multiple imputation for nonresponse in surveys.* New York, NY: Wiley.

Rubin, D. B. (1996). Multiple imputation after 18 + years. *Journal of the American Statistical Association, 91,* 473–489.

Rubin, D. B., & Schafer, J. L. (1990). Efficiently creating multiple imputations for incomplete multivariate normal data. *1990 Proceedings of the Statistical Computing Section, American Statistical Association,* 83–88.

Schafer, J. L. (1997). *Analysis of incomplete multivariate data.* London: Chapman & Hall.

Schafer, J. L., & Schenker, N. (2000). Inference with imputed conditional means. *Journal of the American Statistical Association, 449,* 144–154.

Schafer, J. L., & Yucel, R. M. (2002). Computational strategies for multivariate linear mixed-effects models with missing values. *Journal of Computational and Graphical Statistics, 11,* 437–457.

Scharfstein, D. O., Rotnitzky, A., & Robins, J. M. (1999). Adjusting for nonignorable drop-out using semiparametric nonresponse models. *Journal of the American Statistical Association, 94,* 1096–1120.

Snijders, T. A. B., & Bosker, R. J. (1999). *Multilevel analysis. An introduction to basic and advanced multilevel modeling.* London, UK: Sage Publications Ltd.

SPSS Inc. (2008). *SPSS 17.0 base user's guide* [Computer software]. Chicago, IL: SPSS Inc.

StataCorp LP (2008). *STATA 10 user's guide* [Computer software]. College Station, TX: STATA Press.

Stoop, I. A. L. (2005). *The hunt for the last respondent: Nonresponse in sample surveys.* Rijswijk, The Netherlands: Sociaal en Cultureel Planbureau.

Thomas, N., & Gan, N. (1997). Generating multiple imputations for matrix sampling data analyzed with item response models. *Journal of Educational and Behavioral Statistics, 22,* 425–445.

Van Buuren, S. (2007). Multiple imputation of discrete and continuous data by fully conditional specification. *Statistical Methods in Medical Research, 16,* 219–242.

Van Buuren, S., Boshuizen, H. C., & Knook, D. L. (1999). Multiple imputation of missing blood pressure covariates in survival analysis. *Statistics in Medicine, 18,* 681–694.

Van Buuren, S., Brand, J. P. L., Groothuis-Oudshoorn, C. G. M., & Rubin, D. B. (2006). Fully conditional specification in multivariate imputation. *Journal of Statistical Computation and Simulation, 76,* 1049–1064.

Van Buuren, S., & Groothuis-Oudshoorn, K. (2000). *Multivariate imputation by chained equations: MICE V1.0 user's manual.* (PG/VGZ/00.038 ed.) Leiden, The Netherlands: TNO Quality of Life.

Verbeke, G., & Molenberghs, G. (2000). *Linear mixed models for longitudinal data.* New York, NY: Springer.

Yucel, R. M. (2008). Multiple imputation inference for multivariate multilevel continuous data with ignorable non-response. *Philosophical Transactions of the Royal Society A, 366,* 2389–2403.

Yucel, R. M., Schenker, N., & Raghunathan, T. E. (2006). Multiple imputation for incomplete multilevel data with SHRIMP. Retrieved from: http://www.umass.edu/family/pdfs/talkyucel.pdf (April 12, 2010).

Zaidman-Zait, A., & Zumbo, B. D. (2005). *Multilevel (HLM) models for modeling change with incomplete data: Demonstrating the effects of missing data and level-1 model mis-specification.* Paper presented at the Hierarchical Linear Modeling (SIG) of the American Educational Research Association conference April 2005 in Montreal, Quebec, Canada. Retrieved from: http://educ.ubc.ca/faculty/zumbo/aera/papers/Zaidman_Zait_Zumbo_AERA2005.pdf (April 12, 2010).

Zeger, S. L., & Karim, M. R. (1991). Generalized linear models with random effects: A Gibbs sampling approach. *Journal of the American Statistical Association, 86,* 79–86.

Zhang, D. (2005). *A Monte Carlo investigation of robustness to nonnormal incomplete data of multilevel modeling.* College Station, TX: Texas A&M University.

11

Handling Omitted Variable Bias in Multilevel Models: Model Specification Tests and Robust Estimation

Jee–Seon Kim and Chris M. Swoboda
Department of Educational Psychology,
University of Wisconsin at Madison, Madison, Wisconsin

11.1 INTRODUCTION

Multilevel models allow researchers to examine hypothesized relationships across different units of analysis in a statistically appropriate way, thus permitting more accurate modeling of complex systems. At the same time, the complexity of multilevel models introduces other challenges in statistical modeling, as many assumptions are needed (Goldstein, 2003; Hox, 2002; Raudenbush & Bryk, 2002; Snijders & Bosker, 1999). In particular, there are multiple random effects in multilevel models and it is assumed that all predictors in the model are uncorrelated with all of the random effects. Standard estimation methods for multilevel models such as full information maximum likelihood (FIML), restricted (or residual) maximum likelihood (REML), generalized least squares (GLS), empirical Bayes, and fully Bayesian estimators all assume the independence of predictors from random effects and would yield biased estimates if the assumption is violated.

However, independence between predictors and random effects is prone to be violated in practice. There are three common forms of bias due to correlated effects (Kim & Frees, 2007). First, unobserved effects can lead to *omitted variable bias*. Second, predictors might be measured imprecisely and result in *measurement error* or *error-in-variable bias*. Third, some predictors may not only cause but also be influenced by the outcome variable (two-way causality), yielding *simultaneity bias*. Among these, this chapter focuses on handling correlated effects due to omitted variables, which is a common thread in most observational and quasi-experimental studies in the social and behavioral sciences.

Whereas model specification tests have been one of the most important areas of research in econometrics for decades (Frees, 2004; Hausman, 1978; Wooldridge, 2002), specification issues have been often overlooked in multilevel analysis. We argue that concerns for omitted variables and other specification issues should be routine in multilevel analysis. An omitted variable at one level may yield severe bias at all levels in terms of regression coefficients as well as variance components (Kim, 2009).

This chapter provides a tutorial for the recent statistical developments in Kim and Frees (2006, 2007), which introduced a set of statistical tools for testing the severity of omitted variable bias and for obtaining robust estimators in the presence of omitted variable effects. While these two original articles are more technical, this chapter provides a more conceptual review of the methodology with less emphasis on mathematics. An application of the methods is illustrated with a well-known data set, the National Education Longitudinal Study of 1988 (NELS:88). For details on the procedures and longitudinal data examples, we refer to the original papers.

The rest of the chapter is organized as follows: We define fundamental concepts such as endogeneity and exogeneity in linear models and provide a review of econometric treatment of omitted variables in Section 11.2. Sections 11.3 and 11.4 present omitted variable tests and the *generalized method of moments* (GMM) estimation technique, respectively. Section 11.5 applies the methodologies of the previous two sections to NELS:88. The last section provides a summary and ends with a discussion of related topics in various disciplines.

11.2 BACKGROUNDS

11.2.1 Endogeneity and Omitted Variable Bias

The problem of *endogeneity* in a regression model occurs when a predictor is correlated with the error term in the model. Endogeneity is defined similarly in multilevel modeling but, unlike a regression model, there exist multiple random components in a multilevel model and thus more opportunities for endogeneity to occur.

We consider a linear model as:

$$\mathbf{y} = \mathbf{X}\boldsymbol{\beta} + \boldsymbol{\delta},$$

where $\mathbf{X}$ is the collection of predictors across the levels and $\boldsymbol{\delta}$ is the collection of all random components. If a predictor is correlated with $\boldsymbol{\delta}$, it is an *endogenous* variable. If not, it is an *exogenous* variable. Thus, the *exogeneity* assumption implies that all predictors are uncorrelated with all random components in the model.

As this chapter focuses on linear models, we express the condition for exogeneity as:

$$\mathrm{E}(\mathbf{X}\boldsymbol{\delta}) = \mathbf{0}. \tag{11.1}$$

Variables that do not satisfy this condition are said to be endogenous. Sometimes, a more restrictive assumption $\mathrm{E}(\boldsymbol{\delta}|\mathbf{X}) = \mathbf{0}$ is also used as the exogeneity condition.

To illustrate the problem of omitted variables, consider a "true" model:

$$\mathbf{y} = \mathbf{X}\boldsymbol{\beta} + \mathbf{U}\boldsymbol{\gamma} + \boldsymbol{\delta}, \tag{11.2}$$

where $\mathbf{U}$ are unobserved predictors that affect the outcome. Since $\mathbf{U}$ is unobserved—hence omitted in the analysis—the "fitted" model is

$$\mathbf{y} = \mathbf{X}\boldsymbol{\beta} + \tilde{\boldsymbol{\delta}},$$

where $\tilde{\boldsymbol{\delta}} = \mathbf{U}\boldsymbol{\gamma} + \boldsymbol{\delta}$. The expected value of the least squares estimates for the regression coefficients associated with $\mathbf{X}$ can be shown to be $\boldsymbol{\beta} + (\mathbf{X}'\mathbf{X})^{-1}\mathbf{X}'\mathbf{U}\boldsymbol{\gamma}$. Unless either $\mathbf{X}'\mathbf{U} = \mathbf{0}$ (observed and unobserved predictors are uncorrelated) or $\boldsymbol{\gamma} = 0$ (unobserved predictors do not affect the outcome), $E(\mathbf{X}\tilde{\boldsymbol{\delta}}) \neq 0$ and the least squares estimator of $\boldsymbol{\beta}$ is biased and inconsistent.

In multilevel models, GLS estimators, ML estimators, and empirical Bayes estimators often provide indistinguishable solutions. GLS solutions are sometimes used as starting values for other estimators for complex multilevel models or a large data set. Importantly, all of these estimators rely on the assumption that predictors are uncorrelated with random components. However, this exogeneity assumption is prone to be violated in most observational and quasi-experimental studies, where researchers do not have the ability to control for or collect all the right variables.

The endogeneity problem has been studied extensively in econometrics, mainly in the context of panel data analysis (Frees, 2004; Frees & Kim, 2008; Hsiao, 2003), and statistical tests for omitted variable effects such as the Hausman test (1978) have been steadily used for the past 30 years. In the following section, we review the econometric treatment of omitted variable bias and explain why the methodology for panel data models is not appropriate for more complex multilevel models.

11.2.2 Econometric Treatment and Its Limitations

A panel data model that includes omitted variables can be written as

$$y_{it} = \alpha_i + \mathbf{x}_{it}'\boldsymbol{\beta} + u_i + \varepsilon_{it}, \qquad (11.3)$$

where individual $i = 1,\ldots,n$ is observed over time points $t = 1,\ldots,T_i$. Equation 11.3 includes the outcome variable y_{it} (test score, for example), disturbance term ε_{it}, predictors $\mathbf{x}_{it}$, and coefficient vector $\boldsymbol{\beta}$. The model also contains a latent intercept variable α_i that is constant over time. This latent variable induces a correlation among individual responses over time and serves as a proxy for unobserved time-constant characteristics, such as "ability," that are uncorrelated with the predictors. Without the omitted variable u_i, the model in Equation 11.3 is a two-level random intercept model.

Unlike α_i, u_i may be correlated with one or more of the predictors in $\mathbf{x}_{it}$. Thus, this variable may create bias in the estimates of $\boldsymbol{\beta}$. To mitigate the effects of u_i, one can apply a fixed effects (FE) transformation, sweeping out the time-constant omitted variable. Here, the phrase "sweeping out" refers to the fact that the FE transformation in Equation 11.4 below would remove all time-constant variables. To see the impact of the FE transformation strategy, take averages over time in Equation 11.3 to get

$$\bar{y}_i = \alpha_i + \mathbf{x}_i'\boldsymbol{\beta} + u_i + \bar{\varepsilon}_i.$$

Subtracting this from Equation 11.3 yields the transformed model equation:

$$y_{it} - \bar{y}_i = (\mathbf{x}_{it} - \bar{\mathbf{x}}_i)'\boldsymbol{\beta} + \varepsilon_{it} - \bar{\varepsilon}_i. \qquad (11.4)$$

Then the least squares estimator of β is

$$\mathbf{b}_{\text{FE}} = \left(\sum_i \sum_t (\mathbf{x}_{it} - \mathbf{x}_i)'(\mathbf{x}_{it} - \bar{\mathbf{x}}_i)\right)^{-1} \times \left(\sum_i \sum_t (\mathbf{x}_{it} - \bar{\mathbf{x}}_i)'(y_{it} - \bar{y}_i)\right). \qquad (11.5)$$

This estimator is unbiased even in the presence of the time-constant omitted variable u_i. We denote this estimator as a fixed effects (FE) estimator $\mathbf{b}_{\text{FE}}$. One important strength of the FE estimator is that there are relatively few assumptions needed, compared to alternative procedures. For example, instrumental variable estimation requires the analyst to identify a proxy for the omitted variable. Similarly, simultaneous equations modeling requires specifying a model for the latent, unobserved, omitted variables. Although these alternatives are certainly appropriate in many circumstances, they do require additional (and sometimes unavailable) knowledge from the analyst.

Another important advantage is that these procedures are easy to implement. Generally, one can implement the calculations using standard statistical packages while only specifying certain "fixed effects" nuisance parameters, or unwanted "dummy variables." Note that we do not actually introduce these extra unwanted parameters into the model; the estimates are simply a by-product of the estimation procedures when Equation 11.5 is calculated by a computer program that is not specifically designed for the FE estimator.

Equation 11.5 also underscores a major limitation of FE estimation. Note that this estimator only provides estimates for variables that vary by time; if the jth predictor of $\mathbf{x}_{it}$ is constant over time, then the jth row of $(\mathbf{x}_{it} - \bar{\mathbf{x}}_i)$ is identically zero so that $\boldsymbol{\beta}$ is not estimable. This means that while the FE approach removes undesirable u_i, it also sweeps out potentially important $\bar{\mathbf{x}}_i$ such as variables related to family characteristics, teacher qualities, and school environments. This loss of information would be a critical drawback in many applications.

Therefore, if bias due to omitted variables is not significant, one may prefer other options that do not lose information. To evaluate the size of omitted variable bias, we again consider the model in Equation 11.3 but now assume that there are no omitted variables so that $u_i = 0$. Without u_i, predictors are uncorrelated with random effects in the model and one can estimate $\boldsymbol{\beta}$ using standard procedures discussed earlier, including REML, FIML, GLS, and Bayesian estimators. We refer to a resulting estimator using the independence assumption between predictors and random effects as a random effects (RE) estimator, denoted as $\mathbf{b}_{\text{RE}}$. Note that $\mathbf{b}_{\text{RE}}$ would be biased in the presence of omitted variables ($u_i \neq 0$).

The severity of omitted variable bias for a given data set can be examined by comparing FE with RE estimators, as the former is robust to the presence of u_i and the latter is not. Hausman (1978) presented a test for examining the effect of omitted variables by measuring the distance between vectors $\mathbf{b}_{\text{FE}}$ and $\mathbf{b}_{\text{RE}}$:

$$\chi^2_{\text{Hausman}} = (\mathbf{b}_{\text{FE}} - \mathbf{b}_{\text{RE}})' (Var\,\mathbf{b}_{\text{FE}} - Var\,\mathbf{b}_{\text{RE}})^{-1} \times (\mathbf{b}_{\text{FE}} - \mathbf{b}_{\text{RE}}). \tag{11.6}$$

This test statistic follows a chi-square distribution under the null hypothesis of no omitted variables, with degrees of freedom equal to the number of parameters in $\mathbf{b}_{\text{FE}}$. Under the null hypothesis that there exist no omitted variables, both $\mathbf{b}_{\text{FE}}$ and $\mathbf{b}_{\text{RE}}$ are unbiased. In contrast, only $\mathbf{b}_{\text{FE}}$ is unbiased if the null hypothesis is not true. Thus, when the hypothesis is retained, the analyst would choose $\mathbf{b}_{\text{RE}}$ without the loss of information in $\mathbf{b}_{\text{FE}}$. On the other hand, if the distance between the two estimators is statistically significant, this implies that the random effects estimator is biased due to the presence of omitted variable effects, and the robust fixed effects estimator should be used despite its limitations.

The Hausman test is an effective tool for panel data models, but its applicability is limited for multilevel models, because the test was developed for examining omitted variable effects at the second level in two level models, not for three- or higher-level models. If there exist omitted variable effects at a lower level (e.g., teacher level), applying the Hausman test at a higher level (e.g., school level) would yield fallacious results, as both estimators in the test would be biased. Simulation studies have shown that the level-3 (school) fixed effects estimator could be as severely biased as the random effects estimator when there exist level-2 (teacher) omitted variable effects (Kim & Frees, 2006). This is not a problem of the Hausman test itself, but rather is an improper use of the test for which it is not designed.

Therefore, while we recommend the Hausman test for a panel data model, which is mathematically equivalent to a two-level random intercept model, it should be noted that the test is not appropriate for examining more complex models with more than two levels and/or random slopes. Test results can be highly misleading in more general multilevel modeling contexts. On the one hand, retaining the null hypothesis may not imply the absence of omitted variable effects but may reflect that both the fixed and random effects estimators are biased in a similar way. On the other hand, rejecting the null hypothesis may suggest that the fixed and random effects estimators are biased in a different way.

In the rest of the chapter, we explain recent methodology for extending the idea of the Hausman test to be applicable in more general multilevel models. Extensions involve multiple hypotheses and corresponding tests to determine the location and severity of omitted variable effects. Moreover, we present an alternative robust estimation approach that overcomes the loss of information problem of the FE estimator. The methodology and its implications are presented in the context of a model for mathematics achievement in NELS:88. This chapter ends with a discussion of related topics in various disciplines.

11.3 OMITTED VARIABLES IN MULTILEVEL MODELS

A key to understanding the methodology for handling omitted variable bias is to recognize that when we have only one set of estimates, we do not know if those estimates are biased or not. To diagnose bias, we need to consider multiple sets of estimates obtained from different estimators, ideally where some are more robust than others. This section presents notation for a three-level model, defines two FE estimators and the RE estimator in the three-level model, and introduces statistical tests to compare these estimators.

11.3.1 Multilevel Models with Unobserved Variables

Consider three levels of nesting, where the subscript s identifies a school, the subscript t identifies a teacher belonging to school s, and the subscript p denotes a pupil belonging to school s and teacher t. The level-1 model is then written as:

$$y_{stp} = \mathbf{Z}_{stp}^{(1)}\boldsymbol{\beta}_{st}^{(1)} + \mathbf{X}_{stp}^{(1)}\boldsymbol{\beta}_1 + \varepsilon_{stp}^{(1)},$$

where y_{stp} denotes the response variable. Predictors $\mathbf{Z}_{stp}^{(1)}$ and $\mathbf{X}_{stp}^{(1)}$ may be related to the pupil, teacher, or school. Parameters that are constant appear in the $\boldsymbol{\beta}_1$ vector and so we interpret $\mathbf{X}_{stp}^{(1)}\boldsymbol{\beta}_1$ to be part of the "fixed effects" portion of the model. The term $\boldsymbol{\beta}_{st}^{(1)}$ captures latent, unobserved characteristics that are school and teacher specific. We wish to allow for, and test, the possibility that these latent characteristics are related to predictors $\mathbf{Z}_{stp}^{(1)}$ and $\mathbf{X}_{stp}^{(1)}$. For identification purposes, we adopt the usual convention and assume that the disturbance term, $\varepsilon_{stp}^{(1)}$, is independent of the other right-hand variables, $\mathbf{Z}_{stp}^{(1)}$, $\mathbf{X}_{stp}^{(1)}$, and $\boldsymbol{\beta}_1$.

The level-2 model describes the variability at the teacher level and is written as

$$\boldsymbol{\beta}_{st}^{(1)} = \mathbf{Z}_{st}^{(2)}\boldsymbol{\beta}_s^{(2)} + \mathbf{X}_{st}^{(2)}\boldsymbol{\beta}_2 + \varepsilon_{st}^{(2)}.$$

Analogous to the level-1 model, the predictors $\mathbf{Z}_{st}^{(2)}$ and $\mathbf{X}_{st}^{(2)}$ relate to the teacher or school. Constant parameters appear in the $\boldsymbol{\beta}_2$ vector and so we interpret $\mathbf{X}_{st}^{(2)}\boldsymbol{\beta}_2$ to be the fixed effects at the teacher level. The term $\boldsymbol{\beta}_s^{(2)}$ captures latent, unobserved characteristics that are school specific; these latent characteristics may be related to predictors $\mathbf{Z}_{st}^{(2)}$ and $\mathbf{X}_{st}^{(2)}$.

Finally, the level-3 model describes variability at the school level, and is written as

$$\boldsymbol{\beta}_s^{(2)} = \mathbf{X}_s^{(3)}\boldsymbol{\beta}_3 + \boldsymbol{\varepsilon}_s^{(3)}.$$

The variables $\mathbf{X}_s^{(3)}$ may depend on the school. We let $\boldsymbol{\varepsilon}_s^{(3)}$ represent other unobserved characteristics of the school that are not explained by the fixed effects portion, $\mathbf{X}_s^{(3)}\boldsymbol{\beta}_3$.

It is well known that a multilevel model may be written as a linear mixed-effects model for, among other reasons, parameter estimation purposes. For fitting multilevel models as linear mixed-effects models, Singer (1998) provides an insightful overview using SAS PROC MIXED examples. Combining the separate models for the three levels above, the multilevel model can be expressed as a linear-mixed effects model:

$$y_{stp} = \mathbf{Z}_{stp}^{(1)}\mathbf{Z}_{st}^{(2)}(\mathbf{X}_s^{(3)}\boldsymbol{\beta}_3 + \boldsymbol{\varepsilon}_s^{(3)}) + \mathbf{Z}_{stp}^{(1)}(\mathbf{X}_{st}^{(2)}\boldsymbol{\beta}_2 + \boldsymbol{\varepsilon}_{st}^{(2)}) + \mathbf{X}_{stp}^{(1)}\boldsymbol{\beta}_1 + \boldsymbol{\varepsilon}_{stp}^{(1)}. \tag{11.7}$$

Next, define $\mathbf{Z}_{2,stp} = \mathbf{Z}_{stp}^{(1)}$ and $\mathbf{Z}_{3,stp} = \mathbf{Z}_{stp}^{(1)}\mathbf{Z}_{st}^{(2)}$. With this notation, we may summarize all random component terms as $\delta_{stp} = \mathbf{Z}_{3,stp}\,\varepsilon_s^{(3)} + \mathbf{Z}_{2,stp}\,\varepsilon_{st}^{(2)} + \varepsilon_{stp}^{(1)}$. Further, define the $K \times 1$ vector $\boldsymbol{\beta} = (\boldsymbol{\beta}_1', \boldsymbol{\beta}_2', \boldsymbol{\beta}_3')$ and the $1 \times K$ vector $\mathbf{X}_{stp} = (\mathbf{X}_{stp}^{(1)} : \mathbf{Z}_{2,stp}\mathbf{X}_{st}^{(2)} : \mathbf{Z}_{3,stp}\mathbf{X}_s^{(3)})$. With this notation and stacking, we may express the model as

$$\mathbf{y}_s = \mathbf{X}_s\boldsymbol{\beta} + \boldsymbol{\delta}_s, \tag{11.8}$$

which emphasizes that schools are independent units in the model. The exogeneity assumption would be violated if some of $\mathbf{X}_s$ are correlated with $\boldsymbol{\delta}_s$.

It is useful to consider a special and common case of this general three-level model where there are no random slopes, but the intercepts are associated with random coefficients. Using the above notation, this implies $\mathbf{Z}_{stp}^{(1)} = 1$ and $\mathbf{Z}_{st}^{(2)} = \mathbf{1}$. Multilevel models of this kind are referred to as *random intercept models*. We can write the three-level random intercept model as a single equation of the form:

$$y_{stp} = \mathbf{X}_s^{(3)}\boldsymbol{\beta}_3 + \mathbf{X}_{st}^{(2)}\boldsymbol{\beta}_2 + \mathbf{X}_{stp}^{(1)}\boldsymbol{\beta}_1 + \varepsilon_s^{(3)} + \varepsilon_{st}^{(2)} + \varepsilon_{stp}^{(1)}. \tag{11.9}$$

To illustrate omitted variable problems in multilevel models, we denote unobserved effects of schools and teachers as $\mathbf{u}_s^{(3)}$ and $\mathbf{u}_{st}^{(2)}$, respectively, and add them to Equation 11.9:

$$y_{stp} = \mathbf{X}_s^{(3)}\boldsymbol{\beta}_3 + \mathbf{X}_{st}^{(2)}\boldsymbol{\beta}_2 + \mathbf{X}_{stp}^{(1)}\boldsymbol{\beta}_1 + \varepsilon_s^{(3)} + \varepsilon_{st}^{(2)} + \varepsilon_{stp}^{(1)} + \mathbf{u}_s^{(3)} + \mathbf{u}_{st}^{(2)}. \quad (11.10)$$

The latent intercept variables $\varepsilon_s^{(3)}$ and $\varepsilon_{st}^{(2)}$ are uncorrelated with the predictors. By contrast, $\mathbf{u}_s^{(3)}$ and $\mathbf{u}_{st}^{(2)}$ may be correlated with one or more of the predictors in the model and thus their omission may create bias in the estimates of $\boldsymbol{\beta}$. In the next section, we present omitted variable tests to examine bias due to $\mathbf{u}_s^{(3)}$ and $\mathbf{u}_{st}^{(2)}$ in a given data set.

11.3.2 Different Estimators

As mentioned earlier, commonly used multilevel model estimators such as REML, FIML, GLS, and Bayesian estimators assume that $\mathbf{u}_s^{(3)} = \mathbf{u}_{st}^{(2)} = 0$ in Equation 11.10, and each of these estimators can be considered as a random effects (RE) estimator, $\mathbf{b}_{\mathrm{RE}}$, which provides an unbiased solution in the absence of omitted variables but yields biased estimates of model parameters if the assumption is violated. To examine the degree of bias in b_{RE}, we also considered the fixed teacher effects estimator $\mathbf{b}_{\mathrm{FEt}}$, which is robust against the presence of both $\mathbf{u}_s^{(3)}$ and $\mathbf{u}_{st}^{(2)}$, and the fixed school effects estimator $\mathbf{b}_{\mathrm{FEs}}$, which is robust against the presence of $\mathbf{u}_s^{(3)}$ but not $\mathbf{u}_{st}^{(2)}$.

The two FE estimators, $\mathbf{b}_{\mathrm{FEt}}$ and $\mathbf{b}_{\mathrm{FEs}}$, can be obtained by treating the teacher and school identification variables as discrete variables. This may take some time and the output can be long (unless suppressed), if the number of units is large, because the program would estimate nuisance parameters for the dummy variables assigned to the entities. In Section 11.4, we present a direct approach to obtain $\mathbf{b}_{\mathrm{FEt}}$ and $\mathbf{b}_{\mathrm{FEs}}$ without the unnecessary dummy variables. Note that while $\mathbf{b}_{\mathrm{RE}}$ provides estimates at all three levels, $\mathbf{b}_{\mathrm{FEt}}$ only provides estimates for the level-1 variables while $\mathbf{b}_{\mathrm{FEs}}$ provides estimates for the level-1 and level-2 variables. Because of this severe loss of information, one may not choose the FE estimators unless the bias in RE is statistically significant.

11.3.3 Three Types of Omitted Variable Tests

Kim and Frees (2006) presented three omitted variable tests for examining omitted variable bias in multilevel models. They can be recognized as multiple-level, intermediate-level, and highest-level tests, respectively. The names of the tests indicate the locations of potential omitted variables being tested. The *multiple-level* test can be viewed as an "omnibus" test for examining all potential omitted variable effects simultaneously across levels by comparing the most robust estimator to the most efficient estimator. If this test indicates that the most efficient but least robust estimator (i.e., $\mathbf{b}_{\mathrm{RE}}$) is unbiased, no further test may be necessary and we can make inferences based on $\mathbf{b}_{\mathrm{RE}}$.

In most applications, the analyst may have some ideas about omitted variables and want to test one or more levels separately. Omitted variable effects at lower levels can be tested by the *intermediate-level* test, regardless of omitted variable effects at higher levels. In a three-level school–teacher–pupil model, the omitted teacher effects can be tested regardless of omitted school effects. In a four-level model, omitted variable effects at the second and/or third level can be tested regardless of omitted variable effects at the fourth level using the intermediate-level test.

Finally, the *highest-level* test examines omitted variable effects at higher levels, assuming there exist no omitted variable effects at lower levels. In the school–teacher–pupil model, omitted school effects can be tested assuming there exist no omitted teacher effects. In a four-level model, omitted variable effects at the fourth level can be tested assuming there exist no omitted variable effects at the second and third levels. Also, omitted variable effects at the third and fourth level can be tested assuming there exist no omitted variable effects at the second level using the highest-level test.

All three tests can be defined by one test statistic:

$$\chi^2_{\text{OVT}} = (\mathbf{b}_{\text{robust}} - \mathbf{b}_{\text{efficient}})' \times (\text{Var}\,\mathbf{b}_{\text{robust}} - \text{Var}\,\mathbf{b}_{\text{efficient}})^{-1} \times (\mathbf{b}_{\text{robust}} - \mathbf{b}_{\text{efficient}}), \quad (11.11)$$

and a pair of robust and efficient estimators is determined with respect to the hypothesis being tested. For a three level model, the hypotheses and corresponding robust and efficient estimators for the three omitted variable tests can be summarized as follows:

Omitted Variable Test	Hypothesis	Robust Estimator	Efficient Estimator
1. Multiple-level test	$\mathbf{u}_s^{(3)} = \mathbf{u}_{st}^{(2)} = 0$	$\mathbf{b}_{\text{FEt}}$	$\mathbf{b}_{\text{RE}}$
2. Intermediate-level test	$\mathbf{u}_{st}^{(2)} = 0$	$\mathbf{b}_{\text{FEt}}$	$\mathbf{b}_{\text{FEs}}$
3. Highest-level test	$\mathbf{u}_s^{(3)} = 0$	$\mathbf{b}_{\text{FEs}}$	$\mathbf{b}_{\text{RE}}$

In regard to determining what tests would be appropriate for a given model, one may consider the following two properties. First, the omnibus multiple-level test would be recommended in general if omitted variable effects are of concern at all levels. When the multiple-level test is rejected, one can subsequently test each level using the other tests. However, if one suspects omitted variable effects at a particular level, the intermediate-level or highest-level test would be more powerful than the multiple-level test.

Second, it is important to note the asymmetry properties of the second and third tests; that is, while the intermediate-level test is valid regardless of omitted variable effects at higher levels, the highest-level test is only valid without omitted variable effects at lower levels. This occurs because the lower-level fixed effects estimators are robust against higher-level omitted variable effects, but the higher-level fixed effects estimators are not robust against lower-level omitted variable effects. Therefore, the analyst should either assume or test $\mathbf{u}_{st}^{(2)} = 0$ before testing $\mathbf{u}_s^{(3)} = 0$. The three omitted variable tests are shown at the top of Figure 11.1, which provides a flowchart for data analysis procedures for model specification tests and selecting the optimal estimator. Figure 11.1 will be revisited with an empirical example in Section 11.5.

11.4 GENERALIZED METHOD OF MOMENTS (GMM) INFERENCE

The generalized method of moments (GMM) is a general estimation method for statistical models. As its name indicates, it is

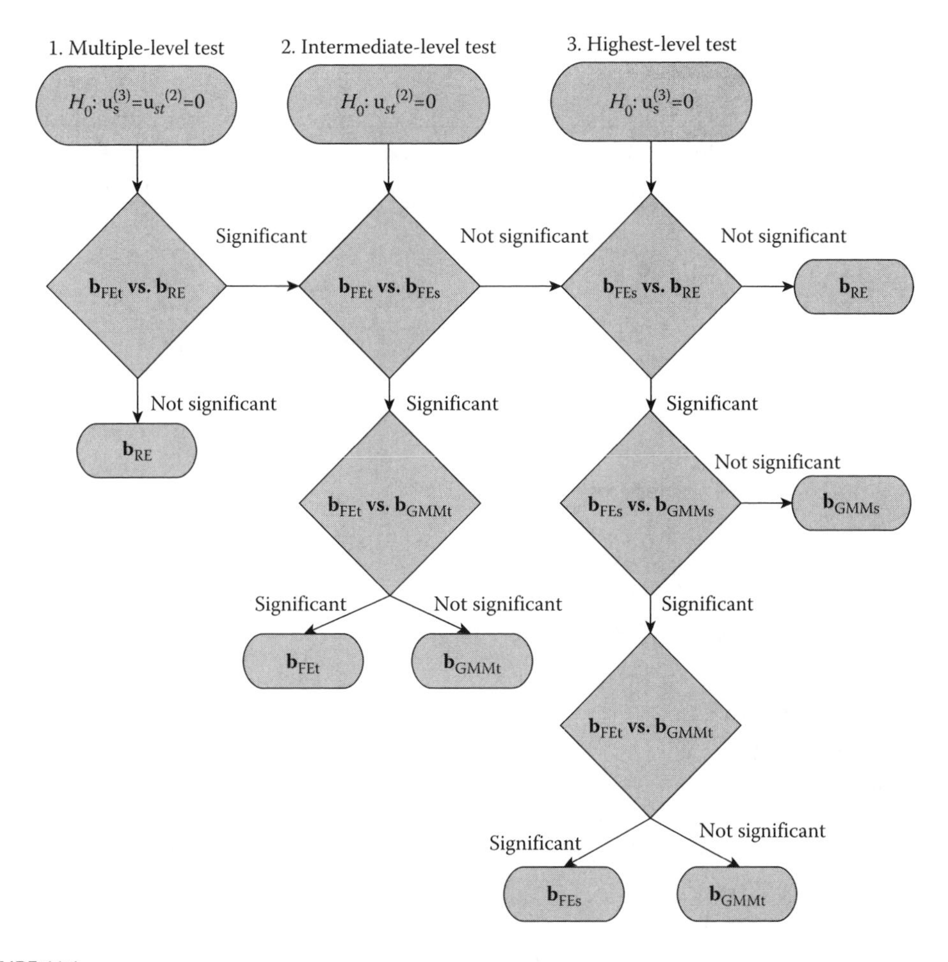

FIGURE 11.1
General guideline for conducting omitted variable tests and selecting the optimal estimator in a three-level model. One may start with any one of the three tests. However, Test 3 is valid when $\mathbf{U}_{st}^{(2)} = 0$. One should assume $\mathbf{U}_{st}^{(2)} = 0$ or conduct Test 2 beforehand.

a generalization of the method of moments (Hansen, 1982) and can also be viewed as an extension of instrumental variable (IV) methods. The area of GMM inference is more technical than other topics in this chapter and involves several statistical concepts such as instrument, projection, transformation, and generalized inverse. For this reason, we seek to convey the conceptual ideas of the methodology rather than its details. For GMM approaches in general, we refer the reader to Chapters 3 and 4 in Hayashi (2000). For their adaptations to multilevel models and formulas for model specification tests and robust estimation, the reader is referred to Kim and Frees (2007).

11.4.1 The GMM Estimator

Exogenous variables that are useful for estimating coefficients in $\boldsymbol{\beta}$ are said to be *instruments*. When all model variables $\mathbf{X}$ are exogenous, we can estimate $\boldsymbol{\beta}$ exclusively based on $\mathbf{X}$. However, if some $\mathbf{X}$ are endogenous, the econometric IV methods require additional nonmodel instrumental variables for consistent estimation.

On the other hand, Kim and Frees (2007) developed a procedure of building "internal" instruments by utilizing the nested structure of hierarchical data that does not require additional "external" variables. In their approach, internal instruments are functions of variables in the model, and various sets of instruments can be defined in multilevel models.

To define the GMM estimator, we recall the stacked version of a linear mixed effects model $\mathbf{y}_s = \mathbf{X}_s\boldsymbol{\beta} + \boldsymbol{\delta}_s$ in Equation 11.8. The outcome $\mathbf{y}_s$ is independent over s with $\mathrm{E}\boldsymbol{\delta}_s = 0$ and $\mathrm{Var}\ \boldsymbol{\delta}_s = \mathbf{V}_s$. As schools are independent units in this model, the variance-covariance matrix $\mathbf{V}$ is a block diagonal matrix, denoted as $\mathbf{V} = blkdiag\ (\mathbf{V}_{1,\ldots,}\mathbf{V}_n)$. Let $\mathbf{W}_s$ be a known weight matrix associated with $\boldsymbol{\delta}_s$. The matrix of instruments, denoted as $\mathbf{H}_s$, are exogenous and thus uncorrelated with the random effects $\boldsymbol{\delta}_s$ in the model such that:

$$E\,\mathbf{H}_s'\mathbf{W}_s\boldsymbol{\delta}_s = 0, \quad s = 1,\ldots,n. \quad (11.12)$$

Let $\mathbf{y} = (\mathbf{y}_1',\ldots,\mathbf{y}_n')'$, $\mathbf{X} = (\mathbf{X}_1',\ldots,\mathbf{X}_n')'$, $\mathbf{H} = (\mathbf{H}_1',\ldots,\mathbf{H}_n')'$ and $\boldsymbol{\delta} = (\boldsymbol{\delta}_1',\ldots,\boldsymbol{\delta}_n')'$. With the matrix of weights $\mathbf{W} = blkdiag\ (\mathbf{W}_{1,\ldots,}\mathbf{W}_n)$, the GMM estimator is defined as:

$$\mathbf{b}_{\mathrm{GMM}} = (\mathbf{X'W}P(\mathbf{H})\mathbf{WX})^{-}\mathbf{X'W}P(\mathbf{H})\mathbf{Wy}, \quad (11.13)$$

where $P(\mathbf{H}) = \mathbf{H}(\mathbf{H'H}) - \mathbf{H'}$ is the projection onto the linear space spanned by the columns of $\mathbf{H}$ and "–" denotes a generalized inverse.

For this GMM estimator, we have introduced weights $\mathbf{W}_s$ to allow for a variance structure, which is usually ignored in IV methods. In that case, one may use the identity for the weight matrix. Alternatively, the weight can be the inverse of the square root of the variance–covariance matrix of the disturbance term ($\mathbf{V}_s^{-1/2}$), thus producing a GLS estimate. As another option, in mixed linear effects modeling (see, for example, Diggle, Heagerty, Liang, & Zeger, 2002), it is customary for analysts to use a weight matrix that approximates $\mathbf{V}_s^{-1/2}$ and then use a robust estimate of standard errors to correct for misspecifications:

$$\widehat{Var\ \mathbf{b}_{\mathrm{GMM}}} = (\mathbf{X'W}P(\mathbf{H})\mathbf{WX})^{-1}\mathbf{X'W}P(\mathbf{H})\mathbf{W} \times (blkdiag(\mathbf{e}_1\mathbf{e}_1',\ldots,\mathbf{e}_n\mathbf{e}_n')) \times \mathbf{W}P(\mathbf{H})\mathbf{WX}'(\mathbf{X'W}P(\mathbf{H})\mathbf{WX})^{-1},$$

determined from the residuals $\mathbf{e}_s = \mathbf{y}_s - \mathbf{X}_s\mathbf{b}_{\mathrm{GMM}}$. We used these *empirical robust* standard errors, also known as Huber–White *sandwich* standard errors (Huber, 1967; White, 1982), instead of model-based standard errors in our applications, as the latter are known to be sensitive to omitted variable effects and may provide severely underestimated standard errors and consequently falsely large effects (Kim & Frees, 2006; Maas & Hox, 2004).

11.4.2 Creating the Estimator Continuum Using Internal Instruments

The formula for the GMM estimator in Equation 11.13 indicates that we obtain different $\mathbf{b}_{\mathrm{GMM}}$ with a different set of instruments $\mathbf{H}$. In fact, we can estimate $\mathbf{b}_{\mathrm{FEt}}$, $\mathbf{b}_{\mathrm{FEs}}$, and $\mathbf{b}_{\mathrm{RE}}$ in previous sections using Equation 11.13, and the calculation of FE estimators is much faster than the dummy variable approach mentioned earlier. More importantly, the GMM estimator provides useful alternative options when the RE estimator is biased and the FE estimators are not desirable. Using the FE, RE, and additional GMM estimators, we can create a continuum of estimators from the most robust but least efficient $\mathbf{b}_{\text{FEt}}$ at one end to the most efficient but least robust $\mathbf{b}_{\text{RE}}$ at the other end. The purpose of building this continuum is to find an estimator that is robust against omitted variables and also as efficient as possible.

The internal instruments **H** can be constructed without nonmodel variables, consisting of functions of model variables **X**. In particular, within-group deviations and group means of **X** can be used as instruments in multilevel models. Note that different groups can be identified in three- or higher-level models. Previously we considered a school–teacher–pupil model, and schools and teachers can thus be considered as groups having different deviations and means. Also, different amounts of information can be used for different estimators. Specifically, $\mathbf{b}_{\text{FEt}}$ is obtained under the idea that all predictors might be endogenous, and thus corresponding instrument $\mathbf{H}_{\text{FEt}}$ is composed of only within-teacher deviations, which are not affected by omitted variables $\mathbf{u}_s^{(3)}$ and $\mathbf{u}_{st}^{(2)}$. Similarly, the instrument $\mathbf{H}_{\text{FEs}}$ for $\mathbf{b}_{\text{FEs}}$ is composed of within-school deviations.[1]

On the other hand, $\mathbf{b}_{\text{RE}}$ is obtained by assuming all predictors are exogenous and $\mathbf{H}_{\text{RE}}$ is composed of within-group deviations as well as the group means of all predictors in the model. Using more information than $\mathbf{b}_{\text{FEt}}$ and $\mathbf{b}_{\text{FEs}}$, $\mathbf{b}_{\text{RE}}$ is more efficient if all predictors are indeed exogenous. Unlike deviations, however, group means are affected by omitted variables and $\mathbf{b}_{\text{RE}}$ would be biased if some predictors are endogenous. Either teachers or schools as groups provide the same $\mathbf{b}_{\text{RE}}$.

There is a third option for the GMM estimator, taking a middle-of-the-road approach by using both deviations and means of exogenous variables and only deviations of endogenous variables. To implement the GMM estimators, therefore, one needs to separate out potentially endogenous variables in the model, and then construct instruments $\mathbf{H}_{\text{GMM}}$ by merging the within-group deviations of all predictors and the group means of exogenous variables. This approach is reasonable because one would not wish to rely on potentially endogenous variables as much as the other exogenous variables for the estimation of model parameters.

For example, if it is of concern that SES, but not MALE, might be endogenous due to unobserved school variables such as district and neighborhood characteristics, only the within-school deviation $(\text{SES}_{stp} - \overline{\text{SES}}_s)$ may be used as an instrument, whereas both school average $\overline{\text{MALE}}_s$ and within-school deviation $(\text{MALE}_{stp} - \overline{\text{MALE}}_s)$ may be included in $\mathbf{H}_{\text{GMMs}}$ for the estimation of the GMM school estimator $\mathbf{b}_{\text{GMMs}}$. Analogously, teacher deviations and means can be used as instruments for the GMM teacher estimator $\mathbf{b}_{\text{GMMt}}$. The idea of the GMM estimator is rooted in the Hausman–Taylor estimator (Hausman & Taylor, 1981) and IV estimators in econometrics (Hsiao, 2003; Wooldridge, 2002).

To summarize, it may be more than is necessary to regard all predictors as endogenous yet too lenient to assume all are exogenous in many applications. The GMM estimator encompasses the two extreme approaches by treating part of the variables as endogenous. In the school–teacher–pupil model, within-teacher deviations and within-school deviations are used as instruments for $\mathbf{b}_{\text{FEt}}$ and $\mathbf{b}_{\text{FEs}}$, respectively. For the corresponding GMM estimators, denoted as $\mathbf{b}_{\text{GMMt}}$ and $\mathbf{b}_{\text{GMMs}}$, teacher means and school means of

[1] The same idea of using deviations to remove unobserved variable u_i from a linear longitudinal model is shown from Equations 11.3 to 11.5.

exogenous variables are additionally used as instruments. Note that different GMM estimators would be obtained for different choices of endogenous variables. When all predictors are considered as exogenous, the GMM estimator is equivalent to the RE estimator. The five estimators can be arranged in the order of $\mathbf{b}_{FEt}$, $\mathbf{b}_{GMMt}$, $\mathbf{b}_{FEs}$, $\mathbf{b}_{GMMs}$, and $\mathbf{b}_{RE}$, with regard to their robustness against omitted variables $\mathbf{u}_s^{(3)}$ and $\mathbf{u}_{st}^{(2)}$.

11.4.3 Comparing Multilevel Model Estimators Using the GMM Tests

After building the estimator continuum, it is of interest to examine if differences among the estimators are statistically significant. The GMM tests for comparing FE, RE, and GMM estimators can be presented as:

$$\chi^2_{GMM} = (\mathbf{b}_2 - \mathbf{b}_1)' (\text{Var}(\mathbf{b}_2 - \mathbf{b}_1))^{-1} (\mathbf{b}_2 - \mathbf{b}_1). \tag{11.14}$$

This test statistic has an asymptotic chi-square distribution with degrees of freedom equal to the rank of Var $(\mathbf{b}_2 - \mathbf{b}_1)$. Hausman and colleagues (Hausman, 1978; Hausman & Taylor, 1981) introduced a test statistic of this form for model specification tests with panel data models. Kim and Frees (2006, 2007) extended the Hausman test to more general multilevel models and also incorporated GMM estimators. As a result, the following five types of tests can be conducted using Equation 11.14:

1. Fixed effects estimator versus random effects estimator: for example, $\mathbf{b}_{FEt}$ versus $\mathbf{b}_{RE}$
2. Two fixed effects estimators: for example, $\mathbf{b}_{FEt}$ versus $\mathbf{b}_{FEs}$
3. GMM estimator versus random effects estimator: $\mathbf{b}_{GMMt}$ versus $\mathbf{b}_{RE}$
4. Two GMM estimators: for example, $\mathbf{b}_{GMMt}$ versus $\mathbf{b}_{GMMs}$
5. Fixed effects estimator versus GMM estimator: $\mathbf{b}_{FEt}$ versus $\mathbf{b}_{GMM}$

To summarize the model specification tests discussed in this chapter, when the model consists of only two levels and no random slopes, the Hausman test in Equation 11.6 is sufficient. For three or more levels, one should consider the multilevel omitted variable tests in Equation 11.11. When the results of the omitted variable tests indicate bias in random effects estimators, one may consider the GMM tests in Equation 11.14. When comparing estimators using the GMM tests, one can also consider each coefficient with a chi-square distribution with one degree of freedom. This is referred to as the *individual coefficient test* or *one-degree-of-freedom test* in Kim and Frees (2007). Individual coefficient tests can be particularly useful in understanding the sources of omitted variable bias.

11.5 AN EXAMPLE WITH NELS:88

The methodology in this chapter can be applied to various forms of multilevel models, including latent growth models for longitudinal data and cross-sectional hierarchical models for organizational data. This section illustrates the implementations of the methods with a three-level model for the National Education Longitudinal Study of 1988 (NELS:88), which is one of the most used large-scale data sets in education.

11.5.1 Selection of Variables and Multiple Imputation

Achievement test scores in the NELS:88 consist of four subjects—mathematics, reading, history, and science—and mathematics is by far the most studied subject among the four. This may be partly due to an understanding that students' mathematics performance is more sensitive to teacher and school effectiveness than other subjects (Shouse & Mussoline, 2002). We also considered the 10th grade mathematics achievement test scores as the outcome variable.

For the selection of predictors, we reviewed the relevant literature extensively to gather variables that are commonly used as predictors in educational production functions. Among many references, we chose most of the variables based on Goldhaber and Brewer (1997), Brewer and Goldhaber (2000), Ehrenberg, Brewer, Gamoran, and Willms (2001), and Rivkin, Hanushek, and Kain (2005) that are hypothesized to be related to student achievement, including teacher and school characteristics as well as student background variables. Wayne and Youngs (2003) surveyed a large number of articles on teacher effectiveness and argued the importance of controlling for prior student achievement. Battistich Solomon, Kim, Watson, and Schaps (1995) and Lee and Smith (1997) also stated that socio-economic status (SES), gender, minority status and some form of prior achievement are necessities in educational production functions. We used the IRT-equated eighth grade mathematics score for controlling for prior achievement.

Table 11.1 shows the predictors at the student, teacher, and school levels and their summary statistics. Note that we compared descriptive statistics based on two different forms of NELS:88; one after listwise deletion ($N = 5278$) by removing all subjects with missing values for the predictors in the model, and the other with imputed observations for the missing values ($N = 7334$) using the multiple imputation procedure implemented in SAS (SAS Institute Inc., 2004; Schafer, 1999). By comparing the percentages of different categories and mean mathematics scores of the variables between the two forms, especially with respect to SES, prior achievement, and minority status, it is apparent that the missing completely at random (MCAR) assumption is far from satisfied in NELS:88 and thus listwise deletion is inappropriate. On the other hand, although the multiple imputation procedure is not assumption free, it requires the considerably weaker assumption of missing at random (MAR) instead of MCAR (Little & Rubin, 2002). Therefore, we used the imputed data set for our analysis in this chapter.

11.5.2 Omitted Variable Tests

We fitted a three-level random intercept model and first obtained the random effects estimator $\mathbf{b}_{\text{RE}}$, assuming no omitted variable effects. This is equivalent to assuming $\mathbf{u}_s^{(3)} = \mathbf{u}_{st}^{(2)} = 0$ in Equation 11.10. As noted earlier, we would not be able to tell if this $\mathbf{b}_{\text{RE}}$ is biased unless we compare it with more robust estimators. Therefore, we also obtained the fixed teacher effects estimator $\mathbf{b}_{\text{FEt}}$, which is robust against the presence of both $\mathbf{u}_s^{(3)}$ and $\mathbf{u}_{st}^{(2)}$. In addition, we obtained the fixed school effects estimator $\mathbf{b}_{\text{FEs}}$ as well, which is robust against the presence of $\mathbf{u}_s^{(3)}$ but not $\mathbf{u}_{st}^{(2)}$.

The three sets of estimates, $\mathbf{b}_{\text{FEt}}$, $\mathbf{b}_{\text{FEs}}$, and $\mathbf{b}_{\text{RE}}$, are listed in Table 11.2. One might question why $\mathbf{b}_{\text{FEt}}$ is not always used if it is robust against both omitted teacher and school effects. The reason is shown in Table 11.2.

TABLE 11.1

Percentages or Means of Predictors and Average Mathematics Scores by Subgroups; Standard Deviations in Parens

Variable	Percentage or Mean		Math Score	
	Listwise Deletion	MI	Listwise Deletion	MI
Student				
N	5278	7334		
Current achievement			45.43 (13.74)	44.35 (13.92)
Prior achievement			37.85 (11.96)	36.97 (12.05)
SES	0.07 (0.79)	0.03 (0.80)		
Female	0.51 (0.50)	0.50 (0.50)	45.02 (13.45)	44.12 (13.58)
Male	0.49 (0.50)	0.50 (0.50)	45.84 (14.02)	44.58 (14.24)
Minority	0.24 (0.43)	0.28 (0.45)	40.43 (13.93)	39.11 (13.75)
Caucasian	0.76 (0.43)	0.72 (0.45)	47.01 (13.30)	46.37 (13.44)
Teacher				
N	2151	3016		
Has a math background	0.28 (0.45)	0.27 (0.44)	46.30 (12.46)	45.51 (12.53)
No math background	0.72 (0.45)	0.73 (0.44)	43.41 (13.24)	42.14 (13.30)
Experienced (3 + years)	0.89 (0.31)	0.89 (0.32)	44.67 (13.09)	43.40 (13.24)
Not experienced (< 3 years)	0.11 (0.31)	0.11 (0.32)	40.93 (12.57)	39.98 (12.38)
Female	0.48 (0.50)	0.49 (0.50)	44.20 (13.06)	42.84 (13.18)
Male	0.52 (0.50)	0.51 (0.50)	44.31 (13.11)	43.21 (13.19)
Minority	0.09 (0.28)	0.10 (0.31)	38.80 (12.57)	37.26 (12.30)
Caucasian	0.91 (0.28)	0.90 (0.31)	44.77 (13.01)	43.69 (13.12)
School				
N	626	859		
Urban	0.36 (0.48)	0.37 (0.48)	45.36 (10.96)	43.33 (11.61)
Rural	0.27 (0.44)	0.26 (0.44)	42.53 (8.82)	41.47 (8.92)
Suburban	0.37 (0.48)	0.37 (0.48)	45.58 (9.89)	44.11 (10.04)
School size/100	11.48 (6.70)	11.85 (6.78)		
% Caucasian/10	7.11 (2.77)	6.87 (2.88)		
% Single parent homes/10	2.83 (1.78)	2.94 (1.79)		
Public school	0.82 (0.39)	0.83 (0.38)	43.02 (9.34)	41.48 (9.64)
Private school	0.18 (0.39)	0.17 (0.38)	52.18 (9.99)	51.21 (10.41)

Note: MI: Multiple imputation.

While $\mathbf{b}_{RE}$ provides estimates for all three levels, $\mathbf{b}_{FEt}$ only provides estimates for the level-1 variables and $\mathbf{b}_{FEs}$ provides estimates for the level-1 and level-2 variables. Because of this severe loss of information, one would not choose $\mathbf{b}_{FEt}$ nor $\mathbf{b}_{FEs}$, unless $\mathbf{b}_{RE}$ is biased.

Figure 11.1 provides a flowchart summarizing different paths for testing omitted variable effects and for finding the optimal estimator. Among the three omitted variable tests at the top of Figure 11.1, we started with the omnibus multiple-level test (Test 1)

TABLE 11.2

Omitted Variable Tests; Empirical Standard Errors in Parens

Variable	$\mathbf{b}_{FEt}$	$\mathbf{b}_{FEs}$	$\mathbf{b}_{RE}$	$\mathbf{b}_{FEt}$ vs. $\mathbf{b}_{RE}$	$\mathbf{b}_{FEt}$ vs. $\mathbf{b}_{FEs}$
Intercept			9.71 (0.77)*		
Prior achievement	0.85 (0.01)*	0.94 (0.01)*	0.95 (0.01)*	−0.10 (0.01)*	−0.09 (0.01)*
SES	0.71 (0.18)*	0.94 (0.15)*	1.11 (0.13)*	−0.40 (0.13)*	−0.23 (0.12)*
Female	0.14 (0.21)	0.17 (0.18)	0.17 (0.17)	−0.04 (0.14)	−0.03 (0.13)
Minority	−0.55 (0.31)	−0.64 (0.25)*	−0.76 (0.23)*	0.21 (0.20)	0.10 (0.18)
Math background		0.71 (0.25)*	0.53 (0.21)*		
Experienced		0.56 (0.36)	0.45 (0.30)		
Female		0.42 (0.25)	0.39 (0.20)*		
Minority		−0.36 (0.46)	−0.42 (0.38)		
Urban			−0.43 (0.28)		
Rural			−0.37 (0.28)		
School size/100			−0.02 (0.02)		
% Caucasian/10			−0.01 (0.05)		
% Single parent homes/10			−0.10 (0.07)		
Public school			−0.93 (0.37)*		

$^*p < 0.05$.
Omitted variable test 1: H_0: No omitted teacher and school effects exist. $\mathbf{b}_{FEt}$ vs. $\mathbf{b}_{RE}$.
$\chi^2 = 143.34$, $df = 4$, $p < 0.01$.
Omitted variable test 2: H_0: No omitted teacher effects exist. $\mathbf{b}_{FEt}$ vs. $\mathbf{b}_{FEs}$.
$\chi^2 = 130.64$, $df = 4$, $p < 0.01$.

for examining omitted school and teacher effects simultaneously. It is generally suggested to start with Test 1 if omitted variable effects are of concern at multiple levels. The other tests can be subsequently conducted as shown in Figure 11.1. One may also start with the intermediate-level test (Test 2), if the analyst wishes to test omitted variable effects at each level separately. However, caution should be made before starting with the highest-level test (Test 3), as it is not valid if there exist omitted variables at lower levels. See Section 11.3 for further discussion on the properties of the three OV tests.

The results of the OV tests are summarized at the bottom of Table 11.2. On the basis of the multiple-level test, it is clear that $\mathbf{b}_{RE}$ is biased. Even with the empirical standard errors in Section 11.1 (as opposed to model-based errors that tend to underestimate variability and may provide falsely large test statistics), the chi-square test statistic is very large (143.34, $df = 4$, $p < 0.01$). Using the individual coefficient tests, we found statistically significant differences for the effects of prior mathematics achievement (0.85 vs. 0.95) and SES (0.71 vs. 1.11). This means the effects of the two variables would be upward biased, if we ignore omitted variable effects and use $\mathbf{b}_{RE}$.

Next, we conducted the intermediate-level test and the results indicate that there exist significant omitted teacher effects ($\chi^2 = 130.64$, $df = 4$, $p < 0.01$) and that $\mathbf{b}_{FEs}$ is also biased. As we found that $\mathbf{b}_{FEs}$ is biased, comparing the biased estimator to another biased estimator

$\mathbf{b}_{RE}$ is not meaningful. Therefore, we did not conduct the highest-level test. This example shows the very reason why the econometric treatment for two-level models discussed in Section 11.2 would fall short in multilevel models. If we had applied the Hausman test to examine omitted school effects at the third level (i.e., OV Test 3), the test would have compared two biased estimators.

Although we have one estimator that is robust to the omitted teacher and school effects, we cannot estimate the effects of teacher-level and school-level variables using $\mathbf{b}_{FEt}$. Sweeping out higher-level effects is a critical limitation of fixed effects approaches and might be one of the main reasons why efficient random effects estimators have been used routinely despite the danger of omitted variable bias. To avoid this problem, some studies selected out students who switched schools, which allow for estimating the effects of teacher and school variables. However, this approach has several serious problems. Most importantly, these "movers" or "switchers" are usually a small proportion of the whole population and often have different characteristics than the rest. Consequently, not only can the estimation be unstable due to a small sample size, but also the findings based on the movers' group may not be generalizable to the majority who did not switch schools. Therefore, instead of relying on biased $\mathbf{b}_{RE}$ or limited information $\mathbf{b}_{FEt}$, we further our analysis to find the GMM estimator.

11.5.3 Searching for the Optimal Estimator

As explained earlier, the RE and FE estimators treat all and none of predictors as exogenous, respectively. On the other hand, the GMM estimator provides a middle ground between the two extreme "all or nothing" approaches and allows for some of predictors to be endogenous and treats the others as exogenous. By making this distinction, the analyst can use both the within- and between-group information of exogenous variables, while using only the within-group information from endogenous variables as instruments (see Section 11.2).

In this chapter, we demonstrate the internal instruments approach that utilizes variables already in the model. If available, external variables can be easily added to the set of instruments. However, it is important that one can obtain the unbiased GMM estimator without relying on additional variables that may not be available. It is also important to identify endogenous variables properly. We expect that the analyst would have some understanding or theory about endogeneity issues in relation to the nature of the topic (e.g., confounding factors) and the data (e.g., data collection process), based on experience and knowledge in the field. The choices of instruments should be guided by this knowledge.

It should be clarified that the purpose of our example is to demonstrate the methodologies rather than to make substantive inferences. Nonetheless, we made efforts to gather proper information and our choice of predictors and endogenous variables are based on the literature in the field. Among the predictors summarized in Table 11.1, prior achievement, SES, and school-level percentage of minority students were suspected to be endogenous. Prior achievement is a concern in regard to factors that would affect both prior and current achievement but are not available in the data set. It is natural to view any value-added type predictor as endogenous because of this inevitable relationship (Alexander, Pallas, & Cook, 1981).

SES would also likely be correlated with numerous factors that influence student achievement. Finally, there is concern that school-level percentage of minority students might be related to omitted school, district, and neighborhood information.

Thus, we hold the three predictors as potential endogenous variables, which means only within-teacher deviations of these variables are used as instruments, whereas both within-teacher deviations and teacher-level means of the rest of the 14 variables are used as instruments **H** in the calculation of $\mathbf{b}_{\text{GMMt}}$ in Equation 11.13. Similarly, within-school deviations of the three endogenous variables are used and both within-school deviations and school-level means of the exogenous variables are used as instruments for $\mathbf{b}_{\text{GMMs}}$. Although we obtained $\mathbf{b}_{\text{GMMs}}$ to complete the estimator continuum, as we already found that $\mathbf{b}_{\text{FEs}}$ is biased in the OV Test 2 above, the corresponding GMM estimator does not carry much value in the current analysis. The results of the five estimators are summarized in Table 11.3. They are ordered from the most robust to the least robust as: $\mathbf{b}_{\text{FEt}}$, $\mathbf{b}_{\text{GMMt}}$, $\mathbf{b}_{\text{FEs}}$, $\mathbf{b}_{\text{GMMs}}$, and $\mathbf{b}_{\text{GLS}}$.

An important question in Table 11.3 is whether $\mathbf{b}_{\text{GMMt}}$ is as robust as $\mathbf{b}_{\text{FEt}}$. If so, we can obtain unbiased estimates for the effects of variables at all levels from $\mathbf{b}_{\text{GMMt}}$. This constitutes the GMM Test 5 in Section 11.3, a test between an FE estimator and a GMM estimator. We found that $\mathbf{b}_{\text{FEt}}$ and $\mathbf{b}_{\text{GMMt}}$ provide almost identical estimates and the test statistic (χ^2) was as small as 0.03. Therefore, we concluded that $\mathbf{b}_{\text{GMMt}}$ is the optimal estimator in our three-level model for the mathematics achievement scores in NELS:88.

While making inferences based on $\mathbf{b}_{\text{GMMt}}$, we also compared $\mathbf{b}_{\text{GMMt}}$ to $\mathbf{b}_{\text{RE}}$ for each coefficient. This comparison may help understand sources of omitted variable bias in $\mathbf{b}_{\text{RE}}$, and the information can be utilized in further research. The detailed results of this comparison are shown in Table 11.4. The test between $\mathbf{b}_{\text{GMMt}}$ and $\mathbf{b}_{\text{RE}}$ results in a large χ^2 value of 150.55 ($p < 0.01$) with 15 degrees of freedom. The individual coefficient tests showed significant differences for many variables including prior achievement, family SES, urbanity, school size, school percentage of Caucasian, and public status, suggesting the omission of some important information in relation to these variables. This is consistent with our literature review and concern over the importance of accounting for individual and family attributes, community dynamics, and school climates in the study of school and teacher effectiveness (e.g., Alexander et al., 1981).

Interestingly, there was a clear pattern in the direction of bias. While $\mathbf{b}_{\text{RE}}$ overestimates the student level variables of prior achievement and family SES, it substantially underestimates several school effects, especially concerning characteristics of schools such as urban status, public status, school size, and ethnicity composition. Based on the robust estimator $\mathbf{b}_{\text{GMMt}}$, the negative effect of students' ethnicity (being minority) became nonsignificant, while several school-level variables, including school size and the percentage of Caucasian, became significant.

We conclude our data analysis example with suggesting several general guidelines for the analysis of other data sets, which can be accompanied by the flowchart in Figure 11.1. First, when $\mathbf{b}_{\text{RE}}$ is biased, one may end the analysis with $\mathbf{b}_{\text{FEt}}$ or $\mathbf{b}_{\text{FEs}}$, especially if it is difficult to identify endogenous variables or it is suspected that the majority of predictors are endogenous. In many applications, however, the GMM estimator

TABLE 11.3

Five Estimators From Most Robust to Most Efficient; Empirical Standard Errors in Parens

Variable	$\mathbf{b}_{FEt}$	$\mathbf{b}_{GMMt}$	$\mathbf{b}_{FEs}$	$\mathbf{b}_{GMMs}$	$\mathbf{b}_{RE}$
Intercept		9.41 (1.57)*		8.97 (0.99)*	9.71 (0.77)*
Prior achievement	0.85 (0.01)*	0.85 (0.01)*	0.94 (0.01)*	0.94 (0.01)*	0.95 (0.01)*
SES	0.71 (0.18)*	0.71 (0.17)*	0.94 (0.15)*	0.94 (0.15)*	1.11 (0.13)*
Female	0.14 (0.21)	0.12 (0.17)	0.17 (0.18)	0.17 (0.17)	0.17 (0.17)
Minority	-0.55 (0.31)	−0.55 (0.31)	−0.64 (0.25)*	−0.68 (0.25)*	−0.76 (0.23)*
Math background		0.58 (0.23)*	0.71 (0.25)*	0.52 (0.21)*	0.53 (0.21)*
Experienced		0.59 (0.31)	0.56 (0.36)	0.47 (0.30)	0.45 (0.30)
Female		0.39 (0.21)	0.42 (0.25)	0.40 (0.20)*	0.39 (0.20)*
Minority		0.16 (0.48)	−0.36 (0.46)	−0.27 (0.40)	−0.42 (0.38)
Urban		0.18 (0.38)		−0.28 (0.30)	−0.43 (0.28)
Rural		−0.54 (0.32)		−0.39 (0.28)	−0.37 (0.28)
School size/100		0.08 (0.03)*		0.04 (0.02)	0.02 (0.02)
% Caucasian/10		0.52 (0.15)*		0.12 (0.08)	0.01 (0.05)
% Single parent homes/10		−0.05 (0.10)		−0.08 (0.07)	−0.10 (0.07)
Public school		−1.94 (0.42)*		−1.07 (0.38)*	−0.93 (0.37)*

*$p < 0.05$.

can be as robust as the corresponding FE estimator and would be preferred. Second, one may choose $\mathbf{b}_{FEs}$ over $\mathbf{b}_{FEt}$, as long as the difference is not significant. Similarly, $\mathbf{b}_{GMMs}$ would be preferred to $\mathbf{b}_{GMMt}$, if both are unbiased. Third, consider the comparison between $\mathbf{b}_{FEs}$ and $\mathbf{b}_{GMMs}$ in the right-most panel. If the difference is significant (i.e., $\mathbf{b}_{GMMs}$ is biased), one may further consider $\mathbf{b}_{GMMt}$ instead of $\mathbf{b}_{FEs}$. Note that $\mathbf{b}_{FEs}$ and $\mathbf{b}_{GMMt}$ are not directly comparable so $\mathbf{b}_{GMMt}$ should be compared to $\mathbf{b}_{FEt}$. If the test between $\mathbf{b}_{FEt}$ and $\mathbf{b}_{GMMt}$ reveals that $\mathbf{b}_{GMMt}$ is biased, one would go back to $\mathbf{b}_{FEs}$ in the previous step. In sum, the flowchart starts with one of the three OV tests and ends with one of the five estimators for a three-level model. There would be a larger number of entities and paths for higher-level models. In our analysis of NELS:88, the starting point was the OV Test 1 and the ending point was $\mathbf{b}_{GMMt}$ in the middle panel.

11.6 SUMMARY AND DISCUSSION

Although few would argue the danger of omitted variable bias, the harmful consequences in data analysis are often overlooked. This is partly because of the lack of statistical methods for handling omitted variables in multilevel models until recently. As in many observational studies the analyst does not have the ability to collect all the “right” variables, it is of great interest to utilize statistical techniques to handle omitted variables as much as possible and ideally obtain unbiased solutions.

This chapter provides an introduction to recent statistical methodology for model specification tests and robust estimation techniques in multilevel models including three types of omitted variable tests (Section 11.3) and GMM estimators (Section 11.4).

TABLE 11.4

Comparing the Teacher-Level GMM Estimator and the Random Effects Estimator

Variable	$\mathbf{b}_{GMMt}$	$\mathbf{b}_{RE}$	Difference	Std Err	Individual Coeff. Test (χ_1^2)
Prior achievement	0.85	0.95	−0.10	0.01	108.70, $p < 0.01$
SES	0.71	1.11	−0.40	0.13	9.85, $p < 0.01$
Female	0.12	0.17	−0.06	0.08	0.42, $p = 0.52$
Minority	−0.55	−0.76	0.21	0.20	1.11, $p = 0.29$
Math background	0.58	0.53	0.06	0.20	0.20, $p = 0.65$
Experienced	0.59	0.45	0.14	0.16	0.73, $p = 0.39$
Female	0.39	0.39	0.00	0.10	0.00, $p = 0.99$
Minority	0.16	−0.42	0.57	0.30	3.55, $p = 0.06$
Urban	0.18	−0.43	0.60	0.27	5.04, $p = 0.02$
Rural	−0.54	−0.37	−0.17	0.16	1.15, $p = 0.28$
School size/100	0.08	0.02	0.06	0.02	11.69, $p < 0.01$
% Caucasian/10	0.52	0.01	0.51	0.13	14.20, $p < 0.01$
% Single parent homes/10	−0.05	−0.10	0.05	0.08	0.44, $p = 0.51$
Public school	−1.94	−0.93	−1.02	0.21	23.72, $p < 0.01$

GMM Test: $\mathbf{b}_{GMMt}$ vs. $\mathbf{b}_{RE}$
$\chi^2 = 150.55$, $df = 15$, $p < 0.01$

It is shown that the versatile GMM technique provides an overarching framework encompassing the well-known random and fixed effects estimators and also offers additional and often more desirable options between the two extremes.

In the three-level model analysis for NELS:88, $\mathbf{b}_{RE}$ turned out to be severely biased. However, we found $\mathbf{b}_{GMMt}$, which is unbiased without losing higher-level information like $\mathbf{b}_{FEt}$. Despite its advantageous properties, an outstanding shortcoming of GMM estimation is that its implementation is cumbersome, as formulas for the GMM estimator in Equation 11.13 and the GMM tests in Equation 11.14 are not utilized in statistical programs. All required formulas are given in Kim and Frees (2007). The SAS IML code written by Frees, Kim, and Swoboda is available by request to the authors of this chapter.

It is well known that multilevel models can be written as linear mixed-effects models. However, this chapter demonstrates that it is critical to retain the multiple-level representation when inspecting omitted variables at different levels. Also, the GMM estimation technique exploits the hierarchical nature of multilevel data and can create internal instruments, so that the researcher is not forced to look for additional variables that were not involved in the original model formulation. For nonhierarchical data (without replications within clusters), one cannot obtain unbiased GMM estimators without external instruments in the presence of endogenous variables.

Additionally, for those who are familiar with IV approaches in econometrics, the GMM methodology in this chapter extends that related work in several important ways. First, the GMM estimator generalizes the original work of Hausman and Taylor (1981) for panel data models to more complex multilevel frameworks. Second, the GMM tests provide a general procedure for directly comparing various types of estimators beyond the FE and RE estimators. Third, empirical standard errors are adapted as opposed to traditional model-based standard errors that are known to underestimate variability when the models are not correctly specified. Finally, the GMM estimator extends the IV estimator by incorporating weights to accommodate the variance structure of a multilevel model and can handle more complex covariance structures in hierarchical data.

As a final note, we recall that Kim and Frees (2006) linked the omitted variable problem to a larger issue of unobserved heterogeneity in the population. Unobserved heterogeneity is a recurrent issue across many disciplines, including econometrics, psychometrics, biostatistics, and sociology. However, the commonality across these literatures has been overlooked, and problems related to unobserved heterogeneity have been acknowledged under various names such as latent classes or finite mixtures, omitted variables, correlated effects, unobserved covariates, measurement error, and confounding variables. In these applications, different assumptions are made about the nature of the unobserved variables (e.g., mutual exclusiveness, independent error terms, time-constant or time-varying variables, parametric or nonparametric mixing distributions) and also different implications of unobserved heterogeneity are emphasized in different disciplines (e.g., impact on causal inference, bias in regression coefficients, collapsibility, etc.). Among the extensive literature dealing with these issues, we refer to Heckman and Singer (1982), Chamberlain (1985), Yamaguchi (1986), Palta and Yao (1991), Vermunt (1997), Frank (2000), McLachlan and Peel (2000), Halaby (2004), and Frees (2004) for further readings.

ACKNOWLEDGMENT

We are thankful to Dan Bolt and David Kaplan for their useful comments.

REFERENCES

Alexander, K. L., Pallas, A. M., & Cook, M. A. (1981). Measure for measure: On the use of endogenous ability data in school-process research. *American Sociological Review, 46,* 619–631.

Battistich, V., Solomon, D., Kim, D.-I., Watson, M., & Schaps, E. (1995). Schools as communities, poverty levels of student populations and students' attitudes, motives and performance: A multilevel analysis. *American Educational Research Journal, 32,* 627–658.

Brewer, D. J., & Goldhaber, D. D. (2000). Improving longitudinal data on student achievement: Some lessons from recent research using NELS:88. In D. Grissmer & M. Ross (Eds.), *Analytic issues in the assessment of student achievement.* Washington DC: NCES.

Chamberlain, G. (1985). Heterogeneity, omitted variable bias, duration dependence. In J. Heckman & B. Singer (Eds.), *Longitudinal analysis of labor market data.* Cambridge, NY: Cambridge University Press.

Diggle, P. J., Heagarty, P., Liang, K.-Y., & Zeger, S. L. (2002). *Analysis of longitudinal data* (2nd ed.). London: Oxford University Press.

Ehrenberg, R. G., Brewer, D. J., Gamoran, A., & Willms, J. D. (2001). Class size and student achievement. *Psychological Science in the Public Interest, 2,* 1–30.

Frank, K. A. (2000). Impact of a confounding variable on a regression coefficient. *Sociological Methods & Research, 29,* 147–194.

Frees, E. W. (2004). *Longitudinal and panel data: Analysis and applications for the social sciences.* Cambridge: Cambridge University Press.

Frees, E. W., & Kim, J.-S. (2008). Panel studies. In T. Rudas (Ed.), *Handbook of probability: Theory and applications* (pp. 205–224). Thousand Oaks, CA: Sage.

Goldhaber, D. D., & Brewer, D. J. (1997). Why don't schools and teachers seem to matter? Assessing the impact of unobservables on educational productivity. *The Journal of Human Resources, 32,* 505–523.

Goldstein, H. (2003). *Multilevel statistical models* (3rd ed.). London, UK: Oxford University Press.

Halaby, C. H. (2004). Panel models in sociological research: Theory into practice. *Annual Review of Sociology, 30,* 507–540.

Hansen, L. P. (1982). Large sample properties of generalized method of moments estimators. *Econometrica, 50,* 1029–1054.

Hausman, J. A. (1978). Specification tests in econometrics. *Econometrica, 46,* 1251–1272.

Hausman, J. A., & Taylor, W. E. (1981). Panel data and unobservable individual effects. *Econometrica, 49,* 1377–1398.

Hayashi, F. (2000). *Econometrics.* Princeton, NJ: Princeton University Press.

Heckman, J. J., & Singer, B. (1982). Population heterogeneity in demographic models. In K. Land & A. Rogers (Eds.), *Multidimensional mathematical demography.* New York, NY: Academic Press.

Hox, J. J. (2002). *Multilevel analysis: Techniques and applications.* Mahwah, NJ: Lawrence Erlbaum Associates.

Hsiao, C. (2003). *Analysis of panel data* 2nd ed. Cambridge, UK: Cambridge University Press.

Huber, P. J. (1967). The behavior of maximum likelihood estimates under non-standard conditions. In *Proceedings of the fifth Berkeley symposium on mathematical statistics and probability.* Berkeley, CA: University of California Press.

Kim, J.-S. (2009). Multilevel analysis: An overview and some contemporary issues. In R. E. Millsap & A. Maydeu-Olivares (Eds.), *Handbook of quantitative methods in psychology* (pp. 337–361). Thousand Oaks, CA: Sage.

Kim, J.-S., & Frees, E. W. (2006). Omitted variables in multilevel models. *Psychometrika, 71,* 659–690.

Kim, J.-S., & Frees, E. W. (2007). Multilevel modeling with correlated effects. *Psychometrika, 72,* 505–533.

Lee, V. E., & Smith, J. B. (1997). High school size: Which works best and for whom? *Educational Evaluation and Policy Analysis, 19,* 205–227.

Little, R. J. A., & Rubin, D. B. (2002). *Statistical analysis with missing data* (2nd ed.). New York, NY: John Wiley.

Maas, C. J. M., & Hox, J. J. (2004). Robustness issues in multilevel regression analysis. *Statistica Neerlandica, 58,* 127–137.

McLachlan, G. J., & Peel, D. (2000). *Finite mixture models.* New York, NY: Wiley.

Palta, M., & Yao, T.-J. (1991). Analysis of longitudinal data with unmeasured confounders. *Biometrics, 47,* 1355–1369.

Raudenbush, S. W., & Bryk, A. S. (2002). *Hierarchical linear models: Applications and data analysis methods* (2nd ed.). Newbury Park, CA: Sage.

Rivkin, S. G., Hanushek, E. A., & Kain, J. F. (2005). Teachers, schools, and academic achievement. *Econometrica, 73,* 417–458.

SAS Institute Inc. (2004). *SAS/STAT 9.1 User's Guide.* Cary, NC: SAS Institute Inc.

Schafer, J. L. (1999). Multiple imputation: A primer. *Statistical Methods in Medical Research, 8,* 3–15.

Shouse, R. C., & Mussoline, L. J. (2002). High risk, low return: The achievement effects of restructuring in disadvantaged schools. *Social Psychology of Education, 3,* 245–259.

Singer, J. (1998). Using SAS PROC MIXED to fit multilevel models, hierarchical models, and individual growth models. *Journal of Educational and Behavioral Statistics, 24,* 323–355.

Snijders, T. A., & Bosker, R. J. (1999). *Multilevel analysis: An introduction to basic and advanced multilevel modeling.* London, UK: Sage.

Vermunt, J. K. (1997). *Log-linear models for event histories.* Thousand Oaks, CA: Sage.

Wayne, A. J., & Youngs, P. (2003). Teacher characteristics and student achievement gains: A review. *Review of Educational Research, 73,* 89–122.

White, H. (1982). Maximum likelihood estimation of misspecified models. *Econometrica, 50,* 1–25.

Wooldridge, J. M. (2002). *Econometric analysis of cross section and panel data.* Cambridge, MA: MIT Press.

Yamaguchi, K. (1986). Alternative approaches to unobserved heterogeneity in the analysis of repeatable events. In B. Tuma (Ed.), *Sociological methodology.* Washington, DC: American Sociological Association.

12

Explained Variance in Multilevel Models

J. Kyle Roberts
Annette Caldwell Simmons School of Education and Human Development, Southern Methodist University, Texas

James P. Monaco
Laboratory for Computational Imaging and Bioinformatics, Rutgers University, New Jersey

Holly Stovall and Virginia Foster
Dedman College, Southern Methodist University, Texas

With the rise of the use and utility of multilevel modeling (MLM), one question has consistently been posed to authors and on listserves: "How much variance does my model explain?" Answering this question within the MLM framework is not an easy task where it is actually possible to explain "negative variance" when the addition of explanatory variables increases the corresponding variance components (Snijders & Bosker, 1999). Because effect size measures previously proposed consider variance at each level, a single measure is needed that helps researchers interpret the strength of the model as a whole. The purpose of this paper is to provide a history of past MLM effect sizes and present three new measures that consider "whole model" effects.

The utility of effect sizes in research interpretation has generated considerable discussion, much of which centers on the role and function of effect sizes, especially concerning the relationship to statistical significance tests (cf. Harlow, Mulaik, & Steiger, 1997). Many authors agree that effect sizes can serve a valuable function to help evaluate the magnitude of a difference or relationship (cf. Cohen, 1994; Kirk, 1996; Schmidt, 1996; Shaver, 1985; Thompson, 1996; Wilkinson & APA Task Force on Statistical Inference, 1999). Their articles, along with current publications (cf., Knapp & Sawilowsky, 2001a; Roberts & Henson, 2002) continue to debate both the use and utility of measures of effect size when considered both in conjunction with and peripheral from statistical significance testing.

One positive thing that has occurred while researchers began to debate the issue of effect size reporting (e.g., Knapp & Sawilowsky, 2001b; Thompson,

2001) is the encouragement of researchers to consider more than just *p*-values before making interpretations as to the noteworthiness (or lack thereof) of a given study. And although it may seem that the field of research follows changes and adopts a new course with the speed and acuteness of a glacier, the fact that this glacier is moving is predicated by the adoption of such language in the *APA Publication Manual* (2001):

> The general principle to be followed, however, is to provide the reader not only with information about statistical significance but also with enough information to assess the magnitude of the observed effect or relationship. (p. 26)

12.1 CATALOG OF EFFECT SIZES IN MLM

A history of effect sizes has been dealt with exhaustively in Huberty (2002), and does not bear repeating here. One thing absent from Huberty's catalog was the use of effect size indices in multilevel analysis. It was wisely absent from Huberty's manuscript, since there is much misconception, and even disagreement, as to the interpretation of these effects. We will quickly list some of the proposed effect size indices for use in MLM and give brief explanations as to their utility.

12.1.1 Intraclass Correlation

Intraclass correlation (ICC) is generally thought of as the degree of dependence of individuals upon a higher structure to which they belong; or, the proportion of total variance that is between the groups of the regression equation. Put more succinctly, it "is the degree to which individuals share common experiences due to closeness in space and/or time" (Kreft & de Leeuw, 1998, p. 9). Hox (1995) explains the ICC as a "population estimate of the variance explained by the grouping structure" (p. 14). The ICC for a two-level model can be represented as:

$$\rho_I = \frac{\tau_0^2}{\tau_0^2 + \sigma^2} = \frac{\sigma_{u0}^2}{\sigma_{u0}^2 + \sigma_{e0}^2}, \qquad (12.1)$$

where the numerator is represented by the variance at the second level of the hierarchy (τ_0^2), and the denominator represents the total variation in the model at both level-2 and level-1 (σ^2).

Although the ICC actually is *not* a measure of the effect size of an MLM model, it bears mentioning here because it sometimes is wrongly thought of as a measure of the "power" or strength of MLM over ordinary least squares (OLS) regression. However, this type of thinking is commonly illustrated in passages like the following:

> Determining the proportion of the total variance that lies systematically between schools, called the intraclass correlation (ICC), constitutes the first step in an HLM analysis. We conduct this analysis with a fully unconditional model, which means that no student or school characteristics are considered. This first step can also indicate whether HLM is needed or whether a single level analytic method is appropriate. *Only when the ICC is more than trivial* (i.e., greater than 10% of the total variance in the outcome) *would the analyst need to consider multilevel methods* [emphasis mine]. Ignoring this step (i.e., assuming an ICC of either 0 or 1) would be inappropriate if the research question were multilevel. Investigation of contextual effects, I argue, is by nature a multilevel question. (Lee, 2000, p. 128)

Roberts (2002) has rightly pointed out that it would be incorrect to interpret this statistic as a measure of the magnitude of difference between OLS and MLM estimates.

12.1.2 Proportion Reduction in Variance

The process of building a multilevel model often begins with a null model (also called the baseline model by Hox, 2002). In this baseline model, just the grand mean is fit in the model such that:

$$y_{ij} = \gamma_{00} + u_{0j} + e_{ij}, \tag{12.2}$$

where, γ_{00} is the model grand mean (or intercept), u_{0j} is the random group-level effect with variance σ^2_{u0}, and e_{ij} is the person level effect with variance σ^2_e. This baseline model's variance estimates serve as a benchmark for determining the R^2 at each level of the hierarchy. By using variance estimates from the null model (e.g., $\sigma^2_{u0|b}$) and variance estimates from the model where all predictors are entered (e.g., $\sigma^2_{u0|m}$), the percentage reduction in variance between the null model and the complete model can be estimated by:

$$R^2_2 = \frac{\sigma^2_{u0|b} - \sigma^2_{u0|m}}{\sigma^2_{u0|b}}, \tag{12.3}$$

for the percentage reduction in level-2 variance and by:

$$R^2_1 = \frac{\sigma^2_{e|b} - \sigma^2_{e|m}}{\sigma^2_{e|b}}, \tag{12.4}$$

for the reduction percentage in level-1 variance, where $|b$ and $|m$ represent the baseline and full models, respectively. This formula is reflected in many forms by Hox (2002, p. 64), or conversely as:

$$R^2_2 = \frac{\tau_{00}(null) - \tau_{00}(full)}{\tau_{00}(null)} \tag{12.5}$$

by Raudenbush and Bryk (2002, p. 74) and Kreft and de Leeuw (1998, p. 118).

Although perceived as a tool for noting the reduction in variance at each level of the model, Hox (2002) and Snijders and Bosker (1999) are quick to caution researchers against directly interpreting this statistic, since it is possible to obtain negative values for R^2 with these formulas when either σ^2_e is a biased estimator or when level-2 predictors are included in the model. This is a difficult concept to grasp, because in normal OLS models, the addition of variables to the model can only help prediction of the dependent variable (raise R^2), not hurt prediction. A negative R^2 value in MLM might be wrongly interpreted to mean that the predictor variables are performing at worse levels than just the grand mean as a predictor.

Negative variance can occur in an example where we use a variable that has almost no variation at one of the levels. Consider the case of a model where we have one single level-1 predictor. It would be safe to assume that the addition of this variable would reduce both the between-and within-groups variance. If we add a group-level predictor, then we could expect that it would reduce only the between-groups variance, not the within-groups variance, ultimately increasing the estimate for the population variance $\hat{\sigma}^2_{u0}$. For example, consider the output in Table 12.1 from a two-level model.

The data in this example were adapted from a hypothetical data set written to illustrate multilevel models (Roberts, 2004). In model

TABLE 12.1

Illustration of Negative Variance With the Addition of a Level-1 Predictor

	Estimate	
Model Formula	σ_e^2	σ_{u0}^2
M0: $science_{ij} = \gamma_{00} + u_{0j} + e_{ij}$	1.979	23.923
M1: $science_{ij} = \gamma_{00} + \gamma_{10}(ses) + u_{0j} + e_{ij}$	0.651	80.890

M1, just a single level-1 predictor is included in the model with variance estimates of $\hat{\sigma}_e^2 = 0.651$ and $\hat{\sigma}_{u0}^2 = 80.923$. If we are to consider this in terms of Equation 12.4, we could see that we actually would be decreasing the variance explained (or increasing $\hat{\sigma}_{u0}^2$) at the second level with the introduction of this predictor (hence negative variance explained between the two models).

Although the amount of variance explained is noteworthy at level-1 ($R_1^2 =$ (1.979 – .0651)/1.979 = 0.671), the amount of variance explained at the second level is actually –2.381(R_2^2 = (23.923 – 80.890)/23.923 = –2.381). Not only is this number troubling, but it is counter-intuitive to the way most researchers think about the effectiveness of a model. If we were to interpret this model without previous knowledge of multilevel models, we might be inclined to say that the addition of the predictor "ses" is a worse predictor of "math achievement" at the second level than if we had no predictor at all, being that it explains –238% variance!

12.1.3 Explained Variance as a Reduction in Mean Square Prediction Error

Snijders and Bosker (1999) argue for a slightly different approach to computing R^2 values in multilevel models by computing the model's associated mean square prediction error. The R^2 for level-1 is then computed as one minus the combined variance at both levels for the full model divided by the combined variance for the null model, or:

$$R_1^2 = 1 - \frac{\text{var}\left(Y_{ij} - \sum_h \gamma_h X_{hij}\right)}{\text{var}(Y_{ij})} = 1 - \frac{\hat{\sigma}^2(full) + \hat{\tau}_0^2(full)}{\hat{\sigma}^2(null) + \hat{\tau}_0^2(null)}, \tag{12.6}$$

where Y_{ij} is the outcome variable, γ_h represents the coefficient for outcome variable X_{hij} for all h variables, $\hat{\sigma}^2$ is an estimate of the variance at the first level, and $\hat{\tau}_0^2$ is an estimate of the variance at the second level.

The level-2 R^2 is then found by dividing the $\hat{\sigma}^2$ by the group cluster size (B), or by the average cluster size for unbalanced data, such that:

$$R_2^2 = 1 - \frac{\text{var}\left(\bar{Y}_{.j} - \sum_h \gamma_h \bar{X}_{h.j}\right)}{\text{var}(\bar{Y}_{.j})} = 1 - \frac{\hat{\sigma}^2(full)/B + \hat{\tau}_0^2(full)}{\hat{\sigma}^2(null)/B + \hat{\tau}_0^2(null)}. \tag{12.7}$$

In this formula, it is easy to see that the R^2 estimate at level-2 is similar to the R^2 for level-1, having just reduced the level-1 variance to represent an average variance for each group. Although this estimation differs from the previous definition of R^2 (Equations 12.3 and 12.4), it is still possible to obtain "negative" values for R^2.

Using Table 12.1, R^2 at level-1 is:

$$R_1^2 = 1 - \frac{0.651 + 80.890}{1.979 + 23.923} = -2.148,$$

and for level-2 is:

$$R_2^2 = 1 - \frac{0.651/10 + 80.890}{1.979/10 + 23.923} = -2.356.$$

Once again this solution is extremely problematic, as we have again obtained negative values for R^2.

12.1.4 Pooling as a Measure of Explained Variance

Like Snijders and Bosker (1999), Gelman and Pardoe (2006) propose a method that defines a measure of explained variance (R^2) at each level of the model. (Here the term level "corresponds to the separate variance components rather than the more usual measurement scales.") However, their measure does *not* require fitting multiple models and can be interpreted similarly to the classical R^2 in that it is the proportion of variability explained by the linear predictors. However, it is important to note that since the proposed R^2 is computed at each level, it does not directly measure the overall predictive accuracy of the model.

They define the explained variance at the data level as:

$$R^2 = 1 - \frac{\hat{\sigma}_y^2 / n}{\hat{\sigma}_\alpha^2 / (n+1) + \hat{\sigma}_y^2 / n},$$

where $\hat{\sigma}_y^2$ is an estimate of the variance of y_j given $y_j \sim N(\alpha_j, \hat{\sigma}_y^2)$ and $\hat{\sigma}_\alpha^2$ is an estimate of the variance of α_j given $\alpha_j \sim N(\mu_\alpha, \hat{\sigma}_\alpha^2)$.

Gelman and Pardoe also define a summary measure for the average amount of pooling (λ) at each level of the model. A low measure of pooling (< 0.5) suggests a higher degree of within-group information than population-level information. The opposite is true for a high measure of pooling. Note that this pooling factor is clearly related to the number of effective parameters. This pooling factor may be defined at the data level as:

$$\lambda = 1 - \frac{n/(n+1)\hat{\sigma}_\alpha^2 + \hat{\sigma}_y^2 / n}{\hat{\sigma}_\alpha^2 + \hat{\sigma}_y^2 / n},$$

and at the group level as:

$$\lambda = 1 - \frac{\hat{\sigma}_\alpha^2}{\hat{\sigma}_\alpha^2 + \hat{\sigma}_y^2 / n}.$$

Together these two measures clarify the role of predictors at the different levels and are very useful for understanding the behavior of complex multilevel models. These measures shed insight into model fit at each level as well as the extent to which the errors are pooled toward their common prior distribution.

12.1.5 Whole Model Explained Variance

Xu (2003) proposes three measures that could be used to estimate the amount of variation being explained by predictors in a linear mixed effects model. To obtain all three of these measures both a null and full model must be fit to the data. The first of these three measures, r^2, uses maximum likelihood to directly estimate the variance components under these two models. An empirical Bayes approach is taken to estimate the random effects, and a second measure, R^2, is obtained by comparing the residuals of the full and null models. The third measure, ρ^2, is based on the idea of explained randomness and relies on the Kullback-Leibler gain. Xu defines the following formulae where:

$$r^2 = 1 - \frac{\hat{\sigma}^2}{\hat{\sigma}_0},$$

$$R^2 = 1 - \frac{RSS}{RSS_0},$$

and

$$\rho^2 = 1 - \frac{\hat{\sigma}^2}{\hat{\sigma}_0} \exp\left(\frac{RSS}{N\hat{\sigma}^2} - \frac{RSS_0}{N\hat{\sigma}_0^2}\right),$$

where $\hat{\sigma}^2$ is an estimate of the residual variance at the lowest level (given by Xu as ε_{ij}) with all predictors present, and $\hat{\sigma}_0^2$ is an estimate of the residual variance given only the clustering. *RSS* is then defined as the residual sum of squares such that $\hat{\sigma}^2 \approx RSS/(N-df)$.

The performance of the three measures, r^2, R^2, and ρ^2, was assessed by Xu through simulation, and the results suggest that r^2 gives accurate estimates of the true amount of variation explained. R^2 and ρ^2 are good estimates when the cluster size is large, but they overestimate if the cluster sizes are too small with R^2 over estimating the truth slightly more than ρ^2. Thus, all three of these measures seem well suited for summarizing the predictive ability of a linear mixed effects model.

12.2 DISTANCE MEASURES FOR CALCULATING R^2 IN MULTILEVEL MODELS

In turning our thoughts to multiple regression, the multiple R^2 can be thought of as the correlation between a function of the predictor variables and the dependent variable. Another way to think about this value is that it is the correlation between the dependent variable, y, and $\hat{y}$, the values of the dependent variable predicted from the independent variables. This association can be seen in Figure 12.1 where x_1, x_2, and x_3 are all predictors of y.

Therefore we could write a formula to represent Figure 12.1 as:

$$R^2_{y(x1,x2,x3)} = R^2_{y\hat{y}}. \tag{12.8}$$

If we are to think theoretically about this formula, we can describe $\hat{y}$ as simply any given person's predicted y score based on the weights derived for each independent variable. The distance between an individual's predicted score, $\hat{y}$, and their observed score, y, would simply be thought of as an error, or variance unaccounted for.

Although this is a relatively simple formula, it can be applied easily to MLM, if we are to think of MLM as belonging to the General Linear Model of statistics. If we consider that the point of any analysis is to try to produce a series of coefficients that closely approximates an individual's original score, then we can see that in MLM, the $\hat{y}$ is simply any predicted value based on a set of regression coefficients derived from the MLM model, and the error term is simply the distance between $\hat{y}$ and y. In the

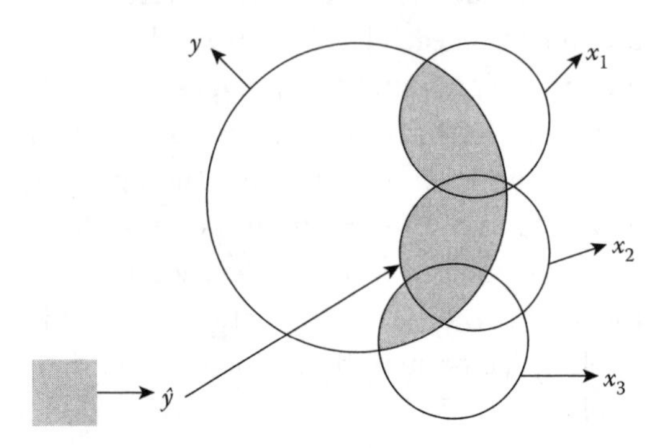

FIGURE 12.1
Graphical representation of correlation between y and predictor variables.

case of hierarchical linear modeling, these weights are derived through maximum likelihood estimates of the fixed effects, with the individual estimates being the product of empirical Bayesian estimates.

Although it would seem that a researcher could simply correlate these $\hat{y}$ and y values to obtain an estimate of R^2, we must remember and maintain in MLM that we wish to honor the procedure by which the estimates were obtained. In OLS regression, we can typically compute the total variance in the model as:

$$\sum_{i=1}^{n}(Y_i - \bar{Y})^2. \qquad (12.9)$$

In a multilevel model, we must remember that we typically define the null model as having both fixed effects (the grand mean for the dependent variable) and random effects (the variation of each group's mean around that grand mean). In defining our model in this manner, we cannot simply compute the total variance in the same manner as it is computed in Equation 12.9. Instead, the total variance must be the predicted values of y when only the grand mean is used as a predictor, or:

$$\tilde{y}_{ij} = \gamma_{00}(cons) + u_{0j}. \qquad (12.10)$$

Notice that the random estimate for individuals has been left out of Equation 12.10. By doing this, the grand mean value of the dependent variable is being estimated for the entire model and the random level-2 deviate u_{0j} (also known as β_{0j} for each group). We then can compute the total amount of variance in this model as:

$$\sigma'^2_{total} = \sum_{j=1}^{J}\sum_{i=1}^{n_j}(y_{ij} - \tilde{y}_j)^2 = e^2_{ij}, \qquad (12.11)$$

where $\tilde{y}_j$ is the random estimate for group j and y_{ij} is the original outcome score for person i in group j. This number is the same value as the sum of the square of all of the residual values e_{ij}, but is distinctly different from the variance estimate from the unbiased estimator of σ^2 that contains a correction factor for the $Q + 1$ regression parameters such that:

$$\hat{\sigma}^2 = \sum e^2_{ij} / (n - Q - 1). \qquad (12.12)$$

As noted in the OLS model, the error variance in a model can be viewed as the distance between $\hat{y}$ and y. Likewise, in MLM after the $\hat{y}$ are calculated using the full model, the error variance for the total model can be explained as:

$$\sigma'^2_{error} = \sum_{j=1}^{J}\sum_{i=1}^{n_j}(y_{ij} - \hat{y}_{ij})^2, \qquad (12.13)$$

where $\hat{y}_{ij}$ is the predicted value for person i in group j based on the full model:

$$\hat{y}_{ij} = \gamma_{00} + \gamma_{q0}X_{ij} + \gamma_{0q}W_j + u_{0j} + e_{ij}, \qquad (12.14)$$

and W_j is a level-2 predictor. Once these values are computed, an R^2 value may be computed as:

$$1 - \frac{\sigma'^2_{error}}{\sigma'^2_{total}}. \qquad (12.15)$$

12.2.1 R^2 Measure that Incorporates a Gaussian Probability Density Function

So far, these formulas are not very dissimilar from the previously proposed estimators of variance explained, with the difference being that they do not use the unbiased estimator for variance. However, consider if we were to aggregate this formula to the level-2 grouping structure such that we gain an R^2 value for each level-2 group and then average across all groups. Doing so would further enhance the above formulas such that the estimate of variance explained would be defined by:

$$R_j^2/k, \tag{12.16}$$

where k is the number of level-2 groups and,

$$R_j^2 = 1 - \frac{e_i^2(full)}{e_i^2(null)}, \tag{12.17}$$

with e_i^2 representing the measure of each residual for the ith person in each distinct group j for both the full model and the null model. Although this would represent an "average" R^2 for the entire model by producing a mean R^2 based on each group's R^2, it does not take into account that the original estimation method used to produce these values was based on a probability of inclusion in the model from maximum likelihood estimates. Unless each group has exactly the same sample size and the same probability of selection, it would not follow to use Equations 12.16 and 12.17 to solve for a model R^2.

By inserting the Gaussian probability density function into the above equations, we could gain a measure of R^2 as a function of the probability of inclusion of the given value assuming the model. Doing so would modify Equation 12.11 such that:

$$\sigma_{total}^2 = \frac{\sum_{ij} p(y_{ij})(e'_{ij})^2}{\sum_{ij} p(y_{ij})}, \tag{12.18}$$

where e'_{ij} is an estimate of the residual for each i person in the jth group in the null model, and:

$$p(y_{ij}) = p(d_{ij} \mid s_j)p(s_j) \tag{12.19}$$

where $p(d_{ij} \mid s_j)$ is the probability of the person i, given that they belong to the jth group, and $p(s_j)$ is the probability of group j, given the entire sample of level-2 units. In extrapolating Equation 12.19 further and applying the probability density function from a Gaussian distribution, it can be shown that:

$$p(y_{ij}) = \frac{1}{\sqrt{2\pi\sigma_{ij}^2 \cdot \sigma_j^2}} \bullet \exp\left[-\frac{(e'_{ij})^2}{2\sigma_{ij}^2}\right] \bullet \exp\left[-\frac{(u'_j)^2}{2\sigma_j^2}\right], \tag{12.20}$$

with σ_{ij}^2 representing the variance of the i individuals around their jth group mean for the null model, σ_j^2 is the variance of β_{0j} around γ_{00}, and e'_{ij} and u'_j are the residual scores for the level-1 and level-2 estimates, respectively.

As would follow from Equation 12.13, the model error could be thought of as:

$$\sigma_{error}^2 = \frac{\sum_{ij} p(y_{ij})(e''_{ij})^2}{\sum_{ij} p(y_{ij})}, \tag{12.21}$$

where e''_{ij} is an estimate of the residual for each i person in the jth group in the full model, and:

$$p(y_{ij}) = \frac{1}{\sqrt{2\pi\sigma_{ij}^2 \cdot \sigma_j^2}} \bullet \exp\left[-\frac{(e''_{ij})^2}{2\sigma_{ij}^2}\right] \bullet \exp\left[-\frac{(u''_j)^2}{2\sigma_j^2}\right], \tag{12.22}$$

with σ^2_{ij} representing the variance of the i individuals around their jth group mean for the full model, σ^2_j is the variance of β_{0j} around γ_{00}, and e''_{ij} and u''_j are the residual scores for the level-1 and level-2 estimates, respectively. The final estimate of variance explained could then be derived from combining Equations 12.18 and 12.21:

$$1-\frac{\sigma^2_{error}}{\sigma^2_{total}}. \qquad (12.23)$$

The strength of a measure such as this is twofold. First, it puts what would normally be a complicated interpretation of a model into a palatable form for the less-informed researcher who might be reading an MLM analysis. Although it often seems that the goal of many statistical concepts is to confuse the graduate student (e.g., the multiplicity of effect sizes currently available for use), in doing so, we only confuse the future researcher, and likewise, future research.

Second, it allows the researcher, outside of the Akaike Information Criterion *AIC* (Akaike, 1987) or the Bayesian Information Criterion *BIC* (Schwarz, 1978), with a single statistic to interpret just how well a model is performing. Since the goal of most research is to find variables that fully describe the variation in the dependent variable, a measure like this could potentially prove very useful in helping the researcher make judgments about the effectiveness of a MLM model.

12.2.2 Group Initiated R^2 based on Weighted Least Squares

In addition to alternative ways of computing an effect size mentioned above, another type of effect size can be conceived through maximum likelihood methods. In a typical multilevel ANOVA, the grand estimate for the slope coefficient is simply the weighted least squares estimator (or maximum likelihood estimate) γ_{00} where:

$$\hat{\gamma}_{00}=\sum \Delta_j^{-1}\bar{Y}_{\bullet j}\Big/\sum \Delta_j^{-1}, \qquad (12.24)$$

and Δ_j is the sum of the two variance components Var(u_{0j}) and Var($\bar{e}_{\bullet j}$; see Raudenbush & Bryk, 2002, p. 40 for further discussion). Put simply, the grand estimate for the mean of all of the groups (γ_{00}) is the sum of all of the group means ($\bar{Y}_{\bullet j}$) after applying the precision parameter (Δ_j^{-1}) and then dividing by the sum of the precision parameters. The effect of these precision parameters on the grand estimate is to apply more weight to the groups that are measured with more precision (cf., more level-1 units). This formula for grand estimates also could be applied to the idea of an R^2 effect size measure.

In typical OLS regression, the multiple R^2 can be expressed as:

$$R^2=1-\frac{\sum_{i=1}^{n}(\hat{y}_i-y_i)^2}{\sum_{i=1}^{n}(y_i-\bar{y})^2}, \qquad (12.25)$$

where y_i is any given individual's score on the dependent variable, $\hat{y}_i$ is that individual's predicted score from the linear regression equation, and $\bar{y}$ is the mean of all individuals on the dependent variable. For any given group within a set of level-2 units, Equation 12.25 could be considered the mathematical equivalent of:

$$R_j^2=1-\frac{1}{n_j}\sum_{i=1}^{n_j}\frac{(\hat{y}_{ij}-y_{ij})^2}{\sigma^2_{ij}}, \qquad (12.26)$$

where n_j is the number of people in group j and σ^2_{ij} is the variance of the individuals in group j. For simplification purposes, we will define the latter part of Equation 12.26 as being an error term corresponding to a normalized error for a given group. In representing this with the term E_i, Equation 12.26 can be thought of as:

$$R^2_j = 1 - \frac{1}{n_j}\sum_{i=1}^{n_j} E_i. \qquad (12.27)$$

This would mean that the total weighted least squares normalized error for all groups could be thought of as:

$$E_j = \sum_{j=1}^{J} p(s_j)E_i, \qquad (12.28)$$

where $p(s_j)$ is the probability of group j existing given the sample of all j groups such that:

$$p(s_j) = \frac{1}{\sqrt{2\pi\sigma^2_j}} \bullet \exp\left[-\frac{(u'_j)^2}{2\sigma^2_j}\right], \qquad (12.29)$$

with u'_j being the residual score for group j around the mean of all groups and σ^2_j the variance of j groups' means around the grand mean of the dependent variable.

In performing analyses that are not multilevel in nature, we could simply compute the R^2 for each group, and then average these values across all groups to obtain an average group R^2 based on the sample of groups that we drew from the greater population of all level-2 units. As was previously stated, this makes little sense, however, since in multilevel modeling we are producing an equation for each second-level group based on weighted least squares estimators. It seems appropriate, then, to produce an entire model R^2 that is also weighted for the probability of the group from which the estimate was drawn. In expanding Equation 12.27 to include all groups and also reflect the need to use a weighted estimator, R^2_T could be thought of as:

$$R^2_T = 1 - \frac{1}{\sum_{j=1}^{J} p(s_j)n_j} E_j, \qquad (12.30)$$

where E_j is the total weighted least squares normalized error for all groups from Equation 12.28. We can further deduce that R^2_T:

$$= 1 - \frac{1}{\sum_{j=1}^{J} p(s_j)n_j} \sum_{j=1}^{J} p(s_j)E_j,$$

$$= 1 - \frac{1}{\sum_{j=1}^{J} p(s_j)n_j} \sum_{j=1}^{J}\left(p(s_j)\sum_{i=1}^{n_j} E_i\right),$$

$$= \frac{\sum_{j=1}^{J} p(s_j)n_j - \sum_{j=1}^{J}\left(p(s_j)\sum_{i=1}^{n_j} E_i\right)}{\sum_{j=1}^{J} p(s_j)n_j},$$

$$= \frac{\sum_{j=1}^{J} p(s_j)n_j\left(1 - 1/n_j\sum_{i=1}^{n_j} E_i\right)}{\sum_{j=1}^{J} p(s_j)n_j}.$$

And since we already have defined R_j^2 in Equation 12.26, we can then interpret:

$$R_T^2 = \frac{\sum_{j=1}^{J} n_j \cdot p(s_j) \cdot R_j^2}{\sum_{j=1}^{J} n_j \cdot p(s_j)}, \qquad (12.31)$$

which, theoretically, is simply the weighted least squares average of all of the R^2 values from each group. What is present in Equation 12.31 is a solution that will produce estimates similar in interpretation to OLS R^2 measures. Equation 12.31 seems considerably more appropriate than Equations 12.15 and 12.23, since it honors both the nesting structure of the data and the fact that the model was derived through weighted least squares estimates.

12.3 CONCLUSIONS

Snijders and Bosker (1999) presented good arguments for instances when Equations 12.6 and 12.7 produce results that yield negative values for R^2 in multilevel models. While it is sometimes helpful to be able to know the variance accounted for at each level of the MLM model, the language with which researchers must refer to these estimates is, at best, confusing to the non-MLM minded researcher. With the further encouragement from editors to begin reporting effect sizes in all research, it is becoming more necessary for researchers using MLM to be able to explain their results in a way that is common with other statistical methods. Although results from a multilevel model probably will need further explanation, it is hoped that the continued development of these models will help in their proliferation.

There is a caution, however, in making these models more accessible. Just because a researcher has the software and programming skills to utilize complicated techniques does not mean that that technique is warranted. With the growth of a likewise complicated field of statistics, MLM, Goldstein (1995) voiced similar concerns:

> There is a danger, and this paper reminds us of it, that multilevel modeling will become so fashionable that its use will be a requirement of journal editors, or even worse, that the mere fact of having fitted a multilevel model will become a certificate of statistical probity. That would be a great pity. These models are as good as the data they fit; they are powerful tools, not universal panaceas. (p. 202)

It is our sincere hope that developing MLM as a more user-friendly field of statistics will improve its utilization and interpretation, but along with this must come responsibility in evaluating such models.

REFERENCES

Akaike, H. (1987). Factor analysis and the AIC. *Psychometrica, 52*, 317–332.

American Psychological Association. (2001). *Publication manual of the American Psychological Association* (5th ed.). Washington, DC: Author.

Cohen, J. (1994). The earth is round ($p < .05$). *American Psychologist, 49*, 997–1003.

Gelman, A., & Pardoe, I. (2006). Bayesian measure of explained variance and pooling in multilevel (hierarchical) models. *Technometrics, 48*(2), 241–251.

Goldstein, H. (1995). Hierarchical data modeling in the social sciences. *Journal of Educational and Behavioral Statistics, 20*, 201–204.

Harlow, L. L., Muliak, S. A., & Steiger, J. H. (Eds.). (1997). *What if there were no significance tests?* Mahwah, NJ: Erlbaum.

Hox, J. (1995). *Applied multilevel analysis.* Amsterdam: TT-Publikaties.

Hox, J. (2002). *Multilevel analysis: Techniques and applications.* Mahwah, NJ: Erlbaum.

Huberty, C. J. (2002). A history of effect size indices. *Educational and Psychological Measurement, 62*(2), 227–240.

Kirk, R. E. (1996). Practical significance: A concept whose time has come. *Educational and Psychological Measurement, 56*, 746–759.

Knapp, T. R., & Sawilowsky, S. S. (2001a). Constructive criticisms of methodological and editorial practices. *The Journal of Experimental Education, 70*, 65–79.

Knapp, T. R., & Sawilowsky, S. S. (2001b). Strong arguments: Rejoinder to Thompson. *The Journal of Experimental Education, 70*, 94–95.

Kreft, I., & de Leeuw, J. (1998). *Introducing multilevel modeling.* Thousand Oaks, CA: Sage.

Lee. V. E. (2000). Using Hierarchical linear modeling to study social contexts: The case of school effects. *Educational Psychologist, 35*(2), 125–141.

Raudenbush, S. W., & Bryk, A. S. (2002). *Hierarchical linear models: Applications and data analysis methods* (2nd ed.). Thousand Oaks, CA: Sage.

Roberts, J. K. (2002). The importance of intraclass correlation in multilevel and hierarchical linear modeling designs. *Multiple linear regression viewpoints, 28*(2), 19–31.

Roberts, J. K. (2004). An introductory primer on multilevel and hierarchical linear modeling. *Learning disabilities: A contemporary journal, 2*(1), 30–38.

Roberts, J. K., & Henson, R. K. (2002). Correction for bias in estimating effect sizes. *Educational and Psychological Measurement, 62*(2), 241–253.

Schmidt, F. (1996). Statistical significance testing and cumulative knowledge in psychology: Implications for the training of researchers. *Psychological Methods, 1*, 115–129.

Schwarz, G. (1978). Estimating the dimension of a model. *Annals of Statistics, 6*, 461–464.

Shaver, J. (1985). Chance and nonsense. *Phi Delta Kappan, 67*, 57–60.

Snijders, T., & Bosker, R. (1999). *Multilevel analysis.* Thousand Oaks, CA: Sage.

Thompson, B. (1996). AERA editorial policies regarding statistical significance testing: Three suggested reforms. *Educational Researcher, 25*(2), 26–30.

Thompson, B. (2001). Significance, effect sizes, stepwise methods, and other issues: Strong arguments move the field. *Journal of Experimental Education, 70*, 80–93.

Wilkinson, L., & American Psychological Association Task Force on Statistical Inference. (1999). Statistical methods in psychology journals: Guidelines and explanation. *American Psychologist, 54*, 594–604. [reprint available through the APA Home Page: http://www.apa.org/journals/amp/amp548594.html]

Xu, R. (2003). Measuring explained variation in linear mixed effects models. *Statistics in Medicine, 22*, 3527–3541.

13

Model Selection Based on Information Criteria in Multilevel Modeling

Ellen L. Hamaker, Pascal van Hattum, Rebecca M. Kuiper, and Herbert Hoijtink
Department Methods and Statistics, Faculty of Social Sciences, Utrecht University, Utrecht, The Netherlands

Model selection consists of comparing two or more models—which represent different theories about reality—and deciding which of these models gives the best description of the current data. The dominant approach to model selection in the social and behavioral sciences is null hypothesis testing (NHT). This approach has received much criticism over the years, ranging from practical issues to more philosophical ones (cf. Raftery, 1995; Wagenmakers, 2007; Weaklim, 2004). Some of the most eminent practical limitations of using NHT for model selection purposes are: its limitation to the comparison of two models; the need for these two models to be nested; and the impossibility of finding evidence for the null model.

An alternative approach to model selection, which does not suffer from these limitations, was started by Akaike (1973), who developed An Information Criterion (AIC, typically referred to as Akaike's Information Criterion). Not long after the introduction of the AIC, the Bayesian Information Criterion (BIC) was introduced by Schwarz (1978), which is also referred to as the Schwarz Information Criterion (SIC). Although the BIC bears a strong resemblance to the AIC, the two measures have quite different roots: While the AIC is derived within the framework of information theory, the BIC is based on the posterior model probability, which is an inherently Bayesian concept. More recently the Deviance Information Criterion (DIC) was introduced by Spiegelhalter, Best, Carlin, and Van der Linde (2002), which is derived within a decision theoretical framework. All three information criteria can be recognized as a penalized log likelihood, which balances model fit in the form of the log likelihood and model complexity in the form of the dimension of the model. Smaller values for an information criterion imply more appropriate models. In general, the information criteria are easy to compute and most software packages report at least one of the information criteria mentioned above.

Despite the advantages of information criteria for model selection and their inclusion in most multilevel modeling (MLM) programs, the literature on model selection in MLM is dominated by NHT (cf. Bryk & Raudenbush, 1992; Hox, 2002; Kreft & de Leeuw, 1998; Longford, 1995; Snijders & Bosker, 1999). What may contribute to this neglect of information criteria is unfamiliarity with these measures and a lack of understanding of the rationale behind them: Their rather simple appearance may give the erroneous impression that these measures were "invented" in an ad hoc fashion, rather than that they are soundly rooted in, for instance, information theory or Bayesian statistics. In addition, it is not always clear how to apply information criteria within the MLM context. Specific questions that may arise here are whether or not the random effects themselves should be penalized (an issue for all three information criteria considered here), and what the actual number of observations is (in case of the BIC).

The aim of this chapter is to provide detailed derivations of the AIC, DIC, and BIC, and to discuss and illustrate their application in the context of MLM. In the following section the basics of model selection using information criteria are presented. This is followed by a section on MLM in which the Laird–Ware model is presented (Laird & Ware, 1982), which is a general way to present multilevel models with two levels. In addition, two inferential focuses in MLM are discussed, which are based on whether or not the random effects are integrated out of the model. In the third section we derive the three information criteria and discuss their application in MLM. In the fourth section we present an empirical illustration, in which we make use of five different software packages and compare multiple nested and nonnested models. We end this chapter with a discussion in which we summarize the most important findings of this chapter and discuss several additional issues that are relevant when using information criteria.

13.1 MODEL SELECTION USING INFORMATION CRITERIA

In comparison to NHT, model selection based on information criteria has three practical advantages. Suppose we are interested in comparing the following four models: M_1: $\theta_1 = 0, \theta_2 = 0, \theta_3 = 0$; M_{2a}: $\theta_1 = 0, \theta_2, \theta_3$; M_{2b}: $\theta_1, \theta_2 = 0, \theta_3$; and M_3: $\theta_1, \theta_2, \theta_3$, where θ_1, θ_2, and θ_3 are parameters of the statistical model at hand. First, comparing nonnested models like M_{2a} and M_{2b} is rather difficult in the context of NHT, because it is not clear which of the two models should serve as the null model (cf. Cox, 1962). In contrast, the use of information criteria is not restricted to nested models, making the comparison of M_{2a} and M_{2b} straightforward.

Second, information criteria allow us to compare multiple models simultaneously. As a result, certain conflicting results, which may arise in the context of NHT due to the kind of comparisons that are made, will not arise here. For instance, suppose testing M_1 against M_{2a} results in $\chi^2_2 = 5.4$, which is not significant, and testing M_{2a} against M_3 results in $\chi^2_1 = 3.2$, which is also insignificant (critical values for $\alpha = .05$ are 5.99 and 3.84, respectively). Then testing M_1 against M_3 must render $\chi^2_3 = 5.4 + 3.2 = 8.6$, which is in fact significant (critical value for $\alpha = .05$ is 7.81). Hence, whether to select M_1 or M_3 using NHT will depend on the *kind of model comparisons that are made.* In contrast, since the use of information criteria allows one to compare multiple models simultaneously, this problem does not exist in this context.

Third, in NHT it is not possible to find evidence for the smallest model—M_1 in the example above. Even if M_1 is not rejected in comparison to any of the other models under which it is nested, this cannot be interpreted as evidence in favor of M_1. In contrast, when using an information criterion, we select the model with the smallest value for this criterion. This may well lead to the conclusion that the smallest model in our set is the best model.

When using information criteria, one first has to specify the models of interest. In addition one has to choose a measure quantifying the support in the data for each of the models. Well known instances of such measures are the AIC (Akaike, 1973), the BIC (Kass & Raftery, 1995; Schwarz, 1978), and the DIC (Spiegelhalter et al., 2002). All three measures consist of a part that represents model (mis)fit and a part that represents the size or dimensionality of the model; that is,

$$\text{IC} = -2\log f(\mathbf{y} \mid \hat{\boldsymbol{\theta}}) + \lambda d, \quad (13.1)$$

where IC denotes information criterion, $f(\mathbf{y} \mid \hat{\boldsymbol{\theta}})$ denotes the likelihood of the data $\boldsymbol{y}$ evaluated at an estimate of the model parameters (i.e., $\hat{\boldsymbol{\theta}}$ is an estimate of $\boldsymbol{\theta}$), λ is the penalty weight, which differs across the various information criteria (i.e., 2 for AIC and DIC, and $\log(n)$ for BIC where n is the sample size), and d is (an estimate of) the size or dimensionality of the model.

Information criteria are based on the idea that models are mere approximations of the truth, so that the issue is not to find the "true" model, but to find the best approximating model given the set of models under consideration (Burnham & Anderson, 2004). Models with smaller values for the quantity in Equation 13.1 are considered better models. When two or more models have about the same value for the criterion at hand, these models are considered equally good candidates. Stated otherwise, model selection can be used to account for model uncertainty (Draper, 1995; Hoeting, Madigan, Raftery, & Volinsky, 1999), and instead of looking for the single best model, it can be used to look for a small set of reasonable models. We return to this issue in the discussion.

While a smaller value for the information criterion at hand indicates a more appropriate model, researchers may feel uncomfortable drawing conclusions based on small differences between two models, which have very large values: What does a difference of three points in AIC or BIC mean, if for instance the smallest value is 20,486 and the second smallest is 20,489? To facilitate the interpretation of information criteria, the values can be transformed, using

$$\text{IC}^* = \exp\left(-\frac{1}{2}\text{IC}\right). \quad (13.2)$$

If one is interested in making a comparison between just two models, one can take the ratio of the transformed criteria of these models. For the AIC this ratio is

$$\begin{aligned}\frac{\mathrm{AIC}_1^*}{\mathrm{AIC}_2^*} &= \exp\left\{-\frac{1}{2}\mathrm{AIC}_1 + \frac{1}{2}\mathrm{AIC}_2\right\} \\ &= \exp\{\log f(\mathbf{y}\,|\,\hat{\boldsymbol{\theta}}_1) \\ &\quad -\log f(\mathbf{y}\,|\,\hat{\boldsymbol{\theta}}_2) - d_1 + d_2\}, \\ &= \frac{f(\mathbf{y}\,|\,\hat{\boldsymbol{\theta}}_1)}{f(\mathbf{y}\,|\,\hat{\boldsymbol{\theta}}_2)}\exp\{d_2 - d_1\}\end{aligned} \tag{13.3}$$

and for the BIC this is

$$\begin{aligned}\frac{\mathrm{BIC}_1^*}{\mathrm{BIC}_2^*} &= \exp\left\{-\frac{1}{2}\mathrm{BIC}_1 + \frac{1}{2}\mathrm{BIC}_2\right\} \\ &= \exp\{\log f(\mathbf{y}\,|\,\hat{\boldsymbol{\theta}}_1) - \log f(\mathbf{y}\,|\,\hat{\boldsymbol{\theta}}_2) \\ &\quad -\frac{1}{2}\log(n)(d_1 - d_2)\}. \\ &= \frac{f(\mathbf{y}\,|\,\hat{\boldsymbol{\theta}}_1)}{f(\mathbf{y}\,|\,\hat{\boldsymbol{\theta}}_2)} n^{\frac{1}{2}(d_2 - d_1)}\end{aligned} \tag{13.4}$$

These ratios can be interpreted in a similar manner as likelihood ratios (Burnham & Anderson, 2002, pp. 77–80), or Bayes factors (Kass & Raftery, 1995; Raftery, 1995). For the values suggested above, this results in exp(–.5·(20486 – 20489)) = 4.48, meaning that the first model is 4.48 times more likely to have generated the data than the second model.

Another transformation that can be used when there are two or more models consists of determining the model weights. Given a set of K models, the weight of model k is

$$w\mathrm{IC}_k = \frac{\mathrm{IC}_k^*}{\sum_{j=1}^{K}\mathrm{IC}_j^*}. \tag{13.5}$$

Note that these model weights add to one.[1] These weights can be based on either the AIC or the BIC, and are referred to as Akaike weights (cf. Burnham & Anderson, 2002, pp. 75–77; Wagenmakers & Farrell, 2004), or Schwarz weights, respectively. Schwarz weights can be interpreted as (an estimate of) the posterior model probability; that is, it is the probability that this model generated the data, given the other models in the set (Raftery, 1995). Similarly, the Akaike weight is interpreted as the weight of evidence in favor of this particular model being the best model, given the set of models (Burnham & Anderson, 2002).

Suppose we have obtained the two values for an information criterion (AIC or BIC) suggested above—20,486 and 20,489—and there is a third model that resulted in a value of 20,501. Then the model weights are .8172 for the first model, .1823 for the second model, and .0005 for the third model. Hence, we can be quite certain the third model did not generate the data. Moreover, the first model is more likely to have generated the data than the second model. These transformations show that it is not *the size of the information criterion* that matters: Rather, it is the *difference between information criteria for competing models* that is of interest.

13.2 MULTILEVEL MODELING

The Laird-Ware linear mixed model is a convenient way of expressing a multilevel

[1] When the values for the IC are very large, this may lead to computational difficulties. To overcome these, one can subtract the same constant from all the IC of each of the K models. For instance, one may subtract the smallest value for the IC that was obtained; that is, IC_{min}. Then, instead of using the IC_j, one uses the $\Delta\mathrm{IC}_j = \mathrm{IC}_j - \mathrm{IC}_{min}$, and the expression in Equation 13.5 becomes: $\omega\mathrm{IC}_k = \Delta\mathrm{IC}_k^* / \Sigma_{j=1}^{k}\,\Delta\mathrm{IC}_j^*$.

model with two levels (Laird & Ware, 1982). In this section we begin with presenting the Laird–Ware model. Next we discuss two different focuses that can be taken when making inferences from multilevel models with two levels.

13.2.1 Laird–Ware Model

Let y_i be the vector with n_i responses from cluster i, with $i = 1, \ldots, m$. Then, the Laird and Ware (1982) model can be written as

$$\mathbf{y}_i = \mathbf{W}_i\boldsymbol{\mu} + \mathbf{V}_i\boldsymbol{\alpha}_i + \boldsymbol{\varepsilon}_i, \qquad (13.6)$$

where $\mathbf{W}_i$ and $\mathbf{V}_i$ are the $n_i \times p$ and $n_i \times q$ matrices with covariates for the fixed and random effects, respectively, $\boldsymbol{\mu}$ is a p-variate vector with unknown fixed effects, $\boldsymbol{\alpha}_i$ is a q-variate vector with unknown random effects, which is normally distributed; that is, $\boldsymbol{\alpha}_i \sim N(\mathbf{0}, \boldsymbol{\Omega})$, and $\boldsymbol{\varepsilon}_i$ is an n_i-variate vector with residuals, which is also normally distributed; that is $\boldsymbol{\varepsilon}_i \sim N(\mathbf{0}, \mathbf{R}_i)$. A further simplification is obtained by assuming that the residual covariance structure does not vary across clusters, and that the residuals are independently and identically distributed, such that $\mathbf{R}_i = \sigma^2\mathbf{I}_{n_i}$, where $\mathbf{I}_{n_i}$ is an identity matrix of size n_i.

The model defined in Equation 13.6 can be used for longitudinal data, where the clusters are made up by individuals and the measurements within the clusters consist of observations made at different occasions. Alternatively, the clusters could be different groups, for instance, schools or hospitals, such that the measurements within a cluster represent individuals belonging to a specific group (i.e., pupils nested in schools or patients nested in hospitals).

13.2.2 Two Inferential Focuses in MLM

Vaida and Blanchard (2005) have argued that researchers dealing with this kind of multilevel data have to choose between two inferential focuses (cf. Spiegelhalter et al., 2002). In the first focus, the clusters are assumed fixed and inferences are made to *other observations from these same clusters*; that is, with respect to $\boldsymbol{\mu}$ and the random effects $\boldsymbol{\alpha}_i$. When dealing with repeated measurements this implies the researcher generalizes to different measurement occasions in the same individuals. When dealing with pupils from different schools it implies the researcher generalizes to different pupils from these same schools. Vaida and Blanchard (2005) refer to this as the *cluster focus*, or the conditional focus, as it is conditional on the random effects.

Alternatively, the clusters may be viewed as a random selection from all clusters in the population. Then the aim is to generalize to the *other clusters from the population* and hence the emphasize is on $\boldsymbol{\mu}$ and $\boldsymbol{\Omega}$, rather than on the actual $\boldsymbol{\alpha}_i$s. This is referred to by Vaida and Blanchard (2005) as the *population focus*, or the marginal focus, as the random effects are integrated out. As a result, in longitudinal data the purpose is to generalize to other individuals from the same population, and for data consisting of pupils in schools the aim is to generalize to other schools from this population.

These different focuses imply different likelihood functions and, consequently, different model dimensions (Vaida & Blanchard, 2005). To show this, we denote the Laird–Ware model as

$$\mathbf{y}_i \mid \boldsymbol{\mu}, \boldsymbol{\alpha}_i, \sigma^2 \sim N(\mathbf{y}_i \mid \mathbf{W}_i\boldsymbol{\mu} + \mathbf{V}_i\boldsymbol{\alpha}_i, \mathbf{R}_i). \qquad (13.7)$$

Let $N = \Sigma_{i=1}^{m} n_i$, and $\mathbf{y} = [\mathbf{y}_1^T, \ldots, \mathbf{y}_m^T]^T$ (meaning $\mathbf{y}$ is a vector with all N observations). Then, conditional on the random effects $\boldsymbol{\alpha} = [\boldsymbol{\alpha}_1^T, \ldots, \boldsymbol{\alpha}_m^T]^T$ the likelihood of the model can be denoted as

$$f(\mathbf{y} \mid \boldsymbol{\mu}, \boldsymbol{\alpha}, \sigma^2) = \prod_{i=1}^{m} f(\mathbf{y}_i \mid \boldsymbol{\mu}, \boldsymbol{\alpha}_i, \sigma^2), \tag{13.8}$$

where

$$f(\boldsymbol{\alpha}_i \mid \boldsymbol{\Omega}) = N(\boldsymbol{\Omega}). \tag{13.9}$$

Note that without Equation 13.9, the model in Equation 13.8 becomes a fixed effects model (i.e., the standard ANOVA model). The conditional likelihood in Equation 13.8 is compatible with the cluster focus, in which the clusters are considered given. For the population focus the random effects are integrated out, such that we get

$$f(\mathbf{y} \mid \boldsymbol{\mu}, \sigma^2, \boldsymbol{\Omega}) = \int f(\mathbf{y} \mid \boldsymbol{\mu}, \boldsymbol{\alpha}, \sigma^2) f(\boldsymbol{\alpha} \mid \boldsymbol{\Omega}) d\boldsymbol{\alpha}, \tag{13.10}$$

where $f(\boldsymbol{\alpha} \mid \boldsymbol{\Omega}) = \Pi_{i=1}^{m} f(\boldsymbol{\alpha}_i \mid \boldsymbol{\Omega})$.

The degrees of freedom for the likelihood functions in Equation 13.8 and Equation 13.10 differ (Vaida & Blanchard 2005). In the population focus, the random effects are integrated out, and the number of degrees of freedom is equal to the number of parameters; that is, $p + 1 + \{q \cdot (q + 1)\}/2$ (p degrees of freedom for the parameters in $\boldsymbol{\mu}$, one degree of freedom for σ^2, and $\{q \cdot (q + 1)\}/2$ degrees of freedom for the unique parameters in $\boldsymbol{\Omega}$). In the cluster focus it is more difficult to determine the effective number of degrees of freedom (Hodges & Sargent, 2001; Vaida & Blanchard, 2005). The number of parameters in the cluster focus is $p + 1 + \{q \cdot (q + 1)\}/2 + m \cdot q$ (i.e., the same parameters as in the population focus, plus an extra $m \cdot q$ number of parameters representing the actual random effects $\boldsymbol{\alpha}_i$). However, since the random effects $\boldsymbol{\alpha}_i$ are tied by the distributional property $f(\boldsymbol{\alpha} | \boldsymbol{\Omega})$, the effective number of degrees of freedom is smaller than this. We elaborate on this issue in the following section.

13.3 THREE INFORMATION CRITERIA

Three popular information criteria are the AIC, DIC, and BIC. Although the appearances of these criteria are rather similar, their derivations are quite different. To elucidate their backgrounds, we begin by presenting the measure of support for each of the three information criteria. Then we give a detailed derivation, such that the reader can see how each information criterion is an approximation of its own measure of support. This is followed by a discussion of their use in MLM. In particular, we discuss the inferential focus that is associated with them. Readers who are not interested in the algebraic details are advised to skip these paragraphs and turn to the summary at the end of this section, in which the three measures and their use in MLM are summarized.

Model selection consists of comparing K models, which can be denoted as $f_1(\cdot|\boldsymbol{\theta}_1)$ to $f_K(\cdot|\boldsymbol{\theta}_K)$. Each model k can be thought of as a set of densities, depending on the values taken on by the parameter vector $\boldsymbol{\theta}_k$. In what follows, we abbreviate $f_k(\cdot|\boldsymbol{\theta}_k)$ to $f(\cdot|\boldsymbol{\theta})$ when possible.

13.3.1 AIC

The AIC is based on minimizing the Kullback-Leibler (K-L) criterion (Bozdogan, 1987; Burnham & Anderson, 2002). Let $p(\cdot)$ be the truth—or the *data generating mechanism*—from which our data $\mathbf{y}$ arise. Then the K-L criterion is the amount of information that is lost when using $f(\cdot|\boldsymbol{\theta})$—our model of interest—to approximate $p(\cdot)$. This criterion is defined as

$$I[p(\cdot), f(\cdot\,|\,\boldsymbol{\theta})] = \mathrm{E}_{p(\mathbf{y})}[\log p(\mathbf{y}) - \log f(\mathbf{y}\,|\,\boldsymbol{\theta})]$$
$$= \int p(\mathbf{y})\{\log p(\mathbf{y}) - \log f(\mathbf{y}\,|\,\boldsymbol{\theta})\}d\mathbf{y}, \tag{13.11}$$

where $\mathrm{E}_{p(\mathbf{y})}$ implies the expectation is taken with respect to the truth $p(\mathbf{y})$. Since for all models $\mathrm{E}_{p(\mathbf{y})}[\log p(\mathbf{y})] = c$, minimizing the expression in Equation 13.11 is equivalent to maximizing the expected log likelihood $\mathrm{E}_{p(\mathbf{y})}[\log f(\mathbf{y}|\boldsymbol{\theta})]$. However, $\boldsymbol{\theta}$ can take on a range of values that makes it impossible to evaluate this expectation directly. Let $\boldsymbol{\theta}^*$ be the vector that maximizes the expected log likelihood (and thus minimizes the K-L criterion) for a particular model $f(\mathbf{y}|\boldsymbol{\theta})$: Then the goal is to find the model with the largest value for $\mathrm{E}_{p(\mathbf{y})}[\log f_k(\mathbf{y}\,|\,\boldsymbol{\theta}_k^*)]$.

Because $\boldsymbol{\theta}^*$ is unknown, we could consider the maximum likelihood estimate obtained in $\mathbf{y}$—which is denoted as $\hat{\boldsymbol{\theta}}_y$—as an estimate of it. However, this would have the undesirable property that the data $\mathbf{y}$ are used twice: first, for obtaining the maximum likelihood estimate and second, for evaluating the fit of the model. To avoid this problem, a hypothetical cross-validation data set $\mathbf{x}$ is introduced, which is also generated by $p(\cdot)$. By entering the maximum likelihood estimates for $\boldsymbol{\theta}$ obtained from $\mathbf{y}$ in the expected log likelihood of $\mathbf{x}$,[2] we get the *estimated relative K-L criterion* $\mathrm{E}_{p(\mathbf{x})}[\log f(\mathbf{x}\,|\,\hat{\boldsymbol{\theta}}_y)]$. Note that this measure is necessarily smaller than the relative K-L criterion $\mathrm{E}_{p(\mathbf{x})}[\log f(\mathbf{x}\,|\,\boldsymbol{\theta})]$, unless $\hat{\boldsymbol{\theta}}_y = \boldsymbol{\theta}^*$.

Finally, to rule out sampling error due to the fact that the maximum likelihood estimate $\hat{\boldsymbol{\theta}}_y$ is based on a single data set, the expectation is taken with respect to $p(\mathbf{y})$, which results in the *expected estimated relative K-L criterion*. Now the aim is to show that

$$\mathrm{E}_{p(\mathbf{y})}[\mathrm{E}_{p(\mathbf{x})}\{\log f(\mathbf{x}\,|\,\hat{\boldsymbol{\theta}}_y)\}] \approx \log f(\mathbf{y}\,|\,\hat{\boldsymbol{\theta}}_y) - d, \tag{13.12}$$

where d is the effective degrees of freedom of the model. Multiplying by −2 results in the expression for the AIC; that is,

$$\mathrm{AIC} = -2\log f(\mathbf{y}\,|\,\hat{\boldsymbol{\theta}}_y) + 2d. \tag{13.13}$$

There are multiple ways to derive the approximation in Equation 13.12 (cf. Bozdogan, 1987; Burnham & Anderson, 2002). The derivation below is based on one of the derivations presented by Burnham and Anderson (2002, pp. 365–368).

13.3.1.1 Derivation of the AIC

Using the expected estimated relative K-L criterion in Equation 13.12 as the point of departure, we begin with a second-order

[2] Note that in Burnham and Anderson (2002) the roles of the observed data set and the hypothetical data set are reversed in that the maximum likelihood estimates from the hypothetical data set are entered in the relative K-L criterion of the observed data set. The current use was chosen such that the similarities and differences between the derivation of the AIC and the DIC become more apparent.

Taylor expansion of $\log f(\mathbf{x}|\hat{\boldsymbol{\theta}}_y)$ around $\boldsymbol{\theta}^*$—the (unknown) vector that maximizes the relative K-L criterion—to obtain

$$\log f(\mathbf{x}|\hat{\boldsymbol{\theta}}_y) \approx \log f(\mathbf{x}|\boldsymbol{\theta}^*) + \left\{\frac{\partial \log f(\mathbf{x}|\boldsymbol{\theta}^*)}{\partial \boldsymbol{\theta}}\right\}^T (\hat{\boldsymbol{\theta}}_y - \boldsymbol{\theta}^*) + \frac{1}{2}(\hat{\boldsymbol{\theta}}_y - \boldsymbol{\theta}^*)^T \left\{\frac{\partial^2 \log f(\mathbf{x}|\boldsymbol{\theta}^*)}{\partial \boldsymbol{\theta} \partial \boldsymbol{\theta}^T}\right\} (\hat{\boldsymbol{\theta}}_y - \boldsymbol{\theta}^*). \quad (13.14)$$

When taking the expectation of this expression with respect to the truth $p(\mathbf{x})$, we can make use of the fact that the expectation of the second term on the right-hand side of Equation 13.14 is equal to zero. For the third term we write

$$\mathrm{E}_{p(\mathbf{x})}\left[\frac{\partial^2 \log f(\mathbf{x}|\boldsymbol{\theta}^*)}{\partial \boldsymbol{\theta} \partial \boldsymbol{\theta}^T}\right] = -\mathbf{I}_{p(\cdot)}(\boldsymbol{\theta}^*). \quad (13.15)$$

Note that the matrix $\mathbf{I}_{p(\cdot)}(\boldsymbol{\theta}^*)$ is not the usual Fisher information matrix, which we denote as $\mathbf{I}(\boldsymbol{\theta}^*)$, because the expectation is taken with respect to the truth $p(\mathbf{x})$, rather than with respect to the fitted model $f(\mathbf{x}|\boldsymbol{\theta})$. Hence, only when $p(\mathbf{x})$ is a special case of the model under consideration $f(\mathbf{x}|\boldsymbol{\theta})$, we have $\mathbf{I}_{p(\cdot)}(\boldsymbol{\theta}^*) = \mathbf{I}(\boldsymbol{\theta}^*)$.

Now the expected estimated relative K-L criterion in Equation 13.12 can be written as

$$\mathrm{E}_{p(\mathbf{y})}[\mathrm{E}_{p(\mathbf{x})}\{\log f(\mathbf{x}|\hat{\boldsymbol{\theta}}_y)\}] \approx \mathrm{E}_{p(\mathbf{y})}[\mathrm{E}_{p(\mathbf{x})}\{\log f(\mathbf{x}|\boldsymbol{\theta}^*)\} + \frac{1}{2}(\hat{\boldsymbol{\theta}}_y - \boldsymbol{\theta}^*)^T(-\mathbf{I}_{p(\cdot)}(\boldsymbol{\theta}^*))(\hat{\boldsymbol{\theta}}_y - \boldsymbol{\theta}^*)] = \mathrm{E}_{p(\mathbf{x})}\{\log f(\mathbf{x}|\boldsymbol{\theta}^*)\} - \frac{1}{2}\mathrm{tr}\{\mathbf{I}_{p(\cdot)}(\boldsymbol{\theta}^*)\Sigma\}, \quad (13.16)$$

where the latter is obtained by applying a standard result for quadratic terms,[3] and $\boldsymbol{\Sigma} = \mathrm{E}_{p(\mathbf{y})}[(\hat{\boldsymbol{\theta}}_y - \boldsymbol{\theta}^*)(\hat{\boldsymbol{\theta}}_y - \boldsymbol{\theta}^*)^T]$ is the large sample covariance matrix of the maximum likelihood estimate of $\boldsymbol{\theta}$.

If $p(\cdot)$ is a special case of the fitted model $f(\cdot|\boldsymbol{\theta})$, then $\mathbf{I}_{p(\cdot)}(\boldsymbol{\theta}^*) = \mathbf{I}(\boldsymbol{\theta}^*) = \boldsymbol{\Sigma}^{-1}$ and thus $\mathrm{tr}\{\mathbf{I}_{p(\cdot)}(\boldsymbol{\theta}^*)\boldsymbol{\Sigma}\} = d$, where d is the number of elements in $\boldsymbol{\theta}$. However, it has been shown that this also holds when the truth is more general than the model, as long as the model is a good approximation of the truth. Moreover, Burnham and Anderson (2002) argue that in case the model is a poor approximation of $p(\cdot)$, the component that represents misfit in the AIC (i.e., $-2\log f(\mathbf{y}|\hat{\boldsymbol{\theta}}_y)$) will be relatively large, such that this model will not be selected anyway.

Hence, we can approximate the expected estimated relative K-L criterion in Equation 13.12 with

$$\mathrm{E}_{p(\mathbf{y})}[\mathrm{E}_{p(\mathbf{x})}\{\log f(\mathbf{x}|\hat{\boldsymbol{\theta}}_y)\}] \approx \mathrm{E}_{p(\mathbf{x})}\{\log f(\mathbf{x}|\boldsymbol{\theta}^*)\} - \frac{1}{2}d. \quad (13.17)$$

To eliminate the unknown vector $\boldsymbol{\theta}^*$ in Equation 13.17, we make use of another second-order Taylor series approximation, but now of $f(\mathbf{x}|\boldsymbol{\theta}^*)$ around $\hat{\boldsymbol{\theta}}_x$—the

[3] Let $\mathbf{A}$ be a symmetric matrix, $\mathrm{E}[\mathbf{y}] = \mathbf{c}$ and $\mathrm{Var}[\mathbf{y}] = \boldsymbol{\Psi}$. Then $\mathrm{E}[\mathbf{y}^T\mathbf{A}\mathbf{y}] = \mathrm{tr}[\mathrm{A}\boldsymbol{\Psi}] + \mathbf{c}^T\mathbf{A}\mathbf{c}$.

maximum likelihood estimate obtained in $\mathbf{x}$—that is

$$\log f(\mathbf{x} \mid \boldsymbol{\theta}^*) \approx \log f(\mathbf{x} \mid \hat{\boldsymbol{\theta}}_x) + \left\{\frac{\partial \log f(\mathbf{x} \mid \hat{\boldsymbol{\theta}}_x)}{\partial \boldsymbol{\theta}}\right\}^T (\boldsymbol{\theta}^* - \hat{\boldsymbol{\theta}}_x) + \frac{1}{2}(\boldsymbol{\theta}^* - \hat{\boldsymbol{\theta}}_x)^T \left\{\frac{\partial^2 \log f(\mathbf{x} \mid \hat{\boldsymbol{\theta}}_x)}{\partial \boldsymbol{\theta} \partial \boldsymbol{\theta}^T}\right\} (\boldsymbol{\theta}^* - \hat{\boldsymbol{\theta}}_x). \tag{13.18}$$

Note that since the second term on the right-hand side of Equation 13.18 is zero, the first term in Equation 13.17 can be approximated by

$$\mathrm{E}_{p(\mathbf{x})}[\log f(\mathbf{x} \mid \boldsymbol{\theta}^*)] \approx \mathrm{E}_{p(\mathbf{x})}[\log f(\mathbf{x} \mid \hat{\boldsymbol{\theta}}_x)] - \frac{1}{2}\operatorname{tr}\{\mathrm{E}_{p(\mathbf{x})}[\hat{\mathbf{I}}(\hat{\boldsymbol{\theta}}_x)(\boldsymbol{\theta}^* - \hat{\boldsymbol{\theta}}_x)(\boldsymbol{\theta}^* - \hat{\boldsymbol{\theta}}_x)^T]\}, \tag{13.19}$$

where $\hat{\mathbf{I}}(\hat{\boldsymbol{\theta}}_x)$ is the negative Hessian of the log likelihood evaluated at $\hat{\boldsymbol{\theta}}_x$. Using the true Fisher information matrix $\mathbf{I}(\boldsymbol{\theta}^*)$ as an approximation of $\hat{\mathbf{I}}(\hat{\boldsymbol{\theta}}_x)$, we can write

$$\begin{aligned} &\operatorname{tr}\{\mathrm{E}_{p(\mathbf{x})}[\hat{\mathbf{I}}(\hat{\boldsymbol{\theta}}_x)(\boldsymbol{\theta}^* - \hat{\boldsymbol{\theta}}_x)(\boldsymbol{\theta}^* - \hat{\boldsymbol{\theta}}_x)^T]\} \\ &\approx \operatorname{tr}\{\mathbf{I}(\boldsymbol{\theta}^*)\mathrm{E}_{p(\mathbf{x})}[(\boldsymbol{\theta}^* - \hat{\boldsymbol{\theta}}_x)(\boldsymbol{\theta}^* - \hat{\boldsymbol{\theta}}_x)^T]\} \\ &= \operatorname{tr}\{\mathbf{I}(\boldsymbol{\theta}^*)\Sigma\} \\ &= d. \end{aligned} \tag{13.20}$$

Using this result in Equation 13.19, and using Equation 13.19 in Equation 13.17, we can write the expected estimated relative K-L criterion in Equation 13.12 as

$$\begin{aligned} &\mathrm{E}_{p(\mathbf{y})}[\mathrm{E}_{p(\mathbf{x})}\{\log f(\mathbf{x} \mid \hat{\boldsymbol{\theta}}_y)\}] \\ &\approx \mathrm{E}_{p(\mathbf{x})}[\log f(\mathbf{x} \mid \hat{\boldsymbol{\theta}}_x)] - d. \end{aligned} \tag{13.21}$$

Finally, since $\mathbf{x}$ and $\mathbf{y}$ were both generated by $p(\cdot)$, we can replace $\mathrm{E}_{p(\mathbf{x})}[\log f(\mathbf{x} \mid \hat{\boldsymbol{\theta}}_x)]$ by $\mathrm{E}_{p(\mathbf{y})}[\log f(\mathbf{y} \mid \hat{\boldsymbol{\theta}}_y)]$. Using $\log f(\mathbf{y} \mid \hat{\boldsymbol{\theta}}_y)$ as an unbiased estimator of this quantity shows that −2 times the expected estimated relative K-L criterion can be approximated by the AIC as defined in Equation 13.13.

13.3.1.2 Applying AIC to MLM

The derivation of the AIC presented above is based on a double expectation, using two independent samples $\mathbf{y}$ and $\mathbf{x}$. As a result, AIC-based model selection is asymptotically equivalent to cross-validation (Burnham & Anderson, 2002; Stone, 1977). However, as Vaida and Blanchard (2005) point out, this cross-validation aspect implies that when the AIC is used in multilevel analyses, researchers have to be clear on what is considered as the independent and identical sample: Is it a sample from the same population with different clusters, or is it a sample of different observations within the same clusters?

If one chooses the population focus, the cross-validation data set $\mathbf{x}$ contains *different clusters* with different random effects. Then the likelihood function that should be considered is the marginal observed likelihood $f(\mathbf{y} \mid \hat{\boldsymbol{\theta}}_y)$, where the parameter vector $\hat{\boldsymbol{\theta}}_y$ contains the p fixed effects $\boldsymbol{\mu}$, the residual variance σ^2, and the $\{q \cdot (q+1)\}/2$ unique elements of $\mathbf{\Omega}$. The penalty d_{mar} associated with the marginal likelihood is equal to the number of parameters in $\hat{\boldsymbol{\theta}}_y$. This AIC is referred to as the *marginal AIC*.

In the cluster focus however, the K-L criterion that needs to be minimized is based on the *conditional likelihood* as defined in Equation 13.8 and the truth $p(\cdot|\upsilon)$, where υ is the vector with true (but unknown) random effects for the clusters under consideration (i.e., the clusters in our data set). This K-L criterion can be expressed as

$$I[p(\cdot \mid \upsilon), f(\cdot \mid \boldsymbol{\alpha}, \boldsymbol{\theta})] = \mathrm{E}_{p(\mathbf{y} \mid \upsilon)}[\log p(\mathbf{y} \mid \upsilon) - \log f(\mathbf{y} \mid \boldsymbol{\theta}, \boldsymbol{\alpha})], \quad (13.22)$$

Again, since $\mathrm{E}_{p(\mathbf{y}|\upsilon)}[\log p(\mathbf{y}|\upsilon)]$ is a constant when we are comparing multiple models, we can focus on maximizing the relative K-L criterion $\mathrm{E}_{p(\mathbf{y}|\upsilon)}[\log f(\mathbf{y}|\boldsymbol{\theta}, \boldsymbol{\alpha})]$. Let $\boldsymbol{\theta}^*$ and $\boldsymbol{\alpha}^*$ be the vectors that maximize this relative K-L criterion. As estimates of these vectors we can use the maximum likelihood estimates $\hat{\boldsymbol{\theta}}_y$ and the empirical Bayes estimates $\hat{\boldsymbol{\alpha}}_y$—also known as the shrinkage estimator—for the random effects.

To avoid having to use the data $\mathbf{y}$ twice, a hypothetical data set $\mathbf{x}$ is introduced, which also arises from $p(\cdot|\upsilon)$. Note that this implies that the cross-validation data set represents observations from the *same clusters*. By entering the estimates from $\mathbf{y}$ into the relative K-L criterion for $\mathbf{x}$, we obtain the estimated relative K-L criterion. Finally, to rule out sampling error due to the fact that the estimates $\hat{\boldsymbol{\theta}}_y$ and $\hat{\boldsymbol{\alpha}}_y$ are based on a single data set, the expectation is taken with respect to $p(\mathbf{y}, \upsilon) = p(\mathbf{y}|\upsilon)\, p(\upsilon)$, such that the *expected estimated relative K-L criterion* is obtained. Vaida and Blanchard (2005) show that this measure can be approximated by

$$\mathrm{E}_{p(\mathbf{y},\upsilon)}[\mathrm{E}_{p(\mathbf{x}|\upsilon)}\{\log f(\mathbf{x} \mid \hat{\boldsymbol{\theta}}_y, \hat{\boldsymbol{\alpha}}_y)\}] \approx \log f(\mathbf{y} \mid \hat{\boldsymbol{\theta}}_y, \hat{\boldsymbol{\alpha}}_y) - d_c. \quad (13.23)$$

Multiplying by –2 results in the *conditional AIC*; that is,

$$cAIC = -2\log f(\mathbf{y} \mid \hat{\boldsymbol{\theta}}_y, \hat{\boldsymbol{\alpha}}_y) + 2d_c. \quad (13.24)$$

The term d_c is the effective number of degrees of freedom associated with the cluster focus. As mentioned in the previous section, the number of degrees of freedom is actually smaller than the number of parameters in the model, due to the fact that the random effects $\boldsymbol{\alpha}_i$ are restricted to come from a particular distribution.

To determine the effective degrees of freedom in the cluster focus, Vaida and Blanchard (2005) propose to rewrite the multilevel model as an ordinary linear regression model, as shown by Hodges and Sargent (2001). From ordinary regression analysis it is known that the dimension of a model is given by the trace of the projection matrix $\mathbf{H}$, for which $\hat{\mathbf{y}} = \mathbf{H}\mathbf{y}$.[4] The purpose of rewriting the multilevel model as an ordinary linear regression model is to find the projection matrix by which to get $\hat{\mathbf{y}}$ from $\mathbf{y}$, such that the dimension of the model can be estimated.

Let $r = m \cdot q$ be the total number of parameters in $\boldsymbol{\alpha}$, then Vaida and Blanchard (2005) propose to rewrite the model in Equation 13.6 as

[4] If $\mathbf{y}$ is regressed on $\mathbf{X}$ through $\mathbf{y} = \mathbf{X}\boldsymbol{\beta} + \boldsymbol{\varepsilon}$, the least squares estimate of $\boldsymbol{\beta}$ is $\hat{\boldsymbol{\beta}} = (\mathbf{X}^T\mathbf{X})^{-1}\mathbf{X}^T\mathbf{y}$. Then the prediction of $\mathbf{y}$ is given by $\hat{\mathbf{y}} = \mathbf{X}\hat{\boldsymbol{\beta}} = \mathbf{X}(\mathbf{X}^T\mathbf{X})^{-1}\mathbf{X}^T\mathbf{y}$. Hence, the projection matrix can be written as $\mathbf{H} = \mathbf{X}(\mathbf{X}^T\mathbf{X})^{-1}\mathbf{X}^T$.

$$\mathbf{Y}=\begin{bmatrix}\mathbf{y}\\\mathbf{0}\end{bmatrix}=\begin{bmatrix}\mathbf{W} & \mathbf{V}\\\mathbf{0} & -\mathbf{I}_r\end{bmatrix}\begin{bmatrix}\boldsymbol{\mu}\\\boldsymbol{\alpha}\end{bmatrix}+\begin{bmatrix}\boldsymbol{\varepsilon}\\\boldsymbol{\alpha}\end{bmatrix}, \quad (13.25)$$

where $\mathbf{0}$ in $\mathbf{Y}$ is an r-variate zero vector making $\mathbf{Y}$ an $(N+r)$-variate vector, $\mathbf{0}$ in $\mathbf{U}$ is a $p \times r$ zero matrix, $\mathbf{W} = [\mathbf{W}_1^T, \ldots, \mathbf{W}_m^T]^T$ is an $N \times p$ matrix with the covariates for the fixed effects, and $\mathbf{V} = \text{diag}[\mathbf{V}_1, \ldots, \mathbf{V}_m]$ is an $N \times r$ block diagonal whose diagonal blocks are the $n_i \times q$ matrices with the covariates for the random effects, and $\mathbf{I}_r$ is an r-dimensional identity matrix. The top N cases in $\mathbf{Y}$ (and $\mathbf{W}$) are referred to as the data cases, while the lower r cases are referred to as the constraint cases. The latter are included to ensure that the random effects are not estimated freely, but actually obey the covariance structure imposed by $\boldsymbol{\Omega}$. To this end the covariance matrix of $\boldsymbol{\eta} = [\boldsymbol{\varepsilon}^T, \boldsymbol{\alpha}^T]^T$ is a block diagonal with $\sigma^2\mathbf{I}_N$ and m times $\boldsymbol{\Omega}$; that is,

$$\begin{aligned} var(\boldsymbol{\eta}) &= \text{E}\{[\boldsymbol{\varepsilon}^T\ \boldsymbol{\alpha}^T]^T[\boldsymbol{\varepsilon}^T\ \boldsymbol{\alpha}^T]\}\\ &= \begin{bmatrix}\text{E}\{\boldsymbol{\varepsilon}\boldsymbol{\varepsilon}^T\} & \text{E}\{\boldsymbol{\varepsilon}\boldsymbol{\alpha}^T\}\\ \text{E}\{\boldsymbol{\alpha}\boldsymbol{\varepsilon}^T\} & \text{E}\{\boldsymbol{\alpha}\boldsymbol{\alpha}^T\}\end{bmatrix} \quad (13.26)\\ &= \begin{bmatrix}\sigma^2\mathbf{I}_N & \mathbf{0}\\ \mathbf{0}^T & \boldsymbol{\Omega}_m\end{bmatrix}, \end{aligned}$$

where $\mathbf{0}$ is an $N \times (r \cdot m)$ zero matrix, and $\boldsymbol{\Omega}_m$ is an $r \times r$ diagonal block matrix with m times the block $\boldsymbol{\Omega}$ on the diagonal.

The expression in Equation 13.29 bears strong resemblance to an ordinary linear regression model. However, the residual terms in $\boldsymbol{\eta}$ are not independently and identically distributed. Hence, we need to find a matrix $\boldsymbol{\Gamma}$ by which to premultiply the expression Equation 13.29 to get an expression in which the residuals are i.i.d. Let $\boldsymbol{\Omega}_m = \sigma^2\boldsymbol{\Phi}_m$, such that $var(\boldsymbol{\eta}) = \sigma^2\text{diag}(\mathbf{I}_N, \boldsymbol{\Phi}_m)$. Now there is some matrix $\boldsymbol{\Delta}$ for which $\boldsymbol{\Phi}_m = (\boldsymbol{\Delta}^T\boldsymbol{\Delta})^{-1}$. Let $\boldsymbol{\Gamma} = \text{diag}(\mathbf{I}_N, \boldsymbol{\Delta})$. Premultiplying $\mathbf{Y}$ with $\boldsymbol{\Gamma}$ results in $\mathbf{Y}$. Premultiplying $\boldsymbol{\eta}$ with $\boldsymbol{\Gamma}$ results in a vector whose covariance matrix is $\sigma^2\mathbf{I}_{N+r}$. Thus we can write

$$\begin{aligned} \mathbf{Y}=\begin{bmatrix}\mathbf{y}\\\mathbf{0}\end{bmatrix} &= \begin{bmatrix}\mathbf{I}_N & 0\\ 0 & \boldsymbol{\Delta}\end{bmatrix}\begin{bmatrix}\mathbf{W} & \mathbf{V}\\ 0 & -\mathbf{I}_r\end{bmatrix}\begin{bmatrix}\boldsymbol{\mu}\\\boldsymbol{\alpha}\end{bmatrix}+\begin{bmatrix}\mathbf{I}_N & 0\\ 0 & \boldsymbol{\Delta}\end{bmatrix}\begin{bmatrix}\boldsymbol{\varepsilon}\\\boldsymbol{\alpha}\end{bmatrix}\\ &= \begin{bmatrix}\mathbf{W} & \mathbf{V}\\ \mathbf{0} & -\boldsymbol{\Delta}\end{bmatrix}\begin{bmatrix}\boldsymbol{\mu}\\\boldsymbol{\alpha}\end{bmatrix}+\begin{bmatrix}\boldsymbol{\varepsilon}\\\boldsymbol{\Delta\alpha}\end{bmatrix} \quad (13.27)\\ &= \mathbf{U}\boldsymbol{\xi}+\boldsymbol{\kappa}, \end{aligned}$$

which is like an ordinary linear regression equation with i.i.d. residuals. Hence, the least squares estimate of $\boldsymbol{\xi}$ is obtained through

$$\hat{\boldsymbol{\xi}}=(\mathbf{U}^T\mathbf{U})^{-1}\mathbf{U}^T\mathbf{Y}=(\mathbf{U}^T\mathbf{U})^{-1}\begin{bmatrix}\mathbf{W}^T\\\mathbf{V}^T\end{bmatrix}\mathbf{y}. \quad (13.28)$$

From this we can find the matrix by which $\mathbf{y}$ is mapped to $\hat{\mathbf{y}}$; that is,

$$\begin{aligned} \hat{\mathbf{y}} &= [\mathbf{W}\ \mathbf{V}]\hat{\boldsymbol{\xi}}\\ &= [\mathbf{W}\ \mathbf{V}](\mathbf{U}^T\mathbf{U})^{-1}\begin{bmatrix}\mathbf{W}^T\\\mathbf{V}^T\end{bmatrix}\mathbf{y} \quad (13.29)\\ &= \mathbf{H}_1\mathbf{y}, \end{aligned}$$

where $\mathbf{H}_1$ is referred to as the "hat" matrix.[5] Hence, the effective number of degrees of freedom of this model is defined as the trace of $\mathbf{H}_1$ (Vaida & Blanchard, 2005).

However, in contrast to the hat matrix in ordinary linear regression analysis, the matrix $\mathbf{H}_1$ is not only a function of the observed values in $\mathbf{W}$ and $\mathbf{V}$, but it also requires us to know σ^2 and $\mathbf{\Omega}$ (since both are needed to derive $\mathbf{\Delta}$). In case that σ^2 is unknown, but $\mathbf{\Phi}_m$ is known, the degrees of freedom d_c become $\text{tr}[\mathbf{H}_1] + 1$. However, when $\mathbf{\Phi}_m$ is also unknown, it is rather complicated to derive the effective degrees of freedom. Therefore, Vaida and Blanchard (2005) propose to use the degrees of freedom associated with σ^2 unknown, but $\mathbf{\Phi}_m$ known. With the function hatTrace() from the R-package lme4, one can estimate the dimensionality of a multilevel model within the cluster focus. However, to obtain the conditional likelihood—which is needed to compute the conditional AIC—some additional programming is needed.

13.3.2 DIC

The DIC is a Bayesian information criterion and is based on Bayesian estimation. The latter consists of combining the likelihood for the data $f(\mathbf{y}|\boldsymbol{\theta})$ with a prior distribution for the parameters $h(\boldsymbol{\theta})$ in order to obtain the posterior distribution for the parameters $g(\boldsymbol{\theta}|\mathbf{y})$. Because the DIC for an approximately normal likelihood with a vague prior is asymptotically equivalent to the AIC (Spiegelhalter et al., 2002), the DIC is sometimes referred to as a Bayesian generalization of the AIC. However, whereas the AIC has an information theoretical background, Spiegelhalter et al. (2002) indicate that the DIC is obtained using an approximate decision theoretical justification.

Point of departure in the derivation of the DIC is the loss, which is defined as $-2\log f(\cdot|\tilde{\boldsymbol{\theta}})$, where $\tilde{\boldsymbol{\theta}}$ is a Bayesian estimate of $\boldsymbol{\theta}$, for instance the mean, mode or median of the posterior distribution $g(\boldsymbol{\theta}|\cdot)$. Because the goal is to favor models for which the loss is expected to be small, the criterion is based on the *expected loss*, also known as the *frequentist loss* (Berger, 1985, pp. 9–10). Making use of two independent samples that arise from $p(\cdot)$—that is, our observed data set $\mathbf{y}$, and a hypothetical cross-validation data set $\mathbf{x}$—the expected loss can be denoted as $E_{f(\mathbf{x}|\boldsymbol{\theta}^*)}[-2\log f(\mathbf{x}|\tilde{\boldsymbol{\theta}}_y)]$. Spiegelhalter et al. (2002) refer to $\boldsymbol{\theta}^*$ as the "pseudotrue" parameter value: It has the same interpretation here as in the derivation of the AIC; that is, it is the parameter vector that minimizes the distance between truth $p(\cdot)$ and the model of interest $f(\cdot|\boldsymbol{\theta})$ in terms of the K-L criterion as defined in Equation 13.11 (see Spiegelhalter et al., 2002, p. 585). Spiegelhalter et al. (2002) indicate that their derivation of the DIC implicitly relies on the assumption that $f(\cdot|\boldsymbol{\theta}^*)$ forms a good approximation of $p(\cdot)$ and they refer to this as the "good model" assumption.

Using $-2\log f(\mathbf{y}|\tilde{\boldsymbol{\theta}}_y)$ as an approximation of the expected loss, we can write

$$E_{f(\mathbf{x}|\boldsymbol{\theta}^*)}[-2\log f(\mathbf{x}\,|\,\tilde{\boldsymbol{\theta}}_y)] = -2\log f(\mathbf{y}\,|\,\tilde{\boldsymbol{\theta}}_y) + c(\mathbf{y},\boldsymbol{\theta}^*,\tilde{\boldsymbol{\theta}}_y), \tag{13.30}$$

[5] Vaida and Blanchard (2005) point out that this matrix is not the usual hat matrix, but only the upper left block of it. The hat matrix itself is $\mathbf{H} = \mathbf{U}(\mathbf{U}^T\mathbf{U})^{-1}\mathbf{U}^T$.

where $c(\mathrm{y}, \boldsymbol{\theta}^{*}, \tilde{\boldsymbol{\theta}}_y)$ is the bias in the approximation, also referred to as the *optimism*. Since a Bayesian perspective is taken, the pseudotrue $\boldsymbol{\theta}^{*}$ is replaced by the random quantity $\boldsymbol{\theta}$ and then, to deal with the fact that $\boldsymbol{\theta}$ is unknown, the posterior expectation of the expected loss is taken (Spiegelhalter et al., 2002, p. 604). Because the posterior expectation of the first term on the right-hand side of Equation 13.30 is equal to $-2\log f(\mathbf{y}|\tilde{\boldsymbol{\theta}}_y)$, this *expected posterior loss* can be written as

$$\mathrm{E}_{g(\boldsymbol{\theta}|\mathbf{y})}\{\mathrm{E}_{f(\mathbf{x}|\boldsymbol{\theta})}[-2\log f(\mathbf{x}\,|\,\tilde{\boldsymbol{\theta}}_y)]\} = -2\log f(\mathbf{y}\,|\,\tilde{\boldsymbol{\theta}}_y) + \mathrm{E}_{g(\boldsymbol{\theta}|\mathbf{y})}\{c(\mathbf{y},\boldsymbol{\theta},\tilde{\boldsymbol{\theta}}_y)\}. \quad (13.31)$$

The DIC is an approximation of the expected posterior loss in Equation 13.31 and its derivation is based on finding an approximation of the posterior expectation of the optimism $\mathrm{E}_{g(\boldsymbol{\theta}|\mathrm{y})}\{c(\mathrm{y}, \boldsymbol{\theta}, \tilde{\boldsymbol{\theta}}_y)\}$. Denoting this approximation as $2d_{DIC}$, the DIC is defined as

$$\mathrm{DIC} = -2\log f(\mathbf{y}\,|\,\tilde{\boldsymbol{\theta}}_y) + 2d_{DIC}. \quad (13.32)$$

Spiegelhalter et al. (2002) provide several exact forms and approximations for the posterior expectation of $c(\mathrm{y}, \boldsymbol{\theta}, \tilde{\boldsymbol{\theta}}_y)$, which apply within specific contexts (i.e., to particular densities and prior distributions). However, the term can also be approximated by making use of Markov Chain Monte-Carlo (MCMC) methods as discussed below.

13.3.2.1 Derivation of the DIC

Before taking the expectation with respect to the posterior $g(\boldsymbol{\theta}|\mathbf{y})$, we first rewrite the optimism term. Let $D(\mathbf{a}|\mathbf{b})$ denote $-2\log f(\mathbf{a}|\mathbf{b})$, such that based on Equation 13.30—and after replacing the pseudotrue $\boldsymbol{\theta}^{*}$ by the random quantity $\boldsymbol{\theta}$—we can express the optimism as

$$\begin{aligned} c(\mathbf{y},\boldsymbol{\theta},\tilde{\boldsymbol{\theta}}_y) &= \mathrm{E}_{f(\mathbf{x}|\boldsymbol{\theta})}[D(\mathbf{x}\,|\,\tilde{\boldsymbol{\theta}}_y)] - D(\mathbf{y}\,|\,\tilde{\boldsymbol{\theta}}_y) \\ &= \mathrm{E}_{f(\mathbf{x}|\boldsymbol{\theta})}[D(\mathbf{x}\,|\,\tilde{\boldsymbol{\theta}}_y) - D(\mathbf{x}\,|\,\boldsymbol{\theta})] \\ &\quad + \mathrm{E}_{f(\mathbf{x}|\boldsymbol{\theta})}[D(\mathbf{x}\,|\,\boldsymbol{\theta}) - D(\mathbf{y}\,|\,\boldsymbol{\theta})] \\ &\quad + D(\mathbf{y}\,|\,\boldsymbol{\theta}) - D(\mathbf{y}\,|\,\tilde{\boldsymbol{\theta}}_y). \end{aligned} \quad (13.33)$$

To approximate the first expectation on the right-hand side of Equation 13.33, we make use of a second-order Taylor expansion of $D(\mathbf{x}|\tilde{\boldsymbol{\theta}}_y)$ around $\boldsymbol{\theta}$; that is,

$$\begin{aligned} D(\mathbf{x}|\tilde{\boldsymbol{\theta}}_y) &\approx -2\log f(\mathbf{x}\,|\,\boldsymbol{\theta}) \\ &\quad -2\left\{\frac{\partial \log f(\mathbf{x}\,|\,\boldsymbol{\theta})}{\partial \boldsymbol{\theta}}\right\}^T (\tilde{\boldsymbol{\theta}}_y - \boldsymbol{\theta}) \\ &\quad -(\tilde{\boldsymbol{\theta}}_y - \boldsymbol{\theta})^T \left\{\frac{\partial^2 \log f(\mathbf{x}|\boldsymbol{\theta})}{\partial \boldsymbol{\theta}\partial \boldsymbol{\theta}^T}\right\} (\tilde{\boldsymbol{\theta}}_y - \boldsymbol{\theta}). \end{aligned} \quad (13.34)$$

Since the first term on the right-hand side of Equation 13.34 is $D(\mathbf{x}|\boldsymbol{\theta})$ we can approximate the first expectation on the right-hand side of Equation 13.33 with

$$\begin{aligned} &\mathrm{E}_{f(\mathbf{x}|\boldsymbol{\theta})}[D(\mathbf{x}|\tilde{\boldsymbol{\theta}}_y) - D(\mathbf{x}\,|\,\boldsymbol{\theta})] \\ &\approx -2\mathrm{E}_{f(\mathbf{x}|\boldsymbol{\theta})}\left[\left\{\frac{\partial \log f(\mathbf{x}\,|\,\boldsymbol{\theta})}{\partial \boldsymbol{\theta}}\right\}^T\right] (\tilde{\boldsymbol{\theta}}_y - \boldsymbol{\theta}) \\ &\quad -(\tilde{\boldsymbol{\theta}}_y - \boldsymbol{\theta})^T \mathrm{E}_{f(\mathbf{x}|\boldsymbol{\theta})}\left[\frac{\partial^2 \log f(\mathbf{x}\,|\,\boldsymbol{\theta})}{\partial \boldsymbol{\theta}\partial \boldsymbol{\theta}^T}\right] (\tilde{\boldsymbol{\theta}}_y - \boldsymbol{\theta}). \end{aligned} \quad (13.35)$$

The first expectation on the right-hand side of Equation 13.35 can be rewritten such that

$$\begin{aligned}
\mathrm{E}_{f(\mathbf{x}|\boldsymbol{\theta})}\left[\frac{\partial \log f(\mathbf{x}|\boldsymbol{\theta})}{\partial \boldsymbol{\theta}^T}\right]
&= \mathrm{E}_{f(\mathbf{x}|\boldsymbol{\theta})}\left[\left\{\frac{1}{f(\mathbf{x}|\boldsymbol{\theta})}\right\}\left\{\frac{\partial f(\mathbf{x}|\boldsymbol{\theta})}{\partial \boldsymbol{\theta}^T}\right\}\right]\\
&= \int\left\{\frac{1}{f(\mathbf{x}|\boldsymbol{\theta})}\right\}\left\{\frac{\partial f(\mathbf{x}|\boldsymbol{\theta})}{\partial \boldsymbol{\theta}^T}\right\} f(\mathbf{x}|\boldsymbol{\theta})d\mathbf{x}\\
&= \int\left\{\frac{\partial f(\mathbf{x}|\boldsymbol{\theta})}{\partial \boldsymbol{\theta}^T}\right\}d\mathbf{x}
\end{aligned} \tag{13.36}$$

$$\begin{aligned}
&= \frac{\partial}{\partial \boldsymbol{\theta}^T}\int f(\mathbf{x}|\boldsymbol{\theta})d\mathbf{x}\\
&= \frac{\partial}{\partial \boldsymbol{\theta}^T}1,
\end{aligned} \tag{13.37}$$

where the step from Equation 13.36 to Equation 13.37 is based on a standard result from score statistics. From this it is clear that the first expectation on the right-hand side of Equation 13.35 is equal to zero.

Hence, the expression in Equation 13.35 can be approximated using

$$\begin{aligned}
&\mathrm{E}_{f(\mathbf{x}|\boldsymbol{\theta})}[D(\mathbf{x}|\tilde{\boldsymbol{\theta}}_y) - D(\mathbf{x}|\boldsymbol{\theta})]\\
&\approx -(\tilde{\boldsymbol{\theta}}_y - \boldsymbol{\theta})^T \mathrm{E}_{f(\mathbf{x}|\boldsymbol{\theta})}\left[\frac{\partial^2 \log f(\mathbf{x}|\boldsymbol{\theta})}{\partial \boldsymbol{\theta}\partial \boldsymbol{\theta}^T}\right] \times (\tilde{\boldsymbol{\theta}}_y - \boldsymbol{\theta})\\
&= \mathrm{tr}\left\{(\tilde{\boldsymbol{\theta}}_y - \boldsymbol{\theta})^T \mathrm{E}_{f(\mathbf{x}|\boldsymbol{\theta})}\left[-\frac{\partial^2 \log f(\mathbf{x}|\boldsymbol{\theta})}{\partial \boldsymbol{\theta}\partial \boldsymbol{\theta}^T}\right] \times (\tilde{\boldsymbol{\theta}}_y - \boldsymbol{\theta})\right\}\\
&= \mathrm{tr}\{\mathbf{I}(\boldsymbol{\theta})(\tilde{\boldsymbol{\theta}}_y - \boldsymbol{\theta})(\tilde{\boldsymbol{\theta}}_y - \boldsymbol{\theta})^T\},
\end{aligned} \tag{13.38}$$

where $\mathbf{I}(\boldsymbol{\theta})$ is the (usual) Fisher information matrix for both $\mathbf{x}$ and $\mathbf{y}$.

As the Bayesian estimate $\tilde{\boldsymbol{\theta}}_y$ we consider the posterior mean $\bar{\boldsymbol{\theta}}_y$. In addition, by making use of the good model assumption, we can replace $\mathbf{I}(\boldsymbol{\theta})$ in Equation 13.38 by the observed Fisher information matrix at the estimated parameter values (i.e., the negative Hessian of the log likelihood evaluated at $\bar{\boldsymbol{\theta}}_y$), which we denote as $\hat{\mathbf{I}}(\bar{\boldsymbol{\theta}}_y) = -\partial^2 \log f(\mathbf{y}|\bar{\boldsymbol{\theta}}_y)/\partial\boldsymbol{\theta}\partial\boldsymbol{\theta}^T$. If we substitute Equation 13.38 in Equation 13.33 and take the expectation with respect to the posterior distribution $g(\boldsymbol{\theta}|\mathbf{y})$, we can approximate the posterior expectation of the optimism with

$$\begin{aligned}
&\mathrm{E}_{g(\boldsymbol{\theta}|\mathbf{y})}[c(\mathbf{y},\boldsymbol{\theta},\bar{\boldsymbol{\theta}}_y)]\\
&\approx \mathrm{tr}\{\hat{\mathbf{I}}(\bar{\boldsymbol{\theta}}_y)\Lambda\}\\
&\quad + \mathrm{E}_{g(\boldsymbol{\theta}|\mathbf{y})}\{\mathrm{E}_{f(\mathbf{x}|\boldsymbol{\theta})}[D(\mathbf{x}|\boldsymbol{\theta}) - D(\mathbf{y}|\boldsymbol{\theta})]\}\\
&\quad + \mathrm{E}_{g(\boldsymbol{\theta}|\mathbf{y})}[D(\mathbf{y}|\boldsymbol{\theta}) - D(\mathbf{y}|\bar{\boldsymbol{\theta}}_y)],
\end{aligned} \tag{13.39}$$

where $\Lambda = E_{g(\boldsymbol{\theta}|\mathbf{y})}[(\bar{\boldsymbol{\theta}}_y - \boldsymbol{\theta})(\bar{\boldsymbol{\theta}}_y - \boldsymbol{\theta})^T]$ denotes the posterior covariance matrix of $\boldsymbol{\theta}$.

Spiegelhalter et al. (2002) suggest that the second term on the right-hand side of Equation 13.39 can be eliminated, based on the following argumentation. Let $f(\mathbf{y})$ be the (fully) marginal likelihood, which is obtained by $f(\mathbf{y}) = \int f(\mathbf{y},\boldsymbol{\theta})d\boldsymbol{\theta}$. Note that $f(\mathbf{y},\boldsymbol{\theta}) = f(\mathbf{y})g(\boldsymbol{\theta}|\mathbf{y}) = h(\boldsymbol{\theta})f(\mathbf{y}|\boldsymbol{\theta})$. We abbreviate $\mathrm{E}_{f(\mathbf{x}|\boldsymbol{\theta})}[D(\mathbf{x}|\boldsymbol{\theta}) - D(\mathbf{y}|\boldsymbol{\theta})]$ to $Q(\mathbf{y},\boldsymbol{\theta})$. Taking the expectation of the second term on the right-hand side of Equation 13.39 with respect to $f(\mathbf{y})$, we can write

$$\begin{aligned}
&\mathrm{E}_{f(\mathbf{y})}(\mathrm{E}_{g(\boldsymbol{\theta}|\mathbf{y})}\{Q(\mathbf{y},\boldsymbol{\theta})\})\\
&= \int f(\mathbf{y})\left(\int g(\boldsymbol{\theta}|\mathbf{y})Q(\mathbf{y},\boldsymbol{\theta})\,d\boldsymbol{\theta}\right)d\mathbf{y}\\
&= \int\int f(\mathbf{y},\boldsymbol{\theta})Q(\mathbf{y},\boldsymbol{\theta})\,d\boldsymbol{\theta}\,d\mathbf{y}\\
&= \int h(\boldsymbol{\theta})\left(\int f(\mathbf{y}|\boldsymbol{\theta})Q(\mathbf{y},\boldsymbol{\theta})\,d\mathbf{y}\right)d\boldsymbol{\theta}\\
&= \mathrm{E}_{h(\boldsymbol{\theta})}(\mathrm{E}_{f(\mathbf{x}|\boldsymbol{\theta})}[D(\mathbf{x}|\boldsymbol{\theta})] - \mathrm{E}_{f(\mathbf{y}|\boldsymbol{\theta})}[D(\mathbf{y}|\boldsymbol{\theta})]).
\end{aligned} \tag{13.40}$$

Clearly, since **y** and **x** are both generated by $p(\cdot)$, the latter is equal to zero. Therefore, Spiegelhalter et al. (2002) drop the second term on the right-hand side of Equation 13.39. In addition, they indicate that they hope this term will cancel out when comparing different models, but indicate this needs further investigation (Spiegelhalter et al., 2002, p. 605). Stone (2002, p. 621) indicates in his commentary to Spiegelhalter et al. (2002) that taking the expectation with respect to $f(\mathbf{y})$ looks "suspicious."

To show that the first term on the right-hand side of Equation 13.39 is approximately equal to the last term on the right-hand side of Equation 13.39, we make use of a second-order Taylor expansion of $D(\mathbf{y}|\boldsymbol{\theta})$ around $\bar{\boldsymbol{\theta}}_y$; that is,

$$D(\mathbf{y}\,|\,\boldsymbol{\theta}) \approx -2\log f(\mathbf{y}\,|\,\bar{\boldsymbol{\theta}}_y) - 2\left\{\frac{\partial f(\mathbf{y}\,|\,\bar{\boldsymbol{\theta}}_y)}{\partial \boldsymbol{\theta}}\right\}^T (\boldsymbol{\theta}-\bar{\boldsymbol{\theta}}_y) -(\boldsymbol{\theta}-\bar{\boldsymbol{\theta}}_y)^T\left\{\frac{\partial^2 \log f(\mathbf{y}\,|\,\bar{\boldsymbol{\theta}}_y)}{\partial\boldsymbol{\theta}\partial\boldsymbol{\theta}^T}\right\}(\boldsymbol{\theta}-\bar{\boldsymbol{\theta}}_y). \tag{13.41}$$

Taking the expectation of Equation 13.41 with respect to the posterior distribution, we can write

$$\begin{aligned}\mathrm{E}_{g(\boldsymbol{\theta}|\mathbf{y})}[D(\mathbf{y}\,|\,\boldsymbol{\theta})] &\approx D(\mathbf{y}\,|\,\bar{\boldsymbol{\theta}}_y) + \mathrm{E}_{g(\boldsymbol{\theta}|\mathbf{y})}\left[\mathrm{tr}\left\{-\frac{\partial^2 \log f(\mathbf{y}\,|\,\bar{\boldsymbol{\theta}}_y)}{\partial\boldsymbol{\theta}\partial\boldsymbol{\theta}^T}(\boldsymbol{\theta}-\bar{\boldsymbol{\theta}}_y)(\boldsymbol{\theta}-\bar{\boldsymbol{\theta}}_y)^T\right\}\right]\\ &= D(\mathbf{y}\,|\,\bar{\boldsymbol{\theta}}_y) + \mathrm{tr}\{\hat{\mathbf{I}}(\bar{\boldsymbol{\theta}}_y)\Lambda\}.\end{aligned} \tag{13.42}$$

Subtracting $D(\mathbf{y}\,|\,\bar{\boldsymbol{\theta}}_y)$ from both sides of Equation 13.42 results in

$$\mathrm{tr}\{\hat{\mathbf{I}}(\bar{\boldsymbol{\theta}}_y)\Lambda\} \approx \mathrm{E}_{g(\boldsymbol{\theta}|\mathbf{y})}[D(\mathbf{y}\,|\,\boldsymbol{\theta})] - D(\mathbf{y}\,|\,\bar{\boldsymbol{\theta}}_y), \tag{13.43}$$

which shows that the first term on the right-hand side of Equation 13.39 is approximately equal to the last term on the right-hand side of Equation 13.39. Substitution of this result in Equation 13.39 results in an expression for the posterior expectation of the optimism; that is,

$$\begin{aligned}&\mathrm{E}_{g(\boldsymbol{\theta}|\mathbf{y})}[c(\mathbf{y},\boldsymbol{\theta},\bar{\boldsymbol{\theta}}_y)]\\ &\quad\approx 2\mathrm{E}_{g(\boldsymbol{\theta}|\mathbf{y})}[D(\mathbf{y}\,|\,\boldsymbol{\theta})] - 2D(\mathbf{y}\,|\,\bar{\boldsymbol{\theta}}_y).\end{aligned} \tag{13.44}$$

From this it follows that the DIC as defined in Equation 13.32 is an approximation of the posterior expected loss as defined in Equation 13.31, with

$$\begin{aligned}d_{DIC} &= \mathrm{E}_{g(\boldsymbol{\theta}|\mathbf{y})}[D(\mathbf{y}\,|\,\boldsymbol{\theta})] - D(\mathbf{y}\,|\,\bar{\boldsymbol{\theta}}_y)\\ &= \overline{D(\mathbf{y}\,|\,\boldsymbol{\theta})} - D(\mathbf{y}\,|\,\bar{\boldsymbol{\theta}}_y).\end{aligned} \tag{13.45}$$

The term d_{DIC} can be easily computed using an MCMC procedure, as is discussed below.

13.3.2.2 Applying DIC to MLM

By entering the expression obtained for d_{DIC} in Equation 13.45 into the DIC defined in Equation 13.32 we get

$$\begin{aligned}\mathrm{DIC} &= D(\mathbf{y}\,|\,\bar{\boldsymbol{\theta}}_y) + 2\{\overline{D(\mathbf{y}\,|\,\boldsymbol{\theta})} - D(\mathbf{y}\,|\,\bar{\boldsymbol{\theta}}_y)\}\\ &= 2\overline{D(\mathbf{y}\,|\,\boldsymbol{\theta})} - D(\mathbf{y}\,|\,\bar{\boldsymbol{\theta}}_y).\end{aligned} \tag{13.46}$$

Hence, the issue becomes to find $\overline{D(\mathbf{y}\,|\,\boldsymbol{\theta})}$ and $D(\mathbf{y}\,|\,\bar{\boldsymbol{\theta}}_y)$. These quantities can be estimated using the results from an MCMC method. Let $\boldsymbol{\theta}^{(1)},\ldots,\boldsymbol{\theta}^{(Q)}$ be Q draws from $g(\boldsymbol{\theta}|\mathbf{y})$ obtained with an MCMC method. Then we can estimate

$$\overline{D(\mathbf{y}\,|\,\boldsymbol{\theta})} \approx \frac{1}{Q}\sum_{q=1}^{Q} -2\log f(\mathbf{y}\,|\,\boldsymbol{\theta}^{(q)}), \quad (13.47)$$

and

$$D(\mathbf{y}\,|\,\bar{\boldsymbol{\theta}}_y) \approx -2\log f(\mathbf{y}\,|\,\frac{1}{Q}\sum_{q=1}^{Q}\boldsymbol{\theta}^{(q)}). \quad (13.48)$$

Note that the number of samples Q does not depend on (observed) sample size, and hence the estimates in Equation 13.47 and Equation 13.48 will converge to their true values if Q is chosen large enough.

The possibility to either maintain the random effects—resulting in a cluster or conditional DIC—or integrate them out—resulting in a population or marginal DIC—was already suggested by Spiegelhalter et al. (2002), and the discussants of their paper. However, when using MCMC methods for MLM, the random effects are typically sampled along with the other parameters in the model, and are thus kept in the model. In that case, use of the Laird-Ware model requires us to know the posterior distribution $g(\boldsymbol{\theta}|\mathbf{y})$ of the parameters in $\boldsymbol{\mu}$, $\boldsymbol{\alpha}$, σ^2, and $\boldsymbol{\Omega}$. It can be shown that this posterior distribution is proportional to the product of the density of the data (i.e., the conditional likelihood in Equation 13.8), the prior distributions of $\boldsymbol{\mu}$, $\boldsymbol{\alpha}$, σ^2, and $\boldsymbol{\Omega}$; that is

$$g(\boldsymbol{\mu}, \boldsymbol{\alpha}, \sigma^2, \boldsymbol{\Omega}\,|\,\mathbf{y}) \propto f(\mathbf{y}\,|\,\boldsymbol{\mu}, \boldsymbol{\alpha}, \sigma^2)h(\boldsymbol{\mu})h(\sigma^2)f(\boldsymbol{\alpha}\,|\,\boldsymbol{\Omega})h(\boldsymbol{\Omega}). \quad (13.49)$$

The distribution at the second level; that is, $f(\boldsymbol{\alpha}|\boldsymbol{\Omega}) = \Pi_i f(\boldsymbol{\alpha}_i|\boldsymbol{\Omega})$, is as defined in Equation 13.9, and is referred to as the prior. The remaining distributions—$h(\boldsymbol{\mu})$ and $h(\sigma^2)$, and the distribution for the level two parameters; that is, $h(\boldsymbol{\Omega})$—are referred to as hyperprior distributions, or simply as the third level. Note that the parameters that define the hyperprior distributions have to be specified by the user. Common choices for the hyperprior distributions, such that they are vague, are: $h(\boldsymbol{\mu}) \sim N(\boldsymbol{\mu}_0, \mathbf{S}_0)$, with large diagonal elements for $\mathbf{S}_0$; $h(\sigma^2) \sim \chi^{-2}(\nu_0, t_0^2)$, with ν_0 chosen small; and $h(\boldsymbol{\Omega}) \sim W^{-1}(\delta_0, \mathbf{T}_0)$, with δ_0 chosen as small as possible (i.e., equal to the number of rows of $\boldsymbol{\Omega}$).

In a recent paper Celeux, Robert, and Titterington (2007) indicated that multilevel models can also be thought of as missing data problems, where the random effects $\boldsymbol{\alpha}_i$ have not been observed. They discussed three different focuses that can be chosen when dealing with missing data: the cluster focus based on $f(\mathbf{y}|\boldsymbol{\mu}, \boldsymbol{\alpha}, \boldsymbol{\Omega}, \sigma^2)$; the population focus based on $f(\mathbf{y}|\boldsymbol{\mu}, \sigma^2, \boldsymbol{\Omega})$; and a novel focus, which they refer to as the joint focus based on $f(\mathbf{y}, \boldsymbol{\alpha}|\boldsymbol{\mu}, \boldsymbol{\Omega}, \sigma^2)$. For each focus Celeux et al. (2007) present multiple formulations of the DIC, showing that within the same focus, one may obtain different values for the DIC, while DICs from different focuses may lead to the same value of the DIC. As of yet it is unclear how these different DICs perform in the context of MLM.

13.3.3 BIC

Thus far we have discussed a frequentist approach to model selection, which is based on minimizing the K-L criterion (AIC), and a Bayesian approach based on minimizing the frequentist loss (DIC). An alternative Bayesian model selection approach consists of choosing the model M_k (from a set of K models), which has the largest posterior model probability $p(M_k|\mathbf{y})$. The *posterior model probability* for model M_K is defined as

$$p(M_k \mid \mathbf{y}) = \frac{f_k(\mathbf{y})p(M_k)}{\sum_{j=1}^{K} f_j(\mathbf{y})p(M_j)}, \quad (13.50)$$

where $f_k(\mathbf{y})$ is the marginal model probability of the data for model M_K and $p(M_K)$ is the prior probability for model M_K. Assuming that the prior model probabilities $p(M_K)$ are equal for all models, finding the model with the largest posterior probability is the same as choosing the model with the largest marginal model probability $f_k(\mathbf{y})$.

For ease of presentation we drop the model subscript k hereafter. It can be shown that −2 times the *log of the marginal model probability of the data* that is, log $f(\mathbf{y})$ can be approximated by the BIC (Congdon, 2005; Raftery, 1995), which is defined as

$$\text{BIC} = -2\log f(\mathbf{y} \mid \hat{\boldsymbol{\theta}}_y) + d\log(n), \quad (13.51)$$

where $-2\log f(\mathbf{y}|\hat{\boldsymbol{\theta}}_y)$ is the measure of misfit, $\hat{\boldsymbol{\theta}}_y$ is the maximum likelihood estimator, and d log(n) is a penalty term based on the dimensionality of the model d and sample size n. The derivation below closely follows Raftery (1995).

13.3.3.1 Derivation of the BIC

The marginal model probability of the data $f(\mathbf{y})$ in Equation 13.50 is obtained by integrating the joint distribution $f(\mathbf{y}, \boldsymbol{\theta})$ over the parameter space $\boldsymbol{\theta}$; that is,

$$f(\mathbf{y}) = \int f(\mathbf{y}, \boldsymbol{\theta})d\boldsymbol{\theta} = \int f(\mathbf{y} \mid \boldsymbol{\theta})h(\boldsymbol{\theta})d\boldsymbol{\theta}, \quad (13.52)$$

where $f(\mathbf{y}|\boldsymbol{\theta})$ is the likelihood of the data and $h(\boldsymbol{\theta})$ is the prior distribution of the parameters $\boldsymbol{\theta}$. Because the marginal model probability of the data is obtained by integrating the parameters out, it is also referred to as the integrated or averaged likelihood, or as the marginal likelihood.[6]

To rewrite the expression in Equation 13.52, we first obtain the second-order Taylor approximation of log $f(\mathbf{y}, \boldsymbol{\theta})$ around the posterior mode $\breve{\boldsymbol{\theta}}$; that is,

$$\begin{aligned}\log f(\mathbf{y},\boldsymbol{\theta}) &\approx \log f(\mathbf{y},\breve{\boldsymbol{\theta}}) + \left\{\frac{\partial \log f(\mathbf{y},\breve{\boldsymbol{\theta}})}{\partial \boldsymbol{\theta}}\right\}^T (\boldsymbol{\theta}-\breve{\boldsymbol{\theta}}) \\ &+ \frac{1}{2}(\boldsymbol{\theta}-\breve{\boldsymbol{\theta}})^T \left\{\frac{\partial^2 \log f(\mathbf{y},\breve{\boldsymbol{\theta}})}{\partial \boldsymbol{\theta}\partial \boldsymbol{\theta}^T}\right\}(\boldsymbol{\theta}-\breve{\boldsymbol{\theta}}). \end{aligned} \quad (13.53)$$

Since $\breve{\boldsymbol{\theta}}$ is the posterior mode, the second term on the right-hand side of Equation 13.53 equals zero. Taking the exponential of

[6] Note that this is not the same as the likelihood denoted in Equation 13.10, which is associated with the population focus: Although the latter is also referred to as the marginal likelihood sometimes—because it is obtained by integrating the random effects out—it still depends on model parameters; in contrast, the marginal model probability of the data used here does not depend on any parameters (i.e., it is completely marginal), and it is not restricted to the context of MLM.

Equation 13.53 and substituting it into Equation 13.52 gives

$$f(\mathbf{y}) \approx \int f(\mathbf{y},\breve{\boldsymbol{\theta}}) \exp\left[\frac{1}{2}(\boldsymbol{\theta}-\breve{\boldsymbol{\theta}})^T \left\{\frac{\partial^2 \log f(\mathbf{y},\breve{\boldsymbol{\theta}})}{\partial\boldsymbol{\theta}\partial\boldsymbol{\theta}^T}\right\}(\boldsymbol{\theta}-\breve{\boldsymbol{\theta}})\right] d\boldsymbol{\theta}$$
$$= f(\mathbf{y},\breve{\boldsymbol{\theta}}) \int \exp\left[-\frac{1}{2}(\boldsymbol{\theta}-\breve{\boldsymbol{\theta}})^T \mathbf{A}(\boldsymbol{\theta}-\breve{\boldsymbol{\theta}})\right] d\boldsymbol{\theta}, \quad (13.54)$$

where $\mathbf{A}$ equals the negative second-order derivative of $\log f(\mathbf{y},\breve{\boldsymbol{\theta}})$, and is referred to as the precision.[7]

The integrand in Equation 13.54 can be rewritten making use of the Laplace methods of integrals. Let d be the number of parameters in $\boldsymbol{\theta}$. Then using the well-known expression for the cumulative multivariate normal distribution function, we can write

$$\int 2\pi^{-d/2} |\mathbf{A}^{-1}|^{-1/2} \exp\left[-\frac{1}{2}(\boldsymbol{\theta}-\breve{\boldsymbol{\theta}})^T \mathbf{A}(\boldsymbol{\theta}-\breve{\boldsymbol{\theta}})\right] d\boldsymbol{\theta} = 1, \quad (13.55)$$

such that

$$\int \exp\left[-\frac{1}{2}(\boldsymbol{\theta}-\breve{\boldsymbol{\theta}})^T \mathbf{A}(\boldsymbol{\theta}-\breve{\boldsymbol{\theta}})\right] d\boldsymbol{\theta} = 2\pi^{d/2} |\mathbf{A}^{-1}|^{1/2}. \quad (13.56)$$

Since $|\mathbf{A}^{-1}|^{1/2} = |\mathbf{A}|^{-1/2}$, using the result from Equation 13.56 in Equation 13.54 gives

$$f(\mathbf{y}) \approx f(\mathbf{y},\breve{\boldsymbol{\theta}})(2\pi)^{\frac{d}{2}} |\mathbf{A}|^{-\frac{1}{2}}. \quad (13.57)$$

Taking the log of the marginal likelihood approximation in Equation 13.57 renders

$$\log f(\mathbf{y}) \approx \log f(\mathbf{y}\,|\,\breve{\boldsymbol{\theta}}) + \log h(\breve{\boldsymbol{\theta}}) + \frac{d}{2}\log(2\pi) - \frac{1}{2}\log|\mathbf{A}|. \quad (13.58)$$

In samples with large n, $\breve{\boldsymbol{\theta}} = \hat{\boldsymbol{\theta}}_y$; that is, the posterior mode is the same as the maximum likelihood estimator, and $\mathbf{A} \approx n\mathbf{J}$, where $\mathbf{J}$ is the expected Fisher information matrix *for one observation*. In that case, $|\mathbf{A}| \approx n^d |\mathbf{J}|$ and substituting this in Equation 13.58 gives

$$\log f(\mathbf{y}) \approx \log f(\mathbf{y}\,|\,\hat{\boldsymbol{\theta}}_y) + \log h(\hat{\boldsymbol{\theta}}_y) + \frac{d}{2}\log(2\pi) - \frac{d}{2}\log(n) - \frac{1}{2}\log|\mathbf{J}|. \quad (13.59)$$

Suppose that the prior $h(\boldsymbol{\theta})$ is a multivariate normal density with mean $\hat{\boldsymbol{\theta}}_y$ and covariance matrix $\mathbf{J}^{-1}$, meaning it contains the same amount of information as a single observation (Raftery, 1995). Then the prior distribution in Equation 13.59 can be written as

$$h(\hat{\boldsymbol{\theta}}_y) = 2\pi^{-d/2} |\mathbf{J}|^{1/2} \exp\left[-\frac{1}{2}(\hat{\boldsymbol{\theta}}_y - \hat{\boldsymbol{\theta}}_y)^T \mathbf{J}(\hat{\boldsymbol{\theta}}_y - \hat{\boldsymbol{\theta}}_y)\right], \quad (13.60)$$

such that

$$\log h(\hat{\boldsymbol{\theta}}_y) = -\frac{d}{2}\log(2\pi) + \frac{1}{2}\log|\mathbf{J}|. \quad (13.61)$$

[7] Note that since this second-order derivative is based on the joint distribution (i.e., of the data and the parameters), it differs from the second-order derivatives encountered in the derivations of the AIC and the DIC, which were based on $f(\mathbf{y}|\boldsymbol{\theta})$.

Substituting this result in Equation 13.59 gives

$$\log f(\mathbf{y}) \approx \log f(\mathbf{y} \mid \hat{\boldsymbol{\theta}}_y) - \frac{d}{2}\log(n). \quad (13.62)$$

Multiplying the expression in Equation 13.62 by −2 results in the BIC defined in Equation 13.51.

13.3.3.2 Applying BIC to MLM

In the derivation of the AIC and DIC a hypothetical cross-validation data set was used, which clearly showed the nature of inferences that can be made when using them. That is, the cross-validation data set was either conditional on the random effects such that inferences can be made to other observations from the same clusters, or it was independent of the random effects such that inferences can be made to observations from other clusters in the same population.

Since the derivation of the BIC is not based on the use of a cross-validation data set, it is less obvious which inferential focus is used. Note however that the BIC is an approximation of $-2\log f(\mathbf{y})$, where $f(\mathbf{y})$ is the marginal probability of the data. Kass and Raftery (1995) indicate that $\log f(\mathbf{y})$ may be viewed as a predictive score, since $\log f(\mathbf{y}) = \Sigma_i \log f(y_i|y_{i-1}, \ldots, y_1)$. This shows that out-of-sample predictions are made *without relying on parameter values.* For this reason, some have suggested that, within the context of MLM, inferences based on the BIC go beyond the population from which the current sample was taken.[8] For example, if one has observed pupils (level 1) in schools (level 2) in a specific country, countries may be thought of as the third level for which only one case is observed. Using the marginal probability of the data—as is done implicitly when using the BIC—implies inferences are made to *observations from other countries,* because not only the cluster effects $\boldsymbol{\alpha}$, but also the parameters that define the current population (i.e., $\boldsymbol{\mu}$ and $\boldsymbol{\Omega}$) are integrated out.

Another issue in using the BIC for MLM is the sample size n, which is used in the penalty term. As a general rule, Raftery (1995) defines n as the scalar that makes the approximation $|\hat{\mathbf{I}}(\breve{\boldsymbol{\theta}}| \approx n^d \, |\mathbf{J}|$, used in the derivation of Equation 13.59, most accurate. In MLM there are several numbers that could be considered as the sample size. First, one could consider it to be equal to the number of clusters m. Second, one could consider the number of responses in each cluster n_i. Third, it could be the total number of responses, represented by $\Sigma_{i=1}^{m} n_i$. In practice, all three "sample sizes" are being used (DeLeeuw, 2004; Vermunt, 2004), but as shown in the illustration below most MLM packages use the third option as n.

13.3.4 Conclusion

In this section the derivations of the AIC, DIC, and BIC were given and their use in the context of MLM has been discussed. The DIC and conditional AIC can be used if the purpose is to make inferences to observations from the same clusters. The marginal AIC can be used if one wishes to make inferences to observations from different clusters that come from the same population. With respect to the BIC it has been suggested

[8] See the WinBUGS Web site http://www.mrc-bsu.cam.ac.uk/bugs/winbugs/dicpage.shtml and the slides by David Spiegelhalter, http://www.mrc-bsu.cam.ac.uk/bugs/winbugs/DIC-slides.pdf

that—despite its similar appearance to the marginal AIC—it can be used for making inferences to observations from different populations.

While the derivations of the three information criteria in this section have emphasized their different backgrounds, several authors have also pointed out connections between these three measures. We already mentioned the asymptotic equivalence between the AIC and the DIC for approximately normal likelihoods with uninformative priors (Spiegelhalter et al., 2002; see also Kuha, 2004). Note that whereas the derivation of the AIC is based on the estimated relative K-L criterion $E_{p(\mathbf{x})}[\log f(\mathbf{x}|\hat{\boldsymbol{\theta}}_y)]$—where the expectation is taken with respect to $p(\cdot)$—the DIC is based on expected loss $E_{f(\mathbf{x}|\boldsymbol{\theta}^*)}[\log f(\mathbf{x}|\tilde{\boldsymbol{\theta}}_y)]$—where the expectation is taken with respect to the best candidate of $f(\cdot|\boldsymbol{\theta})$, which minimizes the K-L criterion $I[p(\cdot), f(\cdot|\boldsymbol{\theta})]$. This implies that if one uses the expected loss to compare a number of models, typically the expectation will be taken with respect to a different density for each model (unless $f_k(\cdot|\boldsymbol{\theta}_k^*) = f_{k'}(\cdot|\boldsymbol{\theta}_{k'}^*)]$, where $k \neq k'$).

With respect to the AIC and the BIC, Burnham and Anderson (2002, pp. 302–305) have shown that the Akaike weights can be thought of as posterior model probabilities in a Bayesian context as defined in Equation 13.50, albeit with different model priors $p(M_k)$ than the ones used in the derivation of the BIC. Moreover, they also show how the BIC can be derived as a non-Bayesian result using the K-L distance. While these are interesting general connections between the AIC and BIC, it is yet unclear how to interpret this in the light of the different inferential focuses that are associated with the (marginal) AIC and the BIC in the context of MLM.

13.4 EMPIRICAL EXAMPLE

To illustrate the use of information criterion for model selection in MLM, we make use of five different software packages: MLwiN, a program specifically developed for MLM; SPSS, which contains the module "Linear mixed-effects modeling;" M*plus*, a flexible structural equation modeling package that also allows for MLM restricted to two levels; R, which is a freely available statistical package and for which we made use of the lme-package for the multilevel analyses; and WinBUGS, which is also freely available and can be used for Bayesian estimation and model evaluation. Note that although SPSS and R offer the option of restricted maximum likelihood, we made use of full maximum likelihood estimation in the first four programs to ensure that the results are comparable.

The data used here are included in MLwiN under the title "tutorial.ws," and consist of a selection from a large data set concerning the examination results from six inner London education authorities. The data include 4,059 students clustered in 65 schools. We will predict the children's normalized exam performance from a number of other variables described below. The main research question in this section is whether the child's gender and/or the gender of the school have an influence on the normalized exam score after correcting for differences in verbal reasoning abilities.

Using i as school index and j as subject index (with $j = 1, \ldots, n_i$), model 1 (M_1) with school as a random effect and verbal reasoning as a covariate can be written as $y_{ij} = \mu_0 + w_{ij,1}\mu_1 + w_{ij,2}\mu_2 + \alpha_i + \varepsilon_{ij}$, where

$w_{ij,1}$ is a dummy variable to indicate whether the verbal reasoning score of individual j in school i is in the top 25% ($w_{ij,1} = 1$) or not ($w_{ij,1} = 0$), and $w_{ij,2}$ is a dummy variable to identify whether the individual's score is in the lowest 25% ($w_{ij,2} = 1$) or not ($w_{ij,2} = 0$). Note that this implies that the reference group consists of the middle 50% (identified by $w_{ij,1} = w_{ij,2} = 0$). We can also present this as the Laird–Ware model,

$$\begin{bmatrix} y_{i1} \\ y_{i2} \\ \cdots \\ y_{in_i} \end{bmatrix} = \begin{bmatrix} 1 & w_{i1,1} & w_{i1,2} \\ 1 & w_{i2,1} & w_{i2,2} \\ \cdots & & \\ 1 & w_{in_i,1} & w_{in_i,2} \end{bmatrix} \begin{bmatrix} \mu_0 \\ \mu_1 \\ \mu_2 \end{bmatrix} + \begin{bmatrix} 1 \\ 1 \\ \cdots \\ 1 \end{bmatrix} [\alpha_i] + \begin{bmatrix} \varepsilon_{i1} \\ \varepsilon_{i2} \\ \cdots \\ \varepsilon_{in_i} \end{bmatrix}. \tag{13.63}$$

The other models consist of extensions of M_1 with additional fixed effects. Since these extensions are straightforward, we will not give their Laird–Ware representation. In M_{2a} the school's gender is included as a fixed effect. To this end two dummies are used, $w_{ij,3}$ and $w_{ij,4}$, which identify students in boys' schools and girls' schools, respectively (i.e., $w_{ij,3} = 1$ and $w_{ij,4} = 0$ implies the child is in a boys' school; $w_{ij,3} = 0$ and $w_{ij,4} = 1$ implies the child is in a girls' school; and $w_{ij,3} = w_{ij,4} = 0$ implies the child is in a mixed school). This results in $y_{ij} = \mu_0 + w_{ij,1}\mu_1 + w_{ij,2}\mu_2 + w_{ij,3}\mu_3 + w_{ij,4}\mu_4 + \alpha_i + \varepsilon_{ij}$. Instead of including the school's gender, M_{2b} consists of including the child's gender as a fixed effect by having a dummy variable $w_{ij,5}$ to identify girls (i.e., $w_{ij,5} = 1$ implies the child is a girl, $w_{ij,5} = 0$ implies it is a boy). Hence, the model becomes $y_{ij} = \mu_0 + w_{ij,1}\mu_1 + w_{ij,2}\mu_2 + w_{ij,5}\mu_5 + \alpha_i + \varepsilon_{ij}$. Note that while M_1 is nested under both M_{2a} and M_{2b}, the latter two are not nested. In M_3 we include both the child's gender and the school's gender as fixed effects, so we have $y_{ij} = \mu_0 + w_{ij,1}\mu_1 + w_{ij,2}\mu_2 + w_{ij,3}\mu_3 + w_{ij,4}\mu_4 + w_{ij,5}\mu_5 + \alpha_i + \varepsilon_{ij}$.

The results for these four models obtained with MLwiN, SPSS, M*plus* and R are given in Table 13.1. As can be seen, all criteria favor M_{2b}; that is, child's gender and not school gender influenced the normalized examination score. The estimates of the parameters in M_{2b} are displayed in Table 13.2. This shows that on average girls score .18 higher than boys on the standardized exam score, after

TABLE 13.1

Results Obtained With MlwiN, SPSS, M*plus*, and R for Four Nonnested Models

Model	$-2\log f(y\mid\hat{\theta}_y)$	d_{mar}	*AIC*	*BIC*
1	9645.2	5	9655.2	9686.7
2a	9638.4	7	9652.3	9696.5
2b	9615.2	6	9627.1	9665.0
3	9611.3	8	9627.3	9677.0

Notes: Data consist of 4059 children clustered in 65 schools.
The dependent variable is normalized exam performance. Model 1 contains a random intercept and two dummies to indicate whether the child's verbal reasoning score is in the top or bottom 25%; in model 2a two dummies are added to indicate whether the child is going to a girls' school, boys' school, or a mixed school; in model 2b, instead of the school's gender, the child's gender is added; in model 3 both the school's gender and the child's gender are included. The BIC uses sample size n = 4059, which is total number of children (number of schools is 65).

controlling for differences in verbal reasoning and school.

Note however that the AIC values of M_{2b} and M_3 are rather close. Transformation of these values to the ratio defined in Equation 13.3 renders $AIC^*_{2b}/AIC^*_3 = \exp(-1/2\{9627.1 - 9627.3\}) = 1.1$; that is, according to the Akaike weight the relative support for M_{2b} and M_3 in the data is about equal. In a similar manner the BIC of both models can be transformed to an approximate Bayes factor using Equation 13.4; that is, $BIC^*_{2b}/BIC^*_3 = \exp(-1/2\{9665.0 - 9677.0\}) = 403.4$, which indicates that the support in the data is about 403 times larger for M_{2b} than for M_3.

The results from WinBUGS for the DIC associated with the cluster focus are presented in Table 13.3. Based on the DIC we conclude that both M_{2b} and M_3 are equally likely to have generated the data. Note that the dimensionality d_{DIC} of all four models is much larger than the dimensionality d_{mar} for the marginal AIC and the BIC, which are based on $f(\mathbf{y}|\boldsymbol{\mu}, \boldsymbol{\Omega}, \sigma^2)$. Since the number of clusters m in this example is 65, it makes sense that the dimensions of the models is a little less than $65 + d_{mar}$.

In sum, this example illustrates the following. First, it shows that both nested and nonnested models can be compared using information criteria. Second, it shows that all models can be compared simultaneously, which makes it possible to select the smallest model as the most appropriate model. Third, this example shows how the ratio of two transformed information criteria can be used to determine how much evidence there is for a particular model in comparison to another model. Note, however, that the ratios can lead to

TABLE 13.2

Parameter Estimates and Their Standard Errors as Obtained Within MLwiN, SPSS, M*plus*, and R for M_{2b}

θ	$\hat{\theta}$	SE
μ_1	−.23	.05
μ_2	.82	.03
μ_3	−.79	.04
μ_5	.18	.03
ω	.10	.02
σ^2	.60	.01

Notes: μ_1 is the mean of boys in the group with the middle 50% of the verbal reasoning score; μ_2 is the difference between children with the highest 25% verbal reasoning scores and children in the middle 50%; μ_3 is the difference between children with the lowest 25% verbal reasoning scores and children in the middle 50%; μ_5 is the difference between the boys and girls; ω is the variance of the random intercept; and σ^2 is the residual variance.

TABLE 13.3

Results Obtained With WinBUGS for Four Nonnested Models Using the Cluster Focus

Model	$\overline{D(y,\theta)}$	$D(y,\bar{\theta}_y)$	d_{DIC}	DIC
1	9497.18	9436.52	60.66	9557.85
2a	9497.05	9436.77	60.27	9557.32
2b	9470.10	9408.94	61.15	9531.25
3	9470.18	9408.85	61.33	9531.52

Notes: Data consist of 4,059 children clustered in 65 schools.
Model 1 contains a random intercept and two dummies to indicate whether the child's verbal reasoning score is in the top or bottom 25%; in model 2a two dummies are added to indicate whether the child is going to a girls' school, boys' school, or a mixed school; in model 2b, instead of the school's gender, the child's gender is added; in model 3 both the school's gender and the child's gender are included.

quite different conclusions, depending on whether they are based on the (marginal) AIC or the BIC.

13.5 DISCUSSION

In this chapter we have discussed model selection in the context of MLM, based on three information criteria (i.e., AIC, BIC, and DIC). In order to provide more insight into the philosophical and mathematical backgrounds of these three information criteria, we have presented a derivation for each of them. In addition, we have discussed the inferential focuses that can be taken when using information criteria in MLM and we have illustrated these by using five different software packages for MLM on an empirical data set.

The question arises what information criterion should be used when involved in MLM. While there is no simple answer to this question, there are several issues one should keep in mind. First, information criteria from different inferential focuses should not be mixed when comparing models. In particular, one should not compare the marginal AIC from one model to the conditional AIC of another model (Vaida & Blanchard, 2005). Second, the researcher should decide whether inferences are to be made to the clusters, the population, or even beyond the (current) population. While most models can be compared using any of these focuses, there is an important limitation: Vaida and Blanchard (2005) point out that if one wishes to compare a random intercept model to a standard ANOVA, one is restricted to the cluster focus, because the standard ANOVA is based on considering the clusters fixed.

Our illustration has shown that by choosing specific software, we may also be limiting our choice of inferential focus: The DIC obtained with WinBUGS is compatible with the cluster focus, while the AIC from MLwiN, SPSS, R, and M*plus* is based on the population focus (i.e., it is the marginal AIC). The BIC that is obtained with these programs appears to be very similar to the marginal AIC, but it is suggested that the focus associated with it goes beyond the current population.

In a broader context—not restricted to MLM—it can be stated that the AIC and the BIC have a different target model. The AIC aims at finding the model that minimizes the K-L criterion and it does not require the true model to be in the set of models. Even if the true model is in the set of models, it may not be the best model in terms of the AIC, because it may not lead to the best predictions of future observations. Moreover, if we are considering a world in which the truth is infinitely dimensional—such that the fitted models are simplifications of the truth—and there are many tapering effects, then the "best" model actually depends on the sample size (Burnham & Anderson, 2002, p. 298; Forster, 2001). Hence, within this framework it makes sense to select more complicated models as sample size increases, which is exactly what AIC does.

In contrast, the BIC aims at selecting the model with the largest posterior probability and it has been shown that asymptotically BIC selects with certainty 1 the true model if it is in the set of models. Hence, what should be considered the best model within the framework of the BIC does not depend on sample size (Burnham & Anderson, 2002), at least not when the true model is among the fitted models.

This difference in target model explains why simulation studies may give conflicting results when comparing the performance of AIC and BIC: Depending on the setup of the study—aspects such as whether or not the true model that generated the data is included in the set of fitted models and the sample sizes that are considered—one may conclude that AIC outperforms BIC or vice versa (Burnham & Anderson, 2002; Forster, 2001; Kuha, 2004; Weaklim, 2004). Hence, Weaklim (2004) stated that the choice between AIC and BIC depends in part on the researcher's belief in the plausibility that certain effects are completely nonexisting (i.e., hypotheses of the form $\theta = 0$). If this is considered likely to be true, the BIC should be preferred, but if one assumes that most effects are never completely nonexisting, it is better to use the AIC, as "The AIC weeds out only those parameters that are so poorly estimated that they detract from the predictive power of the model" (Weaklim, 2004, p. 183).

Another approach to model selection, which is advocated by Kuha (2004), is based on multiple information criteria (possibly in combination with NHT), rather than choosing one measure over the other. If all criteria point to the same model, one can be fairly confident that this is the best model in the set. However, if the measures lead to different conclusions, Kuha (2004) suggests that instead of trying to select a single model, the researcher should select a subset of models from the entire set and present these as the best models. In such cases, and in cases where the best model is not much better than the second best model (as we saw for the AIC in our example), one may also consider the option of multimodel inference, which is also referred to as model averaging (e.g., Burnham & Anderson, 2002; Madigan & Raftery, 1994; Raftery, 1995). This consists of weighting the parameter estimates from different models by their (posterior) model probabilities as obtained in Equation 13.5. However, despite the fact that model averaging has been advocated for quite some time in Bayesian statistics, it is a relatively new topic in frequentistic statistics, and it is likely to take some time before this alternative approach to model selection and prediction becomes accepted by the mainstream.

ACKNOWLEDGMENTS

This work was supported by the Netherlands Organization for Scientific Research (NWO), VENI grant 451-05-012 awarded to E. L. Hamaker, and VICI grant 453-05-002 awarded to H. Hoijtink.

REFERENCES

Akaike, H. (1973). Information theory and an extension of the maximum likelihood principle. In B. N. Petrov & F. Caski (Eds.), *Proceedings of the second international symposium on information theory.* Budapest, Hungary: Akademiai Kaido.

Berger, J. O. (1985). *Statistical decision theory and Bayesian analysis* (2nd ed.). New York, NY: Springer.

Bozdogan, H. (1987). Model selection and Akaike's criterion (AIC): The general theory and its analytic extensions. *Psychometrika, 52,* 345–370.

Bryk, A. S., & Raudenbush, S. W. (1992). *Hierarchical linear models: Applications and data analysis methods* (2nd ed.). Newbury Park, CA: Sage Publications.

Burnham, K. P., & Anderson, D. R. (2002). *Model selection and multimodel inference: A practical information-theoretic approach* (2nd ed.). New York, NY: Springer.

Burnham, K. P., & Anderson, D. R. (2004). Multilevel inference: Understanding AIC and BIC in model selection. *Sociological Methods and Research, 33,* 261–304.

Celeux, G., Robert, C., & Titterington, D. (2007). Deviance information criteria for missing data models. *Bayesian Data Analysis, 1,* 651–706.

Congdon, P. (2005). *Bayesian models for categorical data.* New York, NY: John Wiley.

Cox, D. R. (1962). Further results on tests of separate families of hypotheses. *Journal of the Royal Statistical Society, Series B, 24,* 406–424.

DeLeeuw, J. (2004). Book review. *Journal of Educational Measurement, 41,* 73–77.

Draper, D. (1995). Assessment and propagation of model uncertainty. *Journal of the Royal Statistical Society, Series B, 57,* 45–97.

Forster, M. R. (2001). The new science of simplicity. In A. Zellner, H. A. Keuzekamp, & M. McAleer (Eds.), *Simplicity, inference and modeling: Keeping it sophisticatedly simple* (pp. 83–119). Cambridge, UK: University Press.

Hodges, J. S., & Sargent, D. J. (2001). Counting degrees of freedom in hierarchical and other richly-parameterized models. *Biometrika, 88,* 367–379.

Hoeting, J., Madigan, D., Raftery, A, & Volinsky, C. (1999). Bayesian model averaging. *Statistical Science, 14,* 382–401.

Hox, J. (2002). *Multilevel analysis: Techniques and applications.* Mahwah, NJ: Lawrence Erlbaum.

Kass, R., & Raftery, A. (1995). Bayes factors. *Journal of the American Statistical Association,* 90, 773–795.

Kreft, I., & de Leeuw, J. (1998). *Introducing multilevel modeling.* London, UK: Sage.

Kuha, J. (2004). AIC and BIC: Comparisons of assumptions and performance. *Sociological Methods and Research, 33,* 188–229.

Laird, N. M., & Ware, J. H. (1982). Random-effects models for longitudinal data. *Biometrics, 38,* 963–974.

Longford, N. T. (1995). *Random coefficients models.* New York, NY: Oxford University Press.

Madigan, D., & Raftery, A. E. (1994). Model selection and accounting for model uncertainty in graphical models using Occam's window. *Journal of the American Statistical Association, 89,* 1535–1546.

Raftery, A. E. (1995). Bayesian model selection in social research. *Sociological Methodology, 25,* 111–163.

Schwarz, G. (1978). Estimating the dimension of a model. *The Annals of Statistics, 6,* 461–464.

Snijders, T., & Bosker, R. (1999). *An introduction to basic and advanced multilevel modelling* (1st ed.). London, UK: Sage Publications.

Spiegelhalter, D. J., Best, N. G., Carlin, B. P., & Van der Linde, A. (2002). Bayesian measures of model complexity and fit. *Journal of the Royal Statistical Society, Series B, 64,* 583–639.

Stone, M. (1977). An asymptotic equivalence of choice of model by cross-validation and Akaike's criterion. *Journal of the Royal Statistical Society, Series B, 39,* 39–43.

Stone, M. (2002). Commentary to: Bayesian measures of model complexity and fit. *Journal of the Royal Statistical Society, Series B, 64,* 621.

Vaida, F., & Blanchard, S. (2005). Conditional Akaike information for mixed-effects models. *Biometrika, 92,* 351–370.

Vermunt, J. K. (2004). An EM algorithm for the estimation of parametric and nonparametric hierarchical nonlinear models. *Statistica Neerlandica, 52,* 220–233.

Wagenmakers, E. J. (2007). A practical solution to the pervasive problems of p-values. *Psychonomic Bulletin and Review, 14,* 779–804.

Wagenmakers, E. J., & Farrell, S. (2004). AIC model selection using Akaike weights. *Psychonomic Bulletin and Review, 11,* 192–196.

Weaklim, D. L. (2004). Introduction to the special issue on model selection. *Sociological Methods and Research, 33,* 167–187.

14

Optimal Design in Multilevel Experiments

Mirjam Moerbeek
Department of Methods and Statistics, Utrecht University, Utrecht, The Netherlands

Steven Teerenstra
Department of Epidemiology, Biostatistics and Health Technology Assessment, Radboud University Nijmegen Medical Centre, Nijmegen, the Netherlands

14.1 INTRODUCTION

In the social, behavioral, and biomedical sciences experiments are conducted to evaluate the effect of an experimental treatment condition on particular outcome variables of interest. The randomized controlled trial is generally considered the gold standard for comparing treatment conditions. With this type of trial, subjects are randomly sampled from a population and randomly assigned to treatment conditions. One of the treatment conditions is a control condition existing of either a conventional treatment or no treatment at all. When possible, such a trial is double-blind such that neither the patient nor the experimentalist knows who is assigned to which treatment condition. Data obtained from such trials are often analyzed with traditional linear or logistic regression. These analysis methods provide valid point estimates and statistical tests on the treatment effect if the assumption of independence is met (Moerbeek, van Breukelen, & Berger, 2003).

In trials where humans are the study subjects there is often some degree of interaction between the participants. For instance, this is the case when the subjects in the target population are nested within naturally existing groups, such as schools, families, worksites, or general practices. Measurements on opinion, health, behavior, and attitude of subjects within the same group are likely to be correlated due to mutual influence and group characteristics, such as group policy and the behavior of group leaders. Clustering may also occur in individually randomized trials when subjects are clustered within groups that are established for the purpose of the trial (Lee & Thompson, 2005). An example is a trial where subjects with social phobia are randomized to a control condition consisting of pharmacotherapy or an experimental condition consisting of cognitive behavioral-group therapy. Those randomized to the latter treatment condition are subsequently randomized to different therapy groups. Once a therapy group is established, its members will mutually influence each other and be influenced by the therapist leading the group. Once again, we may expect the response variables of subjects within the same group to be correlated.

The presence of correlated responses complicates the analysis of data obtained from trials with nested data. The correct statistical model to analyze data from such experiments is the multilevel model, which explicitly takes nesting of subjects within groups into account by modeling random effects at the subjects and group level (Goldstein, 2003; Hox, 2002; Raudenbush & Bryk, 2002; Snijders & Bosker, 1999). Ignoring the nested data structure may result in type I or type II errors for the test on treatment effect, and consequently in incorrect conclusions with respect to the effectiveness of the experimental treatment condition. In addition, nesting of subjects within groups has implications for the calculation of the optimal design. Not only the total sample size needs to be calculated, but the optimal allocation of units as well. That is, should we sample many small groups or just a few large groups? Of course, a design with many groups and many subjects per group results in maximal statistical power, but is often not feasible because of limitations on the budget and on the number of subjects that are willing to participate in the trial.

The aim of this chapter is to present sample sizes for trials with nested data such that a desired power level is achieved. It restricts to two treatment conditions (i.e., an experimental and a control condition), two levels of nesting and continuous outcomes. Optimal designs for three levels of nesting and binary outcomes can be found elsewhere (Moerbeek, 2000; Moerbeek & Maas, 2005; Moerbeek, van Breukelen, & Berger, 2000, 2001a). Four types of trials with nested data are considered: cluster randomized trials, multisite trials, pseudocluster randomized trials and trials comparing group and individual treatments. With cluster randomized trials existing groups are randomized to treatment conditions and all subjects within the same group receive the same treatment condition. An example is a school-based smoking prevention intervention where schools are randomized to either the intervention or control condition. With multisite trials both the experimental and the control condition are available in each group that participates in the trial. In the medical sciences this type of trial is often referred to as multi-center clinical trial and is adopted when any single center cannot provide sufficient patients to achieve an acceptable power level. Cluster randomized trials and multisite trials are studied in Section 14.2. Section 14.3 focuses on pseudocluster randomized trials. These are an extension to cluster randomized trials. They were developed to overcome problems that are common when cluster randomized trials are implemented in general practice, namely selection bias and slow recruitment in the general practices that are randomized to the less interesting control condition. Section 14.4 focuses on trials comparing group and individual treatments. With this type of trial, we have nesting of subjects within groups in the group therapy condition but not in the individual therapy condition. For each of these trials, the rationale of the design is given and the advantages and disadvantages are discussed, the multilevel model is presented and the required sample sizes to achieve a desired power level are derived. For cluster randomized trials and the multisite trials we will also consider the optimal number of groups and group size given budgetary constraints. At the end of this chapter some concluding remarks are given.

14.2 CLUSTER RANDOMIZED TRIALS AND MULTISITE TRIALS

14.2.1 Description of Designs

With cluster randomized trials complete groups of subjects are randomized to treatment conditions and all members within a group receive the same treatment, see Donner and Klar (2000) and Murray (1998) for a general introduction. In multisite trials, subjects within a site are randomly assigned to treatment conditions. Figures 14.1 and 14.2 give a graphical representation of these two types of trials. For both types, eight groups with 10 subjects each are displayed. The white subjects represent the control condition and the black subjects the experimental condition. The 10 subjects within a circle are nested within the same group, irrespective of the color with which they are represented. Data on both treatment conditions are available within each group for the multisite trial. For cluster randomized trials, data for only one treatment condition are available within each group, so the other data are missing by design.

Cluster randomized trials are often chosen from political, ethical, administrative, and financial reasons (Gail, Mark, Carroll, & Green, 1996). For instance, a school-based smoking prevention intervention where the experimental condition is

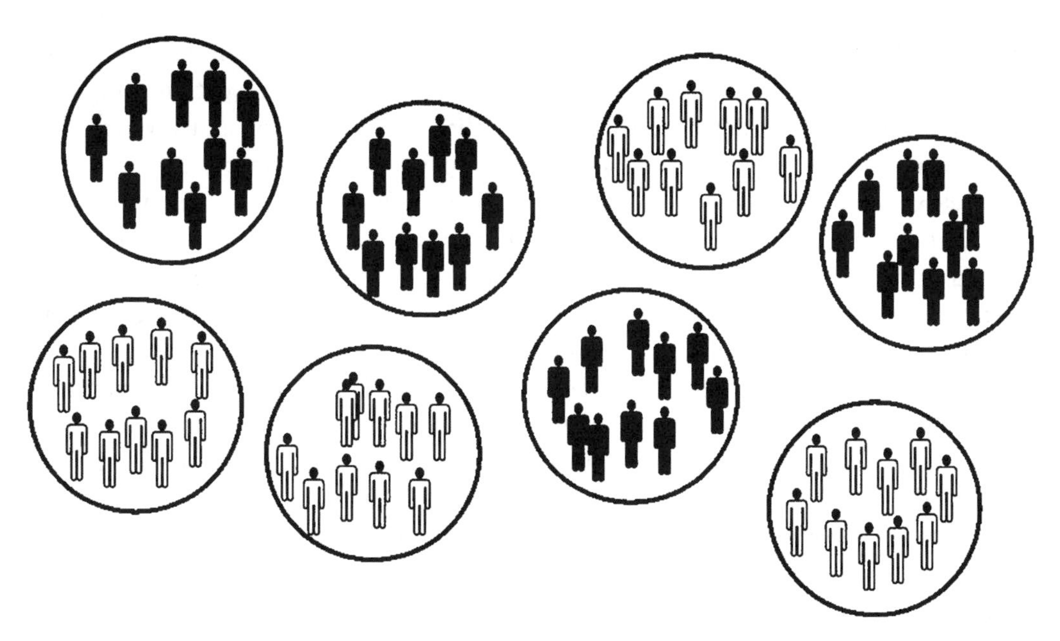

FIGURE 14.1
Graphical representation of a cluster randomized trial.

FIGURE 14.2
Graphical representation of a multisite trial.

delivered to all children within only half of the schools may result in lower travel costs for the health professionals delivering the intervention than a trial where the experimental condition is delivered to only half of the children within each school. In addition, the administrative costs in the first trial may be lower since each child within a school receives the same treatment condition. An example where the choice of the level of randomization may be driven by ethical criteria is a trial on the effects of vitamin-A supplementation on childhood mortality in northern Summatra (Sommer et al., 1986). In such a trial one may choose to randomize complete villages to treatment conditions if it is considered unethical to treat some children within a certain village and withhold the treatment from others. In addition to the reasons given above, cluster randomized trials minimize the probability of control group contamination, which occurs when both the experimental and control condition are available within each group and information on the contents of the experimental condition leaks to the subjects assigned to the control. In some cases, there is no alternative to cluster randomization, such as in community intervention trials where the intervention is designed to affect all members within a community. An example is an intervention using mass media such as television and advertorials in regional or local newspapers.

In contrast to cluster randomized trials, multisite trials are trials where each treatment condition is available within each group. So, randomization to treatment conditions is done at the subject level within each group. Such trials can be shown to be more efficient than cluster randomized trials, and in addition they allow for the estimation of the interaction between treatment and group. Such trials may be a good choice when the magnitude of the treatment effect is expected to vary across groups. On the other hand, they should only be chosen from a practical point of view when the degree of control group contamination is expected to be absent or small. This assumption may be realistic in double blind clinical trials where neither the patient nor the health professional knows who is assigned to which treatment condition, but may not be tenable in trials where the experimental treatment relies on interpersonal communication such as in risk-reduction sessions and peer-pressure groups (Moerbeek, 2005). In other settings, control group contamination may be due to the person delivering the intervention. An example is a trial where general practitioners are given information about new guidelines to reduce unhealthy life styles and both the education and control condition are available in each general practice. It will be difficult for the general practitioner not to let patients in the control condition benefit from the education. In this case, a cluster randomized trial is the only appropriate option.

14.2.2 Multilevel Model

For multisite trials the model that relates treatment condition x_{ij} to outcome y_{ij} of subject i within group j is given by

$$y_{ij} = \beta_{0j} + \beta_{1j}x_{ij} + e_{ij}. \qquad (14.1)$$

The random effect $e_{ij} \sim N(0,\sigma_e^2)$ is the deviate of the outcome of subject i in group j from its group mean. If treatment condition is coded −1 for the control condition and +1 for the experimental condition, then the intercept β_{0j} is the mean outcome in group j and the slope β_{1j} is *half* the difference in mean outcomes between the control and experimental condition in group j.

The intercept and slope may vary across groups, randomly and/or as a function of covariates at the group level. In this chapter we will restrict to models without covariates, but it can be shown that the results also hold for models with covariates at the subject and group level as long as these covariates are uncorrelated with the treatment condition and the treatment condition is coded −1 and +1 (Moerbeek, van Breukelen, & Berger, 2001b). The random intercept is now written as $\beta_{0j} = \beta_0 + u_{0j}$, where β_0 is the overall mean and the random term $u_{0j} \sim N(0,\sigma_{u0}^2)$ is the deviate from this mean for group j. Similarly, $\beta_{1j} = \beta_1 + u_{1j}$, where β_1 is *half* the overall difference in mean outcomes between the control and experimental condition and the random term $u_{1j} \sim N(0,\sigma_{u1}^2)$ is the deviate from this difference in group j. The random effects u_{0j} and u_{1j} may be correlated and their covariance is denoted σ_{u01}. In the remainder of this chapter we assume this correlation to be equal to zero. This does not affect estimation or hypothesis testing of the fixed effects or variances of the random effects as long as the data are balanced and the treatment indicator x_{ij} is centered around zero (Raudenbush, 1993).

Substitution of the expressions for the random intercept and slope into Equation 14.1 results in the single-equation model

$$y_{ij} = \beta_0 + \beta_1 x_{ij} + u_{0j} + u_{1j}x_{ij} + e_{ij}. \qquad (14.2)$$

The random error term at the subject level, e_{ij}, is assumed to be independent from u_{0j} and u_{1j}. As follows from Equation 14.2, the random interaction $u_{1j}x_{ij}$ between treatment and group can be estimated in a multisite trial. That is, the treatment effect is allowed to vary across the groups. The mean treatment effect is given by $\delta = 2\beta_1$, and this parameter is often of particular interest in experiments.

For cluster randomized trials only one treatment condition is available within each group, so the variability of the treatment effect across the groups cannot be estimated. Equation 14.2 reduces to

$$y_{ij} = \beta_0 + \beta_1 x_j + u_j + e_{ij}. \tag{14.3}$$

The variance components σ^2_{u0} and σ^2_{u1} cannot be estimated separately. Instead, their sum $\sigma^2_u = \sigma^2_{u0} + \sigma^2_{u1}$ is estimated. As for multisite itrials, the random effects $u_j \sim N(0, \sigma^2_u)$ and $e_{ij} \sim N(0, \sigma^2_e)$ are assumed to be independent of each other, and the treatment effect is given by $\delta = 2\beta_1$.

For both levels of randomization, the total variance of the outcome variable is given by $\text{var}(y_{ij}) = \sigma^2_{u0} + \sigma^2_{u1} + \sigma^2_e$, and the proportion variance at the group level is given by the intraclass correlation coefficient $\rho = (\sigma^2_{u0} + \sigma^2_{u1}) / (\sigma^2_{u0} + \sigma^2_{u1} + \sigma^2_e)$. This parameter measures the proportion in total outcome variance that is at the group level. It ranges between 0 and 1. If $\rho = 0$ then all variability is at the subject level and the outcomes of persons within the same group are no more correlated than outcomes from different groups. If $\rho = 1$ then subjects within the same group respond identically. The intraclass correlation coefficient may be split up into two parts associated with the random intercept and slope: $\rho = \rho_0 + \rho_1$, where $\rho_0 = (\sigma^2_{u0}) / (\sigma^2_{u0} + \sigma^2_{u1} + \sigma^2_e)$ and $\rho_1 = (\sigma^2_{u1}) / (\sigma^2_{u0} + \sigma^2_{u1} + \sigma^2_e)$. For a multisite trial ρ_1 is the correlation between persons in the same group with different treatments and $\rho_0 + \rho_1$ is the correlation between persons in the same group with the same treatment. For a cluster randomized trial $\rho_0 + \rho_1$ is the correlation between persons in the same group.

In the calculations that follow a balanced design is assumed: the total number of groups is n_2 and the group size is constant and denoted n_1. For multisite trials $\frac{1}{2}n_1$ subjects per group are randomized to the experimental condition and the others are randomized to the control condition, assuming n_1 is even. For cluster randomized trials the number of groups per treatment condition is $\frac{1}{2}n_2$, assuming an even total number of groups. Optimal sample sizes given an unbalanced design are presented elsewhere (Liu, 2003; Van Breukelen, Candel, & Berger, 2007).

For both levels of randomization the treatment effect δ is simply estimated by $\hat{\delta} = \bar{y}_e - \bar{y}_c$, where $\bar{y}_e$ and $\bar{y}_c$ are the mean outcomes in the experimental and control condition, respectively. For cluster randomized trials, the variance of this estimator is given by

$$\begin{aligned} \text{var}(\hat{\delta}) &= 4\frac{\sigma^2_e + n_1\sigma^2_u}{n_1 n_2} \\ &= 4\frac{\sigma^2_e + \sigma^2_u}{n_1 n_2}(1 + (n_1 - 1)\rho). \end{aligned} \tag{14.4}$$

The term $1 + (n_1 - 1)\rho$ is called the design effect and gives the number of times a sample size from a simple random sample needs to be multiplied in a cluster randomized trial to have the same level of efficiency. The design effect is equal to 1 or larger, and increases with the group size and the intraclass correlation coefficient. Already for small values of the intraclass correlation coefficient the design effect may be considerable. For instance, it is equal to 1.95 when $\rho = 0.05$ and $n_1 = 20$, meaning that almost twice as many subjects are needed in a cluster randomized trial than in a simple randomized trial to achieve the same efficiency.

For a multisite trial, a similar formula holds:

$$\begin{aligned} \text{var}(\hat{\delta}) &= 4\frac{\sigma^2_e + n_1\sigma^2_{u1}}{n_1 n_2} \\ &= 4\frac{\sigma^2_e + \sigma^2_u}{n_1 n_2}(1 - \rho_0 + (n_1 - 1)\rho_1). \end{aligned} \tag{14.5}$$

The design effect $1 - \rho_0 + (n_1 - 1)\rho_1$ may be less than, equal to, or larger than 1, depending on the values ρ_0, ρ_1, and n_1. Justifications of the formulae (Equations 14.4 and 14.5) are given in Moerbeek et al., 2000; Raudenbush, 1997; and Raudenbush and Liu, 2000.

The variance of the treatment effect estimator, $\text{var}(\hat{\delta})$, plays a key role in the determination of the optimal allocation of units since this variance is inversely related to the power of the test on treatment effect. The size and significance of the treatment effect are generally the main interest in experiments, therefore the variance of its estimator is chosen as sole optimality criteria in this chapter. Of course, in some settings interest may also lie in other parameters than the treatment effect, for instance the intraclass correlation coefficient. If it turns out that the amount of between-school variability in the outcome variable in a school-based health-promotion intervention is large, then it may be worthwhile to identify the schools for which the intervention performs worst and try to describe these in terms of their school characteristics. The intervention can then be adjusted for this type of school. However, in this chapter the variance of the treatment effect estimator is our main objective in optimizing the number of groups and group size. Optimal designs that combine multiple optimality criteria are not part of this chapter but can be found elsewhere (Moerbeek & Wong, 2002).

The multisite trial has lower variance than the cluster randomized trial with the same group size and number of groups. This is also depicted in Figure 14.3, where the relative efficiency of cluster randomized trials and multisite trials is shown as a function of the intraclass correlation coefficient. The relative efficiency is defined as the ratio of the $\text{var}(\hat{\delta})$ as achieved with a multisite trial and the $\text{var}(\hat{\delta})$ as achieved with a cluster randomized trial. The left plot represents that the case treatment by group interaction is absent (i.e., $\rho_1 = 0$), the right panel is an example where this interaction is present ($\rho_1 = 0.05$). The inverse of the relative efficiency gives the number of times the cluster randomized trial needs to be replicated to be as efficient as the multisite trial.

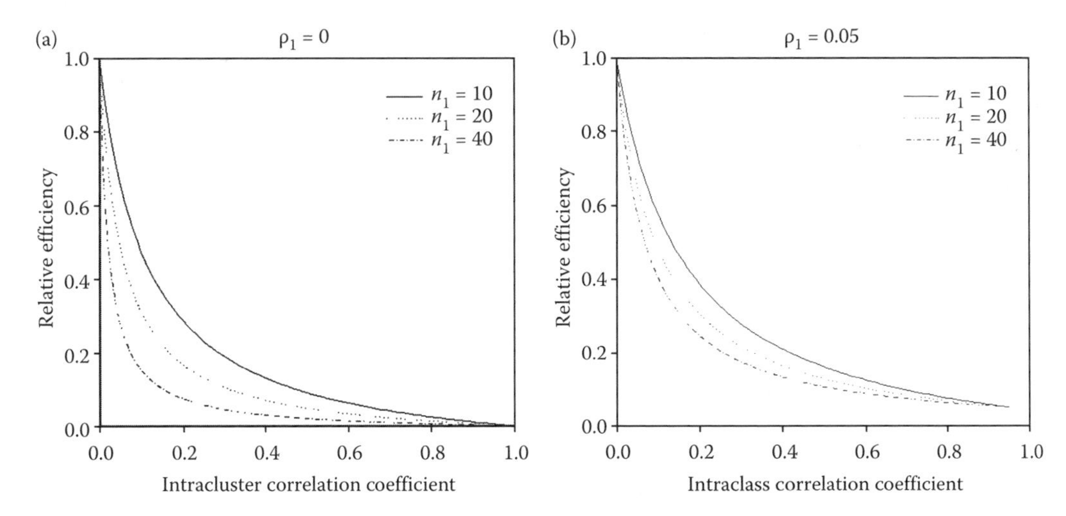

FIGURE 14.3
Relative efficiency of a cluster randomized trial (a) versus a multisite trial (b).

		Unknown truth	
		H_0 true	H_a true
Decision based on statistical test	Retain H_0	Correct decision	Type II error
	Reject H_0	Type I error	Correct decision

FIGURE 14.4
Correct and incorrect decisions in statistical hypothesis testing.

Already for small values of the intraclass correlation coefficient the loss of efficiency may be large. For absent treatment by group interaction the relative efficiency is about 0.66 if $\rho = 0.05$ and $n_1 = 10$. This means that $(0.66^{-1} - 1) \times 100\% = 52\%$ more groups should be added to the cluster randomized trial in order to be as efficient as the multisite trial. Especially for large group sizes and large intraclass correlation coefficients the loss in efficiency may be considerable, meaning that the multisite randomized trial is preferable from a statistical point of view. However, when treatment by group interaction is assumed present, the relative efficiency becomes larger, meaning that the loss in efficiency from using a cluster randomized trial is less severe.

Of course, the choice of the optimal level of randomization should not only be driven by statistical criteria, but by practical considerations as well, as was discussed at the beginning of this section. In many types of trials the experimental condition relies on interpersonal communication and in such trials the risk of control-group contamination may be large. For that reason, the optimal allocation of units is derived for both levels of randomization.

14.2.3 Sample Size and Power

The aim of an experiment is to test whether there is a treatment effect in the population from which the subjects are sampled. This can be done by testing the null hypothesis H_0: $\delta = 0$ against the two-sided alternative H_a: $\delta \neq 0$, where δ is the treatment effect. When appropriate, a one-sided alternative may be used. The test statistic $t = \hat{\delta} / \sqrt{\text{var}(\hat{\delta})}$ has a central t-distribution under the null hypothesis, with n_2–2 degrees of freedom for the cluster randomized trial and $n_2 - 1$ degrees of freedom for the multisite itrial. When the number of groups is large, the standard normal approximation can be used, as will be done in the remainder of this chapter.

In statistical hypothesis testing, two types of errors can be made, see Figure 14.4. A type I error occurs when the null hypothesis is incorrectly rejected. In this case, it is concluded that there is a treatment effect while in reality such an effect does not exist. A type II error is made when the null hypothesis is incorrectly accepted. In this case it is concluded that there is no treatment effect, while in reality such an effect does exist.

The power of a statistical test is the probability to reject the null hypothesis and accept the alternative when a treatment effect does exist. As is obvious, large power levels are desired, and values of 0.8 or 0.9 are generally used in power calculations. The power $1-\gamma$ of a test is related to the type I error rate α, the true treatment effect δ and the variance of the treatment effect estimator $\text{var}(\hat{\delta})$ by the following equation

$$\sqrt{\text{var}(\hat{\delta})} = \frac{\delta}{z_{1-\alpha/2} + z_{1-\gamma}}. \qquad (14.6)$$

Here, $z_{1-\alpha/2}$ and $z_{1-\gamma}$ are the 100(1–α/2)% and 100(1–γ)% standard normal deviates. In case of a one-sided alternative, $z_{1-\alpha/2}$ is replaced by $z_{1-\alpha}$.

The type I error rate α is the probability of incorrectly rejecting the null hypothesis of no treatment effect. Its size may be controlled by the experimenter and should be chosen to reflect the consequences of a type I error. If these consequences are severe, then the type I error should be chosen to be small. It is often set equal to 5%, but values of 10% or 1% may also be appropriate. Larger type I error rates result in larger power levels. This is obvious since the null hypothesis is more likely to be rejected if the type I error rate is large; hence a larger power level is achieved.

The population value of the treatment effect quantifies the true difference between the control and experimental condition with respect to the response variable. As follows from Equation 14.6 a larger treatment effect results in a larger power level. This is obvious, since larger differences between both treatments are easier to detect than small differences. The true value of the treatment effect is generally unknown in the design phase. This introduces a vicious circle: the aim is to conduct an experiment in order to gain insight into the population value of the treatment effect, but in order to design the experiment such that a sufficient power level is guaranteed the true value of this parameter must be known. This problem may be overcome by using the minimal relevant treatment effect in the power calculation. Instead of an absolute treatment effect, one may chose to use a standardized treatment effect, which is defined as the absolute treatment effect divided by the standard deviation of the outcomes. The standardized treatment effect is scale free and a value zero corresponds to the null hypothesis of no treatment effect. Standardized treatment effects of size 0.2, 0.5, and 0.8 are considered small, medium, and large (Cohen, 1992).

The variance of the treatment effect estimator, $\mathrm{var}(\hat{\delta})$, quantifies the error that is made in estimating the treatment effect. As is obvious, larger errors result in lower power levels. The $\mathrm{var}(\hat{\delta})$ can be controlled by the group size n_1 and the number of groups n_2, as follows from the formulae for $\mathrm{var}(\hat{\delta})$ as given in Equations 14.4 and 14.5. By substitution of these equations in Equation 14.6 the required sample sizes can be derived.

In some cases the group size is fixed to n_1. The minimal number of groups n_2 to achieve a power level $1 - \gamma$ in a two-sided test with significance level α is given by

$$n_2 = 4\frac{\sigma_e^2 + n_1\sigma_u^2}{n_1}\left(\frac{z_{1-\alpha/2} + z_{1-\gamma}}{\delta}\right)^2. \qquad (14.7)$$

In other cases the number of groups n_2 is limited by the number of groups willing to participate in the trial. The minimal group size n_1 is then calculated from

$$n_1 = \frac{4\sigma_e^2}{\left(\frac{\delta}{z_{1-\alpha/2} + z_{1-\gamma}}\right)^2 n_2 - 4\sigma_u^2}. \qquad (14.8)$$

These two formulae hold for the cluster randomized trial. In both formulae, the variance component σ_u^2 should be replaced by σ_{u1}^2 if a multisite trial is considered instead of a cluster randomized trial.

Figure 14.5 shows the power to detect a small standardized treatment effect in a cluster randomized trial as a function of the number of groups and the group size in a two-sided test with significance level $\alpha = 0.05$ and an intraclass correlation

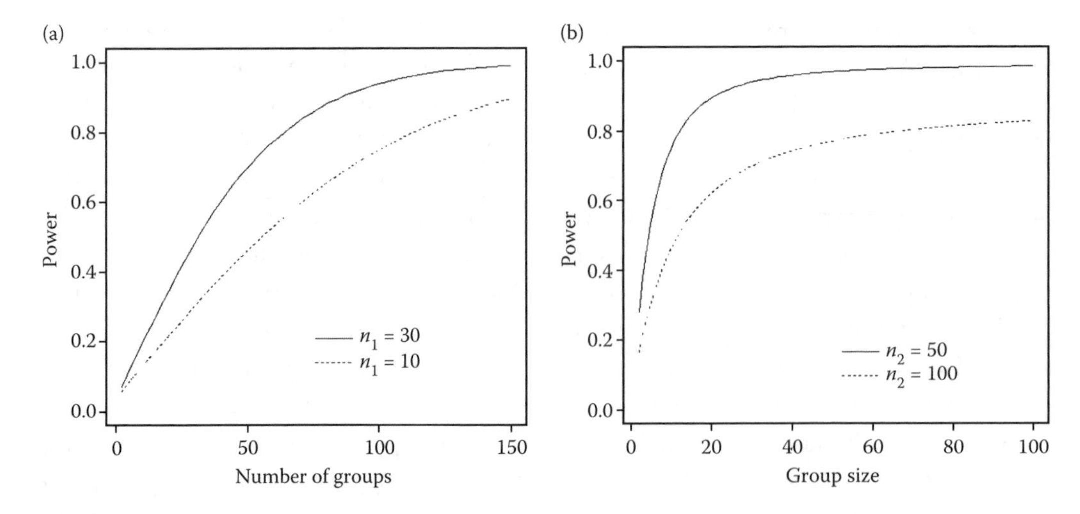

FIGURE 14.5
Power as a function of the number of groups (a) and the group size (b).

coefficient $\rho = 0.05$. As is obvious, power increases with the group size and the number of groups. The left plot in Figure 14.5 shows that the power increases to the value 1 when the number of groups increases and the group size is fixed. The right plot in Figure 14.5 shows that the power increases to a limit not necessarily equal to 1 when the group size increases and the number of groups is fixed. This can be explained by the fact that the number of groups only appears in the denominator of the $\text{var}(\hat{\delta})$ as given by Equation 14.4, whereas the group size appears in both the numerator and the denominator. So, a small number of groups cannot always be compensated by a large group size in order to achieve sufficient statistical power. In such cases one must either increase the number of groups or rely on other methods to improve power, such as matching or prestratification (Murray, 1998; Raudenbush, Martinez, & Spybrook, 2007), the inclusion of covariates (Moerbeek, 2006; Murray & Blitstein, 2003; Raudenbush et al., 2007), or taking repeat measurements (Feldman & McKinlay, 1994; McKinlay, 1994; Murray & Blitstein, 2003).

14.2.4 Optimal Sample Sizes given Budgetary Constraints

In the previous section either the group size or the number of groups was assumed to be fixed in advance and the other was calculated such that a certain power level was achieved. In this section we focus on sample size calculations where neither the group size nor the number of groups is fixed. In particular, we focus on optimal sample sizes given a budgetary constraint. Table 14.1 gives the $\text{var}(\hat{\delta})$, the optimal sample sizes, and $\text{var}(\hat{\delta})$ given these optimal sample sizes.

As follows from Table 14.1, the variance of the treatment effect estimator is a function of the variance components σ_e^2 and σ_{u1}^2 (or the sum σ_u^2 in case of a cluster randomized trial). Furthermore, this variance is a function of the group size n_1 and the number of groups n_2. As is obvious, variance decreases with increasing sample sizes. In practice, sample sizes cannot increase

TABLE 14.1

$\text{var}(\hat{\delta})$. Optimal Sample Sizes, and $\text{var}(\hat{\delta})$ given these Optimal Sample Sizes

Trial	$\text{var}(\hat{\delta})$	Optimal n_1	Optimal n_2	$\text{var}(\hat{\delta})$ Given Optimal n_1, n_2
Cluster randomized trial	$4\dfrac{\sigma_e^2 + n_1\sigma_u^2}{n_1 n_2}$	$\sqrt{\dfrac{c_2\sigma_e^2}{c_1\sigma_u^2}}$	$\dfrac{C}{\sqrt{\dfrac{c_1c_2\sigma_e^2}{\sigma_u^2}} + c_2}$	$4\dfrac{\left(\sqrt{c_1\sigma_e^2} + \sqrt{c_2\sigma_u^2}\right)^2}{C}$
Multisite trial	$4\dfrac{\sigma_e^2 + n_1\sigma_{u1}^2}{n_1 n_2}$	$\sqrt{\dfrac{c_2\sigma_e^2}{c_1\sigma_{u1}^2}}$	$\dfrac{C}{\sqrt{\dfrac{c_1c_2\sigma_e^2}{\sigma_{u1}^2}} + c_2}$	$4\dfrac{\left(\sqrt{c_1\sigma_e^2} + \sqrt{c_2\sigma_{u1}^2}\right)^2}{C}$

without boundaries due to limited financial resources. Therefore, a budgetary constraint is used in the derivation of the optimal allocation of units. This constraint is given by

$$c_1 n_1 n_2 + c_2 n_2 \le C, \qquad (14.9)$$

where the costs to include a person (c_1) are multiplied with the total number of persons, and the costs at the group level (c_2) are multiplied with the total number of groups. These components sum up to the total costs, which should not exceed the budget C that is available for measuring and sampling. For both levels of randomization, the optimal group size n_1 is found by expressing the optimal number of groups n_2 in terms of the costs and budget and group size n_1 using the constraint in Equation 14.9: $n_2 = C/(c_1 n_1 + c_2)$. This expression is then substituted into the $\text{var}(\hat{\delta})$ as given in the second column of Table 14.1 from which the optimal group size n_1 can then be derived.

The optimal allocations of units follow between the third and fourth column of Table 14.1, the $\text{var}(\hat{\delta})$ given the optimal allocation is given in the last column. From Table 14.1 it follows that the number of groups and the group size depend on the variance components as well as the costs at the subject and group level. However, increasing the budget C has an effect on the number of groups but not on the group size. From Table 14.1 we also observe that the optimal group size n_1 increases with the costs ratio c_2/c_1. This is obvious since a design with a few large groups is favored when it is relatively expensive to sample and measure a group. Furthermore, we observe that smaller group sizes are required when the intraclass correlation coefficient is large. This is also obvious since subjects behave more identically when the intraclass correlation is large, thus increasing the group size has a smaller effect on the $\text{var}(\hat{\delta})$ than increasing the number of groups.

The optimal allocation of units as given in Table 14.1 was derived such that the smallest $\text{var}(\hat{\delta})$, and hence largest power, are achieved and the budget C is not exceeded. It can be shown that these sample sizes are also optimal when the budget C to achieve a certain power level is to be minimized.

Figure 14.6 shows the power to detect a small standardized treatment effect in a two-sided test with type I error rate $\alpha = 0.05$. The power is plotted as a function of the group size, the number of groups, and for various budgets. The intraclass correlation coefficient is equal to $\rho = 0.05$, and the costs at the subject and group level are $c_1 = 10$ and $c_2 = 250$. The optimal group size is equal to $n_1 = 22$ and does not depend on the budget. The optimal number of groups is an increasing function of the budget. Power levels of 0.8–0.9 are generally used. For $C = 80{,}000$

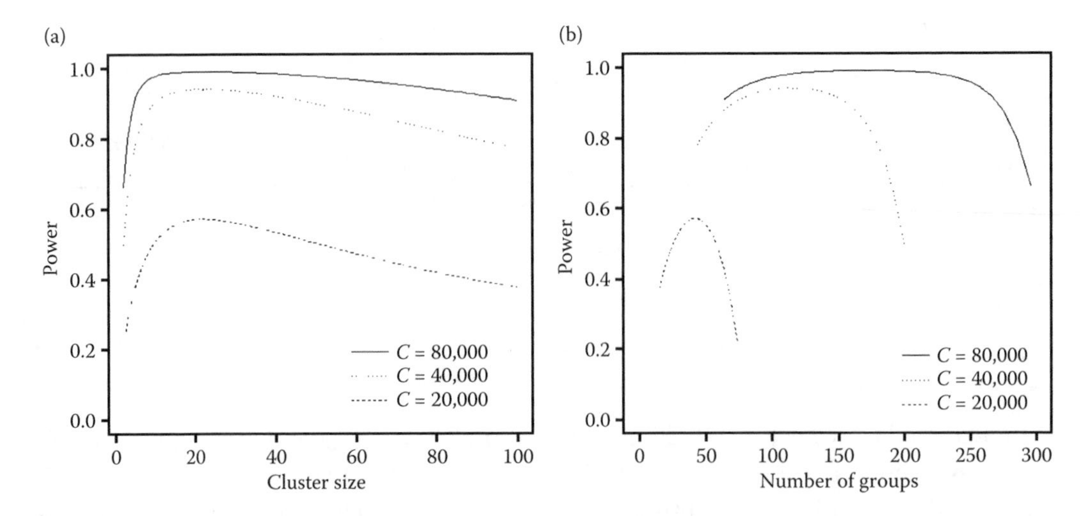

FIGURE 14.6
Power to detect a small standardized treatment effect as a function of the group size (a) and the number of groups (b) for various budgets and $\rho = 0.05$, $c_1 = 10$, $c_2 = 250$.

the maximal power level as achieved with the optimal allocation of units is almost equal to one. In this case, an unnecessarily high sample size is used. For $C = 40{,}000$ the maximal power level is about 0.9 and for $C = 20{,}000$ it is lower than 0.6, which is an unacceptable low power level. With such a small budget, one has to choose between increasing the budget and not conducting the study at all.

Figure 14.6 furthermore shows that the power curves are rather flat around their maximum when the budget is large. In that case, a small deviation from the optimal allocation of units hardly results in a loss of power. This is a useful finding for practical applications where the actual group size is unequal to the optimal group size. For instance, if children are grouped in class sizes of 30 and complete classes are included in a school-based health-promotion intervention study, then the loss in power from using the actual sample size $n_1 = 30$ instead of the optimal sample size $n_1 = 22$ hardly results in a loss of power when the budget C is equal to 40,000 (and $\rho = 0.05$, $c_1 = 10$, $c_2 = 250$).

14.3 PSEUDOCLUSTER RANDOMIZED TRIALS

14.3.1 Description of Design

Pseudocluster randomization combines randomization at the level of the individual and randomization at the level of the cluster. It consists of two steps: first the clusters are randomized into two types, *E* and *C*, and the results of this randomization are not revealed. Then, within each cluster of type *E*, subjects are randomized to either condition in such a way that the majority receives the experimental condition *e* and, therefore, the minority receives the control condition. In the clusters of type *C*, the situation is reversed; that is, the majority of subjects are randomized to the control condition *c* (see Figure 14.7, where the subjects in black and white receive the experimental and control condition, respectively).

Pseudocluster randomization is a compromise between individual randomization and cluster randomization, when neither is preferable over the other. Such a dilemma

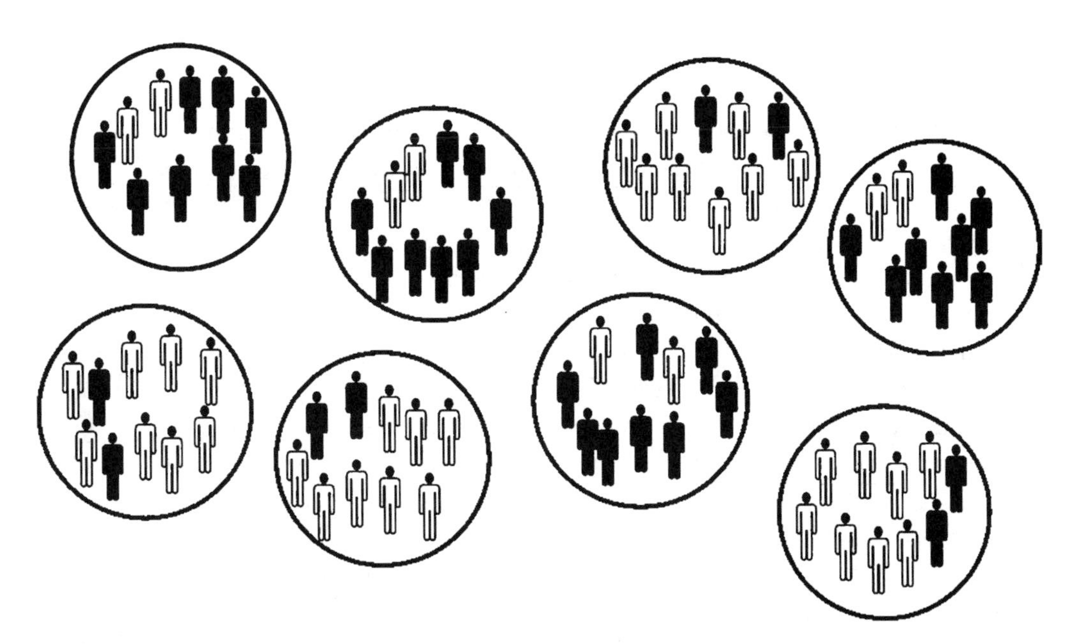

FIGURE 14.7
Graphical representation of a pseudocluster randomized trial.

arises when randomization at the individual level risks contamination, while randomization at cluster level risks selection bias and recruitment problems.

Consider for example a trial that evaluates the effects of a new method for physicians to coach their patients. Randomization on patient (= subject) level is possible, but the physician may consciously or unconsciously mix up elements of the treatments that reduce the contrast between the new coaching method and usual care (contamination). This means that only the contaminated effect can be estimated, which is smaller than the true effect, and that the sample size must be enlarged to have sufficient power for testing this smaller effect (Moerbeek, 2005; Slymen & Hovell, 1997).

To avoid contamination, randomization can be done on physician (= cluster) level (see Section 14.2.2), but this may need a larger sample size (see Figure 14.3). Another disadvantage of cluster randomization is that it is often impossible to recruit patients before randomization of the clusters, for example, because patients are recruited when they consult their physician for a problem. Consequently, the physicians know in advance the treatment their recruited subjects will receive and this may influence their choice of subjects (selection bias), which leads to differences between the treatment groups at baseline (Hahn, Puffer, Torgerson, & Watson, 2005; Jordhoy, Fayers, Ahlner-Elmqvist, & Kaasa, 2002; Puffer, Torgerson, & Watson, 2003). Moreover, advance knowledge of treatment allocation may also influence the rate of recruitment: a physician that is allocated to give a treatment that he or she does not consider to be of much interest, may recruit fewer subjects and produce data of poorer quality (Klar & Donner, 2001; Moore, Summerbell, Vail, Greenwood, & Adamson, 2001).

In a situation as above, pseudocluster randomization can reduce selection bias and contamination, while improving recruitment.

Due to the two-step randomization, the physicians do not know in which type of cluster they are, nor do they know in advance what treatment the next patient will be on. This reduces the chance of selection bias. Moreover, no longer are half of the physicians left with only patients on the "uninteresting" treatment (as in cluster randomization), which may improve recruitment. Finally, in clusters of type E, most patients are on condition e and only a few are on condition c. This means that the contamination of e by c is small. On the other hand, the contamination of c by e may be substantial, but that only affects the few patients on c. Similar considerations apply to clusters of type C.

Upon closer examination, pseudocluster randomization will reduce selection bias to the extent that the predictability of the treatment allocation sequence is reduced. Obviously, the physician can never be sure about which condition will be allocated next, so predictability will always be less than in a cluster randomized design. Nevertheless, the physician may guess how big the chance is of condition e being allocated in his or her cluster. By not revealing the type of cluster the physician is randomized to and by keeping the cluster size small, this effect can be reduced, so that it would be hard for the physician to guess what cluster group he or she is allocated to.

Pseudocluster randomization will reduce contamination if the degree of contamination is proportional to exposure in the following sense: control subjects in a cluster will be less contaminated (or: fewer control subjects will be contaminated) if there are fewer subjects on the experimental condition in that cluster. This is a reasonable assumption if the dissemination of elements of the experimental condition to the control group is a gradual process that depends on the number of subjects on the experimental condition in each cluster. However, it would be an unreasonable assumption if one single patient on the experimental condition will lead to complete contamination of all other patients. For contamination the other way around, which is mostly interpreted as noncompliance, similar considerations apply.

The above two conditions were satisfied in the Dutch EASYcare trial, a pseudocluster randomized study (Melis, van Eijken, & Borm, 2005) on improving the care of elderly with common geriatric problems such as falls or dementia. Usual care by the physician was compared to patient coaching by a specialized geriatric nurse that had expert advice of a geriatrician and regular feedback meetings with the patient's physician. In the EASYcare trial, pseudocluster randomization effectively reduced predictability, as the large majority of physicians thought that a 1:1 randomization ratio was used in their cluster (Melis, Teerenstra, Olde Rikkert, & Borm, 2008). Furthermore, the experimental condition was a complex collaboration of nurse, geriatrician, and physician. Thus, for a physician to copy (elements of) the intervention, not only would he or she have to accumulate passive knowledge through the feedback meetings, but also would he or she need to acquire/learn new skills and attitudes. Therefore, assuming that contamination was proportional to exposure was deemed reasonable. Further evaluation of the performance of pseudocluster randomization in this trial is provided by Melis et al. (2008).

14.3.2 Multilevel Model

The model for the outcome of subject i in cluster j is

$$y_{ij} = \beta_0 + \beta_1 x_{ij} + u_j + e_{ij}, \qquad (14.10)$$

which is similar to the model of the cluster randomized trial (see the formula of Equation 14.3 in Section 14.2.2), except that the treatment indicator x_{ij} varies by subject as there are two conditions in each cluster. The model estimates the effect averaged over all clusters, in particular averaged over both types of clusters. If sufficient information is available (i.e., many clusters are available) the above model can be extended to estimate interaction of the effect by type of cluster and by cluster (so that asymmetrical contamination and variation of the treatment effect over clusters can be assessed).

Assume there are $\frac{1}{2}n_2$ clusters of type *E* and *C*, respectively, and that in each cluster a fraction $f > 0.5$ receives one condition and that the rest (a fraction $(1-f) < 0.5$) receives the other condition; that is,

there are $\frac{1}{2}n_2$ type *E*-clusters of size n_1 with

$$\begin{cases} n_1 f \text{ subjects on treatment } e \text{ ("}Ee\text{ subjects"),} \\ n_1(1-f) \text{ subjects on treatment } c \text{ ("}Ec\text{ subjects"),} \end{cases}$$

there are $\frac{1}{2}n_2$ type *C*-clusters of size n_1 with

$$\begin{cases} n_1(1-f) \text{ subjects on treatment } c \text{ ("}Cc\text{ subjects"),} \\ n_1 f \text{ subjects on treatment } e \text{ ("}Ce\text{ subjects").} \end{cases}$$

Observe that $f = 1$ corresponds to a cluster randomized trial and $f = 0.5$ to a multisite trial. For a pseudocluster randomized trial, the choice of f has to be tailored to each study, but $f = 0.8$ seems to be a good choice in general (Borm, Melis, Teerenstra, & Peer, 2005).

As two types of patients receive the experimental condition *e* (viz., *Ee*- and *Ce*-patients), the effect of condition *e* can be estimated by averaging the mean effect of *Ee*-patients and *Ce*-patients. Let y_j^{Ee} and y_j^{Ce} be the mean outcome of condition *e* in cluster $j = 1, \ldots, \frac{1}{2}n_2$ of type *E* and *C*, respectively. As the *Ce*-patients in a cluster are from a relatively small subset, the estimates y_j^{Ce} may be more contaminated and will at least have less precision than the y_j^{Ee} estimates. Therefore, it may be more efficient to weight the y_j^{Ce} estimates. The mean outcome of condition e is therefore estimated as:

$$\bar{y}_e = \frac{\frac{1}{n_2/2}\sum f y_j^{Ee} + \frac{1}{n_2/2}\sum w(1-f) y_j^{Ce}}{f + w(1-f)}$$

and similarly, the weighted mean outcome of condition *c* is:

$$\bar{y}_c = \frac{\frac{1}{n_2/2}\sum f y_j^{Cc} + \frac{1}{n_2/2}\sum w(1-f) y_j^{Ec}}{f + w(1-f)}.$$

Weights can be chosen such as to maximize the power of the estimator $\hat{\delta} = \bar{y}_e - \bar{y}_c$ (max t-ratio weights). However, calculation of the max t-ratio weights requires knowledge of the (unknown) contamination rates, which limits their applicability at the design and analysis stages. Fortunately, the max t-ratio weights are often close to minimum variance weights (Borm et al., 2005):

$$w = \frac{1 + 2fn_1q}{1 + 2(1-f)n_1q},$$

where $q=\sigma_u^2/\sigma_e^2$ is the ratio of the level 2 and level 1 variance. For this choice of weights, the variance of $\hat{\delta}=\bar{y}_e-\bar{y}_c$ is minimal and equals

$$\text{var}_{\text{min var}} = 4\frac{\sigma_e^2+n_1\sigma_u^2}{n_1 n_2}\cdot\frac{1}{1+4f(1-f)n_1 q}$$
$$=4\frac{\sigma_e^2+\sigma_u^2}{n_1 n_2}\frac{1+(n_1-1)\rho}{1+4f(1-f)n_1\rho/(1-\rho)}. \tag{14.11}$$

In a comprehensive simulation study (Teerenstra, Moerbeek, Melis, & Borm, 2007), various methods of analyzing data from pseudocluster randomized trials were compared including a multilevel model, a covariance pattern model, the generalized estimating equation approach, and a paired t-test. The multilevel model had practically the best power and has the advantage of being readily available in major statistical packages.

14.3.3 Sample Size and Power given Fixed Group Size

The estimator of the treatment effect in the multilevel model is equivalent to the estimator $\bar{y}_e-\bar{y}_c$ for a fixed number of clusters n_2 and a fixed cluster size n_1 (Teerenstra et al., 2007). Therefore, the power can be calculated as in Section 14.2.3 using Equation 14.6 with the substitution of Equation 14.11 for $\text{var}(\hat{\delta})$. This results in the relation

$$n_1 n_2 = 4\frac{\sigma_e^2+n_1\sigma_u^2}{1+4f(1-f)n_1 q}\cdot\left(\frac{z_{1-\alpha/2}+z_{1-\gamma}}{\delta}\right)^2 \tag{14.12}$$

from which either the required cluster size n_1 or number of clusters n_2 can be calculated, given the other (Teerenstra, Melis, Peer, & Borm, 2006). The expected effect δ actually depends on f. For $f=1$, for example, the cluster randomized trial, the contamination will be (ideally) absent and the treatment effect is equal to $\delta_0=\mu_e-\mu_c$, the difference between the expected outcomes in the experimental and control condition in absence of contamination. For $f=0.5$, for example, the multisite trial, cross-exposure of the conditions in each cluster will be maximal, and the contamination of the effect is expected to be maximal. If f is between 0.5 and 1, the expected outcome of condition e in the subset Ee will be $\mu_e - c_{\text{major}}\delta$, where $0\le c_{\text{major}}\le 1$ is the one-sided relative contamination of the majority (the Ee-patients) by the minority (the Ec-patients) in each cluster of type E. In Ce, the expected outcome is $\mu_e - c_{\text{minor}}\delta$. Similarly, the expected outcomes in treatment groups Cc and Ec are $\mu_c + c_{\text{major}}\delta$ and $\mu_c + c_{\text{minor}}\delta$. The expected effect of the weighted estimator $\bar{y}_e-\bar{y}_c$ is

$$\delta=\delta_0-\delta_0\frac{2fc_{\text{major}}+2w(1-f)c_{\text{minor}}}{f+w(1-f)}. \tag{14.13}$$

The relative contamination rates c_{major} and c_{minor} depend on f: when f moves from 0.5 to 1, the contamination c_{major} of the majority will decrease to its minimum while the contamination c_{minor} of the minority will increase to its maximum. However, the total effect of the contamination of majority and minority; that is, the second term in Equation 14.13, depends on the rate of contamination *and* the number of subjects contaminated.

Using Equations 14.12 and 14.13, the relative efficiency of pseudocluster randomization can be compared to that of cluster randomization and individual randomization stratified by cluster (without treatment by cluster interaction).

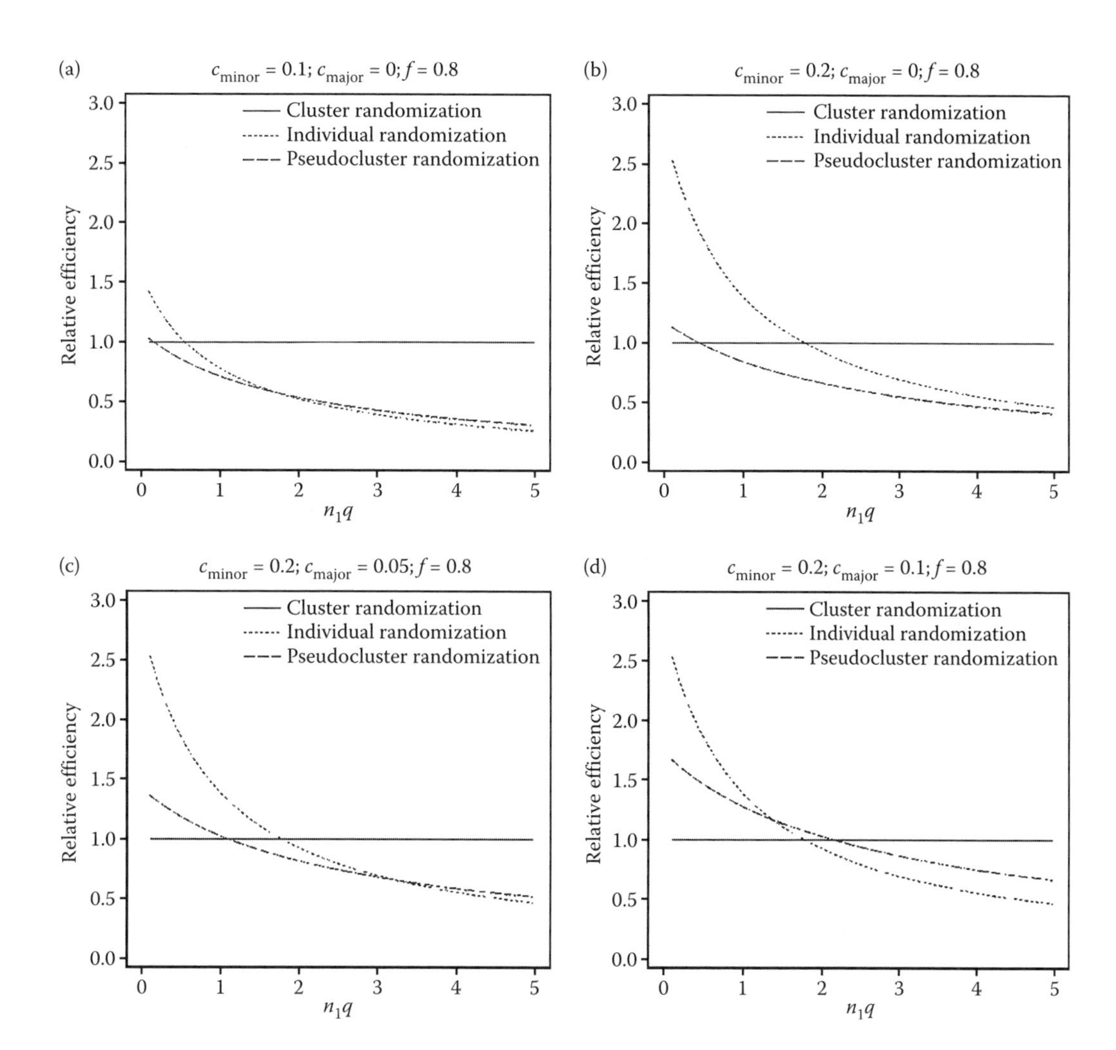

FIGURE 14.8
Relative efficiency of cluster, pseudocluster, and individual randomization for different values of c_{minor} and c_{major} (a–d).

Figure 14.8 shows the relative efficiency of pseudocluster randomization and individual randomization compared to that of cluster randomization (which is set to 1) as a function of $n_1 q = \sigma_u^2/(\sigma_e^2/n_1)$, which is the ratio of the between- and within-variance of the cluster means. The contamination of the majority by a minority in a cluster is set to zero in the upper subplots, while in the lower subplots contamination of the majority by the minority is allowed ($c_{major} = 0.25c_{minor}$ and $c_{major} = 0.5c_{minor}$). In general, cluster randomization is most efficient for small values of $n_1 q$ and individual randomization is better for large values of $n_1 q$. In the range between, pseudocluster generally outperforms both, while its efficiency is close to that of individual randomization for large values of $n_1 q$, especially if in each cluster the contamination of the majority (by the minority) is small compared to the contamination of the minority (by the majority).

In conclusion, the main reasons for applying pseudocluster randomization are methodological (to reduce selection bias compared to a cluster randomized design and/or to reduce contamination compared to an individually randomized design) or practical (to improve recruitment compared to a cluster randomized design). Nevertheless, pseudocluster may also be more efficient than either cluster or individual randomization of which the Dutch EASYcare study is an example (Teerenstra et al., 2006).

14.4 TRIALS COMPARING GROUP AND INDIVIDUAL TREATMENTS

14.4.1 Description of Design

In (pseudo)cluster randomized trials and multisite trials, clustering occurs because individuals are nested within naturally existing groups. Examples of such groups are schools, families, worksites, or general practices. In such cases, the possibility of correlated responses can be easily acknowledged by the experimenter and taken into account in the design and analysis. There can be no excuse for ignoring correlated responses in cluster randomized trials and multisite trials and analysis since computer programs are available for the design (Raudenbush, Bryk, & Congdon, 2004) of such trials and analysis of their data (Bryk, Raudenbush, & Congdon, 1996; Rasbash, Steele, Browne, Goldstein, 2009).

In other types of trials, clustering is less apparent and therefore often not acknowledged or incorrectly ignored. Clustering may occur in individually randomized trials when two types of health professionals are compared, for instance a nurse practitioner and a general practitioner. Patients are randomly assigned to either a nurse practitioner or general practitioner. In general, multiple practitioners are involved and each of them treats more than one patient. Even when there is no interaction among patients being treated by the same practitioner, there can be a practitioner effect. With this type of trial, we have clustering in both arms and we can therefore use sample size formulae for cluster randomized trials as presented in Section 14.3. These can be used when the intraclass correlation coefficient does not vary across the two arms. For more general sample size formulae we refer to Roberts and Roberts (2005).

Another type of an individually randomized trial with clustering effects is a trial comparing group and individual treatments. A graphical representation of this type of trial is given in Figure 14.9. The groups are represented by circles. The subjects represented in black are those in the group therapy and those represented in white are those in the individual therapy. Thus, the trial consists of four groups of size 10 and 40 subjects in the individual therapy. An example is a trial that compares the effectiveness of a smoking cessation group therapy to a control condition that does not offer participation in groups. In this example, subjects are randomly assigned to the group or individual treatment. Those randomized to the group treatment are subsequently randomized to a therapy group. Once these therapy groups are established, the group members within a group will mutually influence each other and be influenced by the therapist leading the group. This results in a group effect; in some groups the members are more likely to quit smoking that in other groups.

The focus of this section is on the latter type of trials. Although such trials are often conducted, there is actually very little research on their design and analysis. This is not surprising since with this type of trial

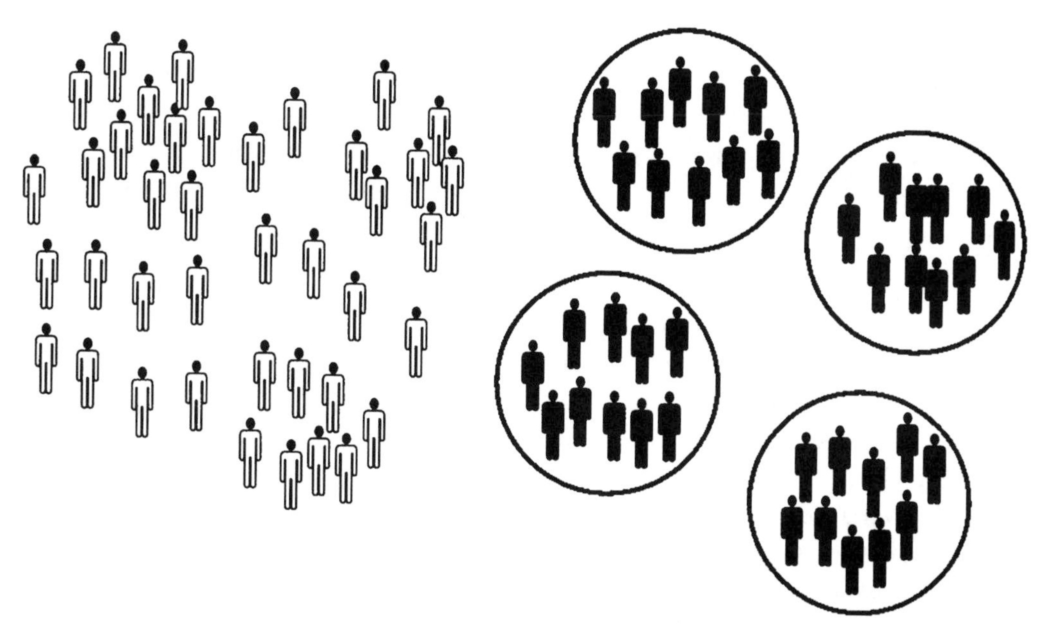

FIGURE 14.9
Graphical representation of a trial comparing group and individual treatments.

we have partially nested data. In the group condition we have subjects nested within groups while in the individual condition subjects are not nested. Experimenters generally tend to ignore this complicated type of data structure and rely on simple t-tests or linear regression analysis and corresponding sample size formulae. This may result in underpowered studies and incorrect conclusions and inflated type I error for the test on treatment effect since the nesting in the group therapy is ignored. Only in the last few years has this type of trial gained attention from statisticians. This section is based on two recent papers (Moerbeek & Wong, 2008; Roberts & Roberts, 2005).

14.4.2 Multilevel Model

The data in trials comparing group and individual treatments are partially nested. We formulate regression equations for each of the two treatments separately and combine these in a single-equation model. We focus on models with continuous outcomes.

For subject $i = 1, \ldots, n$ in the individual condition the outcome variable y_i is related to control (individual) condition x_i by the following equation

$$y_i = \beta_0 + \beta_1 x_i + r_i. \qquad (14.14)$$

Here, $r_i \sim N(0, \sigma_r^2)$ is the random term at the subject level. The error terms are assumed to be independently and identically distributed as the subjects are assumed not to influence each others' outcomes. This assumption is reasonable when there is no mutual influence among subjects and no influence of a therapist treating the subjects. This may be the case, for instance, when the individual treatment consists of a pharmacotherapy treatment or no treatment at all.

For subject $i = 1, \ldots, n_1$ in group $j = 1, \ldots, n_2$ in the experimental (group) condition the relation between outcome y_{ij} and treatment condition x_{ij} is given by

$$y_{ij} = \beta_0 + \beta_1 x_{ij} + u_j + e_{ij}, \quad (14.15)$$

with $u_j \sim N(0,\sigma_u^2)$ and $e_{ij} \sim N(0,\sigma_e^2)$ the random group and individual effect, respectively. These random effects are assumed uncorrelated. As previous, the intraclass correlation coefficient $\rho = \sigma_u^2 / (\sigma_u^2 + \sigma_e^2)$ measures the proportion of variance that is at the group level.

The total variance $\sigma_u^2 + \sigma_e^2$ in the group condition is not necessarily equal to the variance σ_r^2 in the individual condition. Hence, the model allows for heteroscedasticity. The variance ratio is defined as $\theta = (\sigma_u^2 + \sigma_e^2)/\sigma_r^2$. The treatment condition is coded 0 for the individual treatment and 1 for the group treatment. Subsequently, β_0 is the mean outcome in the individual treatment, $\beta_0 + \beta_1$ is the mean outcome in the group treatment and $\delta = \beta_1 - \beta_0$ is the treatment effect.

The two Equations 14.14 and 14.15 can be combined into a single equation by using treatment condition x and its complement $(1 - x)$ as dummy variables:

$$y_{ij} = \beta_0 + \beta_1 x_j + x_j u_j + x_j e_{ij} + (1 - x_j) r_{ij}. \quad (14.16)$$

The computer program MLwiN (Rasbash et al., 2000) allows the incorporation of such dummy variables. It should be noted that the subjects in the individual condition are not nested in a group. This can be solved in the multilevel model by assigning each of them the value $j = k + 1$ to the group indicator, where k is the number of groups in the group condition. In that case, treatment condition only needs to be subscribed by the group indicator j, as is done in Equation 14.16.

The Equation 14.16 can be extended to include covariates. If we assume random assignment of subjects to treatment conditions and large sample sizes, then the covariates are likely to be uncorrelated with treatment condition and in that case the treatment effect is simply estimated by the difference in mean scores of the two treatments: $\hat{\delta} = \bar{y}_e - \bar{y}_c$, where $\bar{y}_e$ and $\bar{y}_i$ are the mean outcomes in the experimental (group) and control (individual) treatment condition, respectively. The variance of this estimator is equal to

$$\begin{aligned} \text{var}(\hat{\delta}) &= \frac{\sigma_e^2 + n_1\sigma_u^2}{n_1 n_2} + \frac{\sigma_r^2}{n} \\ &= \sigma_r^2 \left(\theta \frac{1 + (n_1 - 1)\rho}{n_1 n_2} + \frac{1}{n} \right). \end{aligned} \quad (14.17)$$

This formula holds when all therapy groups have the same size. The first and second terms at the right side are the variances of the mean outcomes in the group and individual condition, respectively. As is obvious, the variance of the mean in the group condition is a function of the design effect $1 + (n_1 - 1)\rho$. The larger the therapy group size and the larger the correlation between members within the same group, the larger the variance of the treatment effect estimator.

14.4.3 Sample Size and Power Given Fixed Therapy Group Size

The power of a trial is inversely related to the variance of the treatment effect estimator. The larger this variance, the smaller the power, and vice versa. The variance of the treatment effect estimator is a function of the therapy group size n_1, the number of therapy groups n_2, and the number of subjects n in the individual condition, see Equation 14.17. The larger these sample sizes, the smaller the variance and hence the larger the power. Of course, these sample sizes cannot increase without boundaries because the number of subjects that is available is often limited. Furthermore, the therapy group size is often fixed in advance. A therapy group is often small to promote dialogue between group members; group sizes between 5 and 15 are reasonable.

In this section sample sizes are presented for fixed therapy group sizes and such that the total number of subjects $n_1 n_2 + n$ is minimized. The optimal sample ratio is given by

$$\frac{n_1 n_2}{n} = \sqrt{\theta((n_1 - 1)\rho + 1)}. \qquad (14.18)$$

From Equation 14.18 we observe that the number of subjects $n_1 n_2$ in the group treatment increases with the variance ratio θ and the design effect $(n_1 - 1)\rho + 1$. This means that more subjects in this condition are required when the variability of the responses in this condition increases, when the therapy group size increases and/or when the correlation between outcomes of subjects within the same therapy group increases. The required number of subjects in the individual condition to achieve a power $1-\gamma$ to detect a unstandardized treatment effect δ in a two-sided test with significance level α is calculated from

$$n = \sigma_r^2\left(\sqrt{\theta((n_1 - 1)\rho + 1)} + 1\right)\left(\frac{z_{1-\alpha/2} + z_{1-\gamma}}{\delta}\right)^2. \qquad (14.19)$$

Figure 14.10 shows the power to detect a small standardized treatment effect as a function of the sample sizes in both conditions in a two-sided test with significance level $\alpha = 0.05$ and an intraclass correlation coefficient $\rho = 0.05$. In both plots, the variance ratio is equal to $\theta = 1$. In the left plot the therapy group size is fixed to $n_1 = 5$, in the right plot it is equal to $n_1 = 10$. For both conditions, a sufficient power level can be achieved provided sample sizes are large enough. However, larger sample sizes are required when the group size increases.

14.5 CONCLUSIONS

This chapter has presented optimal designs for trials with nested data. It was concluded that a multisite trial results in a more efficient design than a cluster randomized trial. Furthermore, multisite trials enable the estimation of the treatment by group interaction. Unfortunately, randomization at the subject level may result in contamination of the control group. On the other hand, cluster randomized trials may be hampered by selection bias and slow recruitment rates in the control condition. Recent research has focused on the development of pseudocluster randomized trials (Borm et al., 2005; Teerenstra et al., 2006, 2007) that aim to overcome these problems.

It should be mentioned that the calculations of the optimal designs were restricted to balanced designs, since those designs have maximal statistical power. This means that the groups are of equal size and that the number of groups per condition for cluster randomized trials or the number of subjects per condition per group for multisite trials does not vary. In practice this assumption is often not tenable. Group sizes vary by nature and by drop-out and nonresponse. Recent research has shown that the relative efficiency of using

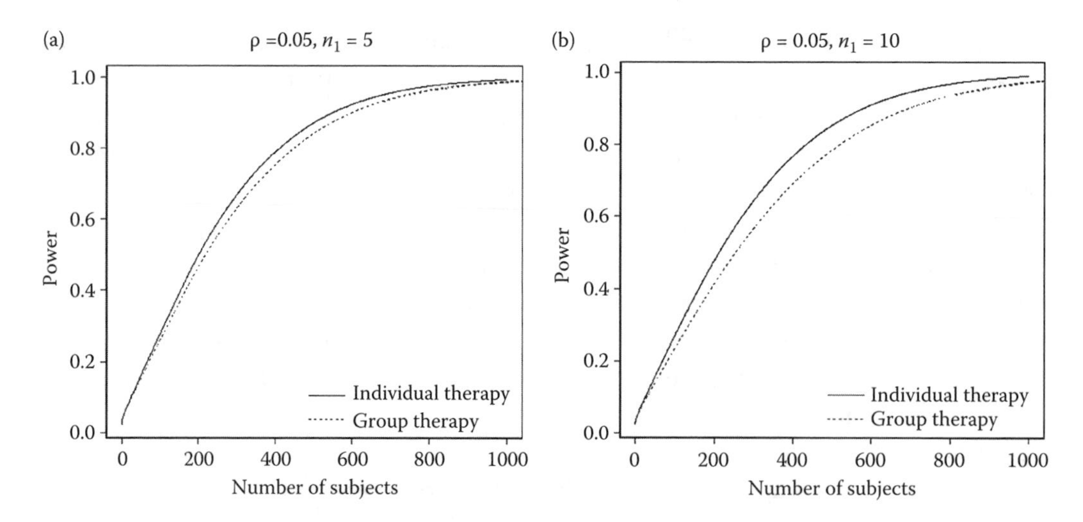

FIGURE 14.10
Power as a function of the sample sizes in both treatment conditions; $n_1 = 5$ (a) and $n_1 = 10$ (b).

varying cluster sizes instead of fixed cluster sizes is generally above 0.9 (Van Breukelen et al., 2007). This means that at most 11% more clusters are required in order to achieve the same power level as a trial with fixed cluster sizes. Furthermore, unequal costs between treatment conditions may be a reason to use unbalanced designs (Liu, 2003). With such trials, less subjects and groups are located to the more expensive treatment condition. Furthermore, there may be other nonstatistical reasons to use an unbalanced design, such as ethics, the need to gain additional information on the new treatment, and avoiding loss of power from dropout or crossover (Dumville, Hahn, Miles, & Torgerson, 2006).

Finally, it should be noted that the optimal sample sizes depend on the true value of the intraclass correlation coefficient. This value is often unknown in the design phase and an educated guess based on expert knowledge or estimates as published in the literature must be used. In the last decade a number of papers that present intraclass correlation coefficient estimates were published in the literature, see Hedges and Hedberg (2007) and Murray, Varnell, and Blitstein (2004) and the references therein. Recent research by the first author of this chapter and coworkers shows that the relative efficiency of a design for a group randomized trial with budgetary restrictions hardly drops below 0.9 when the prior estimate of the intraclass correlation coefficient departs at almost 75% from the true intraclass correlation coefficient. This can be corrected by sampling 11% more clusters (Korendijk, Moerbeek, & Maas, in press). If one is not confident that this requirement can be fulfilled, one has to rely on robust optimal designs, such as maximin designs. A maximin optimal design is a design that maximizes the minimal relative efficiency over a range of plausible values of the intraclass correlation coefficient. In other words, it selects the best worse-case scenario among all possible designs. See Berger and Tan (2004), and Ouwens, Tan, and Berger (2002) for an application to repeated measures designs. Other types of robust optimal designs are Bayesian optimal designs and designs with internal pilots. Bayesian optimal designs pose a prior distribution on the intraclass correlation coefficient and sample a large number of times from this prior distribution in a simulation study. For each sample the required group sizes and number of groups to achieve a specific power level are calculated and summarized in a predictive distribution. One can select the required sample sizes from these distributions such that one is sufficiently confident that the required power level can be achieved. See Spiegelhalter (2001) for an application of Bayesian designs to cluster randomized trials. Designs with internal pilots re-estimate the sample size after data from the pilot have been collected. The estimated sample sizes in the second stage of the trial are based on the estimate of the intraclass correlation coefficient as obtained from the pilot. See Lake, Kammann, Klar, and Betensky (2002) for an application to group randomized trials.

REFERENCES

Berger, M. P. F., & Tan, F. E. S. (2004). Robust designs for linear mixed effects models. *Journal of the Royal Statistical Society Series C Applied Statistics, 53*(4), 569–581.

Borm, G., Melis, R. J. F., Teerenstra, S., & Peer, P. G. (2005). Pseudo cluster randomisation: A treatment allocation method to minimize contamination and selection bias. *Statistics in Medicine, 24,* 3535–3547.

Cohen, J. (1992). A power primer. *Psychological Bulletin, 112*(1), 155–159.

Donner, A., & Klar, N. (2000). Cluster randomization trials. *Statistical Methods in Medical Research, 9*(2), 79–80.

Dumville, J. C., Hahn, S., Miles, J. N. V., & Torgerson, D. J. (2006). The use of unequal randomisation ratios in clinical trials: A review. *Contemporary Clinical Trials, 27*(1), 1–12.

Feldman, H. A., & McKinlay, S. M. (1994). Cohort versus cross-sectional design in large field trials: Precision, sample size, and a unifying model. *Statistics in Medicine, 13*(1), 61–78.

Gail, M. H., Mark, S. D., Carroll, R. J., & Green, S. B. (1996). On design considerations and randomization-based inference for community intervention trials. *Statistics in Medicine, 15*(11), 1069–1092.

Goldstein, H. (2003). *Multilevel statistical models* (3rd ed.). London, UK: Edward Arnold.

Hahn, S., Puffer, S., Torgerson, D. J., & Watson, J. (2005). Methodological bias in cluster randomised trials [On-line]. *BMC Medical Research Methodology, 5*(10). Retrieved from http://www.biomedcentral.com/1471-2288/5/10.

Hedges, L. V., & Hedberg, E. C. (2007). Intraclass correlation values for planning group-randomized trials in education. *Educational Evaluation and Policy Analysis, 29*(1), 60–87.

Hox, J. (2002). *Multilevel analysis. Techniques and applications.* Mahwah, NJ: Erlbaum.

Jordhoy, M. S., Fayers, P. M., Ahlner-Elmqvist, M., & Kaasa, S. (2002). Lack of concealment may lead to selection bias in cluster randomized trials of palliative care. *Palliative Medicine, 16*(1), 43–49.

Klar, N., & Donner, A. (2001). Current and future challenges in the design and analysis of cluster randomization trials. *Statistics in Medicine, 20*(24), 3729–3740.

Korendijk, E. J. H., Moerbeek, M., & Maas, C. J. M. (in press). The Robustness of Designs for Trials with Nested Data against Incorrect Initial Intracluster Correlation Coefficient Estimates. *Journal of Educational and Behavioral Statistics.*

Lake, S., Kammann, E., Klar, K., & Betensky, R. A. (2002). Sample size re-estimation in cluster randomization trials. *Statistics in Medicine, 21*(10), 1337–1350.

Lee, K. J., & Thompson, S. G. (2005). The use of random effects models to allow for clustering in individually randomized trials. *Clinical Trials, 2*(2), 163–173.

Liu, X. F. (2003). Statistical power and optimum sample allocation ratio for treatment and control having unequal costs per unit of randomization. *Journal of Educational and Behavioral Statistics, 28*(3), 231–248.

McKinlay, S. M. (1994). Cost-efficient designs of cluster unit trials. *Preventive Medicine, 23*(5), 606–611.

Melis, R. J. F., Teerenstra, S., Olde Rikkert, M. G. M., & Borm, G. F. (2008). Pseudo cluster randomisation performed well when used in practice. *Journal Clinical Epidemiology, 61*(11), 1169–1175.

Melis, R. J. F., van Eijken, M. I. J., & Borm, G. F. (2005). The design of the Dutch easycare study: A randomised controlled trial on the effectiveness of a problem-based community intervention model for frail elderly people. *BMC Health Services Research, 5*(1), 65.

Moerbeek, M. (2000). *Design and analysis of multilevel intervention studies*. Maastricht, The Netherlands: Maastricht University.

Moerbeek, M. (2005). Randomization of clusters versus randomization of persons within clusters: Which is preferable? *The American Statistician, 59*(1), 72–78.

Moerbeek, M. (2006). Power and money in cluster randomized trials: When is it worth measuring a covariate? *Statistics in Medicine, 25*(15), 2607–2617.

Moerbeek, M., & Maas, C. J. M. (2005). Optimal experimental designs for multilevel logistic models with two binary predictors. *Communications in Statistics—Theory and Methods, 34*(5), 1151–1167.

Moerbeek, M., van Breukelen, G. J. P., & Berger, M. P. F. (2000). Design issues for experiments in multilevel populations. *Journal of Educational and Behavioral Statistics, 25*(3), 271–284.

Moerbeek, M., van Breukelen, G. J. P., & Berger, M. P. F. (2001a). Optimal experimental design for multilevel logistic models. *The Statistician, 50*(1), 17–30.

Moerbeek, M., van Breukelen, G. J. P., & Berger, M. P. F. (2001b). Optimal experimental designs for multilevel models with covariates. *Communications in Statistics, Theory and Methods, 30*(12), 2683–2697.

Moerbeek, M., van Breukelen, G. J. P., & Berger, M. P. F. (2003). A comparison between traditional methods and multilevel regression for the analysis of multi-center intervention studies. *Journal of Clinical Epidemiology, 56*(4), 341–350.

Moerbeek, M., & Wong, W. K. (2002). Multiple-objective optimal designs for the hierarchical linear model. *Journal of Official Statistics, 18*(2), 291–303.

Moerbeek, M., & Wong, W. K. (2008). Sample size formulae for trials comparing group and individual treatments in a multilevel model. *Statistics in Medicine,* 27(15), 2850–2864.

Moore, H., Summerbell, C. D., Vail, A., Greenwood, D. C., & Adamson, A. J. (2001). The design features and practicalities of conducting a pragmatic cluster randomized trial of obesity management in primary care. *Statistics in Medicine, 20*(3), 331–340.

Murray, D. M. (1998). *Design and analysis of group-randomized trials*. New York, NY: Oxford University Press.

Murray, D. M., & Blitstein, J. L. (2003). Methods to reduce the impact of intraclass correlation in group-randomized trials. *Evaluation Review, 27*(1), 79–103.

Murray, D. M., Varnell, S. P., & Blitstein, J. L. (2004). Design and analysis of group-randomized trials: A review of recent methodological developments. *American Journal of Public Health, 94*(3), 423–432.

Ouwens, M. J. N. M., Tan, F. E. S., & Berger, M. P. F. (2002). Maximin D-optimal designs for longitudinal mixed effects models. *Biometrics, 58*(4), 735–741.

Puffer, S., Torgerson, D., & Watson, J. (2003). Evidence for risk of bias in cluster randomised trials: Review of recent trials published in three general medical journals. *British Medical Journal, 327*(7418), 785–789.

Rasbash, J., Steele, F., Browne, W. J., & Goldstein, H. (2009). A user's guide to MLwiN. Version 2.10. Bristol: Centre for Multilevel Modelling. University Of Bristol.

Raudenbush, S. W. (1993). Hierarchical linear models and experimental design. In L. K. Edwards (Ed.), *Applied analysis of variance in behavioral science* (pp. 459–496). New York, NY: Wiley.

Raudenbush, S. W. (1997). Statistical analysis and optimal design for cluster randomized trials. *Psychological Methods, 2*(2), 173–185.

Raudenbush, S. W., & Bryk, A. S. (2002). *Hierarchical linear models. Applications and data analysis methods*. Thousand Oaks, CA: Sage Publications.

Raudenbush, S. W., Bryk, A. S., & Congdon, R. (2004). HLM 6 for Windows [Computer software]. Lincolnwood, IL: Scientific Software International, Inc.

Raudenbush, S. W., & Liu, X. (2000). Statistical power and optimal design for multisite randomized trials. *Psychological Methods, 5*(2), 199–213.

Raudenbush, S. W., Martinez, A., & Spybrook, J. (2007). Strategies for improving precision in group-randomized experiments. *Educational Evaluation and Policy Analysis, 29*(1), 5–29.

Roberts, C., & Roberts, S. A. (2005). The design and analysis of clinical trials with clustering effects due to treatment. *Clinical Trials, 2*(2), 152–162.

Slymen, D. J., & Hovell, M. F. (1997). Cluster versus individual randomization in adolescent tobacco and alcohol studies: Illustrations for designs decisions. *International Journal of Epidemiology, 26*(4), 765–771.

Snijders, T. A. B., & Bosker, R. J. (1999). *Multilevel analysis: An introduction to basic and advanced multilevel modeling*. London: Sage Publications.

Sommer, A., Tarwotjo, I., Djunaedi, E., West, K. P., Loeden, A. A., Tilden, R.,... The Aceh Study Group. (1986). Impact of vitamin A supplementation on childhood mortality. A randomised controlled community trial. *Lancet, 327*(8491), 1169–1173.

Spiegelhalter, D. J. (2001). Bayesian methods for cluster randomized trials with continuous responses. *Statistics in Medicine, 20*(3), 435–452.

Spybrook, J., Raudenbush, S. W., Liu, X. F., & Congdon, R. (2006). *Optimal design for longitudinal and multilevel research: Documentation for the "optimal design" software.* Ann Arbor, MI: University of Michigan.

Teerenstra, S., Melis, R. J. F., Peer, P. G. M., & Borm, G. F. (2006). Pseudocluster randomization dealt with selection bias and contamination in clinical trials. *Journal of Clinical Epidemiolgy, 59*(4), 381–386.

Teerenstra, S., Moerbeek, M., Melis, R. J. F., & Borm, G. F. (2007). A comparison of methods to analyse continuous data from pseudo cluster randomized trials. *Statistics in Medicine, 26*(22), 4100–4115.

Van Breukelen, G. J. P., Candel, M. J. J. M., & Berger, M. P. F. (2007). Relative efficiency of unequal versus equal cluster sizes in cluster randomized and multicentre trials. *Statistics in Medicine, 26*(13), 2589–2603.

Section V

Specific Statistical Issues

15

Centering in Two-Level Nested Designs

James Algina
Department of Educational Phychology, School of Human Development and Organizational Studies in Education,
University of Florida, Gainesville, Florida

Hariharan Swaminathan
Department of Educational Psychology,
University of Connecticut, Storrs, Connecticut

15.1 CENTERING IN TWO-LEVEL NESTED DESIGNS

In this chapter we review basic results about centering independent variables in two-level nested designs. We present the major results in the context of an example in which the available data are math achievement (*MA*), socioeconomic status (*SES*), and the school a student attends. The goal of the data analysis is to demonstrate the effect of centering of the independent variable *SES* on the relationship between *SES* and the dependent variable *MA*. Three types of centering of *SES* are considered:

- Grand mean centering in which the grand mean of *SES* is subtracted from each *SES* score: $SES - \overline{\overline{SES}}$, where $\overline{\overline{SES}}$ is the average of all of the *SES* scores in the sample.
- Group mean centering in which for each student in a school, the school mean of *SES* is subtracted from the *SES* score: $SES - \overline{SES}_j$, where $\overline{SES}_j$ is the average of *SES* score for students who are in the analysis and attended school *j*.
- No centering, in which a mean is not subtracted from *SES*.

In our review, we emphasize that centering, in particular group mean centering, addresses certain potential assumption violations and also affects the interpretation of the fixed effects and variance components in the model. Because of the impact of centering decisions on interpretation, we present recommendations for centering decisions.

As noted above we present the results pertaining to centering in the context of examples. More formal presentation of these results can be found in Kreft, de Leeuw, and Aiken (1995) and Snijders and Berkhof (2008). Other expositions on centering can be found in Enders and Tofighi (2007); Hox (2002); Misangi, LePine, Algina, and Goeddeke (2006); Raudenbush and Bryk (2002); and Snijders and Bosker (1999).

As noted earlier, we present the major results about centering in the context of an example in which data are available for three variables: *MA*, *SES*, and a variable that indicates the school a participant attended. The *SES* data were taken from the High School and Beyond file supplied with HLM 6.0 but were transformed by adding five points to each student's *SES* score. This change was made in order to highlight the effect of centering on intercepts. The values of the dependent variable *MA* were simulated in order to ensure that the models we use to analyze the data are correct. The sample sizes are 7185 students in 160 schools. The model used to simulate the data for our first set of analyses was:

$$\mathcal{E}(MA_{ij}) = 15 + 2.2(SES_{ij}) + 3.8(\overline{SES}_j)$$

and

$$MA_{ij} = \mathcal{E}(MA_{ij}) + u_{0j} + u_{1j}SES_{ij} + \varepsilon_{ij}$$

where

- MA_{ij} is the mathematics achievement score for student *i* in school *j*
- SES_{ij} is the socioeconomic status score for student *i* in school *j*
- $\overline{SES}_j$ is the mean socioeconomic status score in school *j*
- $\mathcal{E}(MA_{ij})$ is the expected value of mathematics achievement conditional on SES_{ij} and $\overline{SES}_j$
- $u_{0j} + u_{1j}SES_{ij} + \varepsilon_{ij}$ is the residual (i.e., $MA_{ij} - \mathcal{E}(MA_{ij})$). The variables u_{0j} and u_{1j} were constants for students within school *j* but varied across schools. The variable ε_{ij} varied over students within a school.

TABLE 15.1

Descriptive Statistics for Math Achievement and Socioeconomic Status

Variable	*M*	*SD*
MA	45.22	7.00
SES	5.00	0.78

Note: $n = 7185$.

In the simulation, the coefficients for *SES* and school means *SES* were set to values similar to those obtained by analyzing the HSB data. The covariance matrix for the variables u_{0j} and u_{1j} was set as

$$\begin{bmatrix} 16.00 & \\ -2.50 & 0.50 \end{bmatrix}$$

and the variance for ε_{ij} was taken as 37. The variances and covariances above are similar to the values obtained by analyzing the HSB data. Descriptive statistics for the data are presented in Table 15.1.

15.2 SIMPLE LINEAR REGRESSION MODELS

If the data were not multilevel an appropriate analysis would be to use ordinary least squares (OLS) to estimate the simple linear regression model with *MA* as the dependent variable and *SES* as the independent variable:

$$MA_{ij} = \beta_0 + \beta_T SES_{ij} + \varepsilon_{ij}.$$

In this model, the two βs are fixed constants to be estimated and ε_{ij} is a random quantity. Fixed constants, either in regression models or multilevel models, are also called *fixed effects*. Random quantities are called *random effects* and the variance (and covariances) of random effects are called *variance components*. The T subscript in β_T stands for total and indicates that the grouping structure was not taken into account in the analysis.

The OLS estimate of the slope is $\hat{\beta}_T = 3.327$ and its standard error is 0.098.

To examine the effect of centering in this simple model, the independent variable is taken as the deviation of SES from the grand mean $\overline{\overline{SES}}$:

$$MA_{ij} = \beta_0 + \beta_T\left(SES_{ij} - \overline{\overline{SES}}\right) + \varepsilon_{ij}.$$

From Table 15.1, $\overline{\overline{SES}} = 5.00$. The estimated slope is again $\hat{\beta}_T = 3.327$ with a standard error of 0.098. That is, in simple linear regression, grand mean centering does not affect the estimated slope or its standard error.

Although grand mean centering did not affect the estimated slope or its standard error, grand mean centering does affect the estimate of the intercept. When *SES* is used as the independent variable the intercept is 28.587 and when grand mean centered *SES* is used as the independent variable the intercept is 45.222. Why are these estimates so different? When *SES* is not centered the estimated regression equation is

$$\widehat{MA} = 28.587 + 3.327(SES).$$

By definition the intercept is the predicted value when the independent variable is zero. When *SES* is used as the independent variable, zero is an *SES* score that is well below the grand mean of *SES* (5.00). As the standard deviation for *SES* is .78, a *SES* of zero is about 6.4 standard deviations below the grand mean. Thus 28.587 is predicted MA corresponding to an *SES* that is 6.4 standard deviations below the grand mean. When grand mean centered *SES* is used as the independent variable, the estimated regression equation is

$$\widehat{MA} = 45.222 + 3.327\left(SES - \overline{\overline{SES}}\right).$$

Now the intercept, 45.222, is the predicted value of MA when $SES - \overline{\overline{SES}} = 0$; that is, when $SES = 5.00$. So the reason the two intercepts are different is because they are predicted values corresponding to different *SES* scores. Another way to see the relationship between the slopes and the intercepts in the noncentered and the centered models is by noting that

$$MA_{ij} = \beta_0 + \beta_T SES_{ij} + \varepsilon_{ij} = \left(\beta_0 + \beta_T \overline{\overline{SES}}\right) + \beta_T\left(SES_{ij} - \overline{\overline{SES}}\right) + \varepsilon_{ij}.$$

Thus, while the slope remains the same, the intercept in the centered model is $\beta_0 + \beta_T \overline{\overline{SES}}$ and its estimate is 28.585 + 3.327(5) = 45.222.

Is the difference in the intercepts of fundamental importance? No, because any estimate or hypothesis test that can be obtained by using one of the two simple linear regression models can also be obtained by using the other model. If we use the model for noncentered *SES*, the predicted MA score for a student with an *SES* exactly equal to the grand mean is

$$\widehat{MA} = 28.587 + 3.327(5.00) = 45.222$$

and is equal to intercept under grand mean centering. A score of zero on the *SES* scale is equal to a score of 0–5.00 = –5.00 on the grand mean centered *SES* scale. If we use the model for grand mean centered *SES* and calculate the predicted value at the score –5.00 we get

$$\widehat{MA} = 45.222 + 3.327(0 - 5.00) = 28.587$$

Thus, given the results from one model, the intercept for the other model can be obtained by algebra. Models are statistically equivalent if any estimate or hypothesis test that can be obtained by using one of the models can also be obtained by using the other models.

The OLS method makes the important assumption that residuals are independent. This assumption is violated when the data are multilevel in nature as illustrated in Figure 15.1 where an idealized diagram of the data is presented. The large ellipse represents the scatter plot of all the data and the slanted line is the OLS regression line for the model $MA_{ij} = \beta_0 + \beta_T SES_{ij} + \varepsilon_{ij}$. The smaller ellipses represent the scatter plots for six schools. A residual is the deviation of a student's *MA* score from the regression line. The scatter plots for Schools A and B are wholly above the regression line and the residuals for the schools are all positive. The scatter plots for Schools E and F are wholly below the regression line and the residuals for the schools are all negative. The scatter plots for these schools illustrate that the residuals are similar in size within schools. As a result the residuals are statistically dependent, violating an important assumption of the OLS estimation procedure.

15.3 THE FIXED EFFECTS MODEL

One solution to the problem of nonindependent residuals is to use the fixed effects model (see, for example, Baltagi, 2005; Frees, 2004; or Greene, 2007). As applied in the current situation, the fixed effects model specifies a different intercept for each of the 160 schools and a common slope for all schools. The model is

$$MA_{ij} = \beta_{0j} + \beta_W SES_{ij} + \varepsilon_{ij}.$$

The coefficient β_{0j} is the intercept in the jth school. These 160 intercepts are fixed effects that can be estimated but often are not because they are of limited interest. Treating the intercepts as fixed effects is

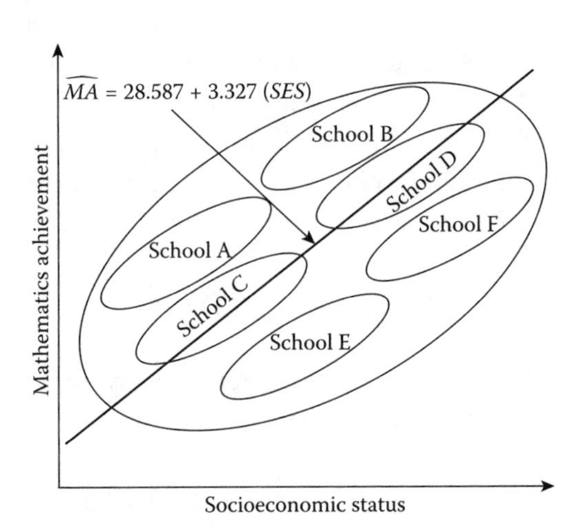

FIGURE 15.1
Depiction of the simple regression model.

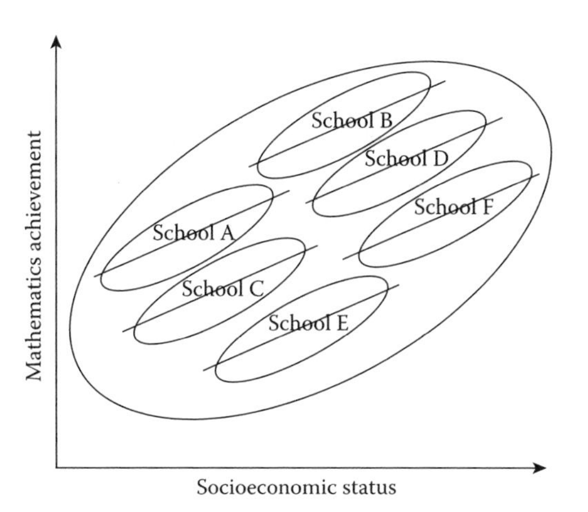

FIGURE 15.2
Depiction of the fixed effects model.

the reason the model is called a fixed effect model. The coefficient β_W is a fixed effect that will be estimated. The W subscript in β_W stands for "within." The fixed effects model is equivalent to the ANCOVA model and is estimated by OLS. As is well known, the slope in the ANCOVA model is assumed to be equal across schools and is a within-school slope; that is, the slope estimates the relationship between *MA* and *SES* within any one of the 160 schools. This is the reason for the W subscript in β_W.

The fixed effects model is depicted in Figure 15.2. Note, for example, the regression line for School A goes through the scatter plot for School A. The residual for a student within a school is defined as the deviation of that student's *MA* score from the regression line for the student's school. In School A, as in all of the other schools in Figure 15.2, there would be both positive and negative residuals and so the problem of nonindependent residuals is addressed. The coefficient $\hat{\beta}_W$ is equal to 2.181 with a standard error of .109. Note that $\hat{\beta}_W$ is approximately .65 the size of $\hat{\beta}_T$. For the purposes of comparing results, coefficients for the various models considered so far and several that will be considered subsequently are presented in Table 15.2.

15.4 THE RANDOM EFFECTS MODEL (RANDOM INTERCEPTS MODEL)

Another possible solution to the problem of nonindependent residuals is to use the random effects model (see, for example, Baltagi, 2005; Frees, 2004; Greene, 2007). The model can be written in the same form as the fixed effects model

$$MA_{ij} = \beta_{0j} + \beta_{SES} SES_{ij} + \varepsilon_{ij}.$$

but (a) β_{SES} is not necessarily equal to β_W, and (b) the β_{0j} are regarded as random quantities. Treating the intercepts as random quantities is the reason the model is called the random effects model. Since β_{0j} is a random variable, it can be expressed as $\beta_{0j} = \gamma_0 + u_{0j}$, where γ_0 is the expected value of β_{0j} and $u_{0j} = \gamma_0 - \beta_{0j}$ is the random error. Substituting $\beta_{0j} = \gamma_0 + u_{0j}$, the model can be written as

$$MA_{ij} = \gamma_0 + \gamma_{SES} SES_{ij} + u_{0j} + \varepsilon_{ij} \qquad (15.1)$$

where $\gamma_{SES} = \beta_{SES}$, and the fixed effects to be estimated are γ_0 and γ_{SES}. Both u_{0j} and

TABLE 15.2

Summary of Results for Selected Models

Model	Equation	Coefficient (Standard Error) for *SES*	Coefficient (Standard Error) for MEAN *SES*
Simple regression	$MA = \beta_0 + \beta_T SES + \varepsilon$	3.327 (0.098)	NA
Simple regression	$MA = \beta_0 + \beta_T(SES - \overline{\overline{SES}}) + \varepsilon$	3.327 (0.098)	NA
Fixed effects	$MA = \beta_{0j} + \beta_W(SES) + \varepsilon$	2.181 (0.109)	NA
Random intercepts	$MA = \gamma_0 + \gamma_{SES} SES + u_0 + \varepsilon$	2.373 (0.107)	NA
Random intercepts	$MA = \gamma_0 + \gamma_{SES}(SES - \overline{\overline{SES}}) + u_0 + \varepsilon$	2.373 (0.107)	NA
Random intercepts	$MA = \gamma_0 + \gamma_W(SES - \overline{SES}_j) + u_{0j} + \varepsilon$	2.181 (0.109)	NA
Intercepts as outcomes	$MA = \gamma_0 + \gamma_W(SES) + \gamma_C \overline{SES}_j + u_{0j} + \varepsilon$	2.181 (0.109)	4.186 (0.400)
Intercepts as outcomes	$MA = \gamma_0 + \gamma_W(SES - \overline{\overline{SES}}) + \gamma_C \overline{SES}_j + u_{0j} + \varepsilon$	2.181 (0.109)	4.186 (0.400)
Intercepts as outcomes	$MA = \gamma_0 + \gamma_W(SES - \overline{SES}_j) + \gamma_B \overline{SES}_j + u_{0j} + \varepsilon$	2.181 (0.109)	6.367 (0.385)

ε_{ij} are random effects. The variances of u_{0j} and ε_{ij} are the variance components. In the random effects model, the variance of u_{0j} is also the variance of the school specific intercept β_{0j}. The estimated coefficient for *SES* is $\hat{\gamma}_{SES} = 2.373$.

The term random effects model comes from the econometric literature. In the multilevel modeling literature, the random effects model is often called the random intercepts model. The model is depicted in Figure 15.3. As for the simple linear regression model, a single regression line for all schools is estimated. A residual is the deviation of a student's MA score from the regression line. Just as in the depiction for the simple linear regression model in Figure 15.1, the scatter plots for Schools A and B are wholly above the regression line and the residuals for the schools are all positive and the scatter plots for Schools E and F are wholly below the regression line and the residuals for the schools are all negative. Thus it would seem that the random effects model has the same problem as the simple linear regression model. However, in the random intercepts model the residual has two components $u_{0j} + \varepsilon_{ij}$ whereas the residual in the simple regression model has only one component. Because u_{0j} is equal for all students within a school, the inclusion of u_{0j} in the random intercepts model accounts for the similarity of residuals within schools and therefore the random intercepts model is more appropriate for the data than is the simple linear regression model.

Just as in the simple linear regression model, with the random intercepts model *SES* can be grand mean centered:

$$MA_{ij} = \gamma_0 + \gamma_{SES}(SES_{ij} - \overline{\overline{SES}}) + u_{0j} + \varepsilon_{ij}$$

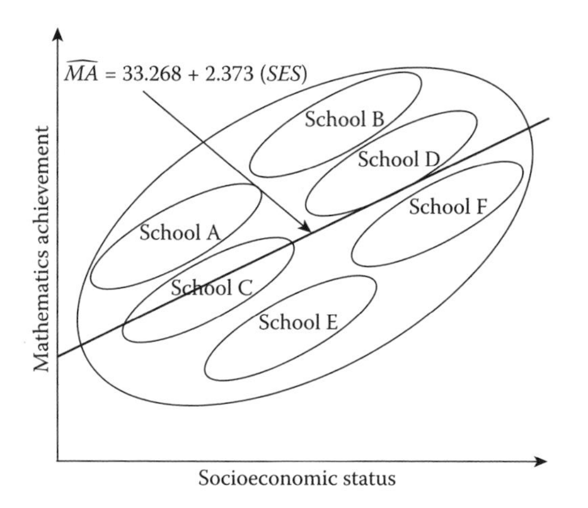

FIGURE 15.3
Depiction of the random effects model.

Grand mean centering does not affect $\hat{\gamma}_{SES}$ because

$$\begin{aligned} MA_{ij} &= \gamma_0 + \gamma_{SES} SES_{ij} + u_{0j} + \varepsilon_{ij} \\ &= \left(\gamma_0 + \gamma_{SES} \overline{\overline{SES}}\right) + \gamma_{SES}\left(SES_{ij} - \overline{\overline{SES}}\right). \\ &\quad + u_{0j} + \varepsilon_{ij}. \end{aligned}$$

As can be seen from the preceding equation, the intercepts when *SES* is not centered and when *SES* is grand mean centered are different and for the same reason presented in connection with the simple linear regression model. Because the two versions of the model are statistically equivalent, the intercept for each model can be calculated from the results for the other model. Similarly, while the estimated variance component for u_{0j}, which is the estimated variance for β_{0j}, will not be the same for the two models, the variance component for u_{0j} for one model can be obtained from the results for the other. These statements apply to all statistically equivalent models considered in this section and in the next section, and will not be repeated in each case.

15.5 INTERCEPTS AS OUTCOMES MODEL

The random intercepts model presented in Equation 15.1 assumes that the random effects (i.e., u_{0j} and r_{ij}) are uncorrelated with *SES*. Because $\beta_{0j} = \gamma_0 + u_{0j}$, the random intercepts model also assumes that β_{0j} is uncorrelated with *SES*. You can envision investigating this assumption by considering a plot with *SES* on the horizontal axis and β_{0j} on the vertical axis. If the assumption is met, the scatter plot would have a zero slope. Furthermore, if the assumption is met, a scatter plot with β_{0j} on the vertical axis and school mean *SES* (i.e., $\overline{SES}_j$) on the horizontal axis would also have a zero slope. That is, assuming β_{0j} is uncorrelated with *SES* implies the assumption that β_{0j} is uncorrelated with $\overline{SES}_j$.

We can write the random intercepts model as a two-level model

Level-1:

$$MA_{ij} = \beta_{0j} + \beta_{SES} SES_{ij} + \varepsilon_{ij}$$

Level-2:

$$\beta_{0j} = \gamma_0 + u_{0j}$$

$$\beta_{SES} = \gamma_{SES}.$$

Since β_{SES} is fixed, it does not have an error component; changing the notation from β_{SES} to $\beta_{SES} = \gamma_{SES}$ is merely for the purpose of conformity. Substituting the level-2 models into the level-1 model shows that the two-level model yields the random intercepts model in Equation 15.1. If we think the data violate the assumption that β_{0j} is uncorrelated with $\overline{SES}_j$ or wish to test the assumption, we can add school mean *SES* to the model for β_{0j}:

Level-1:

$$MA_{ij} = \beta_{0j} + \beta_{SES}SES_{ij} + \varepsilon_{ij}$$

Level-2:

$$\beta_{0j} = \gamma_0 + \gamma_C \overline{SES}_j + u_{0j}$$

$$\beta_{SES} = \gamma_W,$$

and our model becomes

$$MA_{ij} = \gamma_0 + \gamma_W \left(SES_{ij}\right) + \gamma_C \overline{SES}_j + u_{0j} + \varepsilon_{ij}. \tag{15.2}$$

This is an example of an intercepts as outcomes model.

In Equation 15.2 a noncentered level-1 variable and the mean of the level-1 variable are included in the model. The mean is called a contextual or compositional variable (Raudenbush, 1989) because it is a measure of the context of the group. The coefficient for the contextual variable is called a context coefficient and the subscript *C* on γ_C denotes context. Note that with the introduction of $\overline{SES}_j$ as an independent variable in the model, the subscript to the coefficient for *SES* has been changed from *SES* to *W*, signaling that including $\overline{SES}_j$ in the model changes the interpretation of the coefficient for *SES*. Estimates of the parameters are $\hat{\gamma}_W = 2.181$ and $\hat{\gamma}_C = 4.186$. (The standard errors are 0.109 and 0.400, respectively.) Note that the coefficient for *SES* is exactly the same in the fixed effects model and in the intercepts as outcomes model.

An alternative model using grand mean centering for *SES* is

$$MA_{ij} = \gamma_0 + \gamma_W \left(SES_{ij} - \overline{SES}\right) + \gamma_C \overline{SES}_j + u_{0j} + \varepsilon_{ij}.$$

Since

$$\begin{aligned} MA_{ij} &= \gamma_0 + \gamma_W \left(SES_{ij}\right) + \gamma_C \overline{SES}_j + u_{0j} + \varepsilon_{ij} \\ &= \left(\gamma_0 + \gamma_W \overline{SES}\right) + \gamma_W \left(SES_{ij} - \overline{SES}\right) \\ &\quad + \gamma_C \overline{SES}_j + u_{0j} + \varepsilon_{ij}, \end{aligned}$$

the grand mean centered model is statistically equivalent to the model in Equation 15.2 and the estimates of γ_W and γ_C are the same for the two models. The intercept in the grand mean centered model is, however, $\gamma_0 + \gamma_W \overline{SES}$.

15.6 THE RANDOM INTERCEPTS MODEL WITH GROUP MEAN CENTERING

Recall that β_{0j} in the random intercepts model (see Equation 15.1) is assumed to be uncorrelated with *SES* and this assumption implies that β_{0j} is uncorrelated with school mean *SES*. Estimating the intercepts as outcomes model (see Equation 15.2) is one way to address this implication. Another way to address the assumption is by group (school) mean centering *SES* in the random intercepts model. To school mean center *SES*, we replace *SES* in

Equation 15.1 by $SES_{ij} - \overline{SES}_j$. The assumption now is that β_{0j} is uncorrelated with the school means of $SES_{ij} - \overline{SES}_j$. Because the school mean for $SES_{ij} - \overline{SES}_j$ must be zero for each school, the correlation between β_{0j} and the school-mean-centered variable will be zero. Thus by school mean centering *SES* we have ensured that the assumption is met.

The random intercepts model in which *SES* is group mean centered is

$$MA_{ij} = \gamma_0 + \gamma_W (SES_{ij} - \overline{SES}_j) + u_{0j} + \varepsilon_{ij}.$$

The estimated coefficient for *SES* is $\hat{\gamma}_W = 2.181$ with standard error 0.109 exactly the same as for the fixed effects model and the intercepts as outcomes model. However, $\hat{\gamma}_W$ is different than $\hat{\gamma}_{SES}$ that was obtained by using the random intercepts model with noncentered *SES* or with grand mean centered *SES*. This is because while the uncentered model is

$$MA_{ij} = \gamma_0 + \gamma_{SES} SES_{ij} + u_{0j} + \varepsilon_{ij},$$

the group mean centered model is

$$MA_{ij} = \gamma_0 + \gamma_W (SES_{ij} - \overline{SES}_j) + u_{0j} + \varepsilon_{ij}$$
$$= \gamma_0 + \gamma_W SES_{ij} + (-\gamma_W)\overline{SES}_j + u_{0j} + \varepsilon_{ij}.$$

The group mean centered model is equivalent to a model that has two independent variables *SES* and $\overline{SES}_j$ and, in which, the coefficients for the two variables have opposite signs that are equal in absolute value. Because of this, the uncentered model is not statistically equivalent to the group mean centered model. Hence, for the random intercepts model, while the uncentered and the grand mean centered models are statistically equivalent producing identical slope parameters (and equivalent intercepts), these two models are not statistically equivalent to the random intercepts model with school mean centering.

15.7 THE INTERCEPTS AS OUTCOMES MODEL WITH GROUP MEAN CENTERING

As an alternative to the intercepts as outcomes model without centering or with grand mean centering, a model with school mean centering can be used:

$$MA_{ij} = \gamma_0 + \gamma_W (SES_{ij} - \overline{SES}_j) + \gamma_B (\overline{SES}_j) + u_{0j} + \varepsilon_{ij}. \quad (15.3)$$

This model is statistically equivalent to the uncentered or the grand mean centered intercepts as outcome model. In comparing the uncentered model with the group mean centered model in Equation 15.3, we see that

$$MA_{ij} = \gamma_0 + \gamma_W (SES_{ij} - \overline{SES}_j) + \gamma_B (\overline{SES}_j) + u_{0j} + \varepsilon_{ij}$$
$$= \gamma_0 + \gamma_W SES_{ij} + (\gamma_B - \gamma_W)(\overline{SES}_j) + u_{0j} + \varepsilon_{ij}.$$

Thus the coefficient of the group centered *SES* is the same as the coefficient of uncentered *SES*; the coefficient of $\overline{SES}$ in the group centered model is equal to $\gamma_B - \gamma_W$ in the uncentered model. As in the other two intercepts as outcomes models, the coefficient for *SES* is the within-school coefficient and is estimated to be $\hat{\gamma}_W = 2.181$ just as for the other two models. The coefficient for school mean *SES*, however, is not the context effect (γ_C). Instead it is γ_B, the between groups coefficient.

To help understand the difference between γ_C and γ_B consider the means as outcomes model

$$MA_{ij} = \gamma_0 + \gamma_B(\overline{SES}_j) + u_{0j} + \varepsilon_{ij}. \quad (15.4)$$

In this model the coefficient for the *SES* variable is a slope relating school specific *MA* means to school specific *SES* means. This is a between school relationship and therefore the coefficient for mean *SES* is denoted by γ_B; for the current example $\hat{\gamma}_B = 6.365$. Similarly the coefficient for mean *SES* in the intercepts as outcomes model with group mean centering is a between-school relationship; the coefficient for mean *SES* is also denoted by γ_B and the estimated coefficient is $\hat{\gamma}_B = 6.367$. (The estimates of γ_B in the means as outcomes model and the intercepts as outcomes model are not necessarily equal, but will be similar.) Earlier we found that $\hat{\gamma}_W = 2.181$ and $\hat{\gamma}_C = 4.186$. We can see that $\hat{\gamma}_C$ is equal to $\hat{\gamma}_B - \hat{\gamma}_W = 4.186$. Moreover, as was seen earlier, $\gamma_C = \gamma_B - \gamma_W$. That is, the context effect is simply the difference between the between-school effect and the within-school effect.

Recall that the models that include γ_C (i.e., the context effect) are

$$MA_{ij} = \gamma_0 + \gamma_W(SES_{ij}) + \gamma_C\overline{SES}_j + u_{0j} + \varepsilon_{ij},$$

and

$$MA_{ij} = \gamma_0 + \gamma_W(SES_{ij} - \overline{\overline{SES}}) + \gamma_C\overline{SES}_j + u_{0j} + \varepsilon_{ij}.$$

The coefficient γ_C asks whether including school mean *SES* in the model is necessary if individual *SES* (in its non centered or grand mean centered forms) is also in the model. If $\gamma_C = 0$, it is not necessary to include school mean *SES* in the model and one can use the random intercepts model

$$MA_{ij} = \gamma_0 + \gamma_{SES}(SES_{ij}) + u_{0j} + \varepsilon_{ij}.$$

Also if $\gamma_C = 0$, then $\gamma_B = \gamma_W$ and

$$MA_{ij} = \gamma_0 + \gamma_W(SES_{ij} - \overline{SES}_j) + \gamma_B(\overline{SES}_j) + u_{0j} + \varepsilon_{ij} \quad (15.5)$$

simplifies to

$$MA_{ij} = \gamma_0 + \gamma_{SES}(SES_{ij}) + u_{0j} + \varepsilon_{ij},$$

implying it is unnecessary to include school mean *SES* in Equation 15.5. It can be shown that when $\gamma_B = \gamma_W$, $\hat{\gamma}_{SES}$ estimates the common coefficient and $\hat{\gamma}_{SES}$ has a smaller sampling variance than does $\hat{\gamma}_W$ or $\hat{\gamma}_B$.

According to Raudenbush and Bryk (2002), $\hat{\gamma}_{SES}$ is a weighted average of $\hat{\gamma}_B$ and $\hat{\gamma}_W$. This makes it clear that if $\gamma_B \neq \gamma_W$, then the random intercepts model without centering or with grand mean centering should not be used because it averages coefficients that provide information about two different aspects of the relationship between *MA* and *SES*.

The means as outcomes model in Equation 15.4 includes $\overline{SES}_j$ but does not include $SES - \overline{SES}_j$, whereas the intercepts as outcomes model with school mean centered *SES* in Equation 15.3 includes both variables. Nevertheless the estimate of the coefficient for $\overline{SES}_j$ (i.e., $\hat{\gamma}_B$) will be similar for the models. Why is the estimate of γ_B largely unaffected by the inclusion of $SES - \overline{SES}_j$ in the intercepts as outcomes model? Because $SES-\overline{SES}_j$ and $\overline{SES}_j$ are uncorrelated. If $SES-\overline{SES}_j$ is plotted against $\overline{SES}_j$, then at any point on the $\overline{SES}_j$ axis, the mean of $SES-\overline{SES}_j$ must be zero and so the plot will have a slope that is equal to zero. (This claim assumes that the participants who have a score on $SES-\overline{SES}_j$ are the same participants for whom $\overline{SES}_j$ was calculated.)

Rather than including $\overline{SES}_j$ as an independent variable we may want to include a level-2 variable that is not a mean of a level-1 independent variable. Following Susser (1994), we refer to a variable that is not a mean of a level-1 independent variable as an *integral* variable. Just as $SES-\overline{SES}_j$ and $\overline{SES}_j$ are uncorrelated, $SES-\overline{SES}_j$ will be uncorrelated with any level-2 integral variable. For example, suppose we want to study the relationship between the schools' disciplinary climate (*DC*) and *MA*. The correlation between *DC* and $SES-\overline{SES}_j$ will be zero. As a result the coefficient for *DC* will be similar in the following models (these analyses use the actual HSB data):

Means as outcomes model:

$$MA_{ij}=\gamma_0+\gamma_{DC}^{B}DC_j+u_{0j}+\varepsilon_{ij} \qquad (15.6)$$

$$\widehat{MA}=12.59+(-1.49)DC$$

Intercepts as outcomes model:

$$MA_{ij}=\gamma_0+\gamma_W(SES-\overline{SES}_j)+\gamma_{DC}^{B}DC_j+u_{0j}+\varepsilon_{ij} \qquad (15.7)$$

$$\widehat{MA}=12.59+2.19(SES-\overline{SES}_j)+(-1.49)DC$$

where the superscript in γ_{DC}^{B} indicates a between-schools coefficient. So, including group mean centered *SES* in the model did not materially affect the estimate of γ_{DC}^{B}. In essence, by group mean centering *SES* in Equation 15.7 we have failed to control for *SES*. Alternatively we can say that we have controlled for the deviation of *SES* from school mean *SES*, but not for school mean *SES*. Regardless of how we describe the control, we do not know whether the relationship of *MA* to *DC* would be the same if we had a more complete control of *SES*. In our opinion models like that in Equation 15.7 should not be used unless the researcher wants to estimate the within-school relationship for the level-1 variables and a between-school relationship for the level-2 variables.

More complete control of *SES* can be achieved by including school mean *SES* in the model:

$$MA_{ij}=\gamma_0+\gamma_W(SES-\overline{SES}_j)+\gamma_{\overline{SES}}^{B}\overline{SES}_j+\gamma_{DC}DC_j+u_{0j}+\varepsilon_{ij}, \qquad (15.8)$$

$$\widehat{MA}=12.63+2.19(SES-\overline{SES}_j)+3.11\overline{SES}+(-0.69)DC.$$

The coefficient γ_{DC} does not have a *B* superscript because γ_{DC} is not a between-school coefficient when both $SES-\overline{SES}_j$ and $\overline{SES}_j$ are included in the model. The coefficient γ_{DC} does not have a *C* superscript because we reserve the term context effect for the coefficient of the mean of a level-1 variable (e.g., $\overline{SES}_j$) when the model also includes the level-1 variable and that variable is not centered or is grand mean centered. Either of the other two types of centering will provide a statistically equivalent model.

15.8 OLS ESTIMATION REVISITED

Just as school mean *SES* can be included in multilevel models, it can be included in a single-level regression model:

$$MA_{ij} = \beta_0 + \beta_W SES_{ij} + \beta_C \overline{SES}_j + \varepsilon_{ij},$$

$$MA_{ij} = \beta_0 + \beta_W \left(SES_{ij} - \overline{SES}\right) + \beta_C \overline{SES}_j + \varepsilon_{ij},$$

and

$$MA_{ij} = \beta_0 + \beta_W \left(SES_{ij} - \overline{SES}_j\right) + \beta_B \overline{SES}_j + \varepsilon_{ij}.$$

For all three models $\hat{\beta}_W = \hat{\gamma}_W = 2.181$. The coefficient $\hat{\beta}_C$ will be similar to $\hat{\gamma}_C$ but not identical (unless the sample size is the same in all schools). Likewise, the coefficients $\hat{\gamma}_B$ and $\hat{\beta}_B$ will be similar (but will be identical if the sample size is the same in all schools).

If the regression models and multilevel models result in the same or similar coefficients, is it necessary to use the multilevel model? In the multilevel models the schools are viewed as having been sampled from a larger group of schools. If this reflects the researcher's point of view, then using $\hat{\gamma}_W$, and/or $\hat{\gamma}_B$ and/or $\hat{\gamma}_C$ will result in standard errors that correctly reflect this view. If the researcher's view is that the schools are a fixed set of schools of interest, then the standard errors for $\hat{\beta}_B$ and $\hat{\beta}_C$ will be correct. The standard error, however, for $\hat{\beta}_W$ will be incorrect. The correct standard error for $\hat{\beta}_W$ should be obtained by using the fixed effects model, but if the total sample size is large the two standard errors will be quite similar.

Finally it can be shown that $\hat{\beta}_T$ has a mathematical relationship to $\hat{\beta}_B$ and $\hat{\beta}_W$. Specifically, if R^2 is the proportion of variance due to the schools in a one-way ANOVA of the *MA* data, then

$$\hat{\beta}_T = R^2 \hat{\beta}_B + (1 - R^2)\hat{\beta}_W.$$

(See, for example, Pedhazur, 1982). That is, $\hat{\beta}_T$ is a proportion of variance weighted average of $\hat{\beta}_B$ and $\hat{\beta}_W$ and will be an inappropriate coefficient if $\beta_W \neq \beta_B$.

15.9 SUMMARY AND RECOMMENDATIONS IN REGARD TO RANDOM INTERCEPTS AND INTERCEPTS AS OUTCOMES MODELS

15.9.1 Models without Level-2 Variables

The random intercepts model, without centering or with grand mean centering, entails the assumption that the school specific intercepts (β_{0j}) are uncorrelated with the independent variable. If this assumption is violated, or equivalently, if the within group (γ_W) and between group (γ_B) coefficients are not equal, the coefficients for the independent variables can be misleading in the random intercepts model without centering or with grand mean centering. If the researcher is only interested in within group coefficients, the random intercepts model with group mean centered independent variables should be used.

15.9.2 Models with Level-2 Variables

If the variables in the model are (a) level-1 variables, and (b) contextual level-2 variables that are means of the level-1 variables and if the researcher is interested in the within group coefficient (γ_W) and either the between group coefficient or the context coefficient (γ_C), any of the intercepts as outcomes models can be used. The models in which the independent variable is not centered or is grand mean centered may have greater utility because these models directly result in a test of $H_0:\gamma_C = 0$. However, if the independent variable is group mean centered $H_0:\gamma_C = 0$ can be tested by testing $H_0:\gamma_B - \gamma_W = 0$.

If the researcher is interested in an integral level-2 variable (e.g., disciplinary climate) in the model then the correct centering depends on the question the researcher plans to address. If the researcher is interested in the within-group effect of the level-1 variables and the between group effect of the level-2 variables, then the level-1 variables should be group mean centered. However, such models should be used with caution because the level-2 variables are investigated without complete control of the level-1 variables. In our opinion it will usually be preferable to include means of the level-1 variables in the model and then any of the three centering options can be used (see Equation 15.7, for example).

15.10 MODIFYING THE RANDOM INTERCEPTS AND INTERCEPTS AS OUTCOMES MODELS: ADDING A RANDOMLY VARYING SLOPE

The random intercepts model and the intercepts as outcomes model entail the assumption that the school specific slope is a constant across schools. Incorrectly making this assumption is not likely to have a strong impact on the coefficients in these models, but it can impact the standard errors and precludes estimation of the variance of the school specific slopes. Both the random intercepts model and the intercepts as outcomes models can be modified by specifying that the school specific slope varies across schools.

15.10.1 Random Regression Coefficients Models

By adding a randomly varying slope to the random intercepts model we obtain the random regression coefficients model. The version of this model without centering *SES* is

Level-1:

$$MA_{ij} = \beta_{0j} + \beta_{1j} SES_{ij} + \varepsilon_{ij}.$$

Level-2:

$$\beta_{0j} = \gamma_0 + u_{0j}$$

$$\beta_{1j} = \gamma_{SES} + u_{1j}.$$

Combined

$$MA_{ij} = \gamma_0 + \gamma_{SES} SES_{ij} + u_{0j} + u_{1j} SES_{ij} + \varepsilon_{ij}. \tag{15.9}$$

As with the random intercepts models, there are two other variations of the random regression coefficients model, created by either grand mean centering or group mean centering the independent variable. When the independent variable is grand mean centered the model is

$$\begin{aligned} MA_{ij} &= \gamma_0 + \gamma_{SES}\left(SES_{ij} - \overline{\overline{SES}}\right) \\ &\quad + u_{0j} + u_{1j}\left(SES_{ij} - \overline{\overline{SES}}\right) + \varepsilon_{i} \end{aligned} \tag{15.10}$$

TABLE 15.3

Summary of Results for Random Regression Coefficients and Intercepts and Slopes as Outcomes Models

Model	Equation	Coefficient (Standard Error) for *SES*	Coefficient (Standard Error) for MEAN *SES*
Random regression coefficients	$MA = \gamma_0 + \gamma_{SES} SES + u_0 + u_1 SES + \varepsilon$	2.375 (0.127)	NA
Random regression coefficients	$MA = \gamma_0 + \gamma_{SES}(SES - \overline{\overline{SES}}) + u_0 + u_1(SES - \overline{\overline{SES}}) + \varepsilon$	2.375 (0.127)	NA
Random regression coefficients	$MA = \gamma_0 + \gamma_{SES}(SES - \overline{SES}_j) + u_0 + u_1(SES - \overline{SES}_j) + \varepsilon$	2.173 (0.127)	NA
Intercepts and slopes as outcomes	$MA = \gamma_0 + \gamma_W SES + \gamma_C \overline{SES}_j + u_0 + u_1 SES + \varepsilon$	2.186 (0.128)	4.115 (0.402)
Intercepts and slopes as outcomes	$MA = \gamma_0 + \gamma_W(SES - \overline{\overline{SES}}) + \gamma_C \overline{SES}_j + u_0 + u_1(SES - \overline{\overline{SES}}) + \varepsilon$	2.186 (0.128)	4.115 (0.402)
Intercepts and slopes as outcomes	$MA = \gamma_0 + \gamma_W(SES - \overline{SES}_j) + \gamma_B \overline{SES}_j + u_0 + u_1(SES - \overline{SES}_j) + \varepsilon$	2.183 (0.127)	6.295 (0.382)

The models in Equation 15.9 and Equation 15.10) are statistically equivalent. Because the intercepts in Equations 15.9 and 15.10 are defined for different values on the *SES* scale (see the discussion of the intercept for the simple regression model when the independent variable is not centered and when it is grand mean centered) the estimated intercepts (33.33 and 45.21), variance components for u_{0j} (29.85 and 5.68) and covariance for u_{0j} and u_{1j} (−4.18 and −0.64) are different for the two models. Nevertheless, the results for one of the models can be obtained from the results for the second. The estimated variance component for u_{1j} will be the same for the two models and is 0.71 for the example.

Each of the models given in Equations 15.9 and 15.10 assumes that the school specific intercept (β_{0j}) is independent of *SES*. If the data analyst does not want to make this assumption, the independent variable can be group mean centered. The resulting model is

$$MA_{ij} = \gamma_0 + \gamma_W(SES_{ij} - \overline{SES}_j) + u_{0j} + u_{1j}(SES_{ij} - \overline{SES}_j) + \varepsilon, \quad (15.11)$$

and is not statistically equivalent to the other two. Results for the three models are shown in Table 15.3. The coefficients for *SES* are equal for the first two models, but not for the third.

Type of centering affects the meaning of the intercepts and as a consequence the intercept (45.11) is different for Equation 15.11 than for Equation 15.9 or Equation 15.10. Similarly the variance component for u_{0j} (10.14) and covariance for u_{0j} and u_{1j} (−0.96) are different for Equation 15.11) than for Equations 15.9 and 15.10. For all three types of centering, the variance component for u_{1j} is the variance of the school specific slope (β_{1j}). Nevertheless the estimate is different for Equation 15.11 than for Equation 15.9 and Equation 15.10. For Equation 15.11 the estimate is 0.63. According to Raudenbush and Bryk (2002) group mean centering results in a better estimate of the variance component for u_{1j}.

Comparing results for the random intercepts (Table 15.2) and random regression coefficients (Table 15.3) models, we see that even with the same kind of centering, the coefficients vary somewhat across the two types of models. However, as in this example, the coefficients are typically fairly similar for the two model types.

15.10.2 Intercepts and Slopes as Outcomes Models

By adding a randomly varying slope to the intercepts as outcomes model we obtain an intercepts and slopes as outcomes model. The version of this model without centering *SES* is

Level-1:

$$MA_{ij} = \beta_{0j} + \beta_{1j} SES_{ij} + \varepsilon_{ij},$$

Level-2:

$$\beta_{0j} = \gamma_0 + \gamma_C \overline{SES}_j + u_{0j}$$

$$\beta_{1j} = \gamma_W + u_{0j},$$

Combined:

$$MA = \gamma_0 + \gamma_W SES + \gamma_C \overline{SES}_j + u_{0j} + u_{1j} SES + \varepsilon_{ij}. \quad (15.12)$$

There are two varieties of the intercepts and slopes as outcomes model. In the first variety, which is the subject of this section, the level-2 variable is a predictor of the intercept only. In the second variety, the subject of a subsequent section entitled "Models with Cross-level Interactions," the level-2 variable is a predictor of the slope.

Observations about the first variety of the intercepts and slopes as outcomes model are similar to the observations about random regression coefficients models. The model in Equation 15.12, with a noncentered level-1 independent variable and the model with a grand mean centered level-1 independent variable,

$$MA = \gamma_0 + \gamma_W \left(SES - \overline{\overline{SES}}\right) + \gamma_C \overline{SES}_j + u_{0j} + u_{1j} \left(SES - \overline{\overline{SES}}\right) + \varepsilon_{ij}, \quad (15.13)$$

are statistically equivalent. The coefficient for *SES* in these models is a within school coefficient and the coefficient for school means *SES* is a context coefficient. When the level-1 independent variable is group mean centered, the model is

$$MA_{ij} = \gamma_0 + \gamma_W \left(SES_{ij} - \overline{SES}_j\right) + \gamma_B \overline{SES}_j + u_{0j} + u_{1j} \left(SES_{ij} - \overline{SES}_j\right) + \varepsilon_{ij} \quad (15.14)$$

and is not statistically equivalent to the other two models. The fact that the three models are not statistically equivalent stands in contrast to the equivalence status of the three intercepts as outcomes model.

Recall that we defined statistically equivalent models as models for which any estimate or test statistic obtained by using one of the models can be obtained by using the other models. Kreft et al. (1995) classified models as equivalent in the fixed effects and/or equivalent in the variance components. Models are equivalent in the fixed effects if the population fixed effects for one model can be expressed as functions of the population fixed effects of the other model. Models are equivalent in the variance components if the population variance components for one model can be expressed as functions of the population variance components of the other model. Models that are statistically equivalent by our definition are equivalent by both of the criteria set forth by Kreft et al. (1995). The models in Equations 15.12, 15.13, and 15.14 are equivalent in the fixed effects. The model in Equation 15.14 is not equivalent to the other models in the variance components.

Despite the fact that γ_W is the same parameter for Equation 15.12 to Equation 15.14, the estimate of γ_W for Equation 15.14 is not necessarily the same as is the estimate of γ_W in Equations 15.12 and 15.13. This occurs because the models are not equivalent in the variance components and the variance components are used in estimating the fixed effects. In the present example $\hat{\gamma}_W$ for Equation 15.14 differs from $\hat{\gamma}_W$ in Equations 15.12 and 15.13 in the third decimal place. This is consistent with our experience that all three models will provide similar estimates of γ_W. With the intercepts as outcomes model, $\hat{\gamma}_C = \hat{\gamma}_B - \hat{\gamma}_W$. This is not necessarily true with the intercepts and slopes as outcomes model (see Table 15.3). Comparing results for the intercepts as outcomes (Table 15.2) and intercepts and slopes as outcomes (Table 15.3) models we see that even with the same kind of centering, the coefficients vary across the two types of models.

In Section 15.7 we discussed models in which a level-2 variable is an integral variable rather than a contextual variable. With the following exceptions, the discussion in Section 15.7 applies to the first variety of intercepts and slopes as outcomes model.

1. The difference in the estimates of γ_{DC}^B in the means as outcomes model (see Equation 15.7) and in the intercepts as outcomes model (see Equation 15.6) is likely to be smaller than the difference in the estimates of γ_{DC}^B in the means as outcomes and in intercepts and slopes as outcomes model:

$$MA_{ij} = \gamma_0 + \gamma_W(SES - SES_j) + \gamma_{DC}^B DC_j + u_{0j} + (SES - SES_j)u_{1j} + \varepsilon_{ij}. \quad (15.15)$$

2. The model in Equation 15.15 is not statistically equivalent to the versions of this model in which the level-1 variable is not centered or it is grand mean centered, though it is equivalent in the fixed effects.

15.11 RECOMMENDATIONS: IN REGARD TO RANDOM REGRESSION COEFFICIENTS AND INTERCEPTS AND SLOPES AS OUTCOMES MODELS

When a researcher is only interested in the within group effect (γ_W), the random regression coefficients model with group mean centering should be used. In addition, according to Raudenbush and Bryk

(2002) using group mean centering provides a better estimate of the variance components for u_{1j}.

When the researcher is interested in the within group coefficient and the context coefficient (γ_C) the intercepts and slopes as outcomes model can be used. The independent variable should either be noncentered or grand mean centered. When the researcher is interested in the within group coefficient and the between group (γ_B) the intercepts and slopes as outcomes model with a group mean centered independent variable can be used.

15.12 MODELS WITH CROSS-LEVEL INTERACTIONS

As noted earlier, the model with *SES* and school mean *SES* as independent variables,

$$MA_{ij} = \gamma_0 + \gamma_W SES + \gamma_C \overline{SES}_j + u_{0j} + u_{1j} SES + \varepsilon_{ij},$$

is an example of the first variety of the intercepts and slopes as outcomes model. One assumption of this model is that the school specific slope β_{1j} is uncorrelated with $\overline{SES}_j$. To address this assumption the following intercepts and slopes as outcomes model, an example of the second variety, can be used:

Level-1:

$$MA_{ij} = \beta_{0j} + \beta_{1j} SES_{ij} + \varepsilon_{ij}$$

Level-2:

$$\beta_{0j} = \gamma_0 + \gamma_2 \overline{SES}_j + u_{0j}$$

$$\beta_{1j} = \gamma_1 + \gamma_3 \overline{SES}_j + u_{1j},$$

Combined:

$$MA_{ij} = \gamma_0 + \gamma_1 SES + \gamma_2 \overline{SES}_j + \gamma_3 (SES \times \overline{SES}_j) + u_{0j} + u_{1j} SES + \varepsilon_{ij}. \tag{15.16}$$

The inclusion of $SES \times \overline{SES}_j$ allows investigation of the cross-level interaction. Alternatives to Equation 15.16 replace *SES* by grand mean centered *SES* or by group mean centered *SES*. As usual, the model with group mean centered *SES* is not statistically equivalent to the other two, which are statistically equivalent. In addition, as shown by Kreft et al. (1995) the model with group mean centered *SES* is not equivalent

TABLE 15.4

Summary of Results for Intercept and Slopes as Outcomes Model with a Cross-Level Interaction

Random Effect for the Slope	Centering for *SES*	Coefficient (Standard Error) for *SES* ($\hat{\gamma}_1$)	Coefficient (Standard Error) for Mean *SES* ($\hat{\gamma}_2$)	Coefficient (Standard Error) for Interaction ($\hat{\gamma}_3$)
Yes	None	2.074 (1.568)	4.056 (1.592)	1.524 (0.314)
	Grand Mean	2.074 (1.568)	11.674 (0.378)	1.524 (0.314)
	Group Mean	1.752 (1.660)	21.157 (0.361)	1.587 (0.332)
No	None	1.943 (1.285)	3.937 (1.381)	1.547 (0.257)
	Grand Mean	1.943 (1.285)	11.673 (0.376)	1.547 (0.257)
	Group Mean	1.672 (1.378)	21.157 (0.361)	1.602 (0.276)

TABLE 15.5

Summary of Simulation Results for Intercept and Slopes as Outcomes Model with a Cross-Level Interaction

Centering for *SES*	Coefficient (Standard Deviation) for *SES* ($\hat{\gamma}_1$)	Coefficient (Standard Deviation) for Mean *SES* ($\hat{\gamma}_2$)	Coefficient (Standard Deviation) for Interaction ($\hat{\gamma}_3$)
None	2.19	3.81	1.50
	(1.43)	(0.72)	(0.29)
Grand mean	2.19	11.32	1.50
	(1.43)	(0.42)	(0.29)
Group mean	2.20	20.84	1.50
	(1.52)	(0.42)	(0.30)

to the other two models in the fixed effects or in the variance components.

To generate data to compare the results of applying the three model variations, we used the following model:

$$\mathcal{E}(MA_{ij}) = 15 + 2.2(SES_{ij}) + 3.8(\overline{SES}_j) + 1.5(SES_{ij} \times \overline{SES}_j)$$

and

$$MA_{ij} = \mathcal{E}(MA_{ij}) + u_{0j} + u_{1j}SES_{ij} + \varepsilon_{ij}.$$

The covariance matrix for the u_{0j} and u_{1j} was again

$$\begin{bmatrix} 16.00 & \\ -2.50 & 0.50 \end{bmatrix}$$

and the variance for ε_{ij} was 37.

The first three lines in the body of tables 15.4 contain results for the three models. Note that the estimate of the coefficient for the product term (γ_3) when *SES* is group mean centered is different than the estimate of γ_3 for the other types of centering and is less similar to 1.5, which is the true value of γ_3. To determine if this finding is specific to this particular set of simulated data, we replicated the simulation 1000 times. The results are reported in Table 15.5 and indicate little if any difference among the three centering methods in the parameter being estimated by $\hat{\gamma}_3$. We also calculated the standard deviation of the difference between the estimate of γ_3 under group mean centering and the estimate under either of the other types of centering. These latter two estimates must be equal. The standard deviation was .09, indicating that estimates of γ_3 were fairly similar across centering methods in most replications.

In our example, centering had a minimal effect on the estimate of γ_3. However, the type of centering used in Equation 15.16 can affect the estimate of γ_3 and when it does, it typically means that an independent variable has been omitted from the model. For example, if the following model is correct for the data

$$\mathcal{E}(MA_{ij}) = \gamma_0 + \gamma_1 SES_{ij} + \gamma_2 \overline{SES}_j + \gamma_4 DC_j + \gamma_5 \overline{SES}_j \times DC_j, \quad (15.17)$$

and if $\overline{SES}_j \times DC_j$ and $\overline{SES}_j(SES_{ij} - \overline{SES})$ are correlated, then in

$$\mathcal{E}(MA_{ij}) = \gamma_0 + \gamma_1(SES_{ij} - \overline{\overline{SES}}) + \gamma_2\overline{SES}_j + \gamma_3\overline{SES}_j(SES_{ij} - \overline{\overline{SES}}), \tag{15.18}$$

γ_3 will not be equal to zero and spurious evidence for the cross-level interaction of school mean *SES* and individual *SES* can emerge. The same conclusion holds for the version of Equation 15.17 in which *SES* is not centered. However, if *SES* is group mean centered the correlation between $\overline{SES}_j \times DC_j$ and $\overline{SES}_j(SES_{ij} - \overline{SES}_j)$ must be zero and, assuming Equation 15.17 is the correct model for the data, γ_3 will be equal to zero when the version of Equation 15.18 with a group mean centered *SES* is used.

As an alternative to using Equation 15.18, the following equation can be used

$$\mathcal{E}(MA_{ij}) = \gamma_0 + \gamma_1 SES_{ij} + \gamma_2\overline{SES}_j + \gamma_3\overline{SES}_j \times SES_{ij} + \gamma_4 DC_j + \gamma_5\overline{SES}_j \times DC_j.$$

Then assuming Equation 15.17 is the correct equation, spurious evidence for the cross-level interaction of school mean *SES* and individual *SES* should not emerge. That is, if the researcher knows what integral variable (i.e., DC_j) is likely to interact with the contextual variable (i.e., $\overline{SES}_j$), then the product term $\overline{SES}_j \times DC_j$ can be included in the model and there should not be spurious evidence that $\gamma_3 \neq 0$. This will be true regardless of which type of centering is used for *X*.

Similarly, Raundenbush (1989; see also, Enders & Tofighi, 2007; & Hoffman and Gavin, 1998) pointed out that if Equation 15.17 is correct, then using

$$\mathcal{E}(MA_{ij}) = \gamma_0 + \gamma_1(SES_{ij} - \overline{\overline{SES}}) + \gamma_4 DC_j + \gamma_6(SES_{ij} - \overline{\overline{SES}})DC_j, \tag{15.19}$$

could result in spurious evidence that $\gamma_6 \neq 0$ because $\overline{SES}_j \times DC_j$ and $(SES_{ij} - \overline{\overline{SES}})DC_j$ will be correlated. The same conclusion applies to the version of Equation 15.19 in which *SES* is not centered. This problem can be overcome by using group mean centering. In addition, as pointed out by Enders and Tofighi (2007), if $H_0: \gamma_6 = 0$ is tested by using

$$\mathcal{E}(MA_{ij}) = \gamma_0 + \gamma_1(SES_{ij} - \overline{SES}_j) + \gamma_2\overline{SES}_j + \gamma_4 DC_j + \gamma_5\overline{SES}_j \times DC_j + \gamma_6(SES_{ij} - \overline{SES}_j)DC_j,$$

then, assuming Equation 15.17 is correct, spurious test results for the test of $H_0: \gamma_6 = 0$ should not emerge. The same conclusion holds regardless of the centering of *SES*. The general principal in these two examples is that in multilevel models as in regression models, it is essential to avoid omitting important variables from the model. Raudenbush and Bryk (2002, Chapter 10) describe a procedure for detecting omitted level-2 variables.

In Table 15.4, the estimate of the coefficient for the SES variable (γ_1) is quite different when *SES* is group mean centered than it is for the other two types of centering, but results in Table 15.5, for the 1000 simulated data sets, indicate little if any difference among the three centering methods in the parameters estimated by $\hat{\gamma}_1$. The standard deviation of the difference between the estimates of γ_1 under group mean centering and the estimate of γ_1 under either of the other type of centering was .50, indicating that the estimate of γ_1 under group mean centering could be quite different than the estimate under the other types of centering. This is consistent with the results in Table 15.4.

In Table 15.4, the estimate of the coefficient for school mean *SES* (γ_2) varies dramatically across the three methods of centering. We can understand these differences by considering the formula for the estimated simple slope for school mean *SES*. The general expression for a simple slope is (a) the coefficient for the variable of interest, which is school mean *SES* in the present example and is equal to $\hat{\gamma}_2$, plus (b) the coefficient of the product term, which is $\hat{\gamma}_3$, multiplied by the other variable in the product term, which will be one of the forms of *SES*. Under group mean centering, the formula for the simple slope for $\overline{SES}_j$ is $\hat{\gamma}_2+\hat{\gamma}_3\left(SES-\overline{SES}_j\right)$.' Substituting a value for $\left(SES-\overline{SES}_j\right)$ in the formula provides an estimate of the slope for $\overline{SES}_j$ among students at that level of $\left(SES-\overline{SES}_j\right)$. For example, if we focus attention on students with an *SES* that is two points above the mean *SES* for their school we find 21.157 + 1.587(2) = 24.331, the simple slope for school mean *SES* among students whose *SES* is two points above the mean *SES* for their school. Based on $\hat{\gamma}_2+\hat{\gamma}_3\left(SES-\overline{SES}_j\right)$, $\hat{\gamma}_2$ is obtained by substituting zero for $\left(SES-\overline{SES}_j\right)$ and is the simple slope for school mean *SES* among students whose *SES* is equal to their school mean *SES*. Thus $\hat{\gamma}_2 = 21.157$ is the simple slope for school mean *SES* among students whose *SES* is equal to their school mean *SES*.

When *SES* is not centered, the simple slope for school mean *SES* is $\hat{\gamma}_2+\hat{\gamma}_3 SES$. Using $\hat{\gamma}_2+\hat{\gamma}_3 SES$, $\hat{\gamma}_2$ is obtained by substituting zero for *SES* and is the slope for school mean *SES* among students whose *SES* is 0. Thus $\hat{\gamma}_2 = 4.056$ is the simple slope for school mean *SES* among students whose *SES* is 0. To obtain the simple slope formula for grand mean centering, replace *SES* in $\hat{\gamma}_2+\hat{\gamma}_3 SES$, by $\left(SES-\overline{\overline{SES}}\right)$ to obtain $\hat{\gamma}_2+\hat{\gamma}_3\left(SES-\overline{\overline{SES}}\right)$. Therefore, under grand mean centering $\hat{\gamma}_2 = 11.674$ is the simple slope for school mean *SES* among students whose $SES-\overline{\overline{SES}}$ is equal to zero or, equivalently, whose *SES* is at the grand mean. These considerations show that $\hat{\gamma}_2$ varies across the centering because $\hat{\gamma}_2$ is a simple slope for school mean *SES* when the *SES* independent variable is zero and the meaning of zero on the *SES* independent variable varies across the three centering methods.

Recall that $\hat{\gamma}_2 = 11.674$ in the model with grand mean centered *SES* and that $\hat{\gamma}_2$ is the simple slope for school mean *SES* among students whose *SES* is at the grand mean. The simple slope for mean *SES* in the model in which SES is not centered is $\hat{\gamma}_2+\hat{\gamma}_3 SES$. We can use this expression to find the simple slope for school mean *SES* among students whose *SES* is at the grand mean by substituting 5 for *SES*. We find 4.056 + 1.524(5) = 11.676, which is within rounding error of the result obtained by using the model in which *SES* was grand mean centered. This illustrates that when models are equivalent, any estimate that can be obtained from one model can also be obtained from the other model.

Comparison of the formula for the simple slope for school mean *SES* under group mean centering $\left[\hat{\gamma}_2+\hat{\gamma}_3\left(SES-\overline{SES}_j\right)\right]$ to either the formula when there is no centering $[\hat{\gamma}_2+\hat{\gamma}_3 SES]$ or the formula when there is grand mean centering $\left[\hat{\gamma}_2+\hat{\gamma}_3\left(SES-\overline{\overline{SES}}\right)\right]$

shows that the nature of the simple slope is quite different under group mean centering. Specifically $\hat{\gamma}_2+\hat{\gamma}_3(SES-\overline{SES}_j)$ estimates the effect of school mean *SES* for students who are a particular distance from their school mean *SES*. For example if $(SES-\overline{SES}_j)=1$ then $\hat{\gamma}_2+\hat{\gamma}_3(SES-\overline{SES}_j)=21.157+1.587(1)=22.744$ and tells us that among students whose *SES* is one point above the school mean *SES*, the effect of school mean *SES* is about 23 points. According to the model, this implication holds regardless of the school's mean *SES*. So if we focus attention on students with an SES of 4.5 in schools with an average *SES* of 3.5 (a low *SES*) or on students with an SES of 7.5 in a school with an average *SES* of 6.5 (a high *SES*), the effect of school mean *SES* will be 23 points. By contrast, according to either of the other models the simple slope for school mean *SES* depends on the student's actual *SES*.

Cross-level interactions can also be investigated by using intercepts and slopes as outcomes models without a random effect for the slope. If *SES* is not centered the model is

$$MA_{ij}=\gamma_0+\gamma_1 SES+\gamma_2\overline{SES}_j$$
$$+\gamma_3(SES\times\overline{SES}_j)+u_{0j}+\varepsilon_{ij}$$

Results for this model, as well as results for variations on this model obtained by grand mean or group mean centering, are presented in the last three lines in the body of table 15.4. The effect of deleting the u_{1j} term on the coefficients is fairly small. But it is known that failing to include a required random effect tends to result in standard errors that underestimate the sampling variability in the estimates. Consistent with this result, the standard errors for the estimates tend to be smaller when the u_{1j} term is not included.

15.13 RECOMMENDATIONS FOR MODELS WITH CROSS-LEVEL INTERACTIONS

One issue is whether or not to include the u_{ij} term in the model. Whenever possible the u_{ij} term should be included. Another issue is the type of centering for the level-1 independent variables. In our simulation, centering had little effect on the interaction effect. These results no doubt reflect the fact that we used the same model to simulate the data and to analyze the data. Nevertheless, our experience is that centering often has little effect on the interaction coefficient. When centering has a strong effect on the estimate of the interaction parameter, it typically means that an independent variable has been omitted from the model. If centering does not have a strong impact on the cross-level interaction (i.e., the interaction is significant both when group mean centering is used and when it is not or is not significant in both cases) then we recommend against group mean centering unless the researcher wants to investigate the simple effect for the level-2 variable defined by how far removed the participant is from the group mean on the level-1 variable and not on the participant's actual score on the level-1 variable. If centering does have a strong impact on the cross-level interaction (i.e., the interaction is not significant when group mean centering is used but is significant with the other two types of centering) then we recommend excluding the cross-level-interaction from the model.

15.14 CENTERING LEVEL-2 VARIABLES

When a level-2 variable is centered, the new model is statistically equivalent to the original model. Therefore any estimate or hypothesis test that can be obtained by using the original model can also be obtained by using the revised model. Consequently centering level-2 variables is a less important issue than is centering level-1 variables. To illustrate let us consider Equation 15.16 and the model obtained by grand mean centering $\overline{SES}_j$:

$$MA_{ij} = \gamma_0 + \gamma_1 SES + \gamma_2(\overline{SES}_j - \overline{\overline{SES}}) + \gamma_3(SES \times [\overline{SES}_j - \overline{\overline{SES}}]) + u_{0j} + u_{1j}SES + \varepsilon_{ij}.$$

Results for Equation 15.16, originally reported in Table 15.4, are $\hat{\gamma}_1 = 2.074$, $\hat{\gamma}_2 = 4.056$, and $\hat{\gamma}_3 = 1.524$. After grand mean centering $\hat{\gamma}_1 = 9.692$, $\hat{\gamma}_2 = 4.056$, and $\hat{\gamma}_3 = 1.524$. Grand mean centering did not affect $\hat{\gamma}_2$ or $\hat{\gamma}_3$ because their interpretation is independent of level-2 centering. Grand mean centering did affect $\hat{\gamma}_1$ because its interpretation is affected by level-2 centering:

- When $\overline{SES}_j$ is the level-2 variable $\hat{\gamma}_1$ is the simple slope for *SES* when $\overline{SES}_j$ is zero
- When $\overline{SES}_j - \overline{\overline{SES}}$ is the level-2 variable $\hat{\gamma}_1$ is the simple slope for *SES* when $\overline{SES}_j$ is equal to the grand *SES* mean

Similarly when the level-1 variable is centered (either grand mean or group mean) in the cross-level interaction model, centering the level-2 variable only affects the estimate of γ_1. In the multilevel models considered in this chapter that do not include cross-level interactions, centering the level-2 variable only affects the intercept.

15.15 MODELS WITH REPEATED MEASURES

As an example of a repeated measures design, consider the data provided with HLM 6.0 in which adolescents were asked about their attitudes toward deviant behavior (*ATT*) and exposure to deviant peers (*EXP*) each year from age 11 to age 15. (For the sake of simplicity we used a version of this data set in which all participants with missing data were eliminated from the data file.) Henceforth we refer to data like those in the *MA* and *SES* example as between-subjects data because all of the variables vary between-subjects. We refer to data like those in the *ATT* and *EXP* example as mixed data because the variables vary within-subjects (over time) and between-subjects. The same issues we developed in the context of the *MA* and *SES* example could be developed in the context of this new example. Rather than repeating the developments we introduce some issues that apply primarily or uniquely in repeated measures designs.

Suppose the researchers are interested in the relationship of *ATT* to *EXP* and the researchers use the model

$$ATT_{ij} = \gamma_0 + \gamma_{EXP} EXP_{ij} + u_{0j} + \varepsilon_{ij}, \quad (15.20)$$

where ATT_{ij} is the attitude for person *j* at age *i*. By using this model the researchers run the risk of confounding the within person relationship between the two variables (how *ATT* changes over time with changes in *EXP*) with the between person relationship (how *ATT* scores vary across people in relation to variation in *EXP* scores across people.) Based on the developments for between-subjects data, the two relationships can be separated by using the model

$$ATT_{ij} = \gamma_0 + \gamma_W EXP_{ij} + \gamma_C \overline{EXP}_j + u_{0j} + \varepsilon_{ij}, \quad (15.21)$$

where $\overline{EXP}_j$ is the average *EXP* score for person *j* over the five ages. Separation of the two aspects of the relationship can also be achieved by using the other centering options. With group mean centering, the *EXP* independent variable becomes $EXP_{ij} - \overline{EXP}_j$ and is the deviation of the jth person's *EXP* score at age *i* from that person's mean over the five ages. Thus, group mean centering in a repeated measures design is person mean centering. With grand mean centering, the *EXP* independent variable becomes $EXP_{ij} - \overline{\overline{EXP}}$ and is the deviation of the jth person's *EXP* score at age *i* from the grand mean; that is, the mean over all ages and participants. Applying Equation 15.21 we obtain $\hat{\gamma}_0 = .324$, $\hat{\gamma}_W = .465$, and $\hat{\gamma}_C = .071$. Testing $H_0 : \gamma_C = 0$ tests the same hypothesis that is tested in Hausman's (1978) test to determine if a fixed or random effects model is appropriate for the data. If $H_0 : \gamma_C = 0$ is rejected then Equation 15.21 is preferred over Equation 15.20 and, as noted above, yields $\hat{\gamma}_W$ that is equal to the coefficient one would obtain by using the fixed effects model.

The purpose of including the person mean of *EXP* is to account for the possibility that u_{0j} is correlated with *EXP* in Equation 15.20 or equivalently that the within person intercept $\beta_{0j} = \gamma_0 + u_{0j}$ is correlated with *EXP*. There is another approach to accounting for this possibility that can be used with repeated measures data. The model is

$$ATT_{ij} = \gamma_0 + \gamma_W EXP_{ij} + u_{0j} + \varepsilon_{ij}, \quad (15.22)$$

but the estimation procedure allows for the possibility that u_{0j} is correlated with *EXP* at each occasion (Allison, 2005). Allowing for the possibility that u_{0j} is correlated with *EXP* at each occasion results in estimation of a within person coefficient. This procedure can be implemented in any structural equation modeling program. Appendix A presents an M*plus* program for implementing the method. The code—`a1 a2 a3 a4 a5 (3)`—restricts the variance of the residuals to be equal at all ages to maximize similarity of the results obtained by using Equation 15.21; the restriction is not necessary. The results are $\hat{\gamma}_0 = .324$ and $\hat{\gamma}_W = .465$. The estimate $\hat{\gamma}_W$ and its standard error are equal to the estimate $\hat{\beta}_W$ and its standard error that would be obtained by using the fixed effects model. A likelihood ratio test comparing the model that specifies u_{0j} is correlated with *EXP* at each occasion to one without the specification can be used to select between the two approaches. Without the restriction on the residual variances, the results are $\hat{\gamma}_0 = .319$ and $\gamma_W = .456$.

Estimation of Equation 15.22 allowing for the possibility that u_{0j} is correlated with *EXP* at each occasion can also be implemented in a hierarchical linear modeling program by writing the model as

Level-1:

$$ATT_{ij} = \beta_{0j} + \beta_1 EXP_{ij} + \varepsilon_{ij}$$

Level-2:

$$\beta_{0j} = \gamma_0 + \gamma_{01} EXP_{11j} + \cdots + \gamma_{05} EXP_{15j} + u_{oj}$$

and

$$\beta_1 = \gamma_W.$$

where $EXP_{11j},\ldots,EXP_{15j}$ are the exposure variables at the five ages. The level-2 equation is equivalent to Chamberlin's (1982, 1984) specification for the relationship of u_{0j} to the predictors in the model. The combined model is

$$ATT_{ij} = \gamma_0 + \gamma_W EXP_{ij} + \gamma_{01} EXP_{11j} + \cdots + \gamma_{05} EXP_{15j} + u_{0j} + \varepsilon_{ij}, \tag{15.23}$$

Comparing Equations 15.23 and 15.21 shows that if $\gamma_{01} = \ldots = \gamma_{05} \equiv \gamma$ then Equation 15.23 simplifies to Equation 15.21 with $\gamma_C = 5\gamma$. To estimate Equation 15.23, we used the multivariate linear two-level model with heterogeneous residual variances in HLM 6.0. A screen shot of the program is presented in Appendix B. The results obtained were $\hat{\gamma}_0 = .302$ and $\gamma_W = .465$. The coefficient $\hat{\gamma}_0$ is different in Equation 15.23 and Equation 15.22, but the intercept obtained by using Equation 15.22 can be obtained by revising Equation 15.23 with each of $EXP_{11j},\ldots,EXP_{15j}$ centered around its mean. The hypothesis $H_0{:}\gamma_{01} = \ldots = \gamma_{05} = 0$ can be tested to determine if correlations between the person-specific intercepts and the exposure variable at each age are required in the model. The hypothesis tested in Hausman's test is a special case of $H_0{:}\gamma_{01} = \ldots = \gamma_{05} = 0$ because Hausman's approach assumes $\gamma_{01} = \ldots = \gamma_{05} = \gamma_C/5$.

Another approach to centering in repeated measures designs involves the concept of the cross-sectional and the longitudinal effects (see Diggle, Heagerty, Liang, & Zeger, 2002). Consider the two-level model

Level-1:

$$ATT_{ij} = \beta_{0j} + \beta_{1j}\left(EXP_{ij} - EXP_{11j}\right) + \varepsilon_{ij}$$

Level-2:

$$\beta_{0j} = \gamma_0 + \gamma_X EXP_{11j} + u_{0j}$$

and

$$\beta_{1j} = \gamma_L$$

where the subscript X denotes cross-sectional and the subscript L abbreviates longitudinal. Rather than group mean or grand mean centering *EXP*, the variable $EXP_{ij}-EXP_{11j}$ expresses *EXP* as a deviation from the value of *EXP* at age 11. The combined model is

$$ATT_{ij} = \gamma_0 + \gamma_X EXP_{11j} + \gamma_L\left(EXP_{ij} - EXP_{11j}\right) + u_{0j} + \varepsilon_{ij}.$$

In the level-1 equation, *EXP* is expressed as a deviation from *EXP* at age 11. As a consequence of this new type of centering, the intercept β_{0j} is expected *ATT* when *EXP* is equal to its value at age 11. We can think of β_{0j} as model-implied attitude for the adolescent j at age 11. An adolescent with a high β_{0j} will tend to have a high *ATT* at age 11 and an adolescent with a low β_{0j} will tend to have a low *ATT* at age 11. The coefficient γ_X is the cross sectional effect and measures the effect of *EXP* at age 11 on *ATT* at age 11. The variable $EXP_{ij}-EXP_{11j}$ measures change over time in *EXP*. Thus γ_L measures the effect of changes in *EXP* on *ATT*. The results are $\hat{\gamma}_X = .409$, implying that at age 11 adolescents with more exposure to deviant peers have a more positive attitude toward deviance, and $\hat{\gamma}_L = .504$, implying that as exposure to deviant peers increases (decreases) overtime, attitude toward deviance increases (decreases).

An alternative conceptualization of the model is

Level-1:

$$ATT_{ij} = \beta_{0j} + \beta_{1j}EXP_{ij} + \varepsilon_{ij}$$

Level-2:

$$\beta_{0j} = \gamma_0 + \gamma_1 EXP_{11j} + u_{0j}$$

and

$$\beta_{1j} = \gamma_2.$$

The combined model is

$$ATT_{ij} = \gamma_0 + \gamma_1 EXP_{11j} + \gamma_2 EXP_{ij} + u_{0j} + \varepsilon_{ij}.$$

It follows then that $\gamma_2 = \gamma_L$ is the longitudinal effect, and $\gamma_1 = \gamma_X - \gamma_L$ is the difference between the cross sectional and longitudinal effects. The two models are statistically equivalent. The alternative model clarifies the nature of the assumption in regard to β_{0j}. The model assumes that β_{0j} is uncorrelated with *EXP* at ages 12 to 15. By contrast the models in Equations 15.21 and 15.22 do not make this assumption. The results of the new model are $\hat{\gamma}_2 = \hat{\gamma}_L = .504$ and $\hat{\gamma}_1 = \hat{\gamma}_X - \hat{\gamma}_L = -.095$.

In models considered so far, for both between-subjects and repeated measures data, centering was used to address assumption violations but centering also impacted interpretation of some of the parameters of the model. When the distribution of the level-1 variable is the same for all participants, centering is used to enhance interpretation of the parameters, not to address assumption violations. (The discussion that follows also applies when randomly missing data results in a distribution of the level-1 variable that is not the same for all participants.) For example, in the attitude toward deviant behavior example, attitudes were assessed annually from age 11 to 15 and a level-1 model of interest could be

$$Att_{ij} = \beta_{0j} + \beta_{1j}Age_{ij} + \varepsilon_{ij}, \quad (15.24)$$

where Age_{ij} ranges from 11 to 15. Thus the distribution of age is the same for all participants. The level-2 models are

$$\beta_{0j} = \gamma_{00} + u_{0j}$$

and

$$\beta_{1j} = \gamma_{10} + u_{1j}.$$

At any time point, Age_{ij} is the same for all participants and therefore it is impossible for u_{0j} or u_{1j} to be related to age and, consequently, no need to be concerned that u_{0j} or u_{1j} are related to age. Nevertheless, centering can enhance the interpretation of β_{0j} and therefore of γ_{00}. In Equation 15.24 the intercept β_{0j} is the expected attitude score for person j at age 0, and therefore is not subject to a meaningful interpretation. To make the intercept subject to a meaningful interpretation one of two alternative models, obtained by centering *Age*, might be used. The first model is

$$Att_{ij} = \beta_{0j} + \beta_{1j}(Age_{ij} - 11) + \varepsilon_{ij}. \quad (15.25)$$

Using Age_{ij}–11 as the independent variable centers the data so that zero represents the earliest age at which participants are measured. Then the intercept β_{0j} is the expected attitude score for person j at Age_{ij}–11 = 0, that is, at age 11 and in the combined model

$$Att_{ij} = \gamma_{00} + \gamma_{01}(Age_{ij} - 11) + u_{0j}$$
$$+ u_{1j}(Age_{ij} - 11) + \varepsilon_{ij}.$$

γ_{00} is the average attitude at age 11. The second model is

$$Att_{ij} = \beta_{0j} + \beta_{1j}(Age_{ij} - 13) + \varepsilon_{ij}, \quad (15.26)$$

where 13 is the midpoint of the age distribution. Using (Age_{ij}–13) as the independent variable centers the data so that zero represents the midpoint of the ages at which participants are measured. Now, the intercept β_{0j} is the expected attitude score for person j at Age_{ij}–13 = 0, that is, at age 13 and in the combined model

$$Att_{ij} = \gamma_{00} + \gamma_{01}(Age_{ij} - 13) + u_{0j}$$
$$+ u_{1j}(Age_{ij} - 13) + \varepsilon_{ij},$$

γ_{00} is the average attitude at age 13. The intercepts in Equation 15.25 and Equation 15.26 are different and therefore the variance of u_{0j} is different in these models as are the covariances of u_{0j} and u_{1j}. However, γ_{10} is equal in the two models as is the variance of u_{1j}.

When the distribution of the level-1 variable is the same for all participants (or differs only due to randomly missing data), it is common to investigate polynomial trends in the data. For example, a second degree trend in the data could be investigated by replacing Equation 15.24 by

$$Att_{ij} = \beta_{0j} + \beta_{1j}Age_{ij} + \beta_{2j}Age_{ij}^2 + \varepsilon_{ij}. \quad (15.27)$$

Unfortunately, the meaning of β_{1j} is not the same in Equations 15.24 and 15.27. Similarly the meaning of $\gamma_{10} = \mathcal{E}(\beta_{1j})$ is not the same in Equations 15.24 and 15.27. In Equation 15.24, β_{1j} and γ_{10} are the linear trend for person j and the average linear trend in the data, respectively. In Equation 15.27 β_{1j} is the instantaneous rate of change in attitude at zero years of age and γ_{10} measures the average, over adolescents, instantaneous rate of change in attitude at zero years of age. In order to avoid changes in the meaning of terms as higher order powers are added to the model, orthogonal polynomial variables can be used in place of powers of *Age* variable. Table 15.6 shows the orthogonal polynomial variables for use in the *Age* example. For example if the goal were to investigate linear and quadratic trends in the data, the model would be

$$Att_{ij} = \beta_{0j} + \beta_{1j}Linear_{ij} + \beta_{2j}Quadratic_{ij} + \varepsilon_{ij}.$$

TABLE 15.6

Orthogonal Polynomial Variables for the *Age* Example

	Orthogonal Polynomial Variables			
Age	**Linear**	**Quadratic**	**Cubic**	**Quartic**
11	−2	2	−1	1
12	−1	−1	2	−4
13	0	−2	0	6
14	1	−1	−2	−4
15	2	2	1	1

If the data analysts wanted to add a cubic trend to the model, s/he would use

$$Att_{ij} = \beta_{0j} + \beta_{1j}Linear_{ij} + \beta_{2j}Quadratic_{ij} + \beta_{3j}Cubic_{ij} + \varepsilon_{ij}.$$

In both models, β_{1j} measures the linear trend for person j and β_{2j} measures the quadratic trend for person j. It should be noted that the orthogonal polynomials in Table 15.6 are only appropriate if the values of *Age* are equally spaced or if there are five values for *Age*.

15.16 CONCLUSIONS

In this chapter we have used examples to present the basic results on centering in two-level models. Our approach was to present centering as a method that not only addresses assumption violations but also affects interpretation of parameters. Our presentation was primarily in the context of between-subjects designs. Rather than repeating the developments that apply to both between-subjects designs and repeated measures designs, for repeated measures designs we introduce some issues that apply primarily or uniquely to these designs.

REFERENCES

Allison, P. D. (2005). *Fixed effects regression methods for longitudinal data using SAS®*. Cary, NC: SAS Institute Inc.

Baltagi, B. H. (2005). *Econometric analysis of panel data* (3rd ed.). West Sussex, England: Wiley.

Chamberlin, G. (1982). Multivariate regression models for panel data. *Journal of Econometrics, 18*, 5–46.

Chamberlin, G. (1984). Panel data. In Z. Griliches and M. D. Intriligator (Eds.), *Handbook of Econometrics* (pp. 1247–1318). Amsterdam: North Holland.

Diggle, P., Heagerty, P., Liang, K-Y., & Zeger, S. (2002) Analysis of Longitudinal Data (2nd ed.). New York: Oxford University Press, USA.

Enders, C. K., & Tofighi, D. (2007). Centering predictor variables in cross-sectional multilevel models: A new look at an old issue. *Psychological Methods, 12*, 121–138.

Frees, E. W. (2004). *Longitudinal and panel data*. New York, NY: Cambridge University Press.

Greene, W. H. (2007). *Econometric analysis* (6th ed.). Upper Saddle River, NJ: Prentice Hall.

Hausman, J. A. (1978). Specification tests in econometrics. *Econometrica, 59*, 1251–1271.

Hofmann, D. A., & Gavin, M. B. (1998). Centering decisions in hierarchical linear models: Implications for research in organizations. *Journal of Management, 24*, 623–641.

Hox, J. J. (2002). *Multilevel analysis. Techniques and applications*. Mahwah, NJ: Lawrence Erlbaum Associates.

Kreft, I. G. G., de Leeuw, J., & Aiken, L. S. (1995). The effect of different forms of centering in hierarchical linear models. *Multivariate Behavioral Research, 30*, 1–21.

Misangi, V. F., LePine, J. A., Algina, J., & Goeddeke, F. (2006). The adequacy of repeated measures regression for multilevel research designs. *Organizational Research Methods, 9*, 5–26.

Pedhazur, E. J. (1982). *Multiple regression in behavioral research*. New York, NY: Holt, Rinehart, and Winston.

Raudenbush, S. W. (1989). "Centering" predictors in multilevel analysis: Choices and consequences. *Multilevel Modeling Newsletter, 1*(2), 10–12.

Raudenbush, S. W., & Bryk, A. S. (2002). *Hierarchical linear models: Applications and data analysis methods* (2nd ed.).Thousand Oaks, CA: Sage.

Snijders, T. A. B., & Berkhof, J. (2008). Diagnostic checks for multilevel models. In J. de Leeuw and E. Meijer (Eds.), *Handbook of multilevel analysis* (pp. 141–175). New York, NY: Springer.

Snijders, T. A. B., & Bosker, R. J. (1999). *Multilevel analysis: An introduction to basic and advanced multilevel modeling*. London: Sage.

Susser, M. (1994). The logic in ecological: The logic of analysis. *American Journal of Public Health, 85*, 825–829.

APPENDIX A

M*plus* Program for Estimating the Random Intercepts Model With Correlation Between Exposure and the Random Intercepts

```
title:
Random intercepts model; Residual
correlated with exposure;
data:
file is "g:\7474\longcross.correct\
cross.dat";
Variable:
names are a1-a5 e1-e5;
MODEL:
   u0i by a1@1;
   u0i by a2@1;
   u0i by a3@1;
   u0i by a4@1;
   u0i by a5@1;
   a1 ON e1 (1) u0i;
   a2 ON e2 (1) u0i;
   a3 ON e3 (1) u0i;
   a4 ON e4 (1) u0i;
   a5 ON e5 (1) u0i;
   [u0i@0];
   u0i with e1;
   u0i with e2;
   u0i with e3;
   u0i with e4;
   u0i with e5;
   [a1 a2 a3 a4 a5] (2);
   a1 a2 a3 a4 a5 (3);
output:
sampstat;
res;
```

APPENDIX B

Screen Shot of HLM Program for Estimating the Random Intercepts Model With Correlation Between Exposure and the Random Intercepts.

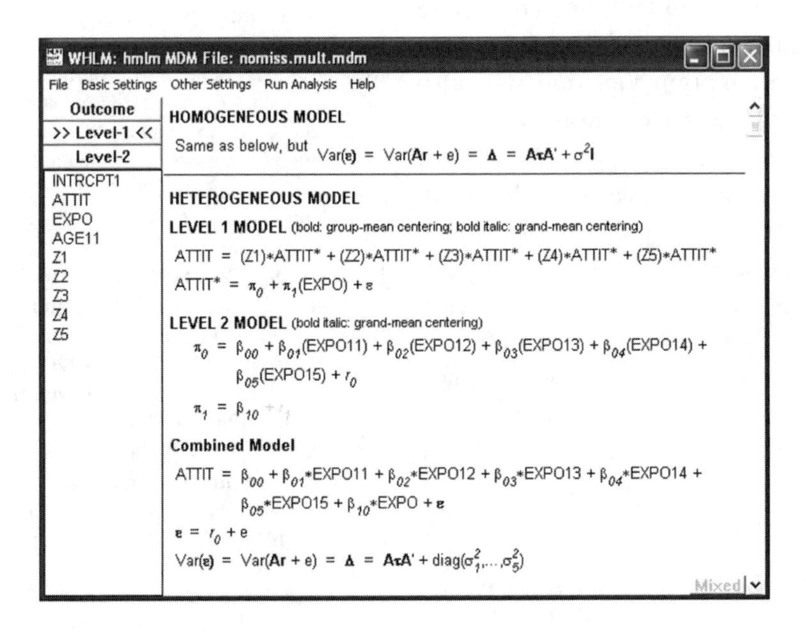

16

Cross-Classified and Multiple-Membership Models

S. Natasha Beretvas
Department of Educational Psychology, The University of Texas at Austin, Austin, Texas

The previous chapters in this book have described models designed to handle purely hierarchical data structures. However, as will be explained in the current chapter, some data structures involve clustered data but do not qualify as *pure* hierarchies. These extensions to the multilevel model are called cross-classified (C-C) models and are commonly encountered in all realms of social science as well as many other types of research. This chapter will provide examples of various cross-classified data structures, ways to assess and depict what factors are cross-classified and to formulate the relevant models. In addition, extensions to cross-classified models (called multiple-membership models) will also be described and explained.

Several multilevel textbooks provide chapters describing cross-classified models (including Beretvas, 2008; Hox, 2002; Raudenbush & Bryk, 2002; Snijders & Bosker, 1999). Hox's and Beretvas's chapters include demonstrations of various multilevel software packages for estimating C-C models. A couple of textbooks include chapters on both cross-classified and multiple-membership models (Goldstein, 2003; Rasbash & Browne, 2001). A report by Fielding and Goldstein (2006) also provides a comprehensive resource offering several examples of cross-classified and multiple-membership data structures. The reader is encouraged to review these chapters and the report to assist in mastery of these complex models.

This first section will build on a simple educational example to explain and demonstrate the distinction between pure hierarchies, cross-classified data sets and structures that are termed multiple-membership structures. This explanation will include the use of tables and figures (in the form of network graphs) to help the reader identify the data's structure. The next section will briefly discuss how cross-classified data sets are typically handled and the effect of inappropriate modeling of cross-classified structures. Next, formulation of both cross-classified and multiple-membership models (as well as combinations thereof) will be presented.

16.1 DISTINCTION BETWEEN PURE AND CROSS-CLASSIFIED DATA STRUCTURES

A very typical example of a hierarchical data structure is the scenario in which individual students are clustered within middle schools. Table 16.1 contains a very small data set used to demonstrate the clustering of students within middle schools. Each letter (A through V) represents a student. Each student only appears once in Table 16.1. The table represents each student's affiliation with only a single middle school although there are multiple students within each middle school. Thus middle school is a clustering variable. Figure 16.1 also depicts the pure clustering of students within middle schools for students in Table 16.1 in a network graph.

In a network graph, each unit of each variable (here, student and middle school) is labeled and lines represent the connections (clusters) among elements across clustered and clustering variables. Network graphs provide a way of identifying data sets' structures. Goldstein (2003) provides an alternative graphical technique for describing a data set's structure, however network graphs seem to provide a better way for *identifying* the type of structure that can be then *summarized* using Goldstein's graphs.

For data to be considered purely clustered, clusters of elements from one level all belong to single elements of a higher level. If it is possible to construct a network graph in which the lines [representing affiliations of lower level units (here, students) with a higher level unit (here, middle schools)] do not cross, then the data's structure can be considered a pure hierarchy.

To extend this first example further, a researcher might recognize that modeling this two-level structure (with students at level-1 clustered within middle schools at level-2) ignores the effect of high school on a student. If each individual high school enrolls students only from a single, associated middle school then it would be impossible to separate out the effects of middle and high schools. If, however, in a better approximation to reality, it could be assumed that each high school took students

TABLE 16.1

Pure Two-Level Clustering of Students (Level-1) within Middle School (Level-2)

Middle School							
1	2	3	4	5	6	7	8
A,B,C	D,E	F,G,H,I	J,K	L,M,N	O,P	Q,R,S,T	U,V

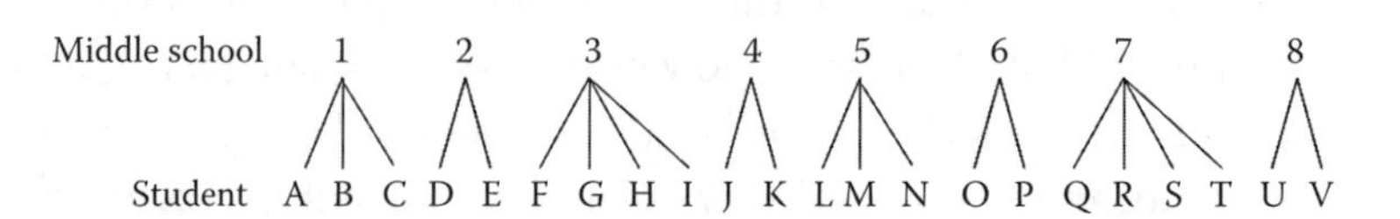

FIGURE 16.1
Network graph of pure two-level clustering of students (level-1) within middle school (level-2).

TABLE 16.2

Pure Three-Level Clustering of Students within Middle Schools within High Schools with Middle School as Column Variable

High School	Middle School 1	2	3	4	5	6	7	8
I	A,B,C	D,E						
II			F,G,H,I	J,K	L,M,N			
III						O,P	Q,R,S,T	
IV								U,V

from its own set of middle schools then the data set could be considered as having three levels. Table 16.2 depicts this pure clustering of students (level-1) clustered within middle schools (level-2) and middle schools as clustered within high school (level-3).

TABLE 16.3

Pure Three-Level Clustering of Students within Middle Schools within High Schools with Middle School as Row Variable

Middle School	High School I	II	III	IV
1	A,B,C			
2	D,E			
3		F,G,H,I		
4		J,K		
5		L,M,N		
6			O,P	
7			Q,R,S,T	
8				U,V

As soon as another clustering variable is being considered, assessment of the structure of a data set becomes more complicated. Depiction of a data set's structure using a table can help identify whether data are purely clustered or not. In Table 16.1, columns represent units of one of the clustering variables (here, middle schools) and rows represent levels of the other clustering variable (here, high school). If the data are a pure hierarchy then either exactly one cell per row or exactly one cell per column (but not both rows and columns) will contain elements in it (Rasbash & Browne, 2001). In this particular example, each column only has one cell with elements in it. For example, Middle School 3 only has elements in the cell that corresponds to High School II. However, in this data set, at least one row (actually, all of them except that of High School IV) contain more than one cell with elements in them. For example, there are students from Middle Schools 6 and 7 who attend High School III. If only one cell per column has all of that column's elements in it, then the column variable is purely nested within the row variable. If, on the other hand, only one cell per row has all of each row's elements within it, then the row variable is purely nested within the column variable.

It does not matter how the table is drawn. In other words, it does not matter what level is used as the column versus the row variable. The data structure in Table 16.2 is also depicted in Table 16.3 in which the Middle School variable is now a row variable and High School is represented by columns. In this depiction of the data's structure, only one cell in each row (i.e., middle school) contains elements and most of the columns have several cells containing elements. Thus we can infer that middle schools are purely clustered within high school.

Figure 16.2 contains a network graph of the data structure contained in Table 16.2 (and, obviously, Table 16.3). For purely clustered data, it is possible to line up each (clustered or clustering) variable's units such that no lines on the graph cross each other. And this holds in Figure 16.2 for both clustering variables. An alternative representation of the data in Figure 16.2 appears in Figure 16.3. The pure clustering of middle school within high school can be modeled more simply by stacking units by level (i.e., student within middle school within high school). None of the lines cross that connect level-1s (students') units with those of level-2 and the same holds for the lines connecting elements of level-2 (middle schools) with level-3 (high schools).

Unfortunately, the pure hierarchy depicted in Tables 16.2 and 16.3 and Figures 16.2 and 16.3 does not always match reality. Not all clustered data structures are pure clusters. In the context of the current example, it is more likely that sets of middle schools do not send their students to a single high school. Instead the data might look more like the example given in Table 16.4. In Table 16.4, students are clustered within middle schools and within high schools as evidenced by some cells in the table containing more than

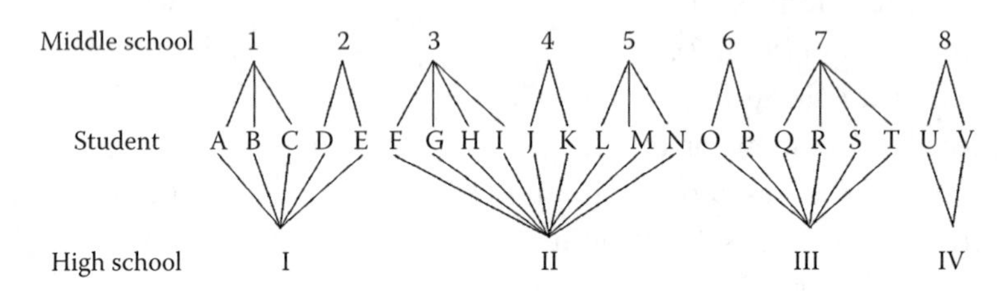

FIGURE 16.2
Network graph of pure three-level clustering of students within middle schools within high schools.

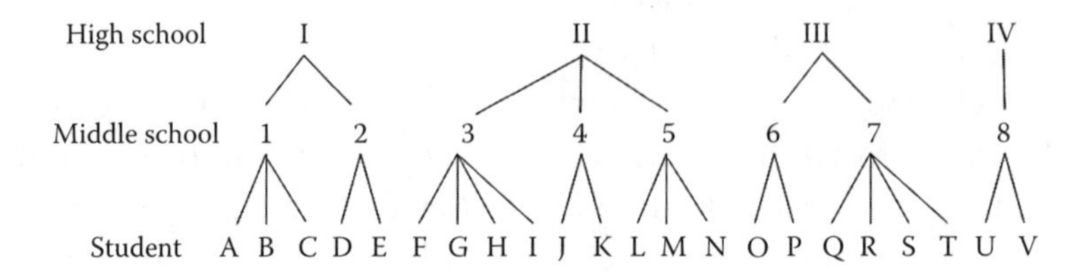

FIGURE 16.3
Alternative network graph of pure three-level clustering of students within middle schools within high schools.

TABLE 16.4

Cross-Classified Dataset Containing Students Cross-Classified by Middle School and High School

High School	Middle School 1	2	3	4	5	6	7	8
I	A,B	D,E						
II	C		F,G,	J,K	L			
III			H,I		M,N	O,P	Q,R	
IV							S,T	U,V

one student. Thus middle school and high school are both clustering variables or "classifications" for students. However, middle schools are not clustered within high schools and high schools are not clustered within middle school. This can be inferred because single cells in rows do not contain all of a row's elements, and vice-versa for columns. The pattern of the data depicted in Table 16.4 involves a *cross-classification* of students by middle and high school. By definition, a cross-classified data structure is inferred because while one set of units (here, students) are clustered within each of the classification variables (middle school and high school), neither of the classification variables is purely clustered within the other.

The crossed lines in the network graph (see Figure 16.4) associated with the contents of Table 16.4 should also lead the reader to infer that students are cross-classified by middle and high school. The cross-classifications in Table 16.4 are alternatively depicted in Figure 16.5. The difference between Figures 16.4 and 16.5 is that in Figure 16.4, students are lined up by their middle school clusters. In Figure 16.5, students are lined up by their high school clusters. In both figures, lines cross thereby supporting the cross-classification of students by middle and high school.

So far, only cross-classified examples with one level of clustering have been discussed. It is, of course, possible to encounter additional levels. For example, in addition to the cross-classification of students by middle and high *school*, the clustering of students within middle school *classrooms* might affect the outcome of interest. The data structure might appear as contained in Table 16.5 and depicted in Figure 16.6. We see in Figure 16.6 the pure clustering of students within middle school classrooms and of these classrooms within middle school. We also see the cross-classification

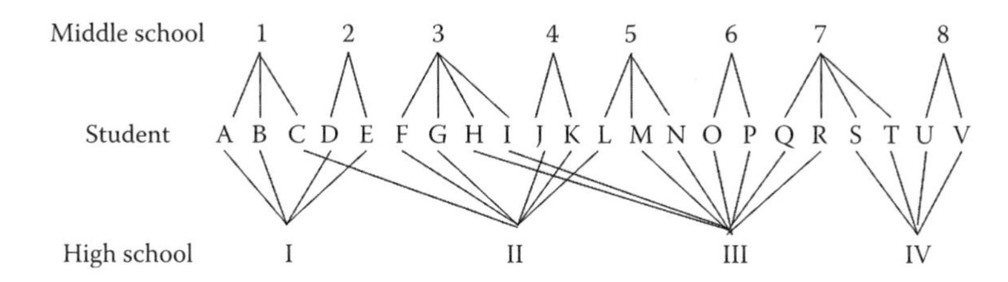

FIGURE 16.4
Network graph depicting cross-classification of students by middle school and high school with students listed by middle school cluster.

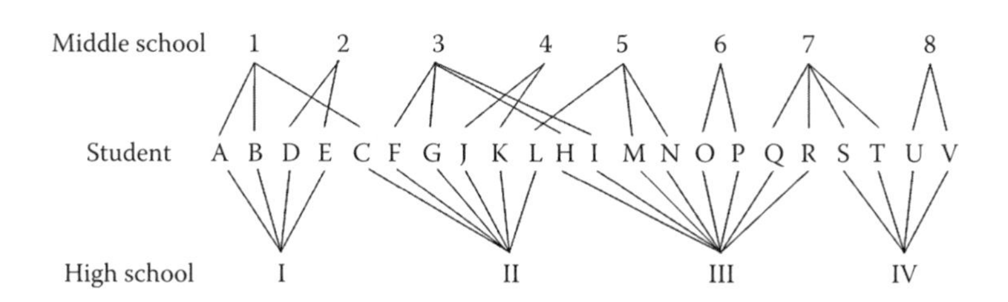

FIGURE 16.5
Network graph depicting cross-classification of students by middle school and high school with students listed by high school cluster.

TABLE 16.5

Data Set Containing Students Nested within Middle School Classrooms and Middle Schools and Cross-Classified by High School

High School	MS 1 Classroom			MS 2 Classroom		MS 3 Classroom		
	a	b	c	d	e	f	g	h
I	A,B	D,E						
II	C		F,G	J,K	L			
III			H,I		M,N	O,P	Q,R	
IV							S,T	U,V

Note: MS = middle school.

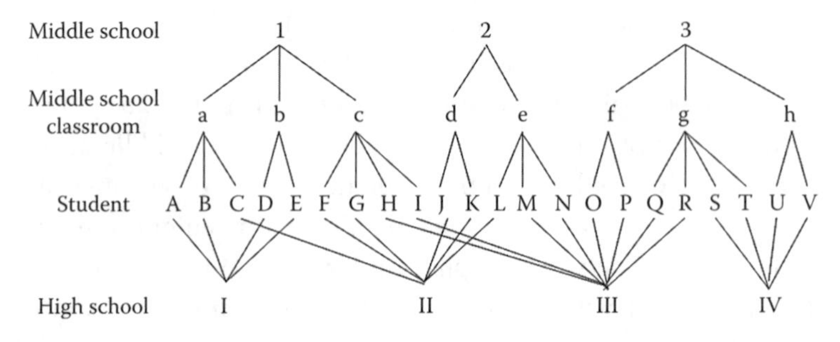

FIGURE 16.6
Network graph depicting clustering of students by middle school and middle school classroom and cross-classification with high school.

of students by middle and high school. We can see in Figure 16.6 that there are two hierarchies with middle school and high school at the highest level of each. We also see that if there is a lower level of clustering within one of the C-C factors, then the crossing of the higher level classifications results in the other C-C factor also being crossed with the lower level clustering variable (Rasbash & Browne, 2001). In the current example, this means that because students are cross-classified by middle and high school, and middle school classrooms are clustered within middle school then students are also cross-classified by middle school classroom and high school.

The three-level data structure in Figure 16.6 depicts two hierarchies with the cross-classification originating at the highest level. It is possible to have a three-level data structure where the cross-classification does not occur at the highest level. For example, variability in student scores on a science achievement test could be attributed to the high school biology and chemistry classes. For this example, it is assumed that students attended only a single high school but that the set of students in a biology class were not exactly the same as the set of students in a corresponding chemistry class. This data structure (see Table 16.6) would consist of students cross-classified by biology and chemistry class where both are purely clustered within high school. Figure 16.7 depicts the cross-classification of students by biology and chemistry classroom. Figure 16.8 depicts the clustering of biology and chemistry classrooms within high

TABLE 16.6

Data Set Containing Students Cross-Classified by Biology and Chemistry Classrooms Clustered within High School

Bio Class	HS 1 Chem. Class		HS 2 Chem. Class	HS 3 Chem. Class	
	C1	C2	C3	C4	C5
B1	A,B,D	F,H			
B2	C,E	G,I,J			
B3			K,L,M,N		
B4				O,Q,R	S,T
B5				P	U,V

Notes: HS = high school; Chem. = Chemistry; Bio = Biology.

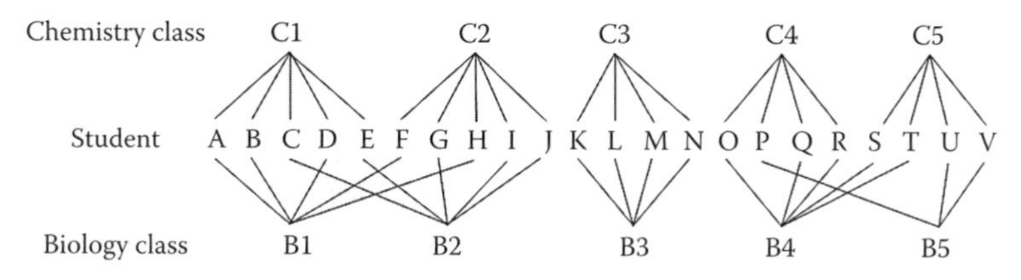

FIGURE 16.7
Network graph depicting cross-classification of students by biology and chemistry classrooms.

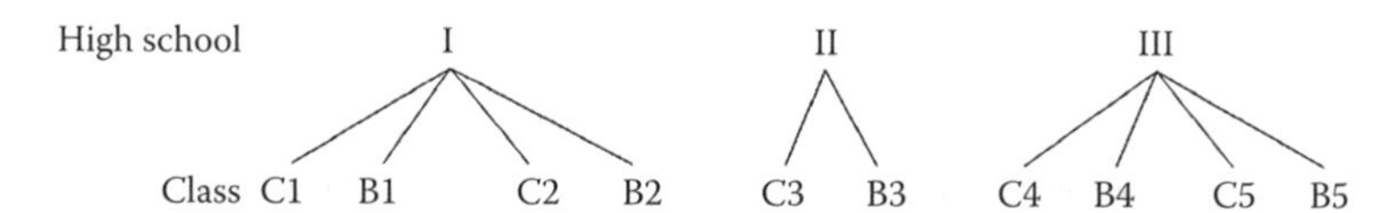

FIGURE 16.8
Network graph depicting clustering of biology and chemistry classrooms within high schools.

school. Figures 16.7 and 16.8 could be combined into one, as presented in Figure 16.9, to represent the cross-classification at level-2 and the clustering of both level-2 C-C factors at level-3. Note that, in this particular data set, students attending high school 2 are not cross-classified by biology and chemistry classroom. However, considering the data set as a whole, there are some students for whom biology and chemistry classrooms are crossed factors.

It is also possible for a cross-classification to occur at level-1. This is manifested by *every* cross-classification cell containing only a single unit within them *by design.* A current, classic example is the cross-classification of item scores on a test by item and student (Goldstein, 2003; Van den Noortgate, De Boeck, & Meulders, 2003). Table 16.7 contains a (very small) sample data set with item scores cross-classified at level-1 by student and item, with students clustered within middle schools. For this very small example, only three of the item scores of the students are depicted. And the sample data set only depicts two students per middle school. Each cell only has a single element within it. (Note that although in the current example all cells have an element within them, it is possible to have

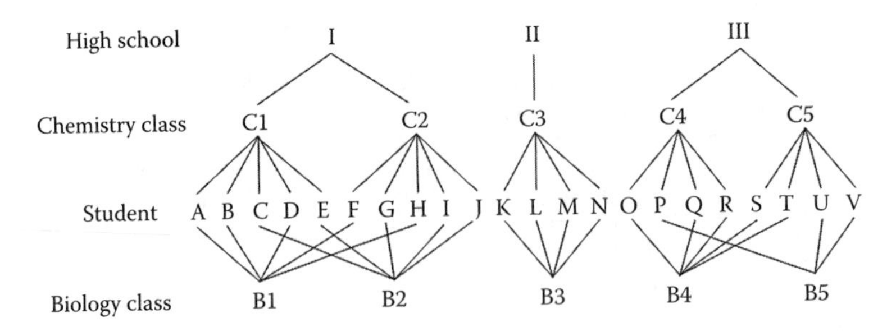

FIGURE 16.9
Network graph depicting clustering of biology and chemistry classrooms within high schools and of cross-classification of students by science classroom type.

TABLE 16.7

Data Set Containing Item Scores Cross-Classified (at Level-1) by Student and Item with Students Clustered within Middle School

	Middle School 1		Middle School 2	
Item	**Student A**	**Student B**	**Student C**	**Student D**
1	${}_1Y_A$	${}_1Y_B$	${}_1Y_C$	${}_1Y_D$
2	${}_2Y_A$	${}_2Y_B$	${}_2Y_C$	${}_2Y_D$
3	${}_3Y_A$	${}_3Y_B$	${}_3Y_C$	${}_3Y_D$

Note: ${}_iY_j$ represents the score on item *i* for person *j*.

crossed level-1 factors with some empty cells).

The single unit per cross-classification cell is occurring by design, not by chance. For example, in Table 16.5 there are two cells that contain only a single element. Only student C attended middle school 1, classroom a and high school II, and only student L attended classroom e in middle school 5 and high school II. However, in Table 16.5, not all cells have only a single element within them and the single-cell occurrences result only by chance, not as a result of the study's design. The data structure from Table 16.7 is depicted in Figure 16.10. The cross-classification is evident from the crossing of the lines connecting items to scores. And if the network graph were depicted with scores listed by item (rather than by student), then the lines connecting students with scores would be crossed.

One further complication in data structures occurs when units are multiple members of one of the classification variables. Returning to the initial example in this section (of students clustered within middle school), students might have attended multiple middle schools. This possibility is demonstrated using the data set in Table 16.8. In the table, students who attended multiple middle schools have their identifier bolded and italicized. In this (very small) data set, three students (A, G, and Q) were members of multiple middle schools. Students A and G attended two middle schools each (middle schools 1 and 2, and 3 and 4, respectively) while student Q attended three middle schools (specifically 6, 7, and 8).

Figure 16.11 contains a network graph representing the data in Table 16.8. In this (multiple membership) network graph, if the student attended multiple middle

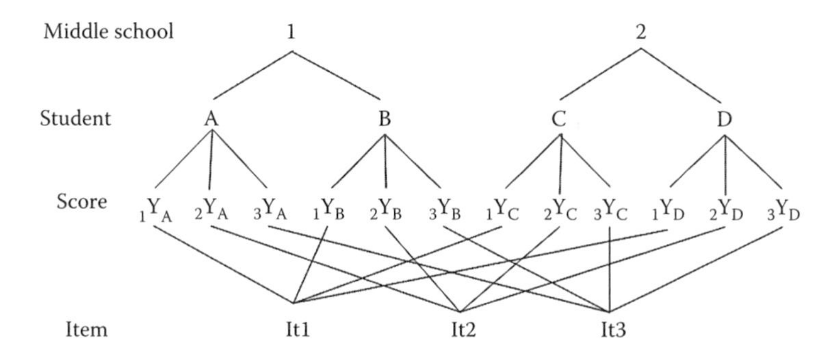

FIGURE 16.10
Network graph depicting level one cross-classification of scores by student and item and clustering of students within middle school.

TABLE 16.8
Multiple-Membership Data Set Containing Students who Attended Multiple Middle Schools

Middle School							
1	**2**	**3**	**4**	**5**	**6**	**7**	**8**
A,B,C	***A***,D,E	F,***G***,H,I	***G***,J,K	L,M,N	O,P,***Q***	***Q***,R,S,T	***Q***,U,V

Note: Bolded and italicized student identifiers represent students who enrolled in more than one middle school.

schools, then hatched lines are used to connect a student with each middle school that the student attended. If, in a network graph, there is more than one line connecting the lowest unit (level-1, here student) with the higher level clustering variable (level-2, here, middle school), then the data structure qualifies as a multiple membership structure. In the single-membership model depicted in Figure 16.1, there is a single line originating from each student to the (middle school) clustering variable. For example, student A is connected *solely* with Middle School 1 (and thus a single line connects student A with middle school 1). In the multiple-membership structure depicted in Figure 16.10, however, two (hatched) lines connect some units of the clustered variable (student) with units of the clustering variable (middle school). For example, there are two lines connecting (student) A with two middle schools (schools 1 and 2).

In multiple-membership network graphs, the lines connecting units that are multiple members to their clustering variable (here, middle school) will cross although the crossing does not necessarily represent a cross-classification. The current example (students as multiple members of middle school) only has a single clustering variable (middle school) and thus cannot be a cross-classified structure. There must be at least two clustering or classification variables for cross-classification to occur. Data sets in which elements are multiple members of a level (or classification) are termed multiple-membership structures. Multiple-membership models can be used to model that the effect of middle school on a student is a function of all middle schools attended.

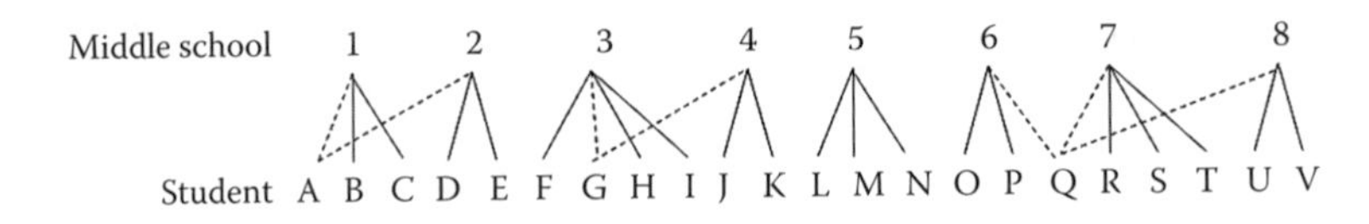

FIGURE 16.11
Network graph depicting multiple-membership data set containing students who attended multiple middle schools.

TABLE 16.9

Three-Level Multiple-Membership Data Set with Level-1 Units (Students) Multiple Members of Level-2 (Middle School) Classification Clustered within Level-3 (High School)

High School	Middle School							
	1	2	3	4	5	6	7	8
I	***A***,B,C	***A***,D,E						
II			F,***G***,H,I	***G***,J,K				
III					L,M,N			
IV						O,P,***Q***	***Q***,R,S,T	***Q***,U,V

Note: Bolded and italicized student identifiers represent students who enrolled in more than one middle school.

Extending the example even further, it is possible that students might have attended multiple middle schools before attending a single high school (see Table 16.9). For example, student G attended both middle school 3 and middle school 4 before attending high school II. Thus, a unit might be a member of multiple units of one clustering variable (middle school) with a higher level of pure clustering (here, of middle schools within high schools). This pure hierarchy is evidenced in Table 16.4 by single cells in each column containing elements while some rows contain more than one nonempty cell. Figures 16.12 and 16.13 are alternative network graphs designed to represent this data structure. They are distinguished by whether high school appears directly connected to students (Figure 16.12) or to middle schools (Figure 16.13). The pure hierarchy of middle schools within high schools is evidenced by no lines crossing in the connection between students and high schools (in Figure 16.12) and thus no lines crossing in the connection between middle and high schools (in Figure 16.13).

It is possible to encounter cross-classified multiple-membership data sets. An example appears in the data set appearing in Table 16.10 and the associated network graphs in Figures 16.14 and 16.15. Table 16.10 clearly contains cross-classified data (with multiple cells containing elements in both rows and columns). The multiple-membership facet (of students within middle schools) is evidenced by some students (A, G, and Q) appearing in multiple middle schools (columns). The crossing of (unhatched) lines in Figures 16.14 and 16.15 also support the existence of a cross-classified data structure. Figure 16.14 contains students listed by middle school clusters. Figure 16.15 contains students listed by high school clusters.

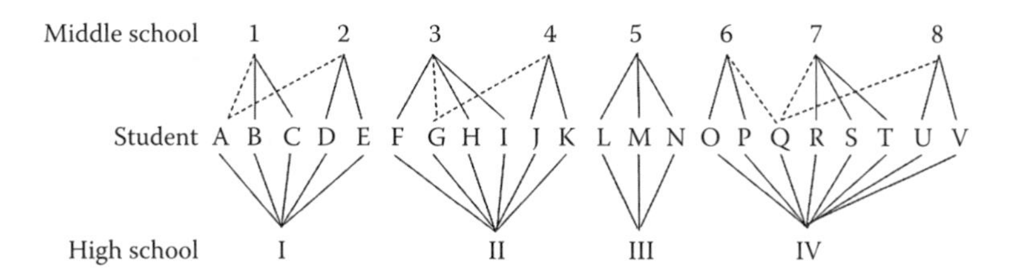

FIGURE 16.12
Network graph of three-level multiple-membership data set with level one units (Students) multiple members of level two (middle school) variable clustered within level three (high school).

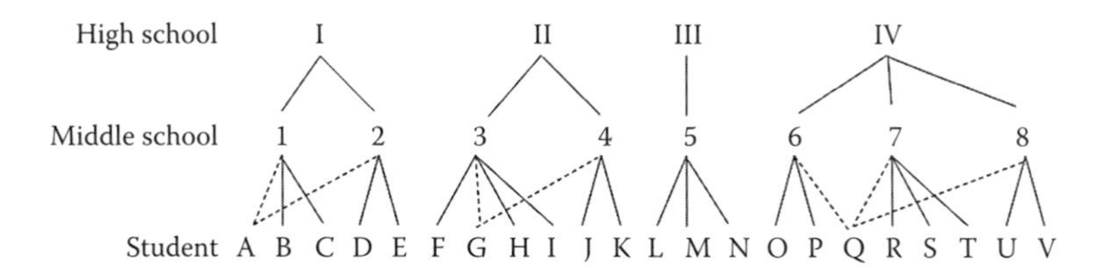

FIGURE 16.13
Alternative network graph of three-level multiple-membership data set with level one units (students) multiple members of level two (middle school) variable clustered within level three (high school).

TABLE 16.10

Students Cross-Classified by Middle School and High School with Some Students Members of Multiple Middle Schools

High School	Middle School 1	2	3	4	5	6	7	8
I	***A***,B	***A***,D,E						
II	C		F,***G***	***G***,J,K	L			
III			H,I		M,N	O,P,***Q***	***Q***,R	***Q***
IV							S,T	U,V

Note: Bolded and italicized student identifiers represent students who enrolled in more than one middle school.

Either depiction leads to crossed lines supporting the cross-classification of students by middle and high schools.

The examples of cross-classified and of multiple-membership data structures that have been described here are of the simplest variety. It is, of course, possible to have more than two cross-classified factors, more than a single multiple-membership variable, more than one or two hierarchies and more than a single level of cross-classifications. In addition, for the purposes of the current chapter, the middle and high school context is being used. The reader is encouraged to think of classifications within their own research context that might correspond with the pattern of structures discussed here. There are some clear and useful additional examples of these models provided in Fielding and Goldstein (2006), Goldstein (2003), and in Rasbash and Browne (2001).

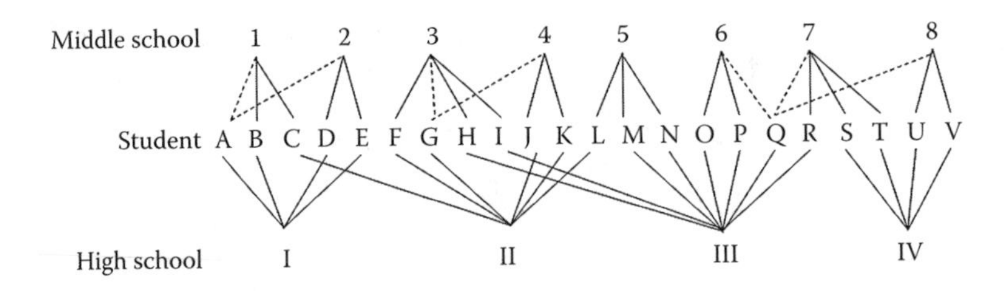

FIGURE 16.14
Network graph of students cross-classified by middle school and high school with some student members of multiple middle schools; sorted by middle schools. Note: Hatched lines connecting a student with a middle school indicates that the student was a member of multiple middle schools.

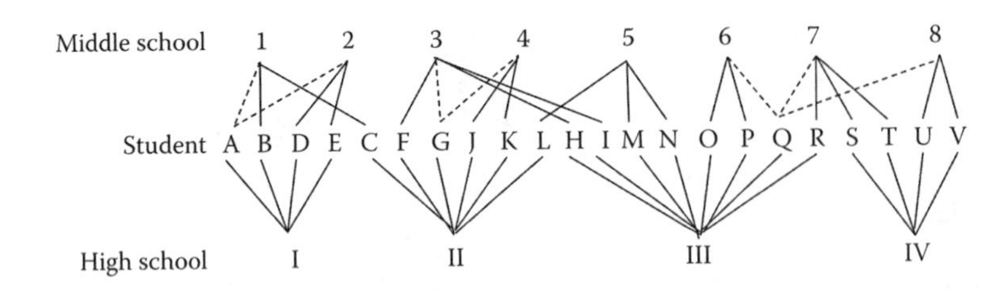

FIGURE 16.15
Network graph of students cross-classified by middle school and high school with some student members of multiple middle schools; sorted by high schools. Note: Hatched lines connecting a student with a middle school indicates that the student was a member of multiple middle schools.

16.2 INAPPROPRIATE MODELING OF CROSS-CLASSIFIED DATA STRUCTURES

Despite the ubiquity of cross-classified data structures, use of C-C random effects modeling is far from commonplace. Estimation of cross-classified models and interpretation of their results is complex. Unfortunately, just as ignoring the pure clustering inherent in many social science data sets can reduce the validity of associated statistical inferences, so will ignoring the cross-classified nature of data sets. Researchers have avoided modeling the C-C nature of their data sets using one of two strategies (Meyers & Beretvas, 2006). The simplest cross-classified data set will be used here to demonstrate these strategies.

As a first strategy, a researcher might simply ignore one of the C-C clustering variables. For example, if students were cross-classified by both middle and high school (as in Table 16.4), then the researcher might ignore the student's middle school membership and instead model the two-level clustering of students within high school. Ignoring a C-C factor will not lead to biased nor inconsistent *fixed effects* estimates (Fielding & Goldstein, 2006). However, as noted by Rasbash and Browne (2001) and Fielding and Goldstein (2006), and empirically supported by others (Meyers & Beretvas, 2006; Raudenbush & Bryk, 2002), ignoring a C-C factor results in an under-specified model that can result in negatively biased standard error estimates. Deflated standard errors will then inflate the Type I error rate of associated statistical tests. Ignoring a C-C factor can also lead to inaccurate variance component estimation (Fielding & Goldstein, 2006; Rasbash & Browne, 2001). A simulation study designed to assess the effect of inappropriately modeling a C-C data structure found that the variance component associated with the ignored C-C factor was instead associated with the nonignored C-C factor (Meyers & Beretvas, 2006). Thus the nonignored C-C factor's variance component will be overestimated.

The data in Table 16.4, in which students are cross-classified by middle and high school, will be used to demonstrate the second strategy. A researcher might recognize that the cross-classification of students by middle and high school seemed unimportant and that for the most part students who attend a set of middle schools go on to attend the same high school. In other words, the data are close to being a pure hierarchy. The researcher could delete from analysis the small sets of students who prevent the data from being a pure hierarchy. Table 16.11 demonstrates the data points (students) that could be deleted from the analysis to provide a data set with a pure hierarchy (i.e., students C, H, I, L, Q, and R would be deleted from analysis). After deletion, the data could be considered a pure clustering of students (level-1) within middle school (level-2) within high school. However, this deletion strategy will result in less power and reduces the generalizability of the results (Meyers & Beretvas, 2006).

Beyond problems with estimation, use of either strategy (either deletion of data or ignoring a C-C factor) results in a loss of information about the ignored factor. In addition, while relationships among characteristics of that factor and the outcome of interest can be modeled, the precision of estimation of this relationship will be biased if the variance component structure is inappropriately modeled. While it is possible to use corrections that adjust standard errors to correct for variance components that are not being explicitly modeled, this restricts the associated research questions that can be explored. For example, when standard errors are adjusted instead of modeling the relevant cross-classified factor, then residuals for that C-C factor cannot be calculated. Currently, residuals can provide important information used, for example, to identify value added by a C-C unit such as school (Fielding & Goldstein, 2006).

The next section will provide formulations for the various cross-classified and multiple-membership models that have been discussed. The section will also briefly demonstrate how characteristics of the different classifications can be explored as predictors of the outcome of interest.

TABLE 16.11

Deletions in the Cross-Classified Data Set from Table 16.4 Resulting in a Pure Hierarchy of Middle School within High School

High	Middle School							
School	1	2	3	4	5	6	7	8
I	A,B	D,E						
II	~~C~~		F,G	J,K	~~L~~			
III			~~H,I~~		M,N	O,P	~~Q,R~~	
IV							S,T	U,V

Note: A double strike through crossing out a student's identifier (letter) indicates a student whose information has been deleted from the analysis.

16.3 FORMULATION OF THE CROSS-CLASSIFIED RANDOM EFFECTS MODEL

16.3.1 Two-Level CCREM with Two C-C Factors

The notation used by Rasbash and Browne (2001) will be adopted here in the formulation of the models. First, a classification is defined as a function that maps the N units at level-1 to a set of M classification units (where $M \leq N$; Rasbash et al., 2000). For purely hierarchical data structures, each level-1 unit is classified by just one classification unit. For a cross-classification to occur, level-1 units are classified by more than one classification but the classification units are not nested within each other.

The simplest example of a cross-classified data structure in which students are cross-classified by middle and high school will be used to present the formulation of the cross-classified random effects model (CCREM) (see Table 16.4 and Figure 16.4 or 16.5). The two C-C factor, two-level CCREM unconditional model for a normally distributed outcome is parameterized as follows:

$$Y_{i(j_1,j_2)} = \beta_{0(j_1,j_2)} + u_{0j_{10}} + u_{00j_2} + u_{00j_1 \times j_2} + e_{i(j_1,j_2)} \quad (16.1)$$

In Rasbash and Browne's (2001) notation, the number of letters in the subscript identifies the number of classifications (here, there are three: student, middle school, and high school). The higher the level of the classification unit, the further to the right the associated subscript will appear. Thus, for example, the level-1 classification unit (here, student) appears as the first subscript letter, i. Subscripts with the same common letter (e.g., "j") appearing in parentheses separated by a comma identify cross-classified factors at the same level. Ordering of subscripts or C-C factors within parentheses is arbitrary. Here, the middle school identifier, j_1, appears before the high school identifier, j_2, but the reverse ordering could also work. This subscripting scheme means that for students cross-classified by middle and high school, $Y_{i(j_1,j_2)}$ represents the score of student i from middle school, j_1 and high school, j_2. The highest number in the subscript of the outcome variable, Y, identifies the number of cross-classified factors at the level (here, two). The lack of a number subscripting the "i" subscript means that the associated unit (here, student) is a member of multiple cross-classifications. Pure clustering of classification units can be interpreted by the lack of a comma separating a set of indices. For example, the following hierarchies exist for the data in Table 16.4: students within middle school (Y_{ij_1}), and students within high school (Y_{ij_2}). It should be noted that while there is a comma between j_1 and j_2 in $Y_{i(j_1,j_2)}$, there is no comma separating the set of level-1 indices (here, just i) from the set of (C-C) level-2 indices (j_1 and j_2).

Four residuals are listed in the unconditional model in Equation 16.1. Each residual, e, is assumed normally distributed with a mean of zero and its own variance, σ_e^2. Covariances among residuals are typically assumed to be zero. Thus, for example, the covariance between level-1 units in different classifications (i.e., here, in different middle and high schools) is assumed to equal zero (Goldstein, 2003). These residuals can be interpreted as follows. The random effects $e_{i(j_1,j_2)}$, u_{00j_1} and u_{00j_2} are associated with each of student i from middle school j_1 and high school j_2, with middle school j_1 and with high school j_1, respectively. The "0" appearing in the subscript of the "*u*"s matches the subscript of the parameter around variability is modeled, here around the intercept, γ_{000}.

The last residual in Equation 16.1, $u_{00j_1 \times j_2}$, represents the random interaction effect between the two C-C factors (middle school and high school). Without sufficiently large within-cell sample sizes, it is hard to separate the variance, σ_e^2, associated with the level-1 classification unit (students) from the variance of this interaction effect (Raudenbush & Bryk, 2002). Thus, this random effect is most typically set to zero when CCREM models are estimated and will be set to zero henceforth in this chapter.

As with purely clustered multilevel models, variance component estimation under the unconditional model can be used to calculate a form of intraclass correlation. The intraclass correlation coefficient (ICC) for one of the cross-classified factors, for example, middle schools, is calculated as follows:

$$\rho_{Y_{i(j_1,j_2)},Y_{i(j_1,j_2')}} = \frac{\sigma^2_{u_{0j_{1}0}}}{\sigma_e^2 + \sigma^2_{u_{0j_{1}0}} + \sigma^2_{u_{00j_2}}}. \quad (16.2)$$

In the current example, the ICC in Equation 16.2 represents the correlation between achievement scores (Y) for students in middle school j_1 who attend different high schools (j_2 and j_2' where $j_2 \neq j_2'$; Raudenbush & Bryk, 2002). The same formula for the ICC for high schools could be calculated and would represent the correlation between students' scores on Y for students who attend the same high school but different middle schools. As with multilevel analyses of purely hierarchical data sets, the ICCs can be used to interpret the proportion of variability attributable to each classification.

Goldstein (2003) cautions that ICCs' values for different classifications can be affected by the number of units within a classification. In Goldstein's example, the ICC for primary schools was three times as large as that of secondary schools and part of the source of that disparity was in the number of units per secondary school being larger than that of primary schools. Keeping this caveat in mind, however, the ICCs for CCREMs can be interpreted in the same way as in traditional multilevel models. Similarly, when predictors are added to the CCREM model, their effect on the ICCs can be explored to assess their relationship with the outcome. Given this does not differ from the use of ICCs with purely hierarchical data, the ICCs will not be mentioned further in this chapter. Readers are encouraged to read Beretvas (2008), Hox (2002), and Raudenbush and Bryk (2002) for more details about cross-classified models' ICCs.

A researcher might be interested in using a student-level descriptor, X, to explain variability in the outcome. For example, the researcher might wish to model gender differences in achievement. Given the complexity of the CCREM's formulation, the levels' equations used in Raudenbush and Bryk (2002) will sometimes be adopted, here, to simplify the explanation of the model. Addition of a level-1 (here, student) descriptor to the unconditional model (in Equation 16.1) results in:

$$Y_{i(j_1,j_2)} = \beta_{0(j_1,j_2)} + \beta_{1(j_1,j_2)} X_{i(j_1,j_2)} + e_{i(j_1,j_2)}, \quad (16.3)$$

for the level-1 equation and at level-2:

$$\begin{cases} \beta_{0(j_1,j_2)} = \gamma_{000} + u_{0j_{1}0} + u_{00j_2} \\ \beta_{1(j_1,j_2)} = \gamma_{100} \end{cases}. \quad (16.4)$$

Equation 16.4 indicates that the relationship between the outcome and the variable, X, is modeled as fixed across middle schools and high schools. As a single model equation, Equations 16.3 and 16.4 become:

$$Y_{i(j_1,j_2)} = \gamma_{000} + \gamma_{100} X_{i(j_1,j_2)} + e_{i(j_1,j_2)} + u_{0j_{10}} + u_{00j_2}. \quad (16.5)$$

As with multilevel models of pure hierarchies, interpretation of the intercept, γ_{000}, changes from the unconditional model to a model including a predictor. In Equation 16.1, γ_{000} is the predicted outcome score averaged across middle and high schools. For the model including a predictor, the intercept becomes the average predicted outcome score for a student with a value of zero on X. Thus, interpretation of the intercept depends on how the predictor (here, gender or X) is centered.

Variability in the intercept across middle and high schools can be explained using middle and high school predictors. For example, a researcher might hypothesize that the middle school characteristic, Z (say, a measure of middle school resources), explains some of the variability in the intercept across middle and high schools. If this is the case, then Equation 16.4 becomes:

$$\begin{cases} \beta_{0(j_1,j_2)} = \gamma_{000} + \gamma_{010} Z_{j_1} + u_{0j_{10}} + u_{00j_2} \\ \beta_{1(j_1,j_2)} = \gamma_{100} \end{cases}. \quad (16.6)$$

Addition of Z to the model again changes interpretation of the intercept so that it represents the predicted score on Y controlling for Z and for X (i.e., for a student with a value of zero on X who attends a middle school with a value of zero on Z). As with the ICCs, interpretation of coefficients with cross-classified models will not be thoroughly reviewed in this chapter because it corresponds exactly with their interpretation in simpler multilevel models. The reader is referred to the chapters by Beretvas (2008), Hox (2002), and Raudenbush and Bryk (2002) for more details about interpreting these coefficients.

In Equation 16.6, the relationship, γ_{010}, between middle school resources, Z, and achievement, Y, is modeled as fixed across high schools. It could instead be modeled as varying across high schools:

$$\beta_{0(j_1,j_2)} = \gamma_{000} + (\gamma_{010} + u_{01j_2}) Z_{j_1} + u_{0j_{10}} + u_{00j_2}. \quad (16.7)$$

A high school descriptor (say, whether the high school is private or public) could also be used to explain variability in the relationship between middle school resources, Z, and the student's achievement score, Y.

Similar to the series of equations demonstrated for the relationship between Z and the intercept, variability in the intercept could also (or alternatively) be explained using a high school descriptor and this relationship modeled as fixed, randomly, or nonrandomly varying across middle schools (see Beretvas, 2008; Raudenbush & Bryk, 2002).

Thus far, all of these equations have modeled the relationship between the level-1 descriptor, X, and the outcome as fixed (as represented by $\beta_{1(j_1,j_2)} = \gamma_{100}$ in Equations 16.4 and 16.6). As with the modeling of the intercept term, $\beta_{0(j_1,j_2)}$, the slope for X, $\beta_{1(j_1,j_2)}$, can be modeled in a number of ways. Residual terms u_{10j_1} and u_{10j_2} could be included in the second equation of the two in Equation 16.4 to model the relationship between X and Y

as varying across middle and high schools, respectively. Either, both or neither residual could be included. For example, if it were hypothesized that the relationship between student's gender and achievement varied across middle schools only, the set of equations in Equation 16.6 (without any predictors of the intercept term) would become:

$$\begin{cases} \beta_{0(j_1,j_2)} = \gamma_{000} + \gamma_{010} Z_{j_1} + u_{0j_10} + u_{00j_2} \\ \beta_{1(j_1,j_2)} = \gamma_{100} + u_{1j_10} \end{cases} \quad (16.8)$$

and the single equation for this model (combining Equations 16.3 and 16.8) would be:

$$\begin{aligned} Y_{i(j_1,j_2)} &= \gamma_{000} + \gamma_{010} Z_{j_1} + u_{0j_10} + u_{00j_2} \\ &\quad + (\gamma_{100} + u_{1j_10}) X_{i(j_1,j_2)} + e_{i(j_1,j_2)} \end{aligned} . \quad (16.9)$$

The researcher might hypothesize that some middle school characteristics, Z, explains some of the variability across middle schools in the gender differences in achievement. In the levels' formulations, Equation 16.9 would then become:

$$\begin{cases} \beta_{0(j_1,j_2)} = \gamma_{000} + \gamma_{010} Z_{j_1} + u_{0j_10} + u_{00j_2} \\ \beta_{1(j_1,j_2)} = \gamma_{100} + \gamma_{110} Z_{j_1} + u_{1j_10} \end{cases} . \quad (16.10)$$

As a single equation, the model then becomes:

$$\begin{aligned} Y_{i(j_1,j_2)} &= \gamma_{000} + \gamma_{010} Z_{j_1} + u_{0j_10} + u_{00j_2} \\ &\quad + (\gamma_{100} + \gamma_{110} Z_{j_1} + u_{1j_10}) X_{i(j_1,j_2)} + e_{i(j_1,j_2)} . \end{aligned} \quad (16.11)$$

We see in Equation 16.11 that the relationship between X (gender) and Y (achievement) is modeled to be moderated by the middle school descriptor, Z.

The set of level-2 equations in Equation 16.11 could be further expanded to include additional predictors of the intercept term, $\beta_{0(j_1,j_2)}$, along with predictors of $\beta_{1(j_1,j_2)}$. However, as with multilevel analyses of pure hierarchies, the more random effects included in a model, the more complex the estimation. A balance needs to be found between correctly specifying the variance components and parsimony.

16.3.2 Three-Level CCREM with Two Hierarchies and C-C at Level Three

The example data structure provided in Table 16.5 and depicted in Figure 16.6 contains two hierarchies. The clustering of students in middle school classrooms nested within middle schools is one three-level hierarchy. In addition, students are clustered within high schools. Middle school and high school are crossed classifications (and thus middle school classroom and high school are also crossed). The unconditional model is:

$$Y_{i(jk_1,k_2)} = \gamma_{0000} + v_{000k_1} + v_{000k_2} + u_{00jk_1} + e_{i(jk_1,k_2)}, \quad (16.12)$$

where $Y_{i(jk_1,k_2)}$ is the score on outcome Y for student i in middle school classroom j of middle school k_1 who attended high school k_2. Each of the four residuals in Equation 16.12 are assumed normally distributed with a mean of zero and its own variance component. As was demonstrated above, characteristics of each classification unit can be added as predictors to the model. For example, it might be hypothesized that there are gender differences in achievement.

The model (represented as levels) becomes, at level-1:

$$Y_{i(jk_1,k_2)} = \pi_{0(jk_1,k_2)} + \pi_{1(jk_1,k)} X_{i(jk_1,k_2)} + e_{i(jk_1,k_2)}, \tag{16.13}$$

at level-2:

$$\begin{cases} \pi_{0(jk_1,k_2)} = \beta_{00(k_1,k_2)} + u_{00jk_1} \\ \pi_{1(jk_1,k_2)} = \beta_{10(k_1,k_2)} + u_{10jk_1} \end{cases}, \tag{16.14}$$

and at level-3:

$$\begin{cases} \beta_{00(k_1,k_2)} = \gamma_{0000} + v_{000k_1} + v_{000k_2} \\ \beta_{10(k_1,k_2)} = \gamma_{1000} \end{cases} \tag{16.15}$$

and as a single equation:

$$Y_{i(jk_1,k_2)} = \gamma_{0000} + (\gamma_{1000} + u_{10jk_1}) X_{i(jk_1,k_2)} + v_{000k_1} + v_{000k_2} + u_{00jk_1} + e_{i(jk_1,k_2)}. \tag{16.16}$$

Equation 16.16 models the relationship between gender, X, and Y as randomly varying across classrooms (due to inclusion of u_{10jk_1} in the level-2 equation for X's slope coefficient, $\pi_{1(jk_1,k_2)}$) and fixed across middle and high schools. A classroom descriptor, C (such as a measure of the teacher's experience), could be modeled as moderating the relationship between gender and achievement so that the equation for the slope of X, $\pi_{1(jk_1,k_2)}$, at level-2 (in Equation 16.16) becomes $\pi_{1(jk_1,k_2)} = \beta_{10(k_1,k_2)} + \beta_{11(k_1,k_2)} C_{jk_1} + u_{10jk_1}$. Thus, we have another coefficient, $\beta_{11(k_1,k_2)}$, which can be modeled as fixed or randomly varying across middle and/or high schools. It could also be modeled as nonrandomly varying with middle and/or high school characteristics used to explain variability in this coefficient across middle and high schools. Again, the reader is warned that inclusion of too many close-to-zero random components can lead to estimation problems.

16.3.3 Three-Level CCREM with Two Hierarchies and C-C at Level Two

Parameterization of the three-level model with cross-classified factors at level-2 will be described using the example depicted in Figures 16.7 and 16.8. In this example student science achievement scores, Y, are cross-classified by biology (j_1) and chemistry (j_2) classrooms that are both clustered within high school (k). Thus, the unconditional model appears as follows, at level-1 (students):

$$Y_{i(j_1,j_2)k} = \pi_{0(j_1,j_2)k} + e_{i(j_1,j_2)k}, \tag{16.17}$$

at level-2 (biology and chemistry classrooms):

$$\pi_{0(j_1,j_2)k} = \beta_{000k} + u_{0j_10k} + u_{00j_2k}, \tag{16.18}$$

and at level-3 (high schools):

$$\beta_{000k} = \gamma_{0000} + v_{000k}, \tag{16.19}$$

and as a single equation:

$$Y_{i(j_1,j_2)k} = \gamma_{0000} + v_{000k} + u_{0j_10k} + u_{00j_2k} + e_{i(j_1,j_2)k}, \tag{16.20}$$

with the expected four variance components associated with the four residuals ($e_{i(j_1,j_2)k}$, u_{0j_10k}, u_{00j_2k}, and v_{000k}). The same assumptions are made about the independence of the residual terms and their (normal) distributions as for the other CCREMs described in this chapter. Similarly, as above, classification factor characteristics can be added to explain variability in the relevant coefficients.

16.3.4 Two-Level CCREM with Cross-Classification at Level One

To demonstrate the formulation of the CCREM when the cross-classification occurs at level-1 (using the example from Table 16.7 and Figure 16.10), it will initially be assumed that item scores are interval-scaled and normally distributed. If this is the case, then the single equation for the unconditional model would be

$$Y_{(i_1,i_2)j} = \gamma_{000} + u_{00j} + e_{i_10j} + e_{0i_2j}, \quad (16.21)$$

where i_1 indexes student, i_2 indexes item, and j indexes middle school. As usual, predictors describing facets of the relevant classification could be included in the model to explain variability identified in a variance component (e.g., at the item, student, or middle school level).

It is more likely that the item score cannot be assumed to be normally distributed. Instead, the item score might be dichotomous (e.g., multiple choice items with scores of zero or one). If this is the case, then the item score could be assumed to follow a binomial distribution (e.g., Bernoulli) and the log-odds of a correct score could instead be modeled with a logit link function in an unconditional model, as follows:

$$\log\left(\frac{p_{(i_1,i_2)j}}{1-p_{(i_1,i_2)j}}\right) = \gamma_{000} + u_{00j} \quad (16.22)$$

where $p_{(i_1,i_2)j}$ is the probability of a score of one on item i_2 for student i_1 at middle school j. Item, student, and middle school predictors can, of course, be added to the model in Equation 16.22 as fixed, random, or randomly varying effects.

It is also possible that the item scores might be ordinal. In this case, models designed for analysis of ordinal outcomes could be used (e.g., a proportional odds model). It should be noted that there are other ways to parameterize this multilevel measurement model (e.g., Kamata, 2001; Williams & Beretvas, 2006) but the C-C random effects version provides a useful context to demonstrate an example of a scenario of a cross-classification occurring at level-1.

16.4 FORMULATION OF THE MULTIPLE MEMBERSHIP MODEL

16.4.1 Two-Level Multiple-Membership Model

As described earlier, it is possible that units of a classification are clustered within a higher level classification and are members of multiple units of that higher level classification. For example, as in Table 16.8 and Figure 16.11, students are clustered within middle schools but some students attended (and are thus multiple members of) more than one middle school. If it is assumed that each middle school contributes to the score of a student, then use of a multiple-membership model should be considered.

Using Rasbash and Browne's (2001) parameterization, the unconditional multiple-membership model would be parameterized as follows:

$$Y_{i\{j\}} = \gamma_{00} + \sum_{h\in\{j\}} w_{ih}u_{0h} + e_{i\{j\}} \quad (16.23)$$

where i indexes the level-1 unit (here, student) that is a member of multiple units of the level-2 classification j (which, in the current example, is middle school), $e_{i\{j\}}$ is the level-1 residual for student i from the set j, $\{j\}$, of middle schools, u_{0h} is the residual associated with level-2 (middle school) unit h, w_{ih} is the weight assigned to level-1 unit i's (student) association with unit h of the j classifications (middle schools) of the student. Each level-1 unit's set of weights sum to one. The maximum number of middle schools attended by students in the data set was three with equal weight assumed assigned for the middle schools that were attended. Thus, using the data in Figure 16.11, Equation 16.23 would be:

$$\begin{aligned} Y_{A\{1\,and\,2\}} &= \gamma_{00} + 0.5u_{01} + 0.5u_{02} + e_{A\{1\,and\,2\}}, \\ Y_{B\{1\}} &= \gamma_{00} + u_{01} + e_{B\{1\}}, \\ Y_{L\{5\}} &= \gamma_{00} + u_{05} + e_{L\{5\}}, \\ Y_{O\{6,7\,and\,8\}} &= \gamma_{00} + 0.3\dot{3}u_{06} + 0.3\dot{3}u_{07} \\ &\quad + 0.3\dot{3}u_{08} + e_{O\{6,7\,and\,8\}} \end{aligned} \tag{16.24}$$

for each of the following students: A, B, L, and O. The student-specific equations listed in Equation 16.24 can be used to demonstrate a couple of facets of the multiple-membership model. In the unconditional model, the score of student i is modeled as a function of the overall average score, γ_{00}, of the level-1 residual, $e_{i\{j\}}$, and of the residuals, u_{0j}s, associated with each of the middle schools of which the student was a member. The "function" of the middle school residuals is the weight. The weights sum to one for each student. In the current example the weight equals one divided by the number of schools attended (i.e., equal weight is assigned to every school attended by a student). Thus if a student attended two schools, the weights for each middle school residual, u_{0j}, is 0.5. If the student only attended a single school, then the weight is one, and so on. Each level-2 (middle school) residual is assumed normally distributed with a mean of zero and a variance of σ^2_u (Goldstein, 2003; Rasbash & Browne, 2001).

The multiple-membership model is a form of two-level cross-classified model (where the level-1 unit is cross-classified by each level-2 unit of which it is a member). In this analogy, the number of cross-classified factors corresponds to the maximum number of level-2 units (middle schools) of which the level-1 unit (student) is a member. In addition, the variance (σ^2_u) of each cross-classified factor is constrained equal and the dummy coding used in model estimation consists of the weights rather than ones (Goldstein, 2003).

The unconditional model is presented in Equation 16.24. Characteristics of each classification can, of course, be added to the model. For example, a researcher might hypothesize that a level-2 (middle school) characteristic, Z, might explain some of the variability in the intercept:

$$Y_{i\{j\}} = \gamma_{00} + \sum_{h \in j} [w_{ih}(\gamma_{01}Z_h + u_{0h})] + e_{i\{j\}}. \tag{16.25}$$

In Equation 16.25, we see that it is assumed that the relationship, γ_{01}, between the middle school characteristic, Z, and the outcome, Y, is assumed constant across middle schools (thus, the single parameter). However, because a student might attend different middle schools that have different values on Z, a weighted average of the Zs across the set of j middle schools for student i is used.

The researcher might also hypothesize that a level-1 (student) characteristic, X, is also related to the outcome. Thus Equation 16.25 becomes:

$$Y_{i\{j\}} = \gamma_{00} + \sum_{h \in j} [w_{ih}(\gamma_{01} Z_h + u_{0h})] + \gamma_{10} X_{i\{j\}} + e_{i\{j\}}. \quad (16.26)$$

Equation 16.26 models the relationship between X and the outcome, Y, as fixed across middle schools. It could instead be modeled as randomly varying across middle schools:

$$Y_{i\{j\}} = \gamma_{00} + \sum_{h \in j} [w_{ih}(\gamma_{01} Z_h + u_{0h})] + \left(\gamma_{10} + \sum_{h \in j} w_{ih} u_{1h}\right) X_{i\{j\}} + e_{i\{j\}}, \quad (16.27)$$

or as nonrandomly varying with part of the variability in the relationship between X and Y explained by the same middle school predictor, Z:

$$Y_{i\{j\}} = \gamma_{00} + \sum_{h \in j} [w_{ih}(\gamma_{01} Z_h + u_{0h})] + \left\{\gamma_{10} + \sum_{h \in j} [w_{ih}(\gamma_{11} Z_h + u_{1h})]\right\} X_{i\{j\}} + e_{i\{j\}}. \quad (16.28)$$

In Equations 16.27 and 16.28, the middle school intercept and slope residuals are distinguished by the zero and one, respectively, appearing as the first character in the subscripts. As with all multilevel models, and mentioned earlier, inclusion of random effects complicates already complex estimation procedures.

16.4.2 Cross-Classified Multiple-Membership Models

This section will provide the parameterization of a cross-classified multiple membership model that could be used to describe relationships among variables for the data structure appearing in Table 16.10 and Figures 16.14 and 16.15. In this data, some students are multiple members of middle schools and are cross-classified by middle school and high school. The associated unconditional model for this pattern of classifications would be:

$$Y_{i(\{j_1\},j_2)} = \gamma_{000} + \sum_{h \in \{j_1\}} w_{ih} u_{0h0} + u_{00j_2} + e_{i(\{j_1\},j_2)} \quad (16.29)$$

where $Y_{i(\{j_1\},j_2)}$ represents the score of student i who attended the set of j_1 middle schools and high school j_2. Equation 16.29 demonstrates the contribution to $Y_{i(\{j_1\},j_2)}$ of high school, u_{00j_2}, the set of middle schools attended, $\sum_{h \in \{j_1\}} w_{ih} u_{0h0}$, and of the individual student, $e_{i(\{j_1\},j_2)}$, over and above the average predicted Y score, γ_{000}. As demonstrated using the simpler multiple-membership model (see Equations 16.23 and 16.25 through 16.27), student, middle school, and high school predictors could be added to this unconditional model and the coefficients modeled as fixed, random, or nonrandomly varying across the units of other classifications.

16.5 CONCLUSIONS

This chapter has provided an introduction to cross-classified and multiple membership models. Some simple examples have been provided offering the link between the data's structure and the resulting model's parameterization. As has been mentioned already, it should be emphasized that further complications are available in terms of additional classifications and predictors. In addition, most of the models demonstrated here have involved univariate, interval-scaled outcomes that are assumed normally distributed. These models can, of course, be used with multivariate outcomes and noninterval-scaled outcomes. As with any sophisticated statistically modeling techniques, a balance should be found between the complexity of the model and the complexity of the necessary estimation procedures and inferences (Raudenbush & Bryk, 2002).

The reader should refer to Hox (2002) and especially Goldstein's (2003) text and his many articles to find detailed information about the estimation procedures and options used for estimating these models. Several software packages can be used to estimate CCREMs although only MLwiN currently provides a direct way of estimating multiple-membership random effects models.

The complexity of the data structures encountered in social science research, the negative effect of inappropriately modeling this complexity, and the availability of software that facilitates estimation of these models behooves the importance of researchers' understanding and increasingly using cross-classified and multiple-membership models.

REFERENCES

Beretvas, S. N. (2008). Cross-classified random effects models. In A. A. O'Connell & D. Betsy McCoach (Eds.), *Multilevel modeling of educational data* (pp. 161–197). Charlotte, SC: Information Age Publishing.

Fielding, A., & Goldstein, H. (2006). *Cross-classified and multiple membership structures in multilevel models: An introduction and review* (Research Report Number 791). Birmingham, UK: Department of Education and Skills, University of Birmingham.

Goldstein, H. (2003). *Multilevel statistical models* (3rd ed.). New York, NY: Hodder Arnold.

Hox, J. (2002). *Multilevel analysis: Techniques and applications*. Mahwah, NJ: Lawrence Erlbaum Associates.

Kamata, A. (2001). Item analysis by the hierarchical generalized linear model. *Journal of Educational Measurement, 38*, 79–93.

Meyers, J. L., & Beretvas, S. N. (2006). The impact of inappropriate modeling of cross-classified data structures. *Multivariate Behavioral Research, 41*, 473–497.

Rasbash, J., & Browne, W. J. (2001). Modeling non-hierarchical structures. In A. H. Leyland and H. Goldstein (Eds.), *Multilevel modeling of health statistics* (pp. 93–105). Chichester, UK: John Wiley & Sons.

Rasbash, J., Browne, W. J., Goldstein, H., Yang, M., Plewis, I., Healy, M. . . . Lewis, T. (2000). *A user's guide to MLwiN* (2nd ed.). London, UK: Institute of Education.

Raudenbush, S. W., & Bryk, A. S. (2002). *Hierarchical linear models: Applications and data analysis methods*. Thousand Oaks, CA: Sage Publications.

Snijders, T. A. B., & Bosker, R. J. (1999). *Multilevel analysis: An introduction to basic and advanced multilevel modeling*. London: Sage Publications.

Van den Noortgate, W., De Boeck, P., & Meulders, M. (2003). Cross-classification multilevel logistic models in psychometrics. *Journal of Educational and Behavioral Statistics, 28*, 369–386.

Williams, N. J., & Beretvas, S. N. (2006). DIF identification using HGLM for polytomous items. *Applied Psychological Measurement, 30*, 22–42.

17

Dyadic Data Analysis Using Multilevel Modeling

David A. Kenny
Department of Psychology, University of Connecticut, Storrs, Connecticut

Deborah A. Kashy
Department of Psychology, Michigan State University, East Lansing, Michigan

Multilevel modeling (MLM) is a method for analyzing hierarchically nested data structures such as over-time data from individuals (i.e., observations are nested within individuals) or group data (i.e., individuals are nested within groups). One way to conceptualize multilevel analyses is as a two-step process: A "lower-level" regression is computed separately for each "upper-level" unit (e.g., the relationship between a person's score on X and the person's score on Y is computed across individuals for each group), and then the lower-level regression estimates are pooled across groups. In the prototypical multilevel case, both the intercepts and slopes from these regressions are treated as random effects. Therefore, in addition to estimating the average intercept and average Y-X slope, both of which are fixed effects, the variance of the intercepts and variance of the slopes are also estimated.

In this chapter we consider multilevel analyses for dyadic data. Dyads are a special case of groups in that the number of individuals nested within each group equals two. From a multilevel perspective, this small group size represents a potential difficulty because, with only two data points, the two-step analysis conceptualization does not work. That is, if there are only two data points, it is not possible to compute a lower-level regression for each dyad because any two data points fall exactly on a line. Nonetheless, as we discuss in this chapter, MLM represents a powerful tool for the analysis of dyadic data, both for simple dyadic designs in which each person is a member of only one dyad as well as for complex designs in which individuals may participate in more than one dyad.

Dyadic data are very common in the social and behavioral sciences. The prototypical dyad study involves heterosexual married couples. However, there are many other possibilities: roommates, dating couples, friends, coworkers, patient and caretaker, siblings, and parent and child. Dyads need not be preexisting but can be created in the laboratory. Members of the dyad need not even interact, as in the case of yoked controls. Moreover, the two observations may not even be from two people; they might be from two animals, two eyes, two arms, or the left and right side of the brain.

Recently Kenny, Kashy, and Cook (2006) have extensively discussed the analysis of dyadic data, and MLM is an important tool in many of the analyses they describe. In this chapter, we review and extend their discussion. We begin with key definitions in dyadic analysis concerning types of dyads, types of dyadic designs, and types of variables. We then consider the use of MLM for the three major types of dyadic designs. Finally, we consider the analysis of over-time data for one of those designs. Most dyadic researchers would benefit from reading the entire chapter even if they have an interest in only one of these designs. Topics that are discussed in one section (e.g., coding of variables, computer syntax, and interpretation of results) are relevant for the other sections of the chapter.

Because we cover many topics, we are limited in the amount of space that can be dedicated to illustrations; however, we have included example data sets, syntax files, and outputs on the website for the book. Here we focus primarily on the syntax of computer applications, but space limitations preclude discussion of all programs, and so we present syntax for use with SAS software (SAS Institute Inc., 2002–2003). We chose SAS because of its flexibility and its ability to link to data transformations and other procedures. However, at times we discuss the use of SPSS (SPSS for Windows Rel. 16.0) and other programs. To help interpret the syntax, we adopt the convention that syntax commands are denoted by upper-case terms whereas variable names are denoted by lower-case terms in bold. We presume that the reader already has some familiarity with the concepts of MLM.

17.1 DEFINITIONS

17.1.1 Distinguishability

One important question in dyadic research and data analysis is whether or not the two dyad members can be distinguished from one another by some variable. In heterosexual dating relationships, dyad members may be distinguishable by gender: Each couple has one man and one woman. In sibling dyads, the two siblings may be distinguished by birth order. In both of these examples, a systematic ordering of the scores from the two dyad members can be developed based on the variable that distinguishes them. However, there are many instances in which there is no such natural distinction. Same-sex roommates or friendship pairs, homosexual romantic partners, and identical twins are all examples of dyads in which the members are typically indistinguishable. If dyad members are indistinguishable or exchangeable, then there is no systematic or meaningful way to order the two scores.

The issue of distinguishability is both conceptual and empirical. The examples of gender in heterosexual couples and birth order in siblings highlight the conceptual component of distinguishability: There must be a categorical variable that can be used to systematically classify dyad members. However, even when dyad members are conceptually distinguishable, they may not be empirically distinguishable. Empirical distinguishability occurs when there are detectable differences between dyad members as a function of the distinguishing variable. That is, if there are no differences in the means, variances, and covariances as a function of the distinguishing variable, then dyad members are not empirically distinguishable and a simpler and more parsimonious model (i.e., the model for indistinguishable dyads) can be estimated.

We shall see that whether dyad members are distinguishable has important implications for analyses, and we discuss procedures that can be used to test distinguishability. If there is no evidence of differences as a function of the distinguishing variable, we recommend that researchers use methods that are appropriate for indistinguishable dyads because such methods allow researchers to pool estimates both within and across dyads, which increases precision and statistical power. Even in cases in which there are compelling conceptual reasons for treating dyad members as distinguishable, one can statistically evaluate whether distinguishability makes a difference.

17.1.2 Typology of Dyadic Designs

In this chapter we discuss variants of three types of dyadic designs at length: the *standard dyadic design,* the *over-time standard dyadic design,* and the *one-with-many design.* We also provide a brief introduction to the Social Relations Model (SRM) design. The factor that differentiates these designs is the number of dyads in which each person participates. In addition, each of these designs can be reciprocal or nonreciprocal. A design is reciprocal when both dyad members provide outcome scores, and a design is nonreciprocal when only one of the two persons is measured on the outcome.

TABLE 17.1

Possible Dyads for the Three Designs With Six People (1 Through 6)

Standard:	{1,2} {3,4} {5,6}
One-with-many:	{1,2} {1,3} {4,5} {4,6}
SRM:	{1,2} {1,3} {1,4} {1,5} {1,6} {2,3} {2,4} {2,5} {2,6} {3,4} {3,5} {3,6} {4,5} {4,6} {5,6}

In the standard design, each person is a member of one and only one dyad. As seen in Table 17.1, the six persons, 1 through 6, form three dyads: {1,2}, {3,4}, and {5,6}. The standard design is by far the most common dyadic research design, and it is typified by studies of marital relationships because (in most studies of marriage at least) each husband is paired with only one wife. For the standard design we consider only reciprocal designs in which both persons are measured on the outcome, because if only one person provides an outcome score the data would not be multilevel.

Likewise, in the over-time standard design, each person is a member of one and only one dyad. However, in this design, both partners are measured at multiple occasions. Most commonly, the two individuals are measured at the same points in time. For example, in a study of the transition to parenthood, both partners' relationship satisfaction might be measured one month prior to birth, and again 1, 3, and 6 months after birth. Thus, in this case individuals are nested within dyads, but time is crossed with individuals. Less commonly, the two partners are measured at different occasions, and so occasions are nested within the individual, and individuals are nested within dyad. We limit our discussion of the over-time standard design to the more common crossed structure.

In the one-with-many design each person is paired with multiple others, but these others are not paired with any other persons. As seen in Table 17.1, for the one-with-many design, the dyads are {1,2}, {1,3}, {4,5}, and {4,6}. Note that the "ones" are persons 1 and 4, and the "many" are 2, 3, 5, and 6. As an example of the one-with-many design, Kashy (1992) asked people to rate the physical attractiveness of each person that they had interacted with over a period of 2 weeks. A second example of the one-with-many design would be having patients rate their satisfaction with their primary care physician (so that there are multiple patients rating the same physician). In this design one person is linked to many others and the others are not linked to each other. In most cases, the one-with-many design is not reciprocal: Data are just from either the "one" or just from the "many."

In a SRM design, each person is paired with multiple others and each of these others is also paired with multiple others. The prototypical SRM design is a round-robin design in which a group of persons rate or interact with each other. As seen in Table 17.1, all possible dyads are formed for the SRM design. For example, in a four-person round-robin design (e.g., persons A, B, C, and D), each person interacts with or rates three partners, and so a four-person round-robin results in a total of six dyads (i.e., AB, AC, AD, BC, BD, and CD). Because round-robin designs are inherently reciprocal, these six dyads generate a total of 12 outcome scores (i.e., the AB dyad generates two scores—A's score with B and B's score with A).

17.1.3 Types of Variables

In multilevel data, variables are traditionally denoted as varying at either the upper level (level 2) or the lower level (level 1), whereas in dyadic data, variables are typically denoted as between-dyads, within-dyads, or mixed. Not surprisingly, these two classification systems are related. Between-dyads variables vary from dyad to dyad, but do not vary within dyads, and so are upper-level variables. For example, in a study of the effects of stress on romantic relationship satisfaction, couples might be randomly assigned to a high stress condition in which they are asked to discuss a difficult problem in their relationship, or they could be assigned to a low stress condition in which they are asked to discuss a current event. For this example, the level of stress would be a between-dyads variable because both dyad members are at the same level of induced stress such that some dyads would be in the high stress condition and others would be in the low stress condition.

Alternatively, both within-dyads and mixed variables vary from person to person within the dyad, and so both of these types of variables can be viewed as lower-level variables from an MLM perspective. The two scores of a within-dyads variable differ between the two members within a dyad, but when averaged across the two dyad members, each dyad has an identical average score. A prototypical within-dyads variable is gender in heterosexual couples in that every couple is comprised of both a man and a woman. A less obvious example of a within-dyads variable is the proportion of housework done by two roommates. With this variable, the average of the two proportions always equals .50, yet within each dyad the amount of housework varies across the two roommates.

Mixed variables vary both within-and between-dyads such that the two partners' scores can differ from each other and there are differences in the dyad averages from dyad to dyad. Age is an example of a mixed independent variable in marital research because the two spouses' ages may differ from one another and some couples may be older on average than others. Because they can differ across the two dyad members, mixed variables fall into the lower-level classification for MLM.

17.2 STANDARD DESIGN

In Table 17.2, we present a fictitious data set based on 10 pairs of same-sex roommates that we use as an example. In this data set the outcome variable is a measure of the person's college adjustment at the end of the first semester (i.e., *Y*; in the table and syntax denoted as **adjust**), and the key predictor is a mixed variable that is an estimate of the average number of days per week each person drank during that semester (**act_drink**). Because the study includes only same-sex roommates, **gender** is a between-dyads variable that is represented as *Z* in this example. Note that the data, syntax, and outputs can be downloaded from http://davidakenny.net/dyadmlm/downloads.htm.

17.2.1 Indistinguishable Dyads

In the standard design, each person is paired with only one other person, and both dyad members are measured on the same variables. As a running example, we use the data presented in Table 17.2 in which

TABLE 17.2

Data set for the Fictitious Dyadic Study of Roommates

Dyad	Person	(Y) Adjust	(X) Act_Drink	Part_Drink	(Z) Gender	Citizen	D1	D2
1	1	5.5	0.2	−0.8	1	1	1	0
1	2	7.0	−0.8	0.2	1	−1	0	1
2	1	3.5	−0.8	0.2	−1	1	1	0
2	2	5.5	0.2	−0.8	−1	−1	0	1
3	1	2.5	2.2	3.2	1	1	1	0
3	2	0.3	3.2	2.2	1	−1	0	1
4	1	5.0	−1.8	−0.8	−1	1	1	0
4	2	5.5	−0.8	−1.8	−1	−1	0	1
5	1	3.0	0.2	1.2	1	1	1	0
5	2	2.0	1.2	0.2	1	−1	0	1
6	1	6.5	−0.8	−0.8	1	1	1	0
6	2	0.0	−0.8	−0.8	1	−1	0	1
7	1	5.5	−1.8	−0.8	1	1	1	0
7	2	6.0	−0.8	−1.8	1	−1	0	1
8	1	7.0	−0.8	0.2	−1	1	1	0
8	2	4.8	0.2	−0.8	−1	−1	0	1
9	1	4.5	1.2	0.2	−1	1	1	0
9	2	5.8	0.2	1.2	−1	−1	0	1
10	1	6.8	−0.8	1.2	−1	1	1	0
10	2	7.5	1.2	−0.8	1	−1	0	1

Note: act_drink and part_drink have been grand-mean centered around their mean of 1.80.

each person provides an outcome score, Y, a score on a lower-level predictor variable, X (note that X could be either mixed or within-dyads but the X depicted in Table 17.2 is mixed), and a score on an upper-level (i.e., between-dyads) predictor variable, Z.

If one takes the two-step analysis perspective for MLM, the level-1 model for person i in dyad j with a single lower-level predictor variable, X, would be

$$Y_{ij} = b_{0j} + b_{1j}X_{ij} + e_{ij},$$

where b_{0j} represents the predicted Y when X equals zero for person i in dyad j, and b_{1j} represents the coefficient that estimates the relationship between X and Y for dyad j. In the example, assuming that X, the person's drinking score (i.e., **act_drink**), has been grand-mean centered, b_{0j} represents the predicted college adjustment when drinking is average, and b_{1j} represents the change in adjustment as drinking increases by one day.[1] The second step of the analysis involves treating the slopes and intercepts from the first-step analyses as outcome variables in two regressions. For these level-2 analyses, the regression coefficients from the first step are assumed to be a function of a dyad-level predictor Z, and the equations would be

$$b_{0j} = a_0 + a_1Z_j + d_j$$

$$b_{1j} = c_0 + c_1Z_j.$$

The first level-2 equation treats the first-step intercepts as a function of the Z variable, and its form is similar to the standard MLM case that specifies that the first-step intercepts are comprised of both fixed and random effects components. Specifically, a_0 estimates the grand mean (assuming that X and Z were either effect coded or grand-mean centered), a_1 estimates the overall effect of Z on Y, and d_j represents that part of the intercepts for dyad j that is not explained by Z (also called the residual). The variance of these residuals captures the nonindependence of the Y scores for the two dyad members. The proportion of variance of these residuals, $s_d^2/(s_d^2 + s_e^2)$, measures the level of nonindependence and is commonly called the *intraclass correlation.*

The second level-2 equation treats the first-step slopes as a function of the Z variable. In this model c_0 estimates the average effect of X on Y and c_1 estimates the effect of Z on the X-Y relationship (i.e., the interaction between X and Z). This equation reflects the one major restriction that is necessary to apply MLM to dyads that need not be made for groups with more than two members: The first-step slopes are not allowed to vary randomly from dyad to dyad, and therefore are comprised of only fixed effects. This is because dyads do not have enough lower-level units (i.e., dyad members) to allow the slopes to vary randomly from dyad to dyad.

One additional modification of the standard MLM formulation is useful in dyadic data analysis. In the standard MLM formulation, nonindependence is modeled as a variance, but an alternative is to treat the scores from the two dyad members as repeated measures such that each dyad member would have an error and the errors would be correlated. Such a formulation models the nonindependence between dyad members as a covariance rather than a variance. This is particularly important when the outcome measure is structured such that when one dyad member has a higher score, the other person's score tends to be

[1] Because the number of days drinking variable has a meaningful zero value, it is unnecessary to grand-mean center. Ordinarily, to have meaningful zero values, it is necessary to grand-mean center.

lower (e.g., variables involving compensation, competition, division of a resource). Nonindependence in dyadic data is often negative and negative nonindependence can be captured by a covariance, but not by a variance. Thus, the standard formulation can be problematic and we strongly urge researchers to use the repeated measures formulation rather than the random intercept approach when analyzing dyadic data. Note that if the nonindependence is positive, then this covariance in the residuals equals the variance of the intercepts (s_d^2) described earlier.

The SAS syntax for a MLM that specifies this basic dyadic model in which the outcome variable is adjustment, the lower-level predictor is the person's drinking, the upper-level predictor is gender, and negative nonindependence is possible, would be:

```
PROC MIXED COVTEST;
CLASS dyad;
MODEL adjust = act_drink gender
   act_drink* gender/S
   DDFM = SATTERTH;
REPEATED/TYPE = CS SUBJECT = dyad;
```

The COVTEST option in the PROC MIXED statement requests that the random effects in the model be tested for statistical significance, thereby providing a test of the partial intraclass correlation for the outcome (partialling out the effects of the person's *X*). The CLASS statement defines **dyad** as a classification variable. The MODEL statement specifies that **adjust** is a function of the person's *X*, **act_drink**, the person's value on *Z*, **gender**, and the *XZ* interaction; the S (or SOLUTION) option requests that SAS print out the estimated fixed-effect coefficients, and the DDFM = SATTERTH option requests that the Satterwaithe approximation be used to compute the degrees of freedom. Finally, the REPEATED statement is used to model the residual variance and covariance, and the SUBJECT = **dyad** option specifies that individuals at the same level of **dyad** are related. The TYPE = CS option requests a residual structure known as compound symmetry, which constrains the residual variances to be equal across the dyad members and specifies that there is a covariance between the residuals as well. The equal variance constraint is particularly important because the dyad members are indistinguishable, and so their residuals are sampled from the same underlying population.

The comparable SPSS syntax is:

```
MIXED
adjust WITH act_drink gender
/FIXED = act_drink gender act_drink*
   gender
/PRINT = SOLUTION TESTCOV
/REPEATED = person | SUBJECT(dyad)
   COVTYPE(CS).
```

The **person** variable in the last syntax statement arbitrarily denotes the two dyad members as a "1" and "2."

Using these models with the example data set (i.e., Y = **adjust**, X = grand-mean centered **act_drink**, and Z = **gender**), $a_0 = 5.24$ and is an estimate of the grand mean for adjustment, and $a_1 = -0.38$ and estimates the effect of gender on adjustment. Given that gender is coded men = 1 and women = –1, this value suggests that women's average college adjustment scores are higher than men's. The average effect that a person's drinking has on his or her college adjustment is $c_0 = -0.51$, indicating that each one unit increase in average weekly drinking corresponds to a predicted decrease of 0.51 points on adjustment. The gender difference in the relationship between drinking and adjustment is $c_1 = -0.63$. These coefficients together suggest that the strong negative relationship between drinking and college adjustment is primarily true of men, because the coefficients predicting adjustment from drinking would be –1.14 for men and 0.12 for women. Finally, the residual variance is $s_e^2 = 1.653$, and the residual covariance (which in this case would equal the variance of the intercepts) is estimated at 0.855. So the partial intraclass correlation that estimates the similarity in the two roommates' adjustment scores after controlling for the effects of drinking and gender is $r_I = .52$, indicating that adjustment scores for roommates are quite similar.

In sum, the MLM model for the standard dyadic design with indistinguishable dyads, one *X* variable, and one *Z* variable has four fixed effects and two random effects (variation in the intercepts and error variance). Although it may seem that constraining the slope variance to zero might bias the other multilevel estimates, this is not the case. Instead, variance in the slopes is modeled as one component of the error variance. Tests of the null hypotheses are not biased when the slopes do in fact vary across dyads.

One common practice in MLM is to compute the mean of the level-1 predictor variable and use it as a level-2 predictor. This can only be done if the level-1 variable is a mixed variable, because within-dyads variables do not vary at level 2. For example, in addition to using a person's drinking as an *X* variable to predict his or her adjustment, the average drinking score for the dyad can be used as a *Z* variable in the level-2 equations. This would result in an estimate of (a) whether a person who drinks more is lower in adjustment, and (b) whether living in a room in which both roommates drink more on average moderates the effect of a person's drinking on his or her adjustment.

For dyadic data we recommend a somewhat different approach. Instead of using the dyad average of the *X* variable as a level-2 predictor variable, we suggest including both the person's *X* and his or her roommate's *X* as level-1 predictors of the person's adjustment. In the drinking example, this would allow us to estimate both the effect of a person's own drinking on his or her college adjustment as well as the effect of the roommate's drinking on the person's adjustment. This approach of using one's own and partner's *X* as predictors has been called the Actor–Partner Interdependence Model (APIM; Kenny, Kashy, & Cook, 2006). Note that in Table 17.2 each person's drinking score variable appears twice in the data file, once in the person's own record as an actor effect (i.e., **act_drink,** denoted in the following MLM equations as *XA*) and again in the partner's record as a partner effect (i.e., **part_drink** denoted in the following MLM equations as *XP*). Data files of this format are sometimes called *pairwise* data sets. The lower-level model that depicts both actor and partner effects is:

$$Y_{ij} = b_{0j} + b_{1j}XA_{ij} + b_{2j}XP_{ij} + e_{ij}.$$

The upper-level models, assuming that there is one upper-level predictor variable, *Z*, would be:

$$b_{0j} = a_0 + a_1Z_j + d_j$$

$$b_{1j} = c_0 + c_1Z_j$$

$$b_{2j} = h_0 + h_1Z_j.$$

The new parameters in these upper-level models, h_0 and h_1, can be interpreted as the average effect that the partner's *X* has on the person's *Y*, and the degree to which *Z* moderates the relationship between the partner's

X and the person's Y, respectively. Thus, the partner effect, h_0, would estimate the effect of the roommate's drinking on the person's college adjustment, and h_1 would estimate gender differences in the partner effect (e.g., having a roommate who drinks heavily may have more impact on men's adjustment than on women's). The SAS syntax for a basic actor–partner model that does not include a dyad-level (i.e., level 2) predictor would be:

```
PROC MIXED COVTEST;
CLASS dyad;
MODEL adjust = act_drink
   part_drink /S DDFM = SATTERTH;
REPEATED/TYPE = CS SUBJECT = dyad;
```

and the corresponding SPSS syntax would be:

```
MIXED
adjust WITH act_drink part_drink
/FIXED = act_drink part_drink
/PRINT = SOLUTION TESTCOV
/REPEATED = person | SUBJECT(dyad)
   COVTYPE(CSR).
```

To extend these models so that they include a Z variable (e.g., **gender**), the MODEL statement in SAS would be extended as follows:

```
MODEL adjust = act_drink part_drink
   gender gender*act_drink gender*
   part_drink/ S DDFM = SATTERTH;
```

and a similar change would be used for SPSS. The coefficients from these APIM models that include gender interactions using the example data set are $a_0 = 5.27$, $a_1 = -0.39$, $c_0 = -0.42$, $c_1 = -0.60$, $h_0 = -0.05$, $h_1 = -0.18$, and $r_I = .53$. The effects for actor are quite similar to those already described. The new partner effect results suggest that having a roommate who drinks more often has a negative effect on men's adjustment (combining h_0 and h_1 for men, the partner effect coefficient is −0.23), but not women's (the coefficient is 0.13).

One additional feature of the actor–partner model is that actor and partner effects can interact to create a new level-2 variable. We can form the interaction in the usual way by computing a product; in the roommate example this interaction might suggest that when both roommates drink a great deal, their adjustment scores are especially low. Alternatively, it may be more appropriate to form the interaction by computing the absolute difference between the person's X and the partner's X scores to create a measure of dissimilarity. Such an interaction might indicate that dissimilarity (i.e., when one person drinks a great deal but the other person does not) has a particularly detrimental effect on the two roommates' adjustment scores. We refer the reader to Chapter 7 of Kenny, Kashy & Cook (2006) for a discussion of these interactions.

17.2.2 Distinguishable Dyads

The standard MLM approach works well when dyad members are indistinguishable, and the model can be adapted to handle cases in which dyad members are distinguishable as well. We outline three different strategies for handling distinguishable dyads. As a running illustration, we amend our example by noting that for each roommate pair, one person is an international student and the other is a U.S. citizen. Thus, **citizen** (or C) is a within-dyads variable that can be used to systematically distinguish between the two roommates. In the data set in Table 17.2, the **citizen** variable is coded such that U.S. students are coded as 1 and non-U.S. students are coded as −1.

The first strategy is identical to the one presented above for handling indistinguishable dyads, but a coded variable (using either 1 and 0 dummy coding or 1 and –1 effect coding as is the case for the citizenship variable) is added to the model to code for the distinguishing variable. For the example, we would add the C variable into the level-1 equation with actor and partner effects (we initially do not include gender to simplify the model):

$$Y_{ij} = b_{0j} + b_{1j}XA_{ij} + b_{2j}XP_{ij} + b_{3j}C_{ij} + e_{ij}.$$

Additionally, we would likely want to allow for interactions between the other variables in the model and the distinguishing variable. For the example, such interactions would specify that the actor and partner effects may differ as a function of citizenship, and are included by multiplying the distinguishing variable times each level-1 variable in the model:

$$Y_{ij} = b_{0j} + b_{1j}XA_{ij} + b_{2j}XP_{ij} + b_{3j}C_{ij} + b_{4j}XA_{ij}C_{ij} + b_{5j}XP_{ij}C_{ij} + e_{ij}.$$

The level-2 equations that include a dyad-level predictor, Z, would then be:

$$b_{0j} = a_0 + a_1Z_{1j} + d_j$$

$$b_{1j} = c_0 + c_1Z_{1j}$$

$$b_{2j} = h_0 + h_1Z_{1j}$$

$$b_{3j} = k_0 + k_1Z_{1j}$$

$$b_{4j} = m_0 + m_1Z_{1j}$$

$$b_{5j} = p_0 + p_1Z_{1j}.$$

In these models, k_0 estimates the average citizenship difference on Y or college adjustment, k_1 estimates the degree to which Z (gender of the two roommates) moderates the citizenship difference on Y, m_0 estimates the degree to which actor effects differ as a function of the person's citizenship, m_1 estimates whether the actor by citizenship interaction varies as a function of Z, and likewise, p_0 estimates the degree to which partner effects differ by citizenship and p_1 estimates the degree to which the partner by citizenship interaction varies as a function of Z.

Our discussion of distinguishable dyads thus far presumes that the residual variances are the same for both types of members (this is akin to the homogeneity of variance assumption in ANOVA). However, because the dyads are distinguishable, we would probably want to allow for heterogeneity of variance across levels of the distinguishing variable. In the example, this would allow the residual variances in college adjustment to differ for United States versus international students (perhaps there would be more unexplained variance in adjustment for international students relative to U.S. students).

The SAS syntax below specifies a model that includes actor and partner effects for the mixed predictor (i.e., **act_drink** and **part_drink**), a distinguishing variable, **citizen**, and a between-dyads variable, **gender**. (Note that the distinguishing variable can equivalently be treated as a class variable or simply as a dichotomous predictor—the only difference is in the appearance of the output.) By including interactions between the distinguishing variable and the actor and partner effects, the model allows the actor and partner effects to differ as a function of the distinguishing variable. Moreover, by changing the TYPE to CSH (heterogeneous compound symmetry) this syntax also allows for heterogeneous variances as a function of the distinguishing variable. Note finally that the model also includes interactions with the upper-level or Z variable (i.e., dyad gender), and so there may be three-way interactions between gender, citizenship, and either actor or partner effects.

```
PROC MIXED COVTEST;
CLASS dyad;
MODEL adjust = citizen gender
   act_drink part_drink citizen*gender
   citizen*act_drink citizen*part_drink
   gender*act_drink gender*
   part_drink citizen*gender*act_drink
   citizen*gender*part_drink / S DDFM
   = SATTERTH;
REPEATED/TYPE = CSH SUBJECT =
   dyad;
```

The corresponding SPSS syntax is:

```
MIXED
adjust WITH citizen gender act_drink
   part_drink
/FIXED = citizen gender act_drink
   part_drink citizen*gender
   citizen*act_drink citizen*part_drink
   gender*act_drink gender*part_drink
   citizen* gender*act_drink
   citizen*gender*part_drink
/PRINT = SOLUTION TESTCOV
/REPEATED = person | SUBJECT(dyad)
   COVTYPE(CSH).
```

A second strategy for the analysis of distinguishable dyads is the two-intercept model, which was originally suggested by Raudenbush, Brennan, and Barnett (1995). We presume again here that each dyad contains one U.S. citizen and one international student. Consider the empty model for member *i* of dyad *j*:

$$Y_{ij} = b_{1j}D_{1ij} + b_{2j}D_{2ij},$$

where D_{1ij} is 1 for the U.S. citizen and 0 for the international student, whereas D_{2ij} is 0 for a U.S. citizen and 1 for an international student. (The correlation between D_1 and D_2 is −1.) In this model there is no intercept, at least not in the usual sense, but rather there are two intercepts, b_1 and b_2. The intercept for U.S. students is estimated as b_1 and the intercept for international students is estimated as b_2. In addition, there is no error term in the model, making this a very unusual model. Importantly, in this model both b_1 and b_2 are random effects, and so the model has a variance–covariance matrix of b_1 and b_2 with three elements: the variance of b_1 or s_1^2 (the error variance for U.S. students), the variance of b_2 or s_2^2 (the error variance for international students), and the covariance between the two or s_{12} (the degree of nonindependence).

If there were any X or Z variables, they are added to the model, but any X variables need to be multiplied by each of the two D dummies. We add here an actor and a partner effect for X:

$$Y_{ij} = b_{1j}D_{1ij} + b_{2j}D_{2ij} + b_{3j}D_{1ij}XA_{ij} + b_{4j}D_{2ij}XA_{ij} + b_{5j}D_{1ij}XP_{ij} + b_{6j}D_{2ij}XP_{ij}.$$

Thus, the actor effects for U.S. students is given by b_{3j} and the actor effects for international students is given by b_{4j} and likewise for the partner effects: b_{5j} and b_{6j}.

With SAS there are two ways to estimate the two-intercept model. The first method is relatively simple. In this method, the distinguishing variable is treated as a classification variable. It is then entered into the MODEL statement, and the NOINT option is used to suppress the intercept. By suppressing the intercept and including the distinguishing variable as a categorical or classification variable, we force the program to compute two intercepts, one for **citizen** = 1 and the other for **citizen** = –1. Similarly, we obtain estimates of separate actor and partner effects by including interactions between the distinguishing variable and the actor and partner variables. The syntax below estimates the two intercept model that also includes separate actor and partner effects for United States versus international students. (Note that for simplicity, we have not included **gender**.)

```
PROC MIXED COVTEST;
CLASS dyad citizen;
MODEL adjust = citizen citizen*
   act_drink citizen*part_drink/S
   DDFM = SATTERTH NOINT;
REPEATED / TYPE = CSH SUBJECT =
   dyad;
```

The comparable syntax is SPSS is:

```
MIXED
adjust BY citizen WITH act_drink
   part_drink
/FIXED = citizen citizen*act_drink
   citizen*part_drink | NOINT
/PRINT = SOLUTION TESTCOV
/REPEATED = person | SUBJECT(dyad)
   COVTYPE(CSH).
```

The more direct, but less simple way to estimate the two-intercept model with SAS is to actually create the two dummy variables and then use them in the syntax. For example, we could define **d1** = 1 if **citizen** = 1 and **d1** = 0 otherwise and **d2** = 1 if **citizen** = –1 and **d2** = 0 otherwise. The SAS syntax would then be:

```
PROC MIXED COVTEST;
CLASS dyad;
MODEL adjust = d1 d2 d1*act_drink
   d2*act_drink d1*part_drink d2*
   part_drink
/S DDFM = SATTERTH NOINT;
RANDOM d1 d2 / SUBJECT = dyad
   TYPE = UN;
PARMS 2, 1, 2, 0.000001 / HOLD = 4;
```

The last statement sets the starting values for the random effects, and it is required because in the two-intercept model, the error variance must be constrained to zero. The specific values in the PARMS statement refer to the variables in the RANDOM statement, and because **d1** and **d2** have a UN or unstructured variance–covariance matrix, there are four random effects being estimated: UN(1,1), which is the variance of the U.S. students' intercepts; UN(2,1), which is the covariance between the United States and international students' intercepts; UN(2,2), which is the variance of the international students' intercepts; and a residual variance. The first three numbers in the PARM statement can be almost any value, as long as the two variances are positive and the covariance is less than the absolute value of the product of the two standard deviations. The last value, 0.000001, specifies that the residual or error variance has a starting value that is virtually, but not exactly, zero (i.e., it is 0.000001), and the HOLD = 4 instructs the program to fix the residual variance, the fourth parameter, to its starting value. So far as we know, there is no way within SPSS to fix the error variance to zero. MLwiN does allow zero error variance whereas HLM allows the fixing of the error variance to a very small value.

Finally, it is sometimes useful to use two of the methods we have described to estimate distinguishable models. In particular, the first method we described provides estimates and tests of the interactions between *X* variables and the distinguishing factor (e.g., the test of whether the effect of one's own drinking differs for U.S. students relative to international students). If such interactions emerge, then estimating and testing the simple slopes separately for each level of the distinguishing variable (e.g., what is the effect of one's own drinking for U.S. students, and what is the effect of one's own drinking for international students) is a natural way to break down the interaction. These estimates and tests of the simple slopes can be provided directly by either of the two-intercept models we have described.

17.2.3 Test of Distinguishability

MLM can be used to test whether conceptually distinguishable dyads are actually empirically distinguishable. Kenny, Kashy, & Cook (2006) present such a test using structural equation modeling. To conduct this test using MLM, two models must be estimated, and both of these models should use maximum likelihood estimation (ML) rather than the typical program default of restricted maximum likelihood (REML). The ML option should be used because the distinguishable model generally differs from the indistinguishable model in its fixed effects. In SAS this is accomplished by adding METHOD = ML to the PROC MIXED statement.

In the first model, dyad members are treated as distinguishable both in terms of their fixed and random effects.

```
PROC MIXED COVTEST METHOD =
   ML;
CLASS dyad;
MODEL adjust = citizen act_drink
   part_drink citizen*act_drink
   citizen*part_drink / S
   DDFM = SATTERTH;
REPEATED/TYPE = CSH SUBJECT =
   dyad;
```

In the second model, dyad members are treated as indistinguishable.

```
PROC MIXED COVTEST METHOD =
   ML;
CLASS dyad;
MODEL adjust = act_drink part_drink
   / S DDFM = SATTERTH;
REPEATED/TYPE = CS SUBJECT =
   dyad;
```

A chi-square difference test can then be computed by subtracting the deviances (i.e., the −2*log likelihood values). For example, if we want to compare the indistinguishable actor–partner model (dropping the between-dyads, or Z, variable for simplicity) with the actor–partner model that treats the dyad members as distinguishable, there are three additional fixed effects (**citizen, citizen*act_drink, citizen*part_drink**) and one additional random effect because heterogeneous compound symmetry (CSH rather than CS) allows the two variances to differ but homogeneous compound symmetry does not. If the χ^2 difference with four degrees of freedom were not statistically significant, the data would be consistent with the null hypothesis that the dyad members are indistinguishable. If, however, χ^2 were statistically significant, then there would be support for the alternative hypothesis that dyad members are distinguishable. In the example, the distinguishable model has a deviance of 60.920 and the indistinguishable model has a deviance of 64.507. Thus, the test of distinguishability is $\chi^2(4) = 3.587$, $p = .46$, and so in the example data set, there is not much evidence that the roommates are empirically distinguished by citizenship (the sample size is very small, resulting in low power for this test).

17.3 OVER-TIME STANDARD DESIGN

Here, we consider the standard design with the complication that each member of the dyad is measured at multiple times. By multiple, we mean more than twice and preferably each dyad member is measured more than five times. In over-time data from dyads, there are three factors that define the structure of the data: time, person, and dyad. Researchers often make the mistake of considering these data to be a three-level nested model in which time points are nested within persons and persons are nested within dyads. As pointed out by Laurenceau and Bolger (2005), the problem is that time and person are usually crossed, not nested. That is, for a given dyad, the time point is the same for the two persons at each time point.

There are two potentially undesirable consequences if the three-level nested model is mistakenly assumed. First the correlation between the two partner's intercepts, r_{cc}, is constrained to be positive because it is estimated as a variance. As we have described, there are several types of outcome measures for which this dyadic correlation would likely be negative. The second consequence of mistakenly conceptualizing crossed over-time dyadic data as a three-level nested structure is that the correlation between the two members' errors at each time, r_{ee}, is assumed to be zero. This correlation measures the time-specific similarity (or dissimilarity) in that part of Y that is not explained by the predictors for the two partners. Such time-specific similarity might arise because of arguments or other events that occur immediately prior to an assessment occasion.

An additional issue with over-time data is the necessity of modeling the nonindependence that arises because variables are measured over time, which is commonly called *autocorrelation* (see e.g., Hillmer, 2001). There is probably no more reliable finding in the social and behavioral sciences than the fact that the best predictor of future behavior is past behavior. Statistically, autocorrelation is the association between a measure taken at one point in time and the same measure taken at another point in time.

We can divide over-time models into two major types. First, there are stochastic models in which a person's or dyad's score is a function of past scores plus a random component. Second, there are deterministic models in which the person or dyad is assumed to be on some sort of trajectory. In this chapter, we focus on linear deterministic growth models. In these models, the explanatory variable is time of measurement, and each person has a slope that estimates his or her rate of change, as well as an intercept that measures the person's level at time zero. As discussed in many treatments of growth models (e.g., Biesanz, Deeb-Sossa, Papadakis, Bollen, & Curran, 2004), choice of time zero is an important one; however, in this chapter we simply assume that time zero is the initial observation. We begin by discussing distinguishable dyads. We then turn our attention to the more complex case of indistinguishable dyads.

17.3.1 Distinguishable Dyads

Consider as a simple example, an over-time study of marital satisfaction (**satisf**) in which satisfaction is measured yearly for 5 years. The data for two couples from this fictitious over-time study are presented in Table 17.3, and the distinguishing variable, **gender**, is coded 1 for husbands and –1 for wives. As can be seen in the table, there are 10 records for each dyad (five for each person) and so the data set is structured in a time-as-unit format, sometimes called a *person-period* data set. Thus if 60 married couples were measured at five times, there would be 600 records.

In this data set **time** is coded as zero at the initial assessment and then increases by one each year. On each data record, we recommend creating one additional variable, what we call **timeid,** which simply equals **time.** We do so because we need two different time variables: one of which is continuous time (**time**) and the other is treated as categorical (**timeid**). One key idea in the analysis of data from the over-time standard design is to use the distinguishing variable (e.g., **gender**) to create two dummy variables that represent the two "classes" of individuals. One dummy variable, what we denote as H in the MLM equations (**husband** in Table 17.3), is set to one when the scores are from the husband and to zero otherwise. The other, what we call W (**wife** in Table 17.3) is set to one when the scores are from the wife and zero otherwise.

TABLE 17.3

Data from Two Couples in a Fictitious Over-Time Study of Marital Satisfaction

Dyad	Time	Person	Husband	Wife	Satisf	Gender
1	0	1	1	0	5	1
1	1	1	1	0	6	1
1	2	1	1	0	8	1
1	3	1	1	0	7	1
1	4	1	1	0	4	1
1	0	2	0	1	4	–1
1	1	2	0	1	5	–1
1	2	2	0	1	3	–1
1	3	2	0	1	6	–1
1	4	2	0	1	7	–1
2	0	1	1	0	8	1
2	1	1	1	0	6	1
2	2	1	1	0	4	1
2	3	1	1	0	7	1
2	4	1	1	0	6	1
2	0	2	0	1	6	–1
2	1	2	0	1	7	–1
2	2	2	0	1	8	–1
2	3	2	0	1	3	–1
2	4	2	0	1	6	–1

The basic idea is that we create a two-level model in which level 1 is time or observation for both persons, and level 2 is the dyad. Through its use of the H and W dummy codes, this single model actually represents two growth curves, one for each member of the dyad. Having the two persons represented by one model allows us to model the nonindependence of the intercepts as a covariance and estimate a time-specific correlation between the residuals as well. The level-1 equation for person i in dyad j at time t would be

$$Y_{ijt} = b_{01j}H + b_{02j}W + b_{11j}HT_t + b_{12j}WT_t + He_{1jt} + We_{2jt}.$$

Note that the only predictor variable is time or T_t. There are four random variables at the level of the dyad: the two intercepts, b_{01j} and b_{02j}, and the two slopes, b_{11j} and b_{12j}. For the example, b_{01j} estimates the husband's satisfaction at the beginning of the study (**time** = 0) and b_{02j} estimates the wife's initial satisfaction. The slopes estimate the degree to which the husband's and wife's satisfaction increases or decreases each year on average.

The variance–covariance matrix for these intercepts and slopes contains four variances and six covariances. Two key covariance parameters for the dyadic growth model are the covariances between the two members' intercepts and two slopes. For the example, the covariance between the intercepts estimates whether husbands and wives are similar in their level of satisfaction at the initial assessment. The covariance between the slopes measures whether the rate of change in a husband's satisfaction is similar to his wife's. The four remaining covariances concern correspondence between the intercepts and slopes, including two within-person covariances (e.g., when wives start the study with lower satisfaction do they change more slowly?) as well as two between-person covariances (e.g., when wives start the study with lower satisfaction do their husbands change more slowly?)

There are also two residual variances, one for husbands and one for wives, and these two residuals may have a covariance. In the example, the correlation between the residuals, what we denoted as r_{ee} earlier, measures the degree to which the husband's and wife's satisfaction scores are especially similar at a particular time point, after taking the intercepts and slopes into account.

Dyadic growth models for distinguishable dyads can be estimated using either SAS or SPSS.

For distinguishable dyads, the SAS code is:

```
PROC MIXED COVTEST;
CLASS dyad timeid gender;
MODEL satisf = husband wife
   husband*time wife*time / NOINT S
   DDFM = SATTERTH ;
RANDOM husband wife husband*time
   wife*time / SUB = dyad TYPE = UN;
REPEATED gender / TYPE = CSH
   SUBJECT = timeid(dyad);
```

Because there is no intercept in the model (NOINT), the program estimates separate intercepts and separate slopes for husbands and wives. Using TYPE = UN in the RANDOM statement specifies that there are no equality constraints on the variance or covariance estimates, and so separate values are estimated across the distinguishing variable (e.g., across husbands and wives). The TYPE = CSH option in the repeated statement allows for different residual variances across the distinguishing variable, and it also specifies that there may be a time-specific correlation between the residuals. It is important to realize that although this specification of the error structure allows for different variances across the distinguishing variable, it does fix the error variances to be the same value at each time. Likewise, it estimates a single time-specific covariance rather than different values for each time point.

The SPSS syntax is:

```
MIXED
satisf BY dyad timeid gender
   WITH husband wife time
/FIXED = husband wife husband*time
   wife*time | NOINT
/PRINT = SOLUTION TESTCOV
/RANDOM husband wife husband*
   time wife*time|SUBJECT(dyad)
   COVTYPE(UNR)
/REPEATED gender | SUBJECT(timeid*
   dyad) COVTYPE(CSH).
```

We did not achieve a solution with the SPSS but did with SAS.

We refer to this model as the fully *saturated model* in the sense that we have treated all of the fixed effects (i.e., the two intercepts and two slopes) as random. In some cases, there can be difficulties in the estimation of multilevel models when either one or more of the variances is small or two or more of the terms are highly collinear. The researcher may need to explore simpler models, for example, a model in which intercepts are random and slopes are fixed or a model in which both dyad members share a common slope or intercept. The fixed and random effects can be constrained to the same value across the distinguishing variable by replacing the MODEL and RANDOM statements with

```
MODEL satisf = time/S DDFM =
   SATTERTH ;
RANDOM INTERCEPT time / SUB =
   dyad TYPE = UN;
```

This basic linear growth model is typically only the starting point of an over-time dyadic analysis. Researchers might want to see whether the levels or rates of change differ as a function of person-level predictors such as personality scores. For instance, we could include the actor's and partner's neuroticism in the analysis, and in our marital satisfaction example we would be able to determine whether the change in wife's satisfaction over time differs depending on the wife's own neuroticism and her husband's neuroticism. Dyad-level predictors such as length of relationship or experimental condition can also be treated as moderators of the effect of time. Moreover, time-varying variables such as employment status at each data collection period can also be included as predictors. Finally, growth models need not be linear, and so nonlinear functions of time such as quadratic time variables can be included in the analysis.

17.3.2 Indistinguishable Dyads

Although the fundamental principle with indistinguishable dyads is that they cannot be systematically distinguished, the analysis of data from an over-time standard design for indistinguishable dyads requires that we create a variable (which we call **person**) that distinguishes between the dyad members. Thus, for each dyad, one member is coded **person** as 1 and the other is coded **person** as 2. Using **person**, we create two dummy variables. One dummy, what we call P1, is set to one when the scores are from person 1 and to zero otherwise; and the other, what we call P2, is set to one when the scores are from person 2 and zero otherwise.

The level-1 model is very similar to that for distinguishable dyads,

$$Y_{ijt} = b_{01j}P1 + b_{02j}P2 + b_{11j}P1T_t + b_{12j}P2T_t + P1e_{1jt} + P2e_{2jt}.$$

However, when dyad members are indistinguishable, a series of equality constraints need to be included for both the fixed and random effects. For example, the two intercepts, b_{01j} and b_{02j}, should be equal, as should the two slopes, b_{11j} and b_{12j}. Indeed, the analysis becomes more complicated because we need to place similar constraints on the four-by-four variance-covariance matrix of slopes and intercepts.

This analysis can be conducted using SAS and MLwiN but not the current versions of SPSS and HLM. To accomplish this analysis, a data set (referred to as the **g** data set) that defines the equality constraints on the variance–covariance matrix must be created. The values in this data set refer specifically to the RANDOM statement in the PROC MIXED procedure. As shown below, the RANDOM statement in SAS syntax for this analysis defines four random effects: The two dummy variables, **p1** and **p2**, represent the two persons' intercepts, and **p1*time** and **p2*time** represent the two persons' slopes. These random effects create a four-by-four variance–covariance matrix where the diagonal represents the variances of the random effects.

Because the dyad members are indistinguishable, a certain structure needs to be imposed on this matrix. Specifically, the intercept variances (**p1** and **p2**) need to be fixed to the same value, and the slope variances (**p1*time** and **p2*time**) need to be fixed to the same value. In addition, the within-person intercept-slope covariances (**p1** with **p1*time** and **p2** with **p2*time**) need to be equated, as do the between-person intercept-slope covariances (**p2** with **p1*time** and **p1** with **p2*time**). There are two other elements in this matrix: the cross-person intercept covariance (**p1** with **p2**) and the cross-person slope covariance (**p1*time** with **p2*time**). Thus, although there are potentially 10 elements in the variance–covariance matrix, there are only six unique parameter estimates due to the equality constraints.

The **g** data set has a very specific format and must include the following variables: PARM, ROW, COL, and VALUE. These variables are linked to the ordering of variables in the RANDOM statement in the PROC MIXED syntax. PARM represents parameter number. Thus, because we are specifying six variance–covariance parameters between the intercepts and slopes (the intercept variance, the slope variance, the intercept covariance, the slope covariance, the within-person intercept-slope covariance, and the between-person intercept-slope covariance), PARM takes on six different values. The nature of the variance–covariance structure for the residuals is specified in the REPEATED statement, and does not play a role in **g**.

The ROW and COL values refer to the variables in the random statement, and so a 1 for ROW represents the effect of the **p1** dummy code (the intercept for **p1**), and having ROW = COL = 1 implies that the estimated parameter is the variance of the intercepts for **p1** (i.e., it is the covariance of **p1**'s intercept with itself). Although the second line in this data set has a PARM value of 1, it also has ROW = COL = 2. This information together specifies that the variance of the intercepts is based on the variance of both the **p1** intercept and the **p2** intercept. The next line identifies the second parameter that is the covariance between the two person's intercepts. The third and fourth lines define the variance of the slopes, and the fifth line defines the covariance between the two slopes. The remaining two parameters are the within person intercept-slope covariance (PARM = 5) and the between person intercept-slope covariance (PARM = 6).

The full SAS syntax for creating the G matrix would be

```
DATA g;
INPUT PARM ROW COL VALUE;
DATALINES;
1 1 1 1
1 2 2 1
2 3 3 1
2 4 4 1
3 1 2 1
4 3 4 1
5 1 3 1
5 2 4 1
6 1 4 1
6 2 3 1
;
```

The SAS syntax for the analysis is then:

```
PROC MIXED COVTEST;
CLASS dyad timeid person
MODEL y = time/ S
   DDFM = SATTERTH ;
RANDOM p1 p2 p1*time p2*time /G
   SUB = dyad TYPE = LIN(6) LDATA = g;
REPEATED person/TYPE = CS
   SUB = timeid(dyad);
```

Note on the REPEATED statement that TYPE is CS rather than CSH. This specifies that the residual variances are the same both across time and across person.

To test for distinguishability we could estimate two models using ML. One allows for full distinguishability and the other allows for indistinguishability. The difference in deviances is a chi-square test that, for the model presented, would have 7 degrees of freedom: the four constraints made in the **g** matrix (two sets of variances equal and two sets of covariances equal), the two equality constraints of the two intercepts and slopes, and the equality of the two error variances.

17.4 ONE-WITH-MANY DESIGN

In the one-with-many design, a person is in multiple dyadic relationships, but each of the person's partners is in a relationship with only that one person. For instance, a doctor might interact with many patients. Alternatively, adolescents might provide information about their relationships with their mothers, fathers, romantic partners, and best friends. We refer to the person who has multiple partners (the "one") as the *focal person* and to the multiple others (the "many") as the *partners*. In the first example, doctors would be focal persons and the patients would be their partners; in the second example the adolescents would be the focal persons and their mothers, fathers, romantic partners, and best friends would be the partners. As is illustrated by these two examples, the partners in the one-with-many design can be distinguishable or indistinguishable.

The one-with-many design is a blend of the standard dyadic and SRM designs; it is similar to the standard design in that each partner is paired with only one focal person, and it is like an SRM design in that the focal person is paired with many partners.

The one-with-many design is nonreciprocal if only one of the two members of each dyad provides outcome scores. Thus, if the adolescents rated their relationship closeness with their mothers, fathers, romantic partners, and friends, but these partners did not rate their closeness with the adolescent, the design would be nonreciprocal. Similarly, if the partners rated their closeness with the adolescent, but the adolescent did not make ratings, it would again be nonreciprocal. On the other hand, if both the focal person and the partners are measured, the design would be reciprocal. We first consider the nonreciprocal design for indistinguishable and distinguishable partners, and then we discuss the reciprocal design.

17.4.1 Nonreciprocal One-With-Many Designs: Indistinguishable Partners

As an example, we use the small data set presented in Table 17.4 from a fictitious study of intimacy in friendships. In this table we see that our data set contains nine focal persons (five women and four men; **focalsex** is coded 1 = men and -1 = women), each of whom report on the intimacy of their friendships (1 = not at all intimate, 9 = very intimate) with varying numbers of male and female friends (**partsex** is also coded 1 = men and –1 = women). The data in Table 17.4 actually include two measures of intimacy: One variable denotes the intimacy scores as rated by the focal person (**f_intimacy**), and the second denotes the intimacy scores as rated by the partners (**p_intimacy**). Thus, the data set is actually reciprocal (although we do not treat it as such at this point in our discussion and so **p_intimacy** will be ignored). Say that one question to be addressed in the study is whether the partners' relationship self-esteem (**rsepart**) predicts the focal person's perceptions of intimacy, and whether such a relationship is moderated by the focal person's gender.

Two variables must be included in the data set to apply MLM to the one-with-many design. First, a variable that identifies the focal person that is involved in each dyad should be created. In this chapter, we use **focalid** to denote the focal person. Second, a variable that identifies the partner that is involved in the dyad must be created, and we use the variable **partid** for this purpose. We assume that both **focalid** and **partid** start at the number one and continue to as many as needed. For **partid**, we have a choice. We can either number them consecutively from 1 to the total number of partners in the study, or for each focal person we use the same numbers. That is, if each focal person has three partners, they would be numbered 1, 2, and 3. When partners are distinguishable, the latter strategy is preferable. When partners are indistinguishable, either method can be used, but as we show, the SAS syntax would need to change.

Data from the one-with-many design are hierarchically structured because partners are tied to a focal person. In some ways, the one-with-many design is more closely linked to standard MLM designs because partners are nested within "groups" that are defined by the focal person. Thus, in the one-with-many design, the upper-level or level 2 is the focal person and the lower-level or level 1 is the partner. The variable X (e.g., partner's relationship self-esteem) is a level-1 variable because it is assumed to vary across partners within focal person. The level-1 equation for partner i with focal person j is:

$$Y_{ij} = b_{0j} + b_{1j}X_{ij} + e_{ij}.$$

TABLE 17.4

Data Set for the Fictitious Indistinguishable One-With-Many Design

Focalid	Partid	F_Intimacy	P_Intimacy	Focalsex	Partsex	RSepart	RSepartc
1	1	5	1	1	1	5	-5.06
1	2	8	6	1	1	4	-6.06
1	3	5	3	1	-1	5	-5.06
1	4	2	5	1	-1	13	2.94
2	1	7	6	-1	1	12	1.94
2	2	6	4	-1	-1	11	0.94
2	3	9	3	-1	-1	14	3.94
3	1	5	5	-1	1	8	-2.06
3	2	4	6	-1	1	12	1.94
3	3	2	7	-1	1	9	-1.06
3	4	6	6	-1	-1	15	4.94
3	5	5	7	-1	-1	17	6.94
4	1	7	3	1	-1	8	-2.06
4	2	5	2	1	-1	7	-3.06
4	3	8	4	1	-1	13	2.94
4	4	5	5	1	1	5	-5.06
5	1	1	6	1	1	12	1.94
5	2	2	6	1	1	16	5.94
5	3	5	7	1	-1	15	4.94
6	1	6	8	-1	-1	17	6.94
6	2	5	5	-1	-1	9	-1.06
6	3	8	5	-1	1	14	3.94
7	1	4	6	1	1	12	1.94
7	2	6	7	1	1	8	-2.06
7	3	7	4	1	1	4	-6.06
7	4	8	4	1	-1	3	-7.06
8	1	6	8	-1	-1	1	-9.06
8	2	7	7	-1	1	8	-2.06
8	3	9	8	-1	1	9	-1.06
9	1	7	7	-1	-1	15	4.94
9	2	5	4	-1	1	10	-0.06
9	3	7	4	-1	1	11	0.94

Note: focalsex and partsex is coded men = 1, women = –1.

To make the intercept, b_{0j}, more interpretable, it is generally advised to center the X variables by subtracting the overall partner mean (i.e., the grand mean computed across all partners in the data set). For the example data, this would involve subtracting the grand mean of the partner's relationship self-esteem, ($M = 10.062$) from each partner's **rsepart** score. The term e_{ij} represents the error or residual for partner i with focal person j. The level-2 models are:

$$b_{0j} = a_0 + a_1Z_j + d_j$$

$$b_{1j} = c_0 + c_1Z_j + f_j,$$

where Z_j is a level-2 variable (e.g., focal person gender, or **focalsex**) such that it takes on the same value for all partners of the same focal person. Unlike the standard design, there can be a random component for the slopes as well as for the intercepts. If there are relatively few partners per focal person, allowing such a variance may not be possible. If k refers to the number of partners per focal person (assumed to be equal only for the purposes of this calculation), to treat *all* the level-1 slopes as random for p lower-level predictor variables, k must be at least $p + 2$. In the small example data set, each person reports on at least three friendships, and so we could (in principle, although we do not do so) include a random slope for one lower-level predictor variable.

Although the slopes can be constrained to be equal for all focal persons (i.e., the slopes would be modeled as a fixed effect component only), it is almost always advisable to allow for the possibility that the intercepts vary across focal persons. As was true for the standard design, the variation of the intercepts models the nonindependence in the data—but in this case the nonindependence refers to similarity in scores for individuals who are paired with the same focal person. We can compute the variance of the intercepts or s_d^2 and the error variance or s_e^2. The ratio of the variance due to the intercept to the total variance, or $s_d^2/(s_d^2 + s_e^2)$, provides an estimate of the intraclass correlation, the measure of nonindependence. When there are lower-level predictor variables, or Xs, the intraclass correlation based on the intercepts is a partial intraclass that represents the proportion of variance due to the focal persons after controlling for the effects of the predictor variable(s).

The interpretation of this measure of nonindependence depends on whether the data come from the focal person or the partners. If the data come from the focal person (e.g., **f_intimacy**), then the variance in the intercepts refers to the consistency in how the focal person sees or behaves with the partners. (It is analogous to the actor variance in an SRM design, see below.) In the example analysis of the focal person's ratings of intimacy, the variance in the intercepts measures the degree to which focal persons tend to report similar levels of intimacy across all of their friends. If the data come from the partners, then the variance in the intercepts refers to the consistency in how the partners see or behave with the focal person. (It is analogous to the partner variance in an SRM design, see below.) If we treated the partner-rated intimacy as the outcome measure, the variance of the intercepts would measure the degree to which friends experience similar levels of intimacy with the focal person.

For the example data, we estimated a model predicting the focal person's intimacy ratings that allowed for random intercepts, random slopes for the effect of the partner's relationship self-esteem, and a covariance between the intercepts and slopes. In this model we treated grand-mean centered relationship self-esteem (**rsepartC**) as a lower-level predictor and focal-person gender as an upper-level predictor. The SAS syntax for this analysis is

```
PROC MIXED COVTEST;
CLASS focalid;
MODEL f_intimacy = rsepartC
   focalsex rsepartC*focalsex/S
   DDFM = SATTERTH;
RANDOM INTERCEPT/SUBJECT =
   focalid;
```

and the corresponding SPSS syntax is

```
MIXED
f_intimacy WITH rsepartC focalsex
```

```
/FIXED = rsepartC focalsex focalsex*
   rsepartC
/PRINT = SOLUTION TESTCOV
/RANDOM INTERCEPT | SUBJECT
   (focalid) COVTYPE(VC).
```

Based on the data in Table 17.4, $a_0 = 5.45$, indicating that average intimacy scores were somewhat above the scale midpoint. The effect of focal-person gender was $a_1 = -0.64$, and given the coding scheme, this indicates that men reported lower average intimacy across friends than did women. The average effect of relationship self-esteem on intimacy was relatively small, $c_0 = -0.05$, but there was evidence of a focal-person gender difference for the effect of relationship self-esteem on intimacy, $c_1 = -0.21$. Thus, women reported higher intimacy with friends who had higher relationship self-esteem, and men reported lower intimacy with friends that had higher relationship self-esteem. The random effects yielded the intercept variance of $s_d^2 = 1.127$, and a residual or error variance of $s_e^2 = 2.500$. Thus, the partial intraclass for intimacy ratings, controlling for the effects of self-esteem is .311, and so there is some evidence that focal persons reported similar levels of intimacy across their friends.

It is important to note that by definition a ratio of the variance of the intercepts to the total variance must be nonnegative, and so this method presumes that the intraclass correlation cannot be negative. Normally, we would not expect the intraclass correlation to be negative for the one-with-many design. However negative intraclass correlations could occur for variables requiring social comparison or compensation across partners; for example, if the outcome variable is structured such that if one partner has a high score, other partners have lower scores, the intraclass might be negative. However, treating nonindependence as a variance precludes the possibility of any negative nonindependence. If a negative intraclass correlation (Kenny, Mannetti, Pierro, Livi, & Kashy, 2002) is likely to occur (e.g., variables for which one partner having a high score constrains other partners to having lower scores), we suggest that partners should be treated as a repeated measure so that nonindependence is modeled as a correlation rather than a variance. This would be accomplished in SAS by substituting the RANDOM statement with a REPEATED statement as follows:

```
REPEATED/SUBJECT =
   focalid TYPE = CS;
```

and similarly for SPSS:

```
/REPEATED partid | SUBJECT(focalid)
   COVTYPE(CS).
```

It is possible to treat the mean of the X variable for each focal person, or M_X, as a predictor in the second-stage of the multilevel analysis. For instance, in the example where X is relationship self-esteem, then the effect of M_X on a focal person's intimacy would estimate whether individuals whose friends have higher self-esteem on average report higher levels of intimacy on average. For the standard design, we suggested using the partner's X as a level-1 predictor (e.g., the partner's drinking), but there is not a direct extension of this approach to the one-with-many design. Nonetheless, as we have shown with the example of focal-person gender, characteristics of the focal person may be relevant predictors.

Perhaps the most interesting case occurs when the same predictor is measured for the focal person as well as the partners. In the example this would mean that we have a measure of the focal-person's relationship self-esteem (e.g., **rsefocal**) in addition to the partners' scores on this variable. In this case the lower-level predictor, or X, would be the partner's relationship self-esteem, and the upper-level predictor, Z, would be the focal-person's relationship self-esteem. In such a case, the effect of focal-person self-esteem on focal-person intimacy would be an actor effect (i.e., If my partners have higher self-esteem on average, do I have higher average intimacy with them?), and the effect of partner self-esteem on focal-person intimacy would be a partner effect (i.e., If my partner has higher self-esteem, do I have higher intimacy scores with that partner?). In a parallel fashion, if the outcome measure is the partner-rated intimacy, then the effect of partner self-esteem on partner-reported intimacy would be an actor effect, and the effect of focal-person self-esteem on partner-reported intimacy would be a partner effect.

17.4.2 Nonreciprocal One-With-Many Designs: Distinguishable Partners

One major difference between the distinguishable and indistinguishable cases is that when partners are distinguishable, both the fixed and the random effects may vary by the distinguishing variable. Differential fixed effects are modeled by including partner role as a predictor in the model. In cases in which the effects of other X or Z variables are examined, interactions between these variables and partner role should be included as well.

Differential random effects as a function of the distinguishing variable can take two forms. The most general format does not place any constraints on the variance–covariance matrix of the random effects. In this model (which we will term the unconstrained random effects model) separate variances are computed for each role and separate covariances are estimated for each combination of roles. Because the covariances can differ across partners, this specification suggests that the focal-person effect may vary across partner roles. The alternative specification, which we refer to as the *constrained random effects model*, estimates a random focal-person effect in the form of an intercept variance, and then allows for differential residual variances for the different partner roles. In effect, this model constrains all of the covariances across partner roles to the same value—which is the variance of the intercepts.

As an example, consider the data in Table 17.5 in which the focal person is an adolescent child and the partners are the child's mother (**partrole** = 1), father (**partrole** = 2), and home-room teacher (**partrole** = 3). The key outcome variable is a measure of the child's cooperativeness, and each partner reports on this measure (**cooperate**; 1 = not at all cooperative, 9 = very cooperative). The data set also includes the child's gender (**focalsex;** boys = 1, girls = –1), which would be an upper-level predictor. Although for simplicity our example does not include any X variables, these could be easily added into the model as main effects and in interactions with partner role or other predictors. Finally, to make the fixed effects results more readily interpretable, we create two dummy coded variables: **teacher** = 1 if **partrole** = 3, and **teacher** = 0 otherwise; **father** = 1 if **partrole** = 2, and **father** = 0 otherwise. This coding scheme makes the mothers' ratings serve as the comparison group.

TABLE 17.5

Data Set From the Fictitious Distinguishable One-With-Many Study

Focalid	Focalsex	Partrole	Cooperate	Teacher	Father
1	1	1	6	−1	−1
1	1	2	5	0	1
1	1	3	4	1	0
2	1	1	4	−1	−1
2	1	2	3	0	1
2	1	3	5	1	0
3	−1	1	6	−1	−1
3	−1	2	6	0	1
3	−1	3	3	1	0
4	−1	1	7	−1	−1
4	−1	2	6	0	1
4	−1	3	6	1	0
5	−1	1	8	−1	−1
5	−1	2	7	0	1
5	−1	3	6	1	0
6	1	1	5	−1	−1
6	1	2	3	0	1
6	1	3	3	1	0
7	1	1	6	−1	−1
7	1	2	4	0	1
7	1	3	5	1	0
8	−1	1	6	−1	−1
8	−1	2	5	0	1
8	−1	3	5	1	0
9	1	1	9	−1	−1
9	1	2	7	0	1
9	1	3	6	1	0
10	−1	1	5	−1	−1
10	−1	2	6	0	1
10	−1	3	4	1	0

Note: focalsex is coded boys = 1, girls = −1; partrole is coded 1 = mothers, 2 = fathers, 3 = teachers.

The following SAS syntax includes the child's gender (**focalsex**) as an upper-level predictor, and partner role (**partrole**) is included as a categorical lower-level predictor. This approach provides overall *F*-tests that test whether there are mean differences as a function of partner role, and whether gender interacts with partner role. In addition to allowing for differential fixed effects, this syntax specifies the unconstrained random effects model in which all of the random effects may differ as a function of partner role:

```
PROC MIXED COVTEST;
CLASS focalid partrole;
MODEL cooperate = partrole focalsex
   partrole*focalsex/S DDFM =
   SATTERTH;
REPEATED partrole/TYPE = UN
   SUBJECT = focalid RCORR;
```

The REPEATED statement estimates the variance–covariance matrix for the different partner roles, and because TYPE is set to UN (unspecified), there are no equality constraints on this matrix. Thus, the variances for each partner role can differ, as can the covariances between the different pairs of roles. Adding RCORR to the REPEATED line gives the correlation matrix across partner roles. The syntax for SPSS is:

```
MIXED
cooperate BY partrole WITH focalsex
/FIXED = partrole focalsex partrole*
   focalsex
/PRINT = SOLUTION TESTCOV
/REPEATED = partrole|SUBJECT
   (focalid) COVTYPE(UNR).
```

The analysis using the small data set in Table 17.5 shows evidence of partner role differences in both the fixed and random effects. For example, the *F*-test of the partner role main effect is $F(2,8) = 13.46$, $p < .01$, and the *F*-test for the interaction between partner role and child gender is $F(2,8) = 3.62$. The effect estimates from these sets of syntax can be difficult to interpret because both SAS and SPSS create their own dummy codes for variables that are treated as categorical (or as factors in SPSS terms). As a result, it can be useful to estimate a second model that uses our dummy coded variables for teachers and fathers. For SAS this would be:

```
PROC MIXED COVTEST;
CLASS focalid partrole;
MODELcooperate=fatherteacherfocal-
   sexfather*focalsexteacher*focalsex/S
   DDFM = SATTERTH;
REPEATED partrole/TYPE = UN
   SUBJECT = focalid RCORR;
```

This analysis indicates that both fathers and teachers rated the children's cooperativeness lower than did mothers (i.e., the coefficient for fathers was $b = -1.0$, $t(8) = 4.26$, $p < .01$ and for teachers was $b = -1.5$, $t(8) = 3.81$, $p < .01$). In addition, it appears that the interaction between partner role and child gender is largely due to fathers ($b = -0.6$, $t(8) = 2.56$, $p = .03$). Thus, given the coding for **focalsex**, this suggests that the fathers tended to see boys as more cooperative than they saw girls.

The random effects from this unconstrained model suggest that the variance for mothers was somewhat larger than those for fathers or teachers (mothers' $s^2 = 2.4$, fathers' $s^2 = 1.65$, teachers' $s^2 = 1.5$). The correlations suggest that the two parents' perceptions were more similar to one another ($r = .88$) than they were to the teacher's ratings (for mothers $r = .62$, and for fathers $r = .46$).

The SAS syntax for the constrained random effects model that allows for a constant level-2 (i.e., focal person) effect, while specifying heterogeneous error variances for the different partner roles would be:

```
PROC MIXED COVTEST;
CLASS focalid partrole;
MODEL cooperate = partrole focalsex
   partrole*focalsex/S DDFM =
   SATTERTH;
RANDOM INTERCEPT/SUBJECT =
   focalid;
REPEATED partrole / TYPE = VC
   SUBJECT = focalid GRP = partrole;
```

This model in essence allows for heterogeneous error variances, but assumes that the covariances between partners are homogeneous, which is captured by the variance of the intercepts.

Finally, as was the case in the standard dyadic design, it is possible to test whether it is statistically useful to distinguish the partners. As before we would run two models, both of which will need to use ML rather than the program default of REML. The first model would be the indistinguishable model in which only *X* and *Z* variables predict the outcome, and the error structure is treated as compound symmetry. The second model would be the distinguishable model, and in addition to the effects of *X* and *Z*, this model would include the main effect of partner role as well as interactions between partner role and the other predictors. In addition, this second model specifies a heterogeneous variance–covariance matrix (either using the unconstrained random effects model or the constrained random effects model as described above). For the example data, we estimated the distinguishable model with unconstrained random effects and found a deviance of 76.1 based on a model estimating 12 parameters (6 random effects and 6 fixed effects). We next estimated the model for indistinguishable partners and found a deviance of 101.6 based on a model estimating 4 parameters (2 random effects and 2 fixed effect). The $\chi^2(8) = 25.5$, $p = .001$, suggesting that the model treating dyad members as distinguishable provides a better fit to the data.

17.4.3 Reciprocal One-With-Many Designs: Indistinguishable Partners

In a reciprocal one-with-many design outcome scores are obtained from both the focal person and the partners. For instance, we might ask doctors and their patients if they are each satisfied with one another. A unique and useful aspect of reciprocal one-with-many designs is that they allow estimation of both generalized and dyadic reciprocity. For the doctor–patient satisfaction study generalized reciprocity measures whether doctors who are on average more satisfied with their patients tend to have patients who are on average more satisfied with them. Dyadic reciprocity measures whether a patient with whom the doctor is especially satisfied is also especially satisfied with the doctor.

Table 17.6 shows the data layout required for the analysis of a reciprocal one-with-many design in which the partners are indistinguishable. This table is based on the data presented in Table 17.4, which was a fictitious study of friendship intimacy. To conduct a reciprocal one-with-many analysis using MLM, there would be two records for each focal person–partner dyad, one that contains the focal person's dyad-specific rating (e.g., the score that was presented as **f_intimacy** in Table 17.4, which is the focal person's rating of intimacy with a partner), and one that contains the partner's dyad-specific rating (e.g., **p_intimacy**, which is the partner's rating of intimacy with the focal person). Thus, there is now only one **intimacy** variable in Table 17.6. Three additional variables, which we call **rater**, **focal**, and **partner**, respectively must be created. These variables specify who generated the data—the focal person or the partner. The **rater** would equal 1 if the data are from the focal person, and it would equal –1 if the data are from the partner; the **focal** variable would be 1 if the outcome was generated by the focal person, and 0 if it was generated by the partner; and the **partner** variable would be 0 if the outcome was generated by the focal person, and 1 if it was generated by the partner.

The MLM equations for a reciprocal design are based on the two-intercept approach in which two dummy variables are created to denote the person that provided the outcome score. Thus, we would

TABLE 17.6

Data Set for the Fictitious Reciprocal Indistinguishable One-With-Many Design

Focalid	Partid	Rater	Focal	Partner	Intimacy	RSepart	Focalsex	Partsex	A_sex	P_sex
1	1	1	1	0	5	5	1	1	1	1
1	1	−1	0	1	1	5	1	1	1	1
1	2	1	1	0	8	4	1	1	1	1
1	2	−1	0	1	6	4	1	1	1	1
1	3	1	1	0	5	5	1	−1	1	−1
1	3	−1	0	1	3	5	1	−1	−1	1
1	4	1	1	0	2	13	1	−1	1	−1
1	4	−1	0	1	5	13	1	−1	−1	1
2	1	1	1	0	7	12	−1	1	−1	1
2	1	−1	0	1	6	12	−1	1	1	−1
2	2	1	1	0	6	11	−1	−1	−1	−1
2	2	−1	0	1	4	11	−1	−1	−1	−1
2	3	1	1	0	9	14	−1	−1	−1	−1
2	3	−1	0	1	3	14	−1	−1	−1	−1
3	1	1	1	0	5	8	−1	1	−1	1
3	1	−1	0	1	5	8	−1	1	1	−1
3	2	1	1	0	4	12	−1	1	−1	1
3	2	−1	0	1	6	12	−1	1	1	−1
3	3	1	1	0	2	9	−1	1	−1	1
3	3	−1	0	1	7	9	−1	1	1	−1
3	4	1	1	0	6	15	−1	−1	−1	−1
3	4	−1	0	1	6	15	−1	−1	−1	−1
3	5	1	1	0	5	17	−1	−1	−1	−1
3	5	−1	0	1	7	17	−1	−1	−1	−1

Note: Rater = 1 if the score was provided by the focal person and rater = −1 if the score was provided by the partner.

have R_1, which is coded 1 if the data are provided by the focal person and 0 if the data came from the partner, and R_2, which is coded 0 if the data are provided by the focal person and 1 if the data are provided by the partner. Using these two variables allows us to specify a model with separate effects for the focal persons and the partners and separate residuals for focal persons' ratings of partners and partners' ratings of the focal persons. The lower-level MLM equation is:

$$Y_{ijk} = b_{01j}R_1 + b_{02j}R_2 + R_1e_{ij1} + R_2e_{ij2}.$$

where i refers to the partner, j refers to the focal person, and k denotes whether the data was provided by the focal person (i.e., $k = 1$) or the partner (i.e., $k = 2$).

The basic level-2 models are:

$$b_{01j} = a_{01} + d_{1j}$$

$$b_{02j} = a_{02} + d_{2j}.$$

The two random variables at the level of the focal person are d_{1j} and d_{2j}, and they each have a variance. The variance in d_{1j} measures the degree to which the focal person rates all partners in a similar way (i.e., the actor effect from the SRM, see below), and the variance in d_{2j} measures the degree to which partners' ratings of the focal person are similar (i.e., the partner effect from the

SRM, see below). These two random effects also have a covariance that measures generalized reciprocity, or whether there is correspondence between how the focal person generally sees his or her partners and how the partners generally see the focal person. In the intimacy example, this covariance would measure whether individuals who report higher intimacy scores across their friends, are rated as higher in intimacy by those friends. The two errors from the lower-level model also have variances and a covariance between them. Although the error variances may not be simply interpreted, the covariance between the two can address an interesting question concerning dyadic reciprocity: If the focal person reports especially high intimacy with a particular friend, does that friend also report especially high intimacy?

Beyond the variance decomposition aspect, the reciprocal design can also incorporate predictor variables for either the focal person, the partners, or both. For example, if a partner-level variable, X, such as the relationship self-esteem measure, **rsepart**, were included, the model would be expanded to:

$$Y_{ijk} = b_{01j}R_1 + b_{02j}R_2 + b_{11j}XR_1 + b_{12j}XR_2 + R_1e_{ij1} + R_2e_{ij2}.$$

This model specifies that intimacy ratings are a function of who makes the rating (e.g., the focal person or the friend; b_{01j} and b_{02j}), and it also specifies that the partner's level of relationship self-esteem may moderate both persons' intimacy ratings (b_{11} and b_{12}).

When the partners are indistinguishable, the SAS syntax to estimate the variance partitioning and covariances (i.e., the model with no partner or focal person predictors) is

```
PROC MIXED COVTEST;
CLASS focalid partid rater;
MODEL intimacy = focal partner/
   NOINT S DDFM = SATTERTH;
RANDOM focal partner / SUBJECT =
   focalid TYPE = UN;
REPEATED rater/SUBJECT = partid
   (focalid) TYPE = UN;
```

Here the coding of **partid** is important. If each partner had a unique identification number, we would just need **partid** for the SUBJECT in the REPEATED statement, but **partid(focalid)** also works. If **partid** is not unique, as is the case in Table 17.6, then **partid(focalid)** must be used. Note that the traditional intercept is suppressed (i.e., NOINT) and so there are two intercepts, one for data from the focal person and the other for data from the partners.

The RANDOM statement results in estimates of the variance in the two intercepts, as well as their covariance. For the **focal** variable this variance estimates the degree to which focal persons differ in their average partner ratings (it is akin to the actor variance in the SRM, see below). The variance in the **partner** intercepts estimates the degree to which there are focal person differences in the average ratings they are given by their partners (this is akin to the partner variance in the SRM, see below). The generalized reciprocity covariance is estimated as the covariance between these two level-2 effects.

The REPEATED statement is necessary to specify the error variances and covariances. Again there are two error variances, one for the partners' ratings of the focal person and another for the focal person's ratings of the partners. This statement specifies that rater is repeated across partners, and because the covariance matrix is unspecified (TYPE = UN), it also estimates a covariance that can be viewed as the dyadic or "error" reciprocity covariance.

The SPSS syntax to estimate the variance partitioning and covariances is

```
MIXED
intimacy BY rater focalid partid
   WITH focal partner
/FIXED = focal partner | NOINT
/PRINT = SOLUTION TESTCOV
/RANDOM focal partner | SUBJECT
   (focalid) COVTYPE(UN)
/REPEATED = rater|SUBJECT(focalid*
   partid) COVTYPE(UN).
```

When there are predictor variables measured for both the focal person and the partners, researchers have two options. In the intimacy example, we have both focal-person and partner gender (see **focalsex** and **partsex** in Table 17.6). The first option would be to keep these two gender variables as they are currently coded, and we would allow each of these to predict data from the focal person and the partners. The model statement in SAS for this analysis would be

```
MODEL intimacy = focal partner focal*
      focalsex focal*partsex partner*
      focalsex partner*partsex / NOINT S
      DDFM = SATTERTH;
```

The second option would be to again code two gender variables, but in this case we would code gender of the actor (i.e., the person doing the rating; **a_sex** in Table 17.6) and gender of the partner (i.e., the person being rated; **p_sex**). Thus, when the data point is focal-person rated intimacy, **a_sex** refers to the focal person's gender and **p_sex** refers to the partner's gender. When the data point is the partner's intimacy, **a_sex** refers to the partner and **p_sex** refers to the focal person. The model statement in SAS for this analysis would be

```
MODEL intimacy = focal partner focal*
      a_sex focal*p_sex partner*a_sex
      partner*p_sex/NOINT S DDFM =
      SATTERTH;
```

While mathematically equivalent, it is advisable to try out both ways of coding to determine the coding method that yields the simpler and more interpretable results. In either case, the researcher should allow the two gender variables to interact with the **focal** and **partner** dummy coded variables. The meanings of those interactions are very different for the two coding systems.

17.4.4 Reciprocal One-With-Many Designs: Distinguishable Partners

Recall that in the one-with-many design with distinguishable partners, the focal-person is paired with a set of partners who fall into different roles. As an example, we might have adolescents as the focal person, and the partners might be their mother, father, romantic partner, and best friend. In the reciprocal design, the adolescents would rate their relationship closeness with each of these partners, and we would also have the mother, father, romantic partner, and best friend rate their relationship closeness with the adolescent, and so the outcome variable would be **close**. In this example the data set would include a partner role variable, **partrole**, which differentiates these four types of partners (e.g., for mothers **partrole** = 1, for fathers **partrole** = 2, for romantic partners **partrole** = 3, and for best friends **partrole** = 4).

When partners are distinguishable, there are two ways to model the distinguishability. We might just treat the data as if partners were totally distinguishable. In this analysis, the number of "variables" would be the number of partners times two because the design is reciprocal. In this case we would allow for intercept differences for the different partners (both from the focal person's perspective and from the partner's perspective), different variances, and different covariances between each pair of variables. In essence, this is the saturated model for treating the design as reciprocal. The SAS code for a model of complete distinguishability is as follows:

```
PROC MIXED COVTEST;
CLASS focalid partrole rater;
MODEL close = focal*partrole
   partner*partrole/NOINT S
   DDFM = SATTERTH;
REPEATED partrole*rater /SUBJECT =
   focalid TYPE = UN;
```

The SPSS syntax is:

```
MIXED
close BY rater focalid partrole
   WITH focal partner
/FIXED = focal*partrole partner*
   partrole | NOINT
/PRINT = SOLUTION TESTCOV
/REPEATED = partrole*rater|
   SUBJECT(focalid) COVTYPE(UN).
```

The other alternative is to estimate separate fixed effects for the different roles, while including a general intercept random effect for the focal person (rather than allowing separate random focal-person intercepts for each role) and likewise including a general intercept random effect for partners. That is, rather than allowing for different intercept random effects for the different types of partners, there is only one focal person intercept and only one partner intercept. Thus the focal person variance measures consistency in the adolescent's ratings of his or her partners and the partner variance measures similarity in the partner's ratings of the adolescent. In such a model, distinguishability would be specified by allowing the error variances and covariances to differ by partner role. The SAS syntax would be

```
PROC MIXED COVTEST;
CLASS focalid partrole rater;
MODEL close = focal*partrole
   partner*partrole/NOINT S
   DDFM = SATTERTH;
RANDOM focal partner/SUBJECT =
   focalid TYPE = UN;
REPEATED rater/SUBJECT =
   partrole(focalid) TYPE = CSH
   GRP = partrole;
```

These MLM have many parameters and they may be slow to converge.

Once again we could test for distinguishability. As before we estimate two models using ML estimation, one of which treats the partners as distinguishable and the other treats the partners as indistinguishable. We would then subtract the deviances of the two models, and that difference has a chi-square distribution given the null hypothesis that dyad members are indistinguishable. The degrees of freedom for the test would be the extra number of parameters in the model when dyad members are treated as distinguishable.

17.5 SOCIAL RELATIONS MODEL DESIGNS

In SRM designs, each person is paired with more than one partner and each partner is also paired with multiple others. A full discussion of how to analyze these complex designs using multilevel modeling is beyond the scope of this chapter. For a more complete discussion of the models, we refer interested readers to Kenny (1994) or Kenny, Kashy, & Cook (2006), and for SAS syntax to use MLM to analyze SRM data, see Kenny and Livi (2009). Our discussion here is intended as a brief introduction to the designs.[2] In addition, like the other dyadic models we have discussed, the SRM can be used for both indistinguishable and distinguishable dyads. Here we limit the discussion to the indistinguishable case and readers should consult Kenny Kashy, & Cook (2006) or Kashy and Kenny (1990) for more detail on the distinguishable case.

The prototypical SRM design is a round-robin design where a group of persons rate or interact with all the other persons in the group (e.g., Persons A, B, C, and D interact and then A rates B, A rates C, and A rates D. Similarly, B rates A, C, and D for a total of 12 dyadic scores). By convention, the person who generates the measurement is called *actor* and the other person is called *partner.* For instance, if we ask people interacting in small groups how much they like one another, the person reporting on the liking is the actor and person being liked is the partner.

[2] In addition to the round-robin design, the SRM can also be estimated using a block design. In a block design, the people are divided into two subgroups and members rate or interact with members of the other subgroup but not with members of their own group. In the symmetric block design the members of the two subgroups are indistinguishable whereas in the asymmetric block design the members of the two subgroups are distinguishable (e.g., one subgroup is comprised of men and the other subgroup is comprised of women). Finally, in the half-block design, we have data from one subgroup with one other subgroup (e.g., men with women but not women with men). The half-block design, unlike the other SRM designs, is not reciprocal: Each person is either an actor or a partner but not both.

17.5.1 The SRM Components

The basic SRM equation is:

$$Y_{ijk} = m_k + a_{ik} + b_{jk} + g_{ijk},$$

where Y_{ijk} is the score for person i rating (or behaving with) person j in group k. In this equation m_k is the group mean, a_{ik} is person i's actor effect, b_{jk} is person j's partner effect, and g_{ijk} is the relationship or actor–partner interaction effect. The terms m, a, b, and g, are random variables and each has a variance: σ_m^2, σ_a^2, σ_b^2, and σ_g^2. The SRM also specifies two different correlations between the SRM components of a variable, both of which can be viewed as reciprocity correlations. At the individual level, a person's actor effect can be correlated with that person's partner effect; this covariance assesses generalized reciprocity, and is denoted as σ_{ab}. If the variable being rated is liking, then this covariance measures whether a person who likes everyone in the group is liked by everyone in the group. At the dyadic level, the two members' relationship effects can be correlated; this covariance assesses dyadic reciprocity and is denoted as σ_{gg}. In the example, this covariance measures whether a person who especially likes a particular partner, is especially liked by that partner. There are then seven SRM parameters, one mean, four variances, and two covariances.

As an example, we might have a group of people who rate one another's intelligence. The meanings of the SRM parameters would be as follows:

μ: The overall mean of rated intelligence

σ_m^2: The variance in average ratings of intelligence across groups.

σ_a^2: The variance in how intelligent a person generally sees others.

σ_b^2: The variance in how intelligent a person is generally seen by others.

σ_g^2: Unique variance in the way a particular individual sees a particular partner (i.e., the relationship variance or actor–partner interaction variance plus error).

σ_{ab}: The covariance between how intelligent a person generally sees others with how intelligent other people generally see that person (i.e., generalized reciprocity).

σ_{gg}: The covariance between how one person uniquely sees another with how that other uniquely sees that person (i.e., dyadic reciprocity).

17.5.2 Analysis of Round-Robin Designs

Because of the complexity of the nonindependence, estimating a MLM from round-robin data is problematic. Sometimes researchers treat actor as level 2, and all of the observations within an actor are treated as level 1. Alternatively, researchers may treat partner as level 2, and all of the observations within a partner are treated as level 1. Ironically, both of these are right and wrong at the same time. Actor and partners are levels, but both need to be considered in one model. Moreover, actor and partner are crossed or cross-classified, not nested. Additionally, the analysis must take into account the correlation between the two scores from members of the same dyad.

To estimate the SRM with round-robin data, the structure of the data set is particularly important. For the analyses we present, the data set is structured such that each record is the response of one person in a dyad (e.g., Person A's rating of Person B's intelligence). As seen in Table 17.7, a data set for a round-robin that includes

TABLE 17.7

Data From Group 1 of a Fictitious SRM Round-Robin Study

Group	Actor	Partner	Dyad	Y	Act_x	Part_x
1	1	2	1	6	5	7
1	1	3	2	4	5	6
1	1	4	3	2	5	5
1	1	5	4	7	5	7
1	2	1	1	7	7	5
1	2	3	5	5	7	6
1	2	4	6	6	7	5
1	2	5	7	7	7	7
1	3	1	2	3	6	5
1	3	2	5	4	6	7
1	3	4	8	6	6	5
1	3	5	9	5	6	7
1	4	1	3	2	5	5
1	4	2	6	4	5	7
1	4	3	8	3	5	6
1	4	5	10	6	5	7
1	5	1	4	7	7	5
1	5	2	7	6	7	7
1	5	3	9	7	7	6
1	5	4	10	6	7	5

five individuals in a group would therefore be comprised of 20 records for each group. (Note that this value is 20, not 25 because self-rating data are treated differently by the SRM.) On each record there would be three identification variables: a group identification variable (e.g., **group**), an identification variable that designates the actor who made the rating (e.g., **actor**), and an identification variable that designates the target of the rating (e.g., **partner**). We shall see that sometimes other variables need to be included in the data set.

In this chapter we present a conventional MLM approach. A more complex approach involving dummy variables can be found in Snijders and Kenny (1999) and Kenny and Livi (2009). In this section because of space limitations, we do not explicitly consider fixed variables (what we denoted as X and Z in prior sections). Such variables would be included in the MODEL statement in SAS and the /FIXED statement in SPSS. Individual-level variables (e.g., the individual's extroversion) would need to be included on all records for which that individual is the rater. Moreover, the researcher should consider including both the actor's variables (e.g., the actor's extroversion) and the partner's variables (e.g., the partner's extroversion) as potential predictors.

Increasingly, multilevel programs can estimate models with cross-classified variables. However, in these models the actor–partner covariance is assumed to be zero, which is a major limitation of this method. This is particularly problematic when variables such as liking are under consideration because generalized reciprocity is a likely component for this variable (i.e., likeable people tend to like others). We describe this approach and detail how both SAS and SPSS can be used to estimate the model. We believe that HLM and MLwiN can also estimate the models in this fashion.

To use the conventional MLM approach, identification numbers need to be assigned to each group, each actor, each partner, and each dyad. For SPSS these values must be unique. For example, if there are 15 five-person groups, then the **group** variable would range from 1 to 15; the values for the **actor** variable would range from 1 to 5 for group 1 and from 6 to 10 for group 2; the values for the **partner** variable would be 1–5 in group 1 and 6 to 10 in group 2; and finally, because there are 10 dyads in a five-person round-robin, the **dyad** variable will range from 1 to 10 for group 1, 11 to 20 for group 2, and so on. Unique identification numbers are not required for SAS.

We first present the syntax for SAS and then for SPSS. Note again that the actor–partner covariance is not modeled. The syntax for SAS is

```
PROC MIXED COVTEST;
CLASS actor partner dyad group;
   MODEL y = /S DDFM = SATTERTH
      NOTEST;
   RANDOM INTERCEPT/TYPE = VC
      SUB = actor(group);
   RANDOM INTERCEPT/TYPE = VC
      SUB = partner(group);
   RANDOM INTERCEPT/TYPE = VC
      SUB = group;
   REPEATED/TYPE = CS SUB =
      dyad (group);
```

The syntax for SPSS is as follows:

```
MIXED
   y BY group
   /PRINT = SOLUTION TESTCOV
   /RANDOM INTERCEPT|SUBJECT
      (group ) COVTYPE(VC)
```

```
/RANDOM INTERCEPT|SUBJECT
    (actor) COVTYPE(VC)
/RANDOM INTERCEPT|SUBJECT
    (partner) COVTYPE(VC)
/RANDOM INTERCEPT |SUBJECT
    (dyad) COVTYPE(VC).
```

In SPSS, the REPEATED statement cannot be used for dyad, and so one must presume that the dyadic covariance is positive.[3] Note also that in SPSS the error variance equals the dyad variance plus the error variance, and the dyadic correlation equals the dyad variance divided by the sum of the dyad variance plus the error variance.

The results from SAS and the SPSS would be different if the reciprocity covariance were negative. In that case, it would be incorrectly estimated as zero by SPSS and properly estimated by SAS.

17.6 CONCLUSIONS

We have shown that MLM can be used to estimate a wide range of dyadic models. However, structural equation modeling can also be a useful tool for the analysis of dyadic data, especially when dyad members are distinguishable and when SRM and one-with-many designs are used.

We made a sharp distinction between distinguishable and indistinguishable dyads, but there are variants in-between. For instance, in the one-with-many design, partner can be partly distinguishable and partly indistinguishable. Consider the example in which some partners are men and others are women. We could distinguish between gender, but within gender the partners would be indistinguishable. We limited our discussion to normally distributed outcome variables and we did not consider models for counts and proportions. Although MLM can now be used to estimate models with such outcome scores, these approaches often do not currently allow for correlated errors. In other words, the approaches do not allow for REPEATED statements, which were an important component of nearly every dyadic design that we discussed.

In some sense dyadic models are simple MLMs. Certainly, the basic model for the standard design is a very simple MLM model. However, as we saw, when we allow for distinguishability and more complex designs that are reciprocal, the analysis can become quite complicated. MLM offers the possibility of being able to estimate a wide range of dyadic models and is becoming an important tool for dyadic researchers.

[3] For SPSS 16 and earlier, tests of variances are two-tailed when they should be one-tailed. Thus, *p* values should be divided by two.

ACKNOWLEDGMENTS

We thank Rob Ackerman and Thomas Ledermann, who provided comments on an earlier draft of this paper.

REFERENCES

Biesanz, J. C., Deeb-Sossa, N., Papadakis, A. A., Bollen, K. A., & Curran, P. J. (2004). The role of coding time in estimating and interpreting growth curve models. *Psychological Methods, 9*, 30–52.

Hillmer, S. (2001). Time series regressions. In D. S. Moskowitz & S. L. Hershberger (Eds.), *Modeling intraindividual variability with repeated measures data: Methods and applications*. Mahwah NJ: Erlbaum.

Kashy, D. A. (1992). Levels of analysis of social interaction diaries: Separating the effects of person, partner, day, and interaction. *Dissertation Abstracts International, 53*(1-B), 608–609.

Kashy, D. A., & Kenny, D. A. (1990). Analysis of family research designs: A model of interdependence. *Communication Research, 17,* 462–483.

Kenny, D. A. (1994). *Interpersonal perception: A social relations analysis.* New York, NY: Guilford.

Kenny, D. A., Kashy, D. A., & Cook, W. L. (2006). *Dyadic data analysis.* New York, NY: Guilford.

Kenny, D. A., & Livi, S. (2009). Thoughts on studying leadership in natural contexts. In F. J. Yammarino & F. Dansereau (Eds.), *Multi-Level Issues in Organizational Behavior and Leadership* (Vol. 8 of Research in Multi-level Issues; pp. 147–191). Bingley, UK: Emerald.

Kenny, D. A., Mannetti, L., Pierro, A., Livi, S., & Kashy, D. A. (2002). The statistical analysis of data from small groups. *Journal of Personality and Social Psychology, 83,* 126–137.

Laurenceau, J.-P., & Bolger, N. (2005). Using diary methods to study marital and family processes. Special issue on methodology in family science. *Journal of Family Psychology, 19,* 86–97.

Raudenbush, S. W., Brennan, R. T., & Barnett, R. (1995). A multilevel hierarchical model for studying psychological change within married couples. *Journal of Family Psychology, 9,* 161–174.

Snijders, T. A. B., & Kenny, D. A. (1999). The social relations model for family data: A multilevel approach. *Personal Relationships, 6,* 471–486.

Author Index

A

B

C

D

E

F

G

H

I

J

K

L

M

N

O

P

Q

R

S

W

X

Y

Z

Subject Index

A

B

F

G

J

K

L

T

U

V

W

X

Z